GUIDE TO GEOGRAPHICAL BIBLIOGRAPHIES AND REFERENCE WORKS IN RUSSIAN OR ON THE SOVIET UNION

Chauncy D. Harris

Samuel N. Harper Distinguished Service Professor of Geography
University of Chicago

Annotated List of 2660 Bibliographies or Reference Aids

THE UNIVERSITY OF CHICAGO
DEPARTMENT OF GEOGRAPHY
RESEARCH PAPER NO. 164

1975

DEDICATED TO

THE 23rd INTERNATIONAL GEOGRAPHICAL CONGRESS

MOSCOW, USSR, JULY 27 - AUGUST 3, 1976

Copyright 1975 by

CHAUNCY D. HARRIS

Library of Congress Catalog Card Number: 74-84784

International Standard Book No. 0-89065-071-3

Library of Congress Cataloging in Publication Data

Harris, Chauncy Dennison, 1914-
    Guide to geographical bibliographies and reference
works in Russian or on the Soviet Union.

    (Research paper - University of Chicago, Department
of Geography ; no. 164)
    Includes bibliographical references and index.
    1. Russia--Description and travel--Bibliography.
2. Reference books--Russia. 3. Geography--
Bibliography. 4. Reference books--Geography.
I. Title. II. Series: Chicago. University. Dept.
of Geography. Research paper ; no. 164.
H31.C514 no. 164 [Z2506] 910s 016.0169147'04 74-84784
ISBN 0-89065-071-3

The cover design map of the world on a star-shaped polar projection is
adapted from the emblem used by Glavnoe Upravlenie Geodezii i Kartografii
on its great world atlases (entries 145, 147, 152, and 153 in this Guide).

Published 1975 by the Department of Geography,
The University of Chicago

Research Papers are available from:

    Department of Geography
    The University of Chicago
    5828 University Avenue
    Chicago, Illinois 60637, USA

Price: $5.00 ($4.00 by series subscription)

The 23rd International Geographical Congress scheduled for Moscow, USSR, July 27-August 3, 1976, the first to be held in Russia or the Soviet Union, heightens interest in the geography of this vast area and in the work of its geographers.

This Guide to Geographical Bibliographies and Reference Works in Russian or on the Soviet Union is offered as an aid to individuals outside the Soviet Union who wish to become informed on the corpus of serious scientific work in geography and related disciplines published in Russian or in other languages of the Soviet Union or dealing with the geography of the Soviet Union. It is a relatively comprehensive inventory of bibliographies published in the Soviet Union on all fields of geography. It includes also a selection of the more important reference materials in Russian of particular geographic value, such as atlases, statistical sources, encyclopedias, handbooks, gazetteers, geographical dictionaries, and biographical directories.

Although the Russian and Soviet geographical literature is rather skimpily treated in Western bibliographies of geography, this literature is vast and of great importance. The Soviet Union itself covers a sixth of the ice-free land surface of the earth and its geography thus forms a significant fraction of the total geography of the globe. It encompasses a larger area than any other single political unit. It contains the largest contiguous area in the world in which the major scientific literature is predominantly in a single language. Soviet regional bibliographies and literature are rich also on the adjacent countries of Europe and Asia. In many fields of physical geography, such as landscape studies, paleogeography, geomorphology, climatology, glaciology, snow-cover and permafrost studies, hydrology, oceanography, soil geography, and bio-geography, Soviet scientists occupy a leading international position and their publications are important to anyone wishing to be well informed on scientific advances in these systematic fields.

This bibliography covers primarily the period 1946-1973. Works from earlier periods have been listed selectively. Publications from 1971-1973 are less complete than for 1946-1970 since full checking for these later years awaits the publication of comprehensive Soviet bibliographical sources. This Guide includes both bibliographies published separately and hidden bibliographies incidental to major monographs, but only if the bibliographies are quite substantial, with at least one hundred references.

Languages of the Soviet Union, other than Russian, are doubtless under-represented in this bibliography because of linguistic, bibliographical, and physical problems: linguistic, because the compiler does not read textual

materials in Soviet languages other than Russian; bibliographical, because
the prime bibliographical sources for the Soviet Union are all in Russian and
provide only the Russian form of a title or in some cases only works with Russian
summaries; and physical, because non-Russian publications are poorly represented
in the library in which he did most of the visual inspection of publications to
be listed in this guide. Thus sometimes a work is listed by a Russian title,
even if the work itself is in another language. Although the vast majority of
the entries are in Russian, the following other languages of the Soviet Union
are represented: Ukrainian, Estonian, Latvian, Lithuanian, Belorussian, Moldavian, Georgian, Armenian, Azerbaijani, Kazakh, Uzbek, and Tajik. To assist the
reader not familiar with the languages of the Soviet Union English translations
of titles are presented in parentheses.

The scope of this guide reflects the fields actively cultivated by Soviet
geographers not any theoretical definition of the core, boundaries, or organization of the field of geography.

In this guide where a work concerns a specific subfield of geography and a
specific region, preference is given to the topical listing.* Thus a geomorphology of the world, of the Soviet Union, or of the Caucasus are all listed under
geomorphology. On the other hand comprehensive geographic treatises of a region
are listed under the region. Works covering several topical subfields are also
listed under the region.

Entries in each section have been arranged in what was considered the most
useful order for that section, but this varies from one part of the bibliography
to another. In general, bibliographies of bibliographies come first, followed by
comprehensive bibliographies, by selective bibliographies, by bibliographies in
topical subfields, and by bibliographies on regional subdivisions. In regional
works the Soviet Union precedes foreign areas. Within the Soviet Union works
are typically listed in a geographical order proceeding from northwest to southeast with the Russian Soviet Federated Socialist Republic coming first, followed
by the Ukraine, then by the other Union republics in the following order: Estonia,
Latvia, and Lithuania in the Baltic; Belorussia and Moldavia in the west; Georgia,
Armenia, and Azerbaydzhan in the Trans-Caucasus; Kazakhstan; and Kirgizia,
Uzbekistan, Turkmenistan, and Tadzhikistan in the Soviet Middle Asia. Bibliographies within each division are typically listed in reverse chronological order,
so that the most up-to-date works come first (except for sequential publications
in a single series in which such an arrangement would be awkward). However, the
most comprehensive and valuable work, even if not the latest, may be listed first.
Occasionally another order is used, as in atlases or statistical handbooks of
individual oblasts, which are listed alphabetically by the names of oblasts.

---

* A partial exception is works on specialized topics on the Georgian SSR, since
these entries were added after the sections on subfields of geography had been
finally typed.

The form of listing for Soviet publications poses some complexities, since a typical monograph has a sponsoring institution, such as the Institute of Geography of the Academy of Sciences of the USSR, several authors, perhaps one or two editors, and a title. Since both Soviet and foreign bibliographical sources and libraries may list a given work under any one of these three categories--institution, author or editor, or title--all have been included in the information in each entry in this guide, so that the user will have clues for alternative listings in library catalogues. In general, however, preference in this guide has been for the author listing, except in standard reference works such as atlases or encyclopedias, listed by title, or in certain collected volumes in which the number of authors and editors is very large and the institution clearly provides the best identification.

The user of this volume needs to bear in mind that it is not a list of the most important works in Soviet geography but rather a list of bibliographies of geographic work. Thus a major work with no bibliography or a slight bibliography would not be listed here, whereas a minor contribution with a substantial bibliography would be. A simple bibliographic list without interpretation or annotation would be listed, whereas a review of the field without a specific bibliography might not be.

The transliteration scheme of the Library of Congress has been used for all bibliographical entries. Place names, however, have been spelled in accordance with the system of the United States Board on Geographical Names.

A number is assigned to each entry. These numbers are recorded in the right-hand margin beside each title. In addition the entry numbers occurring on each page are listed at the bottom of that page.

The index has been made rather full to provide multiple entries into the information in this Guide. It includes authors, titles, sponsoring institutions, subjects, and places in a single alphabetical order. References are to entry numbers not pages.

End paper maps indicate the location of the major regions (inside front cover) and of administrative and statistical areas of the Soviet Union at the oblast level (inside back cover).

The compiler wishes to thank Miss Violet Milicevic for the typing and re-typing of the manuscript, Christopher Müller-Wille for drafting the maps, and Shawn E. Bryan for compilation of the index, and the Center for International Studies of the University of Chicago for support.

Gratitude is expressed to the following individuals who made fruitful suggestions for improvements in organization or for additions, corrections, or deletions:

ALIEV, G. A. Institute of Geography, Baku

DAVITAIA, F. F. Institute of Geography, Tbilisi

KONSTANTINOV, O. A.   Herzen Pedagogical Institute, Leningrad

POKSHISHEVSKII, V. V.   Institute of Ethnography, Moscow

SALISHCHEV, K. A.   Moscow State University, Moscow

ARMSTRONG, Terence E.   Scott Polar Research Institute, Cambridge

BRIEND, A.   Service de Documentation et de Cartographie Géographiques du CNRS, Paris

DEMKO, George J.   Ohio State University, Columbus, Ohio

FISCHER, Dora.   Osteuropa Institut der Freien Universität, Berlin

FRENCH, R. A.   University College, London

FUCHS, Roland J.   University of Hawaii, Honolulu

GIBSON, James R.   York University, Downsview, Ontario

HORECKY, Paul L.   Slavic and Central European Division, Library of Congress, Washington, D.C.

HOOSON, David.   University of California, Berkeley, California

JACKSON, W. A. Douglas.   University of Washington, Seattle, Washington

JENSEN, Robert G.   Syracuse University, Syracuse, New York

LEVERSEDGE, Francis M.   University of Victoria, Victoria, British Columbia

LYDOLPH, Paul E.   University of Wisconsin at Milwaukee, Milwaukee, Wisconsin

MATHIESON, R. S.   University of Sydney, Sydney, New South Wales

NORTH, Robert N.   University of British Columbia, Vancouver, British Columbia

SHABAD, Theodore.   American Geographical Society, New York, N. Y.

SIMMONS, J. S. G.   All Souls College, Oxford.

J. S. G. Simmons was particularly valuable on organizational matters, G. A. Aliev and F. F. Davitaia on works in Azerbaijani and Georgian, and James R. Gibson on historical geography.

The following were helpful in the utilization of library resources:

PETROV, M. P.   Leningrad State University, Leningrad

LOSHKOMOEVA, N. M.   Librarian, Geographical Society of the USSR, Leningrad

ORLOVA, A. I.   Librarian, Institute of Geography of the Academy of Sciences of the USSR, Moscow

LAŠKA, Vaclav.   Slavic Bibliographer, Regenstein Library, University of Chicago, Chicago, Illinois

PANKIW, Halyna.   Assistant Slavic Bibliographer, Regenstein Library, University of Chicago, Chicago, Illinois

The compiler, of course, must accept full responsibility for the conception of this Guide and for the selection, form, accuracy, and organization of entries.

CHAUNCY D. HARRIS
University of Chicago

TABLE OF CONTENTS

LIST OF MAPS

PART I. GENERAL BIBLIOGRAPHICAL AIDS

BIBLIOGRAPHIES OF BIBLIOGRAPHIES

Specialized Bibliographies of Geographic Bibliographies

FIZICHESKAIA geografiia: annotirovannyi perechen' otechestvennykh bibliografii
izdannykh v 1810-1966 gg. (Physical geography: annotated list of native bibliog-
raphies published 1810-1966). M. N. Morozova, E. A. Stepanova, comps. V. V.
Klevenskaia, ed. and L. G. Kamanin, scientific consultant. Moskva, Izdatel'stvo
"Kniga," 1968. 309 p. (Gosudarstvennaia Biblioteka SSSR imeni V. I. Lenina.
Institut Geografii Akademii Nauk SSSR. Sektor Seti Spetsial'nykh Bibliotek
Akademii Nauk SSSR).
                                                                              1

Comprehensive bibliography of bibliographies in Soviet physical geography
1810-1966, with 1,336 entries. Includes sections on general geographical bibliographies, bibliographies on history of geography and geographical exploration, and
bibliographies of works of native geographers and travelers (339 entries), in
addition to physical geography, which includes general physical geography, systematic fields of physical geography (geomorphology, climatology, hydrology,
glaciology, geography of soils, plant geography, zoogeography, and regional
physical geography (997 entries). Indexes of names and of places.

Covers mainly works in Russian with only a selection of materials in other
languages of the USSR.

GAVRILOVA, Serafima Abramovna; PARKHOMENKO, Iia Ivanovna; and KUZNETSOVA, N. K.
Bibliografii po geografii, vyshedshie v SSSR (1946-1961 gg.) (Bibliographies of
geography, published in the USSR 1946-1961), Geograficheskii sbornik (Institut
Nauchnoi Informatsii Akademii Nauk SSSR). [1]. Moskva, VINITI, 1963, p. 32-45.
                                                                              2

266 separate bibliographies of geography published in the Soviet Union 1946-
1961, organized systematically: geography as a whole and cartography, history of
geography, biobibliography, physical geography, conservation and transformation
of nature, economic geography, geography of the USSR as a whole, regional bibliographies of the RSFSR and its individual regions, union republics of the Soviet
East (as a whole), union republics: Armenia, Belorussia, Georgia, Kazakhstan,
Kirgizia, Latvia, Lithuania, Turkmenistan, Uzbekistan, Ukraine, and regional geography of foreign countries.

KLEVENSKAIA, Valentina Vasil'evna. Spisok bibliograficheskikh istochnikov po geografii, rassmotrennykh na seminare (List of bibliographical sources for geography
examined in a seminar). Biblioteki SSSR: opyt raboty (Libraries of the USSR:
experience from the field), no. 4. Moskva, 1955, p. 29-39 (Gosudarstvennaia
Biblioteka SSSR imeni Lenina. Nauchno-metodologicheskii Kabinet Biblioteko-
vedeniia).
                                                                              3

147 entries on the principal bibliographies and reference works in Russian on
geography divided into six sections: (1) current inventory-registering (comprehensive) bibliographies, (2) recommending (selected) bibliographies, (3) summary
retrospective bibliographies (bibliographical aids for geography, geographical
literature in bibliographies for related branches of knowledge, geographical
literature in regional bibliographies, and geographical literature in universal
bibliographies), (4) geographical dictionaries, (5) biobibliographies, and (6)
periodical and article indexes. Review of principal Russian and Soviet sources
for geography up to 1955.

GAVRILOVA, Serafima Abramovna, and GORIANINA, A. I. Vtorichnaia informatsiia po geografii za rubezhom. Chast' I. Geografiia v tselom, kartografiia, fiziko-geograficheskie nauki, biogeografiia (Secondary information on geography abroad. Part 1. Geography as a whole, cartography, physical-geographic sciences, bio-geography), Geograficheskii sbornik (Institut Nauchnoi Informatsii Akademii Nauk SSSR), v. 2, Moskva, VINITI, 1966, p. 57-103.   4

    62 bibliographical sources arranged alphabetically by authors, followed by listing and discussion of 114 series that constitute secondary sources of biblio-graphical data or abstracts, arranged systematically.

GAVRILOVA, Serafima Abramovna, and GORIANINA, A. I. Vtorichnaia informatsiia po geografii za rubezhom. Chast' II. Ekonomicheskaia geografiia v tselom i otdel'-nye eё otrasli (Secondary information on geography abroad. Part 2. Economic geography as a whole and its individual branches), Geograficheskii sbornik (Vsesoiuznyi Institut Nauchnoi i Tekhnicheskoi Informatsii), v. 3, Moskva, VINITI, 1969, p. 5-56.   5

    115 bibliographical sources, arranged alphabetically by name of author, followed by 156 series that constitute sources of bibliographical data or abstracts in economic geography, arranged systematically.

ZAKHAROVA, L. N., and MYSHEVA, I. A. Vtorichnaia informatsiia po geografii za rubezhom. Chast' III. Meditsinskaia geografiia. Izdaniia SShA (Secondary in-formation on geography abroad. Part 3. Medical geography: publications in the United States), Geograficheskii sbornik (Vsesoiuznyi Institut Nauchnoi i Tekhnicheskoi Informatsii), v. 3. Moskva, VINITI, 1969, p. 57-83.   6

    45 reviews, abstracts, or bibliography series plus 76 medical journals in the United States of potential interest in medical geography, each section arranged alphabetically by titles.

GAVRILOVA, Serafima Abramovna, and GORIANINA, A. I. Vtorichnaia informatsiia po geografii za rubezhom. Chast' IV. Stranovedenie: Afrika (Secondary information on geography abroad. Part 4. Regional geography: Africa), Geograficheskii sbornik (Vsesoiuznyi Institut Nauchnoi i Tekhnicheskoi Informatsii), v. 4, Moskva, VINITI, 1970, p. 55-84. Bibliography, p. 67-84.   7

    115 bibliographical sources followed by 68 series that provide sources of bibliographical information, abstracts, or similar material. Each arranged alphabetically by name of author or title of series.

ZAKHAROVA, L. N., and MYSHEVA, I. A. Vtorichnaia informatsiia po geografii za rubezhom. Chast' V. Meditsinskaia geografiia v otdel'nykh stranakh Latinskoi Ameriki (Secondary information on geography abroad. Part 5. Medical geography in individual countries of Latin America), Geograficheskii sbornik (Vsesoiuznyi Institut Nauchnoi i Tekhnicheskoi Informatsii), v. 4, Moskva, VINITI, 1970, p. 85-94.   8

    45 medical and related journals of Latin America, arranged in alphabetical order of titles.

MURZIN, I. Materialy dlia spiska bibliograficheskikh ukazatelei literatury po geografii otdel'nykh mestnostei (Material for a list of bibliographic guides for the geography of individual localities), Gosudarstvennoe Geograficheskoe Obshche-stvo, Izvestiia, v. 66, no. 2 (1934), p. 299-305.   9

    162 general bibliographies of continents, of individual countries, and of localities, from the 19th century and up to 1934 in the 20th century.

General Bibliographies of Soviet Bibliographies

Current

BIBLIOGRAFIIA sovetskoi bibliografii (Bibliography of Soviet bibliography).
(Vsesoiuznaia Knizhnaia Palata). 1939, 1946- . Moskva, Izdatel'stvo "Kniga,"
1941, 1948- . Annual.   10

The annual bibliography of bibliographies published in the Soviet Union,
either as separate works or as parts of books or serials. Bibliographies of
geography are listed under "Estestvennye nauki. Matematika" (Natural sciences
and mathematics) in the section, "Geologo-geograficheskie nauki" (Geological-
geographical sciences) under the heading "Obshchie voprosy.Geografiia. Kraevedenie"
(General questions of geography. Regional studies).

The most complete and most convenient annual record of Soviet bibliographies
in the field of geography, listing some 1,894 bibliographies of geography in
Russian and other languages of the USSR for the 26 years, 1939, 1946-1970.

The page location and number of entries for geographical bibliographies in
each annual volume are as follows:

1939, p. 218-220	(48 entries)	1958, p. 79- 83	(93 entries)
1946, p. 46- 47	(29)	1959, p. 92- 97	(104)
1947, p. 62- 65	(61)	1960, p. 92- 97	(107)
1948, p. 41- 44	(72)	1961, p. 97-102	(99)
1949, p. 60- 63	(52)	1962, p. 72- 76	(101)
1950, p. 40- 42	(79)	1963, p. 79- 83	(100)
1951, p. 42- 44	(61)	1964, p. 72- 75	(74)
1952, p. 42- 44	(50)	1965, p. 71- 73	(73)
1953, p. 38- 41	(59)	1966, p. 166-169	(76)
1954, p. 43- 45	(47)	1967, p. 148-150	(64)
1955, p. 51- 54	(57)	1968, p. 77- 79	(71)
1956, p. 58- 63	(98)	1969, p. 87- 90	(64)
1957, p. 65- 69	(87)	1970, p. 71- 73	(68)

Some other sections with bibliographies of geographic interest, with page
numbers (and number of entries) in the 1970 annual volume are:

"Kraevedcheskaia bibliografiia" (regional bibliographies for regions in the
Soviet Union), p. 12-15 (66 entries);
"Geodeziia. Topografiia. Kartografiia" (geodesy, topography, and cartog-
raphy), p. 56-57 (15 entries).
"Geologo-geograficheskie nauki," (geological-geographical sciences), p. 71-
103 (914 entries), with subdivisions: "Obshchie voprosy. Geografiia. Kraevedenie"
(General questions, geography, regional studies), p. 71-73 (68 entries), noted
above; "Geologiia. Petrografiia. Mineralogiia. Geofizika. Geokhimiia" (geology,
petrography, mineralogy, geophysics, and geochemistry), p. 73-98 (676 entries);
"Gidrologiia. Meteorologiia. Klimatologiia" (hydrology, meteorology, and climatology), p. 98-103 (170 entries).
"Biologicheskie nauki" (biological sciences), p. 104-126 (661 entries).
"Agrobiologiia. Obshchee rastenievodstvo. Pochvovedenie" (Agricultural
biology, general crop cultivation. Soil science), p. 197-204 (194 entries).

The total number of bibliographies listed in the 1970 annual volume was
7,021,arranged systematically. List of journals with regular sections on bibliography. List of reviews of bibliographical guides. Index of names of authors.
Index of titles (for works without listed authors). Index of places.

SVODNYI ukazatel' bibliograficheskikh spiskov i kartotek, sostavlennykh bibliotekami
Sovetskogo Soiuza v ... godu: obshchestvennye nauki, khudozhestvennaia literatura,
iskusstvo (Union list of bibliographical lists and card indexes compiled by
libraries of the Soviet Union in _____: social sciences, humanities, and art).
Moskva, Izdatel'stvo "Kniga." (Gosudarstvennaia Biblioteka SSSR imeni V. I.
Lenina).   11

    1960. 1961. 109 p.
    1961. 1962.   88 p.
    1962. 1963.   95 p.
    1963. 1964. 108 p.
    1964. 1965. 100 p.
    1965. 1966.   94 p.
    1966. 1967. 104 p.
    1967. 1968.   96 p.
    1968. 1969. 112 p.
    1969. 1970. 109 p.
    1970. 1971. 108 p.
    1971. 1973.   87 p.
    1972. 1973.   88 p.

SVODNYI katalog bibliograficheskikh rabot, vypolnennykh v Sovetskom Soiuze:
estestvennye i fiziko-matematicheskie nauki. (Union catalogue of bibliographical
works produced in the Soviet Union: natural and physical-matematical sciences).
Leningrad. Quarterly. (Akademiia Nauk SSSR. Biblioteka Akademii Nauk SSSR).   11

    List of published, planned, and unpublished bibliographies compiled in the
Soviet Union on the physical and biological sciences, classified by disciplines.
Major sections are natural sciences as a whole, physical-mathematical sciences,
chemistry, earth sciences, and biology. Section 4. Earth sciences is divided
into: 4.1, geodesy and cartography; 4.2 geophysics; 4.3 geology, and 4.4 geographical sciences (4.4.1. general section, physical geography, applied geography.
4.4.2. regional geography). Author index. Index of compiling institutions.

                                    Retrospective

GOSUDARSTVENNAIA bibliografiia SSSR: spravochnik (State bibliography of the USSR:
handbook). 2nd ed., Moskva, Izdatel'stvo "Kniga," 1967. 111 p. (Vsesoiuznaia
Knizhnaia Palata).   12

    Description of the ten key comprehensive all-Union state bibliographical
series and of the bibliographical series of each of the Union republics. Full
bibliographical details of coverage.

AKADEMIIA Nauk SSSR. Biblioteka Akademii Nauk SSSR. Ukazatel' osnovnykh
otechestvennykh bibliografii i spravochnykh izdanii po estestvennym i fiziko-
matematicheskim naukam (Guide to basic native bibliographies and handbooks published on natural and physical-mathematical sciences). R. L. Baldaev, comp.
A. I. Mankevich, ed. Leningrad, Biblioteka Akademii Nauk SSSR, 1966. 385 p.   13

    1,960 entries, organized by fields. Bibliography for geographical sciences.
p. 159-189, entries 892-1,085 (194 entries) are arranged by general questions,
regional geography, physical geography, regional physical geography, hydrology,
and regional studies. Earth sciences, p. 117-189 include geodesy and cartography,
geophysical sciences, and geological sciences, as well as geography. Biological
sciences, p. 189-244. Handbooks for geography, p. 291-298, entries 1,786-1,845
(60 entries). Index of names and titles.

CIRPICHĒVA, Iraida Konstantinovna. Bibliografiia v pomoshch' nauchnoi rabote:
metodicheskoe i spravochnoe posobie (Bibliography as an aid to scientific work:
methodological and reference aids). P. N. Berkov, ed. Leningrad, 1958. 480 p.
(Gosudarstvennaia Publichnaia Biblioteka imeni M. E. Saltykova-Shchedrina).   14

The standard Russian guide to bibliography and bibliographies and reference
works. Part 1 discusses the techniques of bibliographical work. Part 2, the
core of the book, consists of textual comments and listing of the basic bibliographical sources of information on the literature of all branches of learning;
chapter 1 on Soviet literature and chapter 2 on foreign literature. Part 3 includes the basic bibliographical sources by individual disciplines, grouped by
the social sciences, natural sciences, technology, agriculture, and medicine.

Geography is included in the section, Geologo-geograficheskie nauki (geological and geographical sciences), p. 347-365. It is especially valuable for
its coverage of basic bibliographical sources in Russian.

A German translation, Handbuch der russischen und sowjetischen Bibliographie:
die Allgemeinebibliographien, Fachbibliographien und Nachschlagewerke Russlands,
Leipzig, Verlag für Buch-und Bibliothekswesen, 1962, 225 p. (Bibliothekswissen-
schaftliche Arbeiten aus den Ländern der Volksdemokratie in deutscher Übersetzung,
Reihe B, Band 5) omits the discussion of bibliographical techniques,of prerevolutionary Russian bibliographies, and of foreign bibliographies.

OKUROVA, M. V. ed. Obshchie bibliografii russkikh knig grazhdanskoi pechati 1708-
1955: annotirovannyi ukazatel' (General bibliography of books in Russian secular
type, 1708-1955: an annotated bibliography). P. N. Berkov, ed. 2nd ed. Leningrad,
1956. 283 p. (Gosudarstvennaia Publichnaia Biblioteka imeni M. E. Saltykova-
Shchedrina). (1st ed. 1944).   15

Carefully selected and extensively annotated bibliography of 60 general
bibliographies of books, in chronological order. Folded survey table indicates
periods covered and main characteristics of the bibliographies. Indexes of
authors and titles and bibliography of works consulted.

CURRENT BIBLIOGRAPHIES

Current Geographical Bibliographies

REFERATIVNYI zhurnal: geografiia. 1954- . 12 numbers a year. Vsesoiuznyi Institut Nauchnoi i Teknicheskoi Informatsii, Baltiiskaia ulitsa 14, Moskva, USSR. 11 series either bound together as a combined volume (svodnyi tom), or issued separately. 1954-1955 geography combined with geology. Some sections have author index in each no. Annual author index (separate volume) and annual subject and geographic index (another separate volume). The subject index is in effect a series of separate indexes for the separate series.   16

The most massive and detailed geographical bibliography in any language. Extensive abstracts. Fullest coverage by any geographical bibliography of technical publications in other fields. Fullest coverage by any geographic bibliography of most fields of physical geography. Detailed reporting of works in Russian but extensive coverage of materials in all other languages. Each section edited by specialist. Signed abstracts.

The various series, indicated by Russian letters, and the main subdivisions are as follows:

А. Теоретические и общие вопросы географии
    Общий раздел
    Теоретические вопросы общей физической географии и ландшафтоведения
    Теоретические вопросы экономической географии и страноведения
    География мирового хозяйства. Международное разделение труда
    География хозяйства и международное сотрудничество социалистических стран
    География крупных регионов, охватывающих разные части света
    Топонимика
    Методика преподавания географии
    Общие руководства, справочная литература. Новые издания

A. Theoretical and general questions of geography
    General section
    Theoretical questions of general physical geography and landscape studies
    Theoretical questions of economic geography and regional geography
    Geography of the world economy. International division of labor
    Geography of the economy and international collaboration of socialist countries
    Geography of large regions covering different continents
    Toponymics
    Methods of teaching geography
    General handbooks, reference literature, new editions

М. Картография

M. Cartography

А. Антропогеновый период. Геоморфология суши и морского дна
    Антропогеновый период
    Неотектоника
    Геоморфология

G. The Quaternary period. Geomorphology of the land and the sea bottom
    Quaternary period
    Neotectonics
    Geomorphology

В. Океанология. Гидрология суши.
    Гляциология
    Океанология
    Гидрология
    Гляциология

V. Oceanography. Hydrology of the land.
    Glaciology
    Oceanography
    Hydrology
    Glaciology

Б. Метеорология и климатология
  Общий раздел
  Метеорологические приборы, методы
    наблюдений и обработки
  Физическая метеорология
  Динамическая и синоптическая
    метеорология
  Климатология
  Прикладная метеорология и
    климатология

B. Meteorology and climatology
  General section
  Meteorological instruments, methods
    of observation and analysis
  Physical meteorology
  Dynamic and synoptic
    meteorology
  Climatology
  Applied meteorology and
    climatology

Д. Биогеография. География почв
  Фитогеография суши
  Зоогеография суши
  Биогеография внутренних
    водоемов и пещер
  Биогеография моря
  География эндопаразитов
  Фенология
  География почв

D. Biogeography. Geography of soils
  Plant geography of the land
  Animal geography of the land
  Biogeography of internal water
    bodies and caves
  Biogeography of the sea
  Geography of internal parasites
  Phenology
  Geography of soils

Л. Охрана природы и воспроизводство
    природных ресурсов. Краеведение
  Охрана природы и воспроизводство
    природных ресурсов
  Краеведение. Туризм

L. Conservation and regeneration of
    natural resources. Local studies
  Preservation of nature and the
    regeneration of natural resources
  Local studies. Tourism

Е. География СССР
  Общий раздел
  География природных ресурсов
    и природные условия
    производства
  География населения и населенных
    пунктов
  География народного хозяйства
    и его отраслей
  Районирование страны
  Союзные республики и районы СССР

E. Geography of the USSR
  General section
  Geography of natural resources
    and natural conditions of
    production
  Population and settlement
    geography
  Geography of the economy and its
    branches
  Regionalization of the country
  Union republics and regions of USSR

Ж. География Зарубежной Европы
  Польша
  Чехословакия
  Венгрия
  Румыния
  Болгария
  Албания
  Югославия
  Германская Демократическая
    Республика
  Западный Берлин
  Федеративная Республика Германии
  Австрия. Швейцария
  Бельгия. Нидерланды. Люксембург
  Скандинавские страны
  Исландия. Фарерские острова.
    Западно-Европейский сектор
    Арктики

Zh. Geography of Europe (outside the USSR)
  Poland
  Czechoslovakia
  Hungary
  Romania
  Bulgaria
  Albania
  Yugoslavia
  German Democratic
    Republic
  West Berlin
  Federal Republic of Germany
  Austria. Switzerland
  Belgium. Netherlands. Luxembourg
  Scandinavian countries
  Iceland. Faeroe Islands.
    The West-European sector
    of the Arctic

16

Ж. География Зарубежной Европы
Великобритания. Ирландия
Франция
Страны Пиренейского полуострова
Италия
Греция. Мальта

Zh. Geography of Europe – continued
Great Britain. Ireland
France
Iberian peninsula
Italy
Greece. Malta

И. География Зарубежной Азии и Африки
Зарубежная Азия
Центральная и Восточная Азия
Юго-Восточная Азия
Южная Азия
Западная Азия
Африка
Северная Африка
Западная Африка
Центральная Африка
Восточная Африка
Южная Африка
Острова Южного полушария,
относящиеся к Африке

I. Geography of foreign Asia and of Africa
Foreign Asia
Central and East Asia
Southeast Asia
South Asia
West Asia
Africa
North Africa
West Africa
Central Africa
East Africa
Southern Africa
Islands of the southern hemisphere
belonging to Africa

К. География Америки, Австралии,
Океании и Антарктики
Северная Америка
Латинская Америка
Центральная Америка
Южная Америка
Австралия и Океания
Антарктика

K. Geography of America, Australia,
Oceania, and Antarctica
North America
Latin America
Central America
South America
Australia and Oceania
Antarctica

The number of entries in the year 1973:

Topical and physical		29,536
Theoretical questions of physical and economic geography	2,670	
Cartography	2,351	
Geomorphology. Quaternary period	2,988	
Oceanography. Hydrology. Glaciology.	5,417	
Meteorology and climatology	5,954	
Biogeography. Geography of soils	5,403	
Natural resources. Local studies	4,753	
Regional and economic		12,661
U.S.S.R.	2,349	
Europe	3,857	
Asia and Africa	4,268	
Asia	3,181	
Africa	1,087	
Americas, Australasia, Oceania, and Antarctica		2,187
North America	1,084	
Latin America	789	
Australia and Oceania	285	
Antarctica	29	
Total number of entries, 1973		42,197

LITERATURA po geografii i smezhnym naukam. Ezhemesiachnyi informatsionnyi biulleten'
(izdaetsia s 1969 goda). (Literature on geography and related disciplines: monthly
information bulletin, published since 1969). Moskva. 1969- . monthly.
Akademiia Nauk SSSR. Tsentral'naia Biblioteka po Estestvennym Naukam. [1969-1973
no. 3 as Sektor Seti Spetsial'nykh Bibliotek]. Biblioteka Instituta Geografii.
Annual index volume with name index (Cyrillic and Latin letters separately) and
list of periodicals and other serials.   17

Monthly list of monographs and articles in collected works or periodicals in
geography and related disciplines received by the Library of the Institute of
Geography of the Academy of Sciences of the USSR.

About 4,000 entries per annum including articles from about 150 Soviet periodicals and serials and about 150 from other countries. The annual index for 1971
records about 4,500 Soviet authors and about 1,000 from other countries; the 1970
index about 4,500 Soviet authors and about 1,500 from other countries. The large
number of Soviet authors in relation to the number of entries reflects (1) the
frequency in the Soviet Union in which articles have multiple authors and (2) the
full analysis in this bibliography of many collected Soviet volumes with numerous
articles or chapters by different authors.

The material is organized systematically or regionally. The systematic division is classified by general sections; historical geography; geodesy and cartography; physical geography and landscape studies; conservation and transformation
of nature; geology and hydrogeology; geomorphology and paleogeography; meteorology
and climatology, with a subsection on agricultural meteorology and agricultural
climatology; hydrology of the land; glaciology; snow cover and permafrost; oceanography; soil science and the geography and cartography of soils; biogeography
and biology; agriculture and reclamation; sociology and demography; economic
geography; economics; applied geography, with a subsection on medical geography;
popular literature.

The regional division is classified by the USSR, with subsections on the USSR
as a whole, the RSFSR as a whole, the European part of the RSFSR, the Asiatic part
of the RSFSR, the European part of the USSR, the Caucasus, the Asiatic part of
the USSR; by foreign countries, with subsections on Europe, Asia, Africa, America,
and islands of the Pacific Ocean; oceans and seas; and polar countries. From 1973,
no. 1, each issue contains a list of the specific volumes and numbers of serials
analyzed in that issue.

Annual indexes for 1970 (1971, 50 p.), 1971 (1972, 48 p.), and 1972 (1973,
48 p.) provide an author index for names in Cyrillic and names in Latin letters,
and a list of periodicals and other serials analyzed during the year.

VSESOIUZNYI Institut Nauchnoi i Tekhnicheskoi Informatsii. Geograficheskii sbornik.
Moskva, Proizvodstvenno-izdatel'skii Kombinat VINITI, 1963- . [v. 1.]. 1963.
243 p. Raboty po teoreticheskim i spetsial'nym voprosam nauchno-tekhnicheskoi
informatsii (Works on theoretical and special questions of scientific-technical
information). v. 2. 1966. 307 p. Metody geograficheskogo izucheniia. Ispol'
zovanie prirodnykh resursov (Methods of geographical research. Utilization of
natural resources). v. 3, 1969. 298 p. Nauchnaia informatsiia v geografii.
Teoreticheskie i regional'nye problemy geografii (Scientific information in
geography. Theoretical and regional problems of geography). v. 4, 1970. 330 p.
Obshchie voprosy geografii. Regional'nye problemy geografii (General questions
of geography. Regional problems in geography). v. 1, not numbered, and issued
by Institut Nauchnoi Informatsii Akademii Nauk SSSR. v. 3-4 have supplementary
table of contents in English.   1

Series of review articles on the literature of many fields of geography with
extensive bibliographies.

17 - 18

ITOGI nauki i tekhniki: seriia geografiia (Results of science and technology: geography series). (Vsesoiuznyi Institut Nauchnoi i Tekhnicheskoi Informatsii).
Moskva, VINITI, 1964- .
                                                                                           19
  Surveys of the literature and state of the fields of geography, often with extensive bibliographies. Issued in series:
  Teoreticheskie voprosy geografii (Theoretical questions in geography). 1
(1966).
  Teoreticheskie voprosy fizicheskoi i ekonomicheskoi geografii (Theoretical problems of physical and economic geography). 1- (1972).
  Kartografiia (Cartography). 1- (1962-).
  Geomorfologiia (Geomorphology). 1- (1966-).
  Gidrologiia (Hydrology). 1- (1963-).
  Meditsinskaia geografiia (Medical geography). 1- (1964-).
  Okhrana prirody i vosproizvodstvo prirodnykh resursov (Conservation of nature and reproduction of natural resources). 1- (1968-).
  Geografiia SSSR (Geography of the USSR). 1- (1965-).
  Geografiia zarubezhnykh stran(Geography of foreign countries). 1- (1972-).
  Ekonomgeograficheskaia izuchennost' raionov kapitalisticheskogo mira
(Economic-geography study of the regions of the capitalist world). 1-3 (1964-1966).
  Teoreticheskie i obshchie voprosy geografii (Theoretical and general questions of geography). 1- (1974-).

Current General Soviet Bibliographies

KNIZHNAIA letopis'. Organ gosudarstvennoi bibliografii SSSR. (Book annals. Organ of the state bibliography of the USSR). v. 1- (1907-). weekly. Moskva,
Izdatel'stvo. "Kniga."
                                                                                           20

  Current listing of new books as published, arranged by 31 major classes.
Under section 15, Estestvennye nauki (natural sciences) subsection 4, geologo-geograficheskie nauki. Kraevedenie. (Geological and geographical sciences.
Regional studies), lists most geographical publications, although some in economic and political geography are listed in section 8, Mezhdunarodnye otnosheniia. Politicheskoe i ekonomicheskoe polozhenie zarubezhnykh stran
(International relations: political and economic situation of foreign countries), or section 9, Kommunisticheskoe stroitel'stvo SSSR (Communist construction in the USSR), and on teaching of geography in the schools in section 24, Kultura,
Prosveshchenie. Nauka (Culture. Education. Science).

  Quarterly author, place, and subject indexes help to locate material. This is the fullest listing of Soviet books and is invaluable for bibliographical checking.

  The fullest listing of Soviet publications. In 1970, for example, Knizhnaia letopis' listed 44,545 publications and the supplementary number of Knizhnaia letopis' 36,657; thus Knizhnaia letopis' and its supplement listed 81,202 separate Soviet publications in that year or about 2 1/2 times as many as the 33,777 items in Ezhegodnik knigi SSSR. Indispensable for bibliographical checking.

  The quarterly author, place, and subject indexes (Vspomogatel'nye ukazateli, 1907-) help to locate material.

  The annual guide to series (Ukazatel' seriinykh izdanii) lists under series title the individual volumes of irregular series, such as Itogi nauki, with reference number to full description in the main body of Knizhnaia letopis'.

_____. Dopolnitel'nyi vypusk. (Supplementary number). 1938-1941, 1961- . monthly. Section 15.4 Geologo-geograficheskie nauki. Kraevedenie (Geological-geographical sciences. Regional studies).   21

List of materials issued in smaller editions for limited circle, such as official documents, instructional, teaching, or handbook-informational publications.

Supplement lists dissertation abstracts for both doctoral and candidate degrees, arranged by disciplines. Information includes name of author, title, year, number of pages, and institution at which offered.

EZHEGODNIK knigi SSSR: sistematicheskii ukazatel' (Annual of books of the USSR; classified index). 1941- . annual. Moskva, Izdatel'stvo "Kniga," (Vsesoiuznaia Knizhnaia Palata).   22

Annual cumulation of monographs listed in Knizhnaia letopis', omitting certain categories and thus less complete, but much easier to use. Since 1957 in two volumes, the first covering the humanities and social sciences, the second covering natural sciences and technology; 1943-1944 divided by subject into three issues; 1945-1956 divided into two six-monthly volumes covering all subjects.

The most convenient source for checking Soviet books published since July, 1941.

Geography is in volume 2, chapter 1. Estestvennye nauki. Matematika (Natural sciences and mathematics), section 4. Geologo-geograficheskie nauki (geological and geographical sciences), subsection a. Obshchie voprosy. Geografiia. Kraevedenie (General questions. Geography. Regional studies).

RETROSPECTIVE BIBLIOGRAPHIES

Guides to the Literature of Geography

AKADEMIIA Nauk SSSR. Geograficheskoe Obshchestvo Soiuza SSR. Sovetskaia geografiia:
itogi i zadachi (Soviet geography· accomplishments and tasks). I. P. Gerasimov,
S. V. Kalesnik, O. A. Konstantinov, E. M. Murzaev, K. A. Salishchev, and G. M.
Ignat'ev, eds. Moskva, Geografgiz, 1960. 635 p.   23

SOVIET geography: accomplishments and tasks. A symposium of 50 chapters, contributed
by 56 leading Soviet geographers and edited by a committee of the Geographic
Society of the USSR, Academy of Sciences of the USSR. I. P. Gerasimov, chairman,
G. M. Ignat'yev, secretary, S. V. Kalesnik, O. A. Konstantinov, E. M. Murzayev,
K. A. Salishchev. Translated from the Russian by Lawrence Ecker. English edition
edited by Chauncy D. Harris. New York, American Geographical Society, 1962.
409 p. (Occasional publication no. 1). [Also as entry 2582].   24

    An inventory and appraisal of modern scholarly work in the geographical
sciences in the Soviet Union. Extensive bibliographies at the ends of chapters
provide a valuable introduction to the literature of each subfield of geography
in the USSR. Bibliographies with page references in the English translation are:
Physical geography of the land,.p. 22 (18 entries); physical geography of the seas
and oceans, p. 28 (21); economic geography, p. 38 (24); regional geography, p. 44
(6); cartography, p. 51-52 (61); history of geographical knowledge, p. 60 (22);
climatology, p. 72-73 (21); glaciology, p. 80 (28); permafrost science, p. 86
(21); hydrology of the land, p. 94-95 (40), geomorphology, p. 107 (31); geographic
soil science, p. 116-117 (22); plant geography, p. 127 (30); zoogeography of the
land, p. 139-140 (33); geography of population and settlements, p. 149-150 (22);
geography of industry, p. 155 (20); paleogeography of the ice age, p. 164 (15);
paleogeography of postglacial times, p. 174 (21); heat and water regime of the
earth's surface, p. 179-181 (57); geographic zonality, p. 186-187 (23); zonality
of the ocean, p. 194 (12); snow cover, p. 200 (25); landscape science, p. 204
(5); natural regionalization, p. 209 (18); integrated mapping, p. 213-214 (10);
economic regionalization, p. 223-224 (15); integrated development of productive
forces of economic regions of the USSR, p. 229 (10); economic geography of agriculture, p. 238 (18); transportation geography, p. 248-249 (25); medical geography,
p. 253 (17); origin of geographical names, p. 258 (39); polar lands, p. 273-274
(42); the tayga, p. 280 (21); steppes and deserts, p. 288-289 (12); agricultural
lands, p. 296-297 (27); methods of expeditions, p. 306 (7); aerial photography
and stereophotogrammetry, p. 321 (28); research in physical geography in fixed
stations, p. 329-330 (44); laboratory analysis and experiment, p. 345 (33); nature
preserves, p. 354 (10); methods of research in economic geography, p. 361 (15);
geographic education in secondary schools, p. 369 (9); geographic education in
institutions of higher learning, p. 380-381 (54); tourism and alpinism, p. 389
(8); serials cited in the book, p. 403-406 (150), not in the Russian edition.
Total number of references 1,040 in the Russian edition and 1,190 in the English
edition.

MAR, Igor' Valerianovich; MURZAEV, Edward Makarovich; PREOBRAZHENSKII, Vladimir
Sergeevich; ABRAMOV, Lev Solomonovich; GOKHMAN,Veniamin Maksovich; POKSHISHEVSKII,
Vadim Viacheslavovich; GRATSIANSKII, Andrei Nikolaevich; and MIKHAILOV, Nikolai
Ivanovich. Sovetskaia geograficheskaia literatura v period mezhdu IV i V s"ezdami
Geograficheskogo Obshchestva Soiuza SSR. Materialy V s"ezda Geograficheskogo
Obshchestva SSSR. Leningrad, 1970. 67 p. In English, Soviet geographical lit-
erature (between the fourth and fifth congresses), Soviet geography: review and
translation, v. 12, no. 8 (October, 1971), p. 495-528.   25

A review of geographical publications in the Soviet Union 1965-1970 including principal works on theory and methodology, university and college textbooks, geographical periodicals and serials, information and abstract journals, series on regional geography of the USSR, thematic and special-purpose regional volumes, world regional geography, popular geographic literature, school textbooks, and the organization of publishing activity in geography. For an earlier review see the chapter by V. V. Pokshishevskii in Soviet geography: accomplishments and tasks (English edition, 1962), p. 390-396. (p. 616-626 in the Russian edition, 1960).

GEOGRAFIIA na novykh rubezhakh: rekomendatel'nyi obzor literatury. Moskva, Izdatel'-stvo "Kniga," 1973. 14 p. (Gosudarstvennaia Biblioteka SSSR imeni V. I. Lenina. Novoe v nauke i tekhnike). G. P. Bogatova, comp.   26

Former Current Geographical Bibliographies

Including the USSR

Former current bibliographies covering individual years were published by the Geographical Society in its Vestnik or Izvestiia or in the journal Zemlevedenie. These cover the years 1851-1854, 1858-1880, 1891-1899, 1905, 1915-1916, 1927-1934, 1946-1949, and 1956 and generally include both Russian and foreign works but sometimes only one or the other.

Other earlier current bibliographies were prepared by the Academy of Sciences for 1901-1913 for the International Catalogue of Scientific Literature and 1954-1955 as comprehensive annual bibliographies of the geographic literature of the USSR.

OBOZRENIE russkoi geograficheskoi, etnograficheskoi i statisticheskoi literatury za 1851 [-1854 gg.] (Review of Russian geographical, ethnographic, and statistical literature, 1851-1854). Russkoe Geograficheskoe Obshchestvo, Vestnik. Chasti 4-18, otdel 4 (1852-1856).   27

USTINOV, M. Obozrenie russkoi geograficheskoi literatury 1858 goda (Review of Russian geographical literature in 1858). Russkoe Geograficheskoe Obshchestvo. Vestnik. chast' 26, kniga 5, otdel 4 (1859), p. 1-18; chast' 27, otdel 4 (1859), p. 1-102.   28

MEZHOV, Vladimir Izmailovich. Literatura russkoi geografii, statistiki i etnografii za god 1859-1880. (The literature of Russian geography, statistics, and ethnography 1859-1880). St. Peterburg, Tip. V. Bezobrasova, 1861-1883. 9 v. v. 1 as Bibliograficheskii ukazatel' vyshedshikh v Rossii knig i statei po chasti geografii, topografii, etnografii i statistika. v. 9 as Literatura russkoi geografii, etnografii i statistiki.   29

v. 1...1859-1863. 1861-1864. 112, 166, 166, 132, 163 p. 9,199 entries.
v. 2...1864-1866. 1867. 134, 152, 191 p. 7,505 entries.
v. 3...1867-1868. 1870. 210, 169 p. 6,555 entries.
v. 4...1869-1870. 1873. 178, 241 p. 7,667 entries.
v. 5...1871-1872. 1874. 245, 282 p. 9,675 entries.
v. 6...1873-1874. 1877. 270, 276 p. 10,299 entries.
v. 7...1875-1876. 1878. 324, 326 p. 11,351 entries.
v. 8...1877-1878. 1881. 288, 384 p. 11,086 entries.
v. 9...1879-1880. 1883. 335, 344 p. 10,429 entries.

Comprehensive bibliography on Russian geography covering 22 years in 9 volumes with 83,766 entries, originally published in the Vestnik or Izvestiia of the Geographical Society. Generally in two sections, the first covering general works, the second in turn geography, statistics, mathematical geography, and ethnography. Indexes of names and places, except for v. 1.

GEOGRAFICHESKAIA literatura Rossii Evropeiskoi i Aziatskoi i prilezhashchikh stran,
po dannym Biblioteki Imperatorskogo Russkogo geograficheskogo obshchestva
(Geographical literature of Russia, European and Asiatic, and adjacent countries,
according to information of the Library of the Imperial Russian Geographical
Society), Russkoe Geograficheskoe Obshchestvo, Izvestiia, v. 27-35 (1891-1899). 30

   Russian and foreign Geographical literature on Russia and adjacent lands,
1891-1899, listed in each number, usually at end of issue.

GEOGRAFICHESKAIA literatura po dannym Biblioteki Imperatorskogo Russkogo Geograficheskogo Obshchestva (Geographical literature according to information of the
Library of the Imperial Russian Geographical Society). Russkoe Geograficheskoe
Obshchestvo, Izvestiia, v. 41, no. 4 (1905), p. 806-822; v. 42, no. 1 (1906),
p. 333-352. Bibliografiia russkoi geograficheskoi literatury, v. 42, no. 1
(1906), p. 305-331.   31

   Russian and foreign geographical literature, 1905.

SPISOK knig, postupivshikh v Biblioteku Russkogo Geograficheskogo Obshchestva
(List of books received in the Library of the Russian Geographical Society),
Russkoe Geograficheskoe Obshchestvo, Izvestiia, v. 51-52 (1915-1916).   32

   Russian and foreign geographical literature, 1915-1916, listed in each
number.

BODNARSKII; Mitrofan Stepanovich., and DITMAR, B. P. Bibliografiia zemlevedeniia,
(Bibliography of geography), Zemlevedenie, v. 29 (1927), no. 3/4, p. 33-144,
v. 30 (1928), no. 1-2, p. 119-128, p. 213-216, and no. 4, p. 315-316.   33

   List of current literature appearing in the RSFSR 1927-1929 based mainly
on Knizhnaia letopis' and Letopis' zhurnal'nykh statei. Material organized under
the following headings: mathematical geography; physical geography; geophysics;
anthropology; biogeography; regional geography; history of geography; cartography;
teaching, programs, etc.

BORZOV, Aleksandr Aleksandrovich. Bibliografiia (Bibliography), Zemlevedenie.
v. 32 (1930), no. 1-2, p. 115-128; v. 33 (1931), no. 1/2, p. 154-176, no. 3-4,
p. 339-359; v. 34 (1932), no. 1-2, p. 119-130; v. 35 (1933), no. 3, p. 251-271;
v. 36 (1934), no. 3, p. 303-319.   34

   Lists of current literature, 1930-1934, both Russian and foreign, both
articles and books, based on the systematic catalogue of the Lenin State Library
of the USSR in Moscow.

PISOK geograficheskoi literatury, vyshedshei v SSSR i postupivshei v Biblioteku
Vsesoiuznogo geograficheskogo obshchestva (List of geographical literature, published in the USSR and received by the Library of the Geographical Society of the
USSR), Geograficheskoe Obshchestvo SSSR, Izvestiia, v. 79 (1947), no. 3, p. 373-
380 (252 entries), no. 5, p. 611-615 (124 entries); v. 80 (1948), no. 2, p. 207-
212 (182 entries), no. 3, p. 313-322 (278 entries), no. 5, p. 550-561 (352 entries);
v. 81 (1949), no. 1, p. 127-136 (300 entries), no. 3, p. 364-368 (141 entries),
no. 4, p. 441-449 (266 entries); v. 82 (1950), no. 5, p. 442-448 (193 entries);
v. 86 (1954), no. 4, p. 384-398 (457 entries), no. 5, p. 485-498.   35

   List of geographical works published in the Soviet Union and received by the
Library of the Geographical Society of the USSR 1946-1949 and 1953.

PISOK literatury po geografii i smezhnym distsiplinam, postupivshei v Biblioteku
Geograficheskogo Obshchestva Soiuza SSR (List of literature in geography and
neighboring disciplines, received in the Library of the Geographical Society of
the USSR). (Geograficheskoe Obshchestvo SSSR, Izvestiia, v. 88, no. 1 (1956),
p. 108-126.   36

   About 680 entries for books and periodical publications in Russian, other
languages of the USSR, and foreign languages.

RUSSKAIA bibliografiia po estestvoznaniiu i matematike, sostavlennaia pri Akademii Nauk S.-Peterburgskim biuro mezhdunarodnoi bibliografii (Russian bibliography in science and mathematics, compiled in the Academy of Science by the St. Petersburg bureau of international bibliography). St. Petersburg, Petrograd, Akademiia Nauk, 1904-1917. v. 1-9.   3ʳ

The Russian entries compiled for inclusion in the International Catalogue of Scientific Literature. Covers the years 1901-1913.

One section covers mathematical and physical geography.

BARINOV, Iu. A., KRUGLAKOVSKII, A. N., and SHRAIBER, L. Ia. Geograficheskaia literatura SSSR: bibliograficheskii ezhegodnik (Geographical literature of the USSR: bibliographical yearbook). 1954-1955. Moskva, 1961-1962. (Akademiia Nauk SSSR. Sektor Seti Spetsialnykh Bibliotek). 302 p. 332 p. Rotaprint.   3ⁱ

A comprehensive annual bibliography of the geographic literature of the USSR in the Russian language or in other languages of the USSR if they contain a Russian summary. 2,427 entries for 1954 and 2,903 entries for 1955 including books and articles from periodicals and collected works, organized by geographical disciplines and regions. Sources of titles include publications of the Vsesoiuznaia Knizhnaia Palata, the bibliographical journal, Referativnyi Zhurnal: geografiia i geologiia, current bibliographies in related disciplines, and the basic geographical journals and collections. Alphabetical index of authors, editors, and staff. Covers the years 1954 and 1955 only.

Foreign Literature Only

SPISOK inostrannykh postuplenii Biblioteki Vsesoiuznogo Geograficheskogo Obshchestva (List of foreign literature received by the Library of the Geographical Society of the USSR), Vsesoiuznoe Geograficheskoe Obshchestvo, Izvestiia, v. 69 (1937), no. 1, p. 200-202, no. 2, p. 337-340, no. 3, p. 497-499, no. 4, p. 696-697, no. 5, p. 848-856, no. 6, p. 1,044-1,046; v. 70 (1938), no. 1, p. 145-146, no. 2, p. 330-332, no. 3, p. 441-444, no. 4, p. 668-670, no. 5, p. 818-821; v. 71 (1939), no. 3, p. 475-478, no. 4, p. 646-648, no. 5, p. 791-794, no. 6, p. 950-952, no. 7, p. 1,099-1,102, no. 8, p. 1,273-1,274, no. 9, p. 1,407-1,408; v. 72 (1940), no. 1, p. 123-124, no. 2, p. 296-298, no. 3, p. 450-452, no. 6, p. 874-879; v. 73 (1941), no. 1, p. 149-152, no. 2, p. 498-500; v. 77 (1945), no. 4, p. 251-256; v. 79 (1947), no. 1, p. 97-99, no. 4, p. 502-504; v. 80 (1948), no. 3, p. 322-326, no. 5, p. 562-564; v. 81 (1949), no. 4, p. 449-451; v. 86 (1954), no. 4, p. 398-401, no. 5, p. 499-502.   3

Foreign geographic literature received by the Library of the Geographical Society of the USSR, 1936-1949 and 1954.

SISTEMATICHESKII ukazatel' statei v inostrannykh zhurnalakh. Geologiia, geofizika i geografiia (Systematic guide to articles in foreign journals: geology, geophysics, and geography). nos. 1-12, 1948-1953, Moskva, Izdatel'stvo inostrannoi literatury, 1950-1955. (Vsesoiuznaia Gosudarstvennaia Biblioteka Inostrannoi Literatury). no. 1-2, 1948 (1950) as: Geologiia i geografiia.   4C

Foreign geographic and related literature 1948-1953.

LIBRARY CATALOGUE

USSKOE Geograficheskoe Obshchestvo. Sistematicheskii katalog knig Biblioteki Russkago Geograficheskago Obshchestva. Chast' 1. Knigi na russkom iazyke. v. 1. (classed catalogue of books in the Library of the Russian Geographical Society. Part 1. Books in the Russian language, v. 1). Petrograd, Tipografiia Morskogo Komissariata. 1922. 431 p.   41

About 10,000 entries of books and reprints of articles from periodicals and collected volumes published from the 18th century to 1905, in two parts. Part 1, general geography and related disciplines: history and methodology of geography, mathematical geography, physical geography, biogeography and natural sciences, study of peoples, historical geography and history, statistics, political economy, and finance. Part 2, regional geography, mainly the Russian Empire, organized systematically for the country as a whole or for its regions. Replaces earlier catalogue of the library: Sistematicheskii katalog Biblioteki Russkogo geograficheskogo obshchestva, F. I. Liutsenskii and V. I. Sreznevskii, comps. parts 1-2. St. Peterburg, 1878. Part 1, 42 p. Part 2, 178 p.

PUBLICATIONS OF INSTITUTIONS
OR OF PERSONNEL ASSOCIATED WITH INSTITUTIONS

Geographical Society of the USSR

See also indexes to publications of the Geographical Society of the USSR, 1846-1905, under cumulative indexes to geographical serials [87-90].

For lists of publications and indexes to the publications of numerous individual branches of the Geographical Society of the USSR, see the following sources:

RG, Lev Semenovich. Letopis' Vsesoiuznogo Geograficheskogo Obshchestva za 1845-1945 gg. (Chronicle of the Geographical Society of the USSR, 1845-1945), Geograficheskoe Obshchestvo SSSR, Izvestiia, v. 78, no. 1 (1946), p. 25-90.   42

The basic publications of the Geographical Society of the USSR are given in the text: books, serials, collected volumes, and maps. About 350 items during the first century of the Society.

RG, Lev Semenovich. Spisok glavneishikh izdanii Geograficheskogo Obshchestva (List of the most important publications of the Geographical Society), in his book: Vsesoiuznoe Geograficheskoe Obshchestvo za sto let, 1845-1945. (The All-Union geographical society during a hundred years 1845-1945). Moskva-Leningrad, Izdatel'stvo Akademii Nauk SSSR, 1946. p. 249-252. (Akademiia Nauk SSSR. Nauchno-populiarnaia Seriia).   43

80 entries, selected as the more important publications of the Society in the course of a century.

RECHEN' izdanii Geograficheskogo Obshchestva Soiuza SSR, postupivshikh v Biblioteku Obshchestva za ... (List of publications of the Geographical Society of the USSR received in the Library of the Society for the year ...), 1960- in Geograficheskoe Obshchestva SSSR, Izvestiia, v. 93- (1961-). 1960 ..., v. 93, no. 3 (1961), p. 281-288; 1961 ..., v. 94, no. 5 (1962), p. 451-458; 1962 ..., v. 95, no. 6 (1963), p. 556-578; 1963 ..., v. 96, no. 6 (1964), p. 541-552; 1964..., v. 97, no. 4 (1965), p. 392-399; 1965 ..., v. 98, no. 2 (1966), p. 190-203; 1966 ..., v. 99, no. 4 (1967), p. 338-368; 1967 ..., v. 101, no. 2 (1969), p. 174-190; 1968-1970 ..., v. 104, no. 2 (1972), p. 144-156.   44

VOROB'ĒV, Vladimir Vasil'evich, and VERSHINSKAIA, N. I., comps. Publikatsii sibirskikh i dal'nevostochnykh organizatsii Geograficheskogo Obshchestva SSSR 1945-1963 gg. (sistematicheskii i avtorskii ukazateli). (Publications of the Siberian and Far Eastern organizations of the Geographical Society of the USSR 1945-1963 [systematic and author guides]). Irkutsk, 1966. 166 p. (Akademiia Nauk SSSR. Sibirskoe otdelenie. Institut Geografii Sibiri i Dal'nego Vostoka. Biuro sibirskikh i dal'nevostochnykh organizatsii Geograficheskogo obshchestva SSSR). 45

1495 entries of monographs and articles, arranged by systematic fields of geography. List of publications by branches of the Geographical Society. Alphabetical index of authors, reviews, and names of books without authors.

PUBLIKATSII sibirskikh i dal'nevostochnykh organizatsii Geograficheskogo Obshchestva SSSR. tom 2 (1964-1969 gg.). vypuski 1-2 (Publications of the Siberian and Far Eastern organizations of the Geographical Society of the USSR, v. 2, 1964-1969, nos. 1-2). Irkutsk, 1971. 177 p., 208 p. (Akademiia Nauk SSSR. Sibirskoe otdelenie. Institut Geografii Sibiri i Dal'nego Vostoka. Biuro sibirskikh i dal'nevostochnykh organizatsii Geograficheskogo Obshchestva SSSR). 46

4,391 entries of monographs and articles arranged by systematic fields. Author index.

GEOGRAFICHESKOE Obshchestvo SSSR. Zabaikal'skii Filial. Vestnik nauchnoi informatsii, no. 10. Bibliografiia izdanii filiala, A. I. Sizikov, comp. Chita, Zaibaikal'skii filial Geograficheskogo Obshchestva SSSR, 1971. 54 p. (Akademiia Nauk SSSR. Geograficheskoe Obshchestvo SSSR). 47

Includes bibliography of publications of the Trans-Baykal branch of the Geographical Society of the USSR: general characteristics of scientific-publishing activity 1896-1971; contents of the branch's Vestnik nauchnoi informatsii, nos. 1-9 (1965-1968); Izvestiia Zabaikal'ia, nos. 1-6 (1965-1971); Zabaikal'skii kraevedcheskii ezhegodnik, nos. 1-4 (1967-1970); Zapiski, nos. 18-54 and other publications (1962-1971); and a list of scientific publications of the branch.

VINOGRADOV, Vladimir Nikolaevich. Publikatsii Kamchatskogo otdela Geograficheskogo Obshchestva SSSR, 1941-1969 gg. (Publications of the Kamchatka section of the Geographical Society of the USSR, 1941-1969). Petropavlovsk-Kamchatskii, Kamchatskii Otdel Geograficheskogo Obshchestva SSSR, 1970. 32 p. 48

251 entries. Alphabetical index of authors and titles. Systematic index of authors.

PANKOV, A. V. Sistematicheskii ukazatel' k izdaniiam Uzbekistanskogo Geograficheskogo Obshchestva i ego predshestvennikov, sostavlennyi k 50-letiiu Obshchestva, 1897 po 1947 (Classed guide to the publications of the Uzbekistan Geographical Society and its predecessors, compiled on the 50th anniversary of the Society, 1897 to 1947), Trudy Uzbekistanskogo Geograficheskogo Obshchestva, v. 2 (1948), p. 171-201. 49

323 entries. List of publications used. Author and place indexes.

Academies of Sciences in the USSR

AKADEMIIA Nauk SSSR. Institut Geografii. Biblioteka. Izdaniia Instituta Geografii Akademii Nauk SSSR, 1918-1958: bibliograficheskii ukazatel' (Publications of the Institute of Geography of the Academy of Sciences of the USSR, 1918-1958: a bibliographical guide). E. A. Stepanova and V. A. Sokolova, comps. L. G. Kamanin, ed. Moskva, Akademiia Nauk SSSR. Institut Geografii. Biblioteka, 1959. 169 p. Processed. 50

1,118 entries for publications of the Institute of Geography or by members of its staff, including rotaprint editions and collected works, systematically arranged, and within each division works of a general character first followed by regional works. Index of authors. Chronological list of publications of the Institute.

KADEMIIA Nauk SSSR. Sektor Seti Spetsial'nykh Bibliotek. Biblioteka Instituta Geografii. Izdaniia Instituta Geografii Akademii Nauk SSSR (1959-1969): Bibliograficheskii ukazatel' (Publications of the Institute of Geography of the Academy of Sciences of the USSR, 1959-1969: bibliographic guide). E. S. Steklenkova, V. A. Sokolova, comps. L. G. Kamanin, ed. Moskva, 1972. 218 p. Rotaprint.   51

242 entries organized by systematic fields and regions. Collected works, each a single entry, analyzed in detail. Chronological list of publications, p. 165-204. About 1,350 names of authors, editors, or collaborators listed in the author index, p. 205-218. Valuable as an inventory of work carried on in, published by, or achieved through collaboration with the Institute of Geography in Moscow, the largest group of research geographers in the Soviet Union. Especially strong for physical geography, geomorphology, hydrology, glaciology, soil science and geography of soils, biogeography, and economic geography of the USSR.

KADEMIIA Nauk SSSR. Sibirskoe Otdelenie. Institut Geografii Sibiri i Dal'nego Vostoka. Katalog izdanii (1959-1968 gg.) (Catalogue of publications, 1959-1968). V. S. Kurbatov, comp. Irkutsk, 1968. 20 p.   52

60 entries including collected works, monographs by collaborators, and conferences, published by the Institute of Geography of Siberia and the Far East in Irkutsk in Eastern Siberia.

KADEMIIA Nauk SSSR. Biblioteka Akademii Nauk SSSR. Bibliografiia izdanii Akademii Nauk SSSR: ezhegodnik (Bibliography of publications of the Academy of Sciences of the USSR: annual). Leningrad, Izdatel'stvo Otdela Biblioteki AN SSSR, 1957- . v. 1, 1956 (1957), 342 p.; v. 2, 1957 (1958), 489 p.; v. 3, 1958 (1960), 601 p.; v. 4, 1959 (1961), 752 p.; v. 5, 1960 (1962), 989 p.; v. 6, 1961 (1963), 947 p.; v. 7, 1962 (1964), 901 p.; v. 8, 1963 (1965), 985 p.; v. 9, 1964 (1966), 1005 p.; v. 10, 1965 (1967), 1066 p.; v. 11, 1966 (1969), 893 p.; v. 12, 1967 (1970), 880 p.; v. 13, 1968 (1971), 1082 p.; v. 14, 1969 (1972), 922 p.; v. 15, 1970 (1972), 1041 p. V. A. Vasil'ev, M. V. Voronov, V. V. Komarov, L. A. Churkin, comps. S. P. Luppov, ed.   53

Annual list of books and articles in serial publications of the Academy of Sciences of the USSR, listed by departments, institutes, and societies of the Academy, including the Institute of Geography and the Geographical Society of the USSR. In the 1970 volume, for example, under the chapter on sciences of the earth, p. 333-463 are listed for the Geographical Society of the USSR, p. 352-383, entries 2124-2204 some 81 publications of the Society for this year, including those of its numerous branches (with record of all articles and authors in collected volumes), and for the Institute of Geography, p. 451-456, entries 2640-2677, some 38 publications, including dissertations. Also of value, of course, are publications of institutes in related fields.

ADEMIIA Nauk SSSR. Izdatel'stvo. Katalog knig 1945-1962 gg. (Catalogue of books, 1945-1962). Moskva, Izdatel'stvo "Nauka," 1965. 367 p.   54

Geography, p. 152-158 (181 books), oceanography, p. 258-260 (66 books), permafrost studies, p. 160-162 (66 books), hydrology, p. 162-165, (68 books), volcanology, p. 165-166 (22 books), study of lakes, p. 166-167 (18 books), and geology and mineral deposits, p. 149-152 (67 books).

___. ___. ___ 1963-1967. 1971. 340 p.   55

AKADEMIIA Nauk Gruzinskoi SSR. Tsentral'naia Nauchnaia Biblioteka. Bibliografiia izdanii Akademii nauk Gruzinskoi SSR, 1937-1956 (Bibliography of publications of the Academy of Sciences of the Georgian SSR, 1937-1956). Tbilisi, Izdatel'stvo Akademii Nauk Gruzinskoi SSR, 1959. 945 p. In Russian and Georgian. 56

The geographical sciences, p. 191-215, entries 1502-1699, list 198 publications by fields: general questions; history of geography; cartography; climatology; geomorphology; hydrology; and soil science. Entries are in original language, whether Georgian or Russian, with bibliographical details, including presence of summaries in other languages and number of references in bibliographies. If entry is in Georgian full Russian translation is given.

_____. _____. _____ 1957-1960 gg. 1963. 578 p. 57

Geographical sciences, p. 85-99, entries 660-776, 117 publications under the following headings: general questions and bibliography; historical geography; toponymics; physical geography; climatology; geomorphology; hydrology; speleology; and glaciology.

BLAGOOBRAZOV, Vladimir Alekseevich. Tian'-Shan'skaia fiziko-geograficheskaia stantsiia (Tyan'-Shan' physical-geographic station). Frunze, "Ilim," 1965. 255 p. (Akademiia Nauk Kirgizskoi SSR). 58

426 references on the scientific work of the station in the period 1947-April 1964. Author, subject, and geographical-name indexes. List of 27 publications of the station.

Universities and Pedagogical Institutes

Moscow State University

URUSOVA, N. T., and OPOCHINSKAIA, E. A. Spisok opublikovannykh rabot sotrudnikov Geograficheskogo Fakul'teta Moskovskogo Gosudarstvennogo Universiteta na 1 sentiabria 1957 g. (List of published works by staff members of the Geography Faculty of Moscow State University on September 1, 1957.) K. K. Markov, A. M. Riabchikov, and Iu. G. Saushkin, eds. Moskva, 1957. 196 p. Rotaprint (Moskovskii Gosudarstvennyi Universitet imeni Lomonosova. Geograficheskii Fakul'tet. Otdeleni Geografii Nauchnoi Biblioteki imeni Gor'kogo). 59

About 2,500 entries, arranged alphabetically by authors.

_____. _____ s 1 sentiabria 1957 g. po 1 ianvaria 1959 g. (from September 1, 1957, to January 1, 1959. N. A. Gvozdetskii, and others, eds. 1959. 93 p. 60

About 750 entries, alphabetically by author. Author index.

_____. Spisok opublikovannykh rabot sotrudnikov Geograficheskogo Fakul'teta Moskovskogo Gosudarstvennogo Universiteta s 1 ianvaria 1959 g. po 1 ianvaria 1961 g. (List of published work of members of the Geography Faculty of Moscow State University, January 1, 1959 to January 1, 1961). N. A. Gvozdetskii, ed. Moskva, 1962. 104 p. (Moskovskii Gosudarstvennyi Universitet imeni Lomonosova. Geograficheskii Fakul'tet. Otdel geografii Nauchnoi Biblioteki imeni Gor'kogo). 61

About 1,350 entries for books, articles in periodicals and collected volumes, translations, editorial work, arranged by authors in alphabetical order and by author chronologically. List of earlier publications of new collaborators. List of foreign publications 1959-1960 in which works by members of the faculty are printed. Name index.

_____ . Spisok monografii, uchebnikov, uchebnykh posobii, kursov lektsii, metodicheskikh rabot i nauchno-populiarnykh knig geografov Moskovskogo Universiteta (List of monographs, textbooks, teachers' aids, courses of lectures, methodological works, and scientific-popular books by geographers of Moscow University). In the book: Geografiia v Moskovskom Universitete za 200 let. Spravochnyi material (Geography in Moscow University during 200 years. Handbook material). Moskva, 1955, p. 63-101 (Moskovskii Gosudarstvennyi Universitet imeni Lomonosova).   62

516 entries, 1755-1955. List of persons receiving degrees in geography in Moscow State University.

Leningrad State University

ENINGRADSKII Gosudarstvennyi Universitet. Nauchnaia Biblioteka imeni M. Gor'kogo. Sistematicheskii ukazatel' k periodicheskim izdaniiam S-Peterburgskogo-Leningradskogo universiteta, 1876-1956 (Classed index to the periodical publications of St. Petersburg-Leningrad university, 1876-1956). Leningrad, Izdatel'stvo Leningradskogo Universiteta, 1959. 445 p.   63

5834 entries arranged by fields. Author index. Geographical sciences, p. 226-240, entries 3194-3411, include the following subdivisions: general questions and history; cartography; physical geography; geomorphology, quaternary geology, and glaciology; meteorology and climatology; hydrology and oceanography, hydrochemistry; and economic geography. Within each category entries are arranged alphabetically by name of author.

Other Universities or Pedagogical Institutes

RUSOVA, N. T. Geografiia v izdaniiakh universitetov i pedagogicheskikh institutov (Geography in the publications of universities and pedagogical institutes), Nauchnye doklady vysshei shkoly: geologo-geograficheskie nauki, 1958, no. 3, p. 231-248; 1959, no. 2, p. 229-237; Geografiia i khoziaistvo, no. 6 (1960), p. 87-91, no. 7 (1960), p. 60-64, no. 8 (1960), p. 96-101, no. 9 (1961), p. 85-90, no. 10 (1961), p. 98-103, no. 11 (1961), p. 108-112, no. 12 (1963), p. 94-99.   64

Systematic list of scientific and methodological works, colloquia, conferences, and congresses, abstracts of dissertations, and other material in publications of institutions of higher learning of the USSR, 1957-1960: 604 entries for 1957, 412 for 1958, 342 for 1959, and 284 for 1960.

PUBLIKOVANNYE raboty chlenov kafedr geograficheskogo fakul'teta Moskovskogo Gorodskogo Pedagogicheskogo Instituta imeni V. P. Potëmkina (Publications of members of the chairs of the Geography Faculty of the Potemkin Moscow City Pedagogical Institute). Moskovskii Gorodskoi Pedagogicheskii Institut. Uchënye Zapiski, tom 101; Geograficheskii Fakul'tet, vypusk 6 (1959), p. 269-280.   65

About 250 entries.

ADEL', Kh. S., and SHTRAIMISH, R. I. Trudy geografov Khar'kovskogo Gosudarstvennogo Universiteta im. A. M. Gor'kogo. 1956-1966 gg. (mart). (Publications of the geographers of Khar'kov State University named after Gor'kiy, 1956-1966 [to March]). Materialy Khar'kovskogo Otdela Geograficheskogo Obshchestva Ukrainy, 1968, vypusk 3, p. 164-185.   66

535 entries by authors in alphabetical order.

ORONEZHSKII Gosudarstvennyi Universitet. Fundamental'naia Biblioteka. Trudy nauchnykh sotrudnikov Voronezhskogo Gosudarstvennogo Universiteta: bibliograficheskii ukazatel', 1954-1961 (Publications of scientists of Voronezh State University: bibliography, 1954-1961). Voronezh, Izdatel'stvo Voronezhskogo Universiteta, 1967. 262 p.   67

Physical and economic geography, p. 119-132, entries 1271-1432, lists 162 publications of members of the geography faculty of Voronezh State University, 1954-1961, in alphabetical order by name of author.

PECHATNYE trudy nauchnykh sotrudnikov Biologo-geograficheskogo Instituta pri Irkutskom Gosudarstvennom Universitete za 1923-1949 gg. (Printed works of scientific collaborators of the Biological-geographical Institute at Irkutsk State University 1923-1949), Irkutskii Gosudarstvennyi Universitet. Biologo-Geograficheskii Nauchno-Issledovatel'skii Institut. Izvestiia, v. 11, no. 2 (1950), p. 37-68.   68

353 entries.

KOORITS, V. Bibliografiia trudov prepodavatelei, sotrudnikov i aspirantov Geograficheskogo otdeleniia Tartuskogo Gosudarstvennogo Universiteta (Bibliography of works of teachers, collaborators, and graduate students of the Geographical section of Tartu State University). Tartuskii Gosudarstvennyi Universitet, Uchenye zapiski, no. 237 (1969), Trudy po geografii, no. 6, p. 215-232.   69

About 340 titles of works 1948-1968.

BIBLIOGRAFIIA izdanii Tbilisskogo Gosudarstvennogo Universiteta imeni Stalina 1919-1960 (Bibliography of publications of the Tbilisi State University named for Stalin, 1919-1960). A. A. Kasradze, comp. Tbilisi, Izdatel'stvo Tbilisskogo Gosudarstvennogo Universiteta im. Stalina, 1961. 404 p. In Georgian and Russian.   70

Geography, p. 77-86, entries 587-671, lists 85 publications, entries 587-663 in Georgian, then in Russian translation, with a notation if there is a Russian summary, and entries 664-671 in Russian only with a note if there is a summary in Georgian.

Publishing Houses and Institutes of Scientific Information

GOSUDARSTVENNOE Izdatel'stvo Geograficheskoi Literatury. Katalog knig 1946-1957 gg. (Catalogue of books 1946-1957). I. K. Miachin, comp. and ed. Moskva, Geografgiz, 1958. 224 p.   71

Annotated list of more than 600 books published by the State Publishing House of Geographical Literature, commonly abbreviated as "Geografgiz," 1946-1957, arranged by the following categories: literature on travels and geographical exploration, scientific literature, scientific-popular and mass literature. Replaces catalogue 1946-1954, 1954, 152 p.

_____ . ____ 1957-1961. Moskva, Geografgiz, 1962. 147 p.   72

Annotated list of about 450 works arranged under headings: general problems of geographical sciences, geography of the USSR, geography of foreign countries, history of geography and geographical discoveries, series "tales of nature," series "travels," handbooks, guidebooks and albums, and continuing publications. Alphabetical index.

PARKHOMENKO, Iia Ivanovna, SABUROVA, M. P., and ZAITSEVA, M. D. Publikatsii sotrudnikov otdela Geografii VINITI za 1954-1969 gg. (Publications of collaborators of the section on geography of All-Union Institute of Scientific and Technical Information in the years 1954-1969), Geograficheskii sbornik (Vsesoiuznyi Institut Nauchnoi i Tekhnicheskoi Informatsii), v. 4. Moskva, VINITI, 1970, p. 303-328.   73

375 entries, arranged systematically.

SERIALS

Geographical Serials

Serials of the Geographical Society of the USSR

EOGRAFICHESKOE Obshchestvo SSSR. Seriinye i periodicheskie izdaniia Geografiches-
kogo Obshchestva SSSR za 1917-1966 gg. (Serial and periodical publications of the
Geographical Society of the USSR, 1917-1966), in the book: Geograficheskoe
Obshchestvo Soiuza SSR, 1917-1967, ed. by I. B. Kostrits and D. M. Pinkhenson.
Moskva, Izdatel'stvo "Mysl'," 1968, p. 246-261.   74

    71 serial and periodical publications of the Geographical Society of the
USSR and its sections and branches, in the 50-year period 1917-1967.

ERIINYE i periodicheskie izdaniia Geograficheskogo Obshchestva SSSR: svedeniia
dany na 1 ianv. 1965 g.--izdaniia otdelenii i komissii pri Prezidiume Geograficheskogo Obshchestva SSSR. (Serial and periodical publications of the Geographical
Society of the USSR: information as of January 1, 1965--publications of branches
and of sections of the Presidium of the Geographical Society of the USSR), in
the book: Geograficheskoe Obshchestvo SSSR: Spravochnik (Geographical Society of
the USSR: a handbook). 2nd ed. Leningrad, Geograficheskoe Obshchestvo SSSR,
1965, p. 92-98. Rotaprint. (1st ed., 1962, p. 127-131).   75

    Names of 84 serial and periodical publications of the central organization
and branches of the Geographical Society of the USSR from 1846 to 1964, and names
of 18 rotaprint publications.

Other Geographical Serials of the USSR

ARANOVICH, Vladimir Pavlovich. Materialy k istorii russkoi povremennoi geografi-
cheskoi pechati (Materials on the history of the Russian geographical publications).
Akademiia Nauk SSSR. Komissiia po Izucheniiu Estestvennykh Proizvoditel'nykh
Sil. Geograficheskii Otdel. Trudy, no. 2 (1930), p. 223-248.   76

    Review of periodicals and serial publications containing material on geography,
1726-1930. Appended list of 340 publications (p. 232-248) grouped by: publications
of the Academy of Sciences; publications of the Russian Geographical Society, its
divisions, and subdivisions; publications of scientific institutions, institutions
of higher learning, and naturalist societies in universities; publications of
individuals; and publications of institutions working in fields related to geography.

Foreign Geographical Serials

EKUSHCHIE inostrannye periodicheskie izdaniia po geograficheskim naukam na 1
ianvaria 1969 g. (Current foreign periodical publications on geographical sciences,
on January 1, 1969). E. A. Stepanova, comp. Moskva, 1969. 30 p. (Akademiia Nauk
SSSR. Sektor Seti Spetsial'nykh Bibliotek. Biblioteka Instituta Geografii).   77

    Alphabetical list, partly annotated, of 271 geographical periodicals (strictly
construed to exclude nonperiodical serials) from outside the Soviet Union received
by the Institute of Geography of the Academy of Sciences of the USSR, the Lenin
State Library of the USSR, or the All-Union Institute of Scientific and Technical
Information, all in Moscow, USSR. Data included name, address, frequency, price,
if stated, and general field.

VODNYI spisok inostrannykh periodicheskikh izdanii, poluchaemykh v 1952 godu po
vsei seti bibliotek Akademii Nauk SSSR. Chast' 2. vypusk 5. Geologo-
geograficheskie nauki. (Summary list of foreign periodicals received in 1952
in the network of libraries of the Academy of Sciences of the USSR. Part 2, no.
5. Geological and geographical sciences). E. I. Rofe, comp. Leningrad, 1952.
(Biblioteka Akademii Nauk SSSR).   78

    270 references.   74 - 78

Cumulative Indexes to Geographical Serials

REFERATIVNYI zhurnal: Geografiia. (Reference journal: geography). 1954- monthly.
Moskva. Vsesoiuznyi Institut Nauchnoi i Tekhnicheskii Informatsii.

Avtorskii ukazatel' (Author index).   7

1954-1955, 484 p., as 1955, no. 12; 1956; 1957, 278 p.; 1958, 337 p.; 1959, 392 p.;
1960, 452 p.; 1961, 178 p.; 1962, 139 p.; 1963, 166 p.; 1964, 183 p.; 1965, 140 p.;
1966, 139 p.; 1967, 142 p.; 1968, 159 p.; 1969, 153 p.; 1970, 149 p.; 1971, 183 p.

Author indexes with separate alphabets.

Predmetnyi i geograficheskii ukazatel' (Subject and geographical indexes).   8

1954, 1955, and 1966, Part 1, subjects, 318 p., Part 2, geographical names, 206 p.;
1957, 412 p.; 1958, 381 p.; 1959, 398 p.; 1960, 428 p.; 1961, 501 p.; 1962, 409 p.;
1963, 308 p.; 1964, 239 p.; 1965, 248 p.; 1966, 238 p.; 1967, 232 p.; 1968, 388 p.;
1969, 368 p.; 1970, 386 p.; 1971, 456 p.; 1972, 432 p.

Detailed series of subject indexes divided into sections that vary somewhat
but typically consist of: Theoretical and general questions of geography; Cartography; Geomorphology; Oceanography, hydrology of the land, and glaciology; Biogeography; Geography of soils; Conservation; Local studies; and Regional geography.

The "geographical" index is organized by large regions: the world, the USSR,
Europe, Asia, Africa, America, Australia and Oceania, Antarctica and the ocean,
broken down further by region and topic.

Although not strictly cumulative indexes, the annual two-volume author,
subject, and regional indexes of Referativnyi zhurnal: geografiia are so comprehensive and extensive as to merit special listing. These indexes provide a major
inventory of geographical publication for the period from 1954.

AKADEMIIA Nauk SSSR. Izvestiia. Seriia geograficheskaia. Moskva, 1951- . 6 per
annum.

Sistematicheskii ukazatel' (1951-1966). (Classed index, 1951-1966). E. A.
Stepanova, comp. L. G. Kamanin, ed. Moskva, Izdatel'stvo "Nauka," 1967. 100 p.
(Akademiia Nauk SSSR. Sektor Seti Spetsial'nykh Bibliotek. Institut Geografii).
2,221 entries.   8

Index by systematic fields of articles, notes, and reviews in the leading
Soviet geographical periodical for this period. Table of contents on page 98-100
indicate page locations of systematic fields under which entries are grouped.
Index of names on pages 88-97, includes authors of articles and notes but not of
books reviewed. Information includes number of references in bibliographies of
each article.

AKADEMIIA Nauk SSSR. Izvestiia. Seriia geograficheskaia i geofizicheskaia, 1937-1951

Sistematicheskii ukazatel' statei po geografii, 1937-1951. (Classed index to
articles on geography, 1937-1951). E. S. Steklenkova, comp. L. G. Kamanin, ed.
Moskva, 22 p. rotaprint. (Akademiia Nauk SSSR. Sektor Seti Spetsial'nykh Bibliotek. Biblioteka Instituta Geografii).   8

168 entries organized by systematic fields of physical geography. Author
index includes 117 names. Particularly useful in selecting out geographical
articles from a journal in which geophysics and geography were combined.

MOSKOVSKII Gosudarstvennyi Universitet. Nauchnaia Biblioteka imeni A. M. Gor'kogo.

Sistematicheskii ukazatel' k "Vestniku Moskovskogo universiteta" 1946-1966
(Classed index to the Vestnik (Herald) of Moscow University, 1946-1966). M. K.
Simon, comp. V. L. Birzovich, ed. Moskva, Izdatel'stvo Moskovskogo Universiteta,
1969. 493 p.   8

8,916 entries, arranged by fields. Author index. Geographical sciences,
p. 218-250, entries 4153-4852, include the following subdivisions: general questions and history; physical geography (general questions, physical geography of
the USSR and foreign countries, geomorphology, landscape studies and physical-
geographical regionalization, paleogeography, biogeography, and medical geography);
and economic geography (general questions and history, economic geography of the
USSR, and economic geography of foreign countries). Within each category entries
are listed in chronological order.

OPROSY geografii (Geograficheskoe Obshchestvo SSSR. Moskovskii Filial. Nauchnye
sborniki). Moskva, 1- (1946-). Irregular but several a year.

   Spisok statei, opublikovannykh v sbornikakh "Voprosy geografii" (List of
articles published in the collected volumes "Voprosy geografii): no. 1-20 (1946-
1950) in no. 20 (1950), p. 338-354; no. 21-30 (1950-1952), in no. 30 (1952), p.
342-353; no. 31-40 (1952-1957), in no. 40 (1957), p. 249-261.   84

   Lists of articles arranged by systematic fields and regions. Alphabetical
author indexes. A cumulative index for v. 1-90 (1946-1972) is in preparation.

EMLEVEDENIE; geograficheskii zhurnal (Imperatorskoe Obshchestvo Liubitelei
Estestvoznaniia, Antropologii i Etnologii. Geograficheskoe Otdelenie). Moskva,
v. 1-40 no. 2 (1892-1938. Closed 1938. Succeeded by Zemlevedenie (Moskovskoe
Obshchestvo Ispytatelei Prirody), Moskva, nsv 1- (41-) (1940-). Irregular.

   Ukazatel' statei zametok i retsenzii, pomeshchënnykh v zhurnale "Zemlevedenie"
za 12 let (s 1894 po 1905 gg.) (Index to articles, notes, and reviews located in
the journal "Zemlevedenie" in the 12 years, 1894-1905)., v. 12 (1905), no. 3-4,
supplement, p. 1-63. Also separately, Moskva, I. N. Kushnerev, 1906. 63 p.   85

   About 1,200 entries for the period 1894-1905 arranged by sections: (1) general
and physical geography, (2) regional geography (Russia by regions, p. 7-16, Europe,
Asia, Africa, America, Australia, polar countries), (3) history of geography, (4)
geographical societies, congresses, and museums, (5) necrology and memorials, (6)
pedagogical articles, (7) bibliographical notes divided into (a) Russian books
and periodical publications, p. 28-48, and foreign books and periodical publications, p. 48-59, (8) sessions of the geographical section of the Society, and (9)
alphabetical author index of about 125 authors of articles, p. 62-63.

   Ukazatel' statei zametok i retsenzii, pomeshchennykh v zhurnale "Zemelevedenie"
za 12 let (1906-1917), v. 37 (1935), no. 4, supplement, 35 p.   86

   About 900 entries arranged by the same sections as in the previous cumulative
index with geography of Russia by regions, p. 3-6, bibliography of Russian books
and periodical publications, p. 13-22, bibliography of foreign books and publications, p. 22-24, and author index with 133 names, p. 26.

EOGRAFICHESKOE Obshchestvo SSSR (1845-1916 as Imperatorskoe Russkoe Geograficheskoe
Obshchestvo). Izvestiia. Leningrad. v. 1- (1865-). 6 nos. a year.

   Ukazatel' k izdaniiam Imperatorskago Russkago Geograficheskago Obshchestva i
ego otdelov s 1846 po 1875 god (Index to the publications of the Imperial Russian
Geographical Society and its sections from 1846 to 1875). St. Petersburg, 1886.
144 plus 33 p. Kliuch k ukazateliu (Key to the indexes), with about 1,600 personal
names and 2,000 geographical names.   87

   Primarily a list of publications in chronological order, with contents for
the whole family of serials of the Imperial Russian Geographical Society including its Zapiski 1846-1859, Denkschriften, Geograficheski Izvestiia 1848-1850,
Vestnik, 1851-1860, Zapiski, 1861-1864, Izvestiia, 1865-1874, Godovye otchëty,
1854-1874, individual publications 1848-1875, publications of the sections 1851-
1875, works of expeditions 1853-1875, section on maps 1851-1875, publications of
major regional branches (Caucasus, Siberia, Orenburg, Northwest, and Southwest),
p. 3-144. Separate name index, "Kliuch k ukazateliu," separately paged, divided
into two parts, A. "Imena lichnyia," personal names, p. 1-14, and B. "Imena
geograficheskiia," place names, p. 15-33.

                                                                        84 - 87

Ukazatel'...s 1876 po 1885 god. 1887. 75 plus 20 p. About 800 personal names and about 1,500 geographical names. 8⟨

Ukazatel'...s 1886 po 1895 god. 1896. 190 columns plus 27 p. About 1,200 personal names and about 1,800 geographical names. 8⟨

Ukazatel'...s 1896 po 1905 god. 1910. 334 columns plus 50 p. About 2,000 personal names and 3,500 geographical names. 9⟨

All indexes are similar to the first cumulative index but with increasing number of regional branches reaching in the 1896-1905 cumulative index publications from the following branches: Caucasus, Orenburg, Turkestan, Western Siberia, Eastern Siberia, Amur, Troitsko-Savsk-Kiakhta, Chita, and Vladivostok, and the following sub-branches: Semipalatinsk, Altay, and Krasnoyarsk.

The key cumulative index for the Russian Empire 1846-1905, covering all publications of the Geographical Society, including its specialized sections and its numerous regional branches. Actually more a chronological table of contents than an analyzed cumulated index but does include alphabetical indexes of persons and places.

General Guides to Soviet Serials, Articles, and Reviews

Bibliography of Periodical Bibliographies

MASHKOVA, M. V., and SOKUROVA, M. V. Obshchie bibliografii russkikh periodicheskikh izdanii 1703-1954 i materialy po statistike russkoi periodicheskoi pechati: annotirovannyi ukazatel' (General bibliographies of Russian periodicals 1703-1954 and statistical materials on Russian periodicals: annotated list). P. N. Berkov, ed. Leningrad, Izdanie Gosudarstvennoi Publichnoi Biblioteki im. M. E. Saltykova-Shchedrina, 1956. 139 p. (Gosudarstvennaia Publichnaia Biblioteka imeni M. E. Saltykova-Shchedrina). 9⟨

Detailed descriptions of 62 bibliographies of periodicals with graphic table showing the exact years covered by each.

Inventories of Serials of the USSR

LETOPIS' periodicheskikh izdanii SSSR 1950-1954 gg: bibliograficheskii ukazatel'. (Annual of periodicals in the USSR 1950-1954: bibliographical guide). Moskva, Izdatel'stvo Vsesoiuznoi Knizhnoi Palaty, 1955. 502 p. 9⟨

Geography is in the section on natural history, geography, meteorology, and astronomy, p. 20-28.

_____ 1955-1960: organ gosudarstvennoi bibliografii SSSR. Chast' 1. Zhurnaly, trudy, biulletini (1955-1960: organ of state bibliography of the USSR. Part 1. Journals, transactions, bulletins). 1963. 1,026 p. Indexes. 9⟨

Geography is in the section on general questions of natural sciences, geography, meteorology, hydrology, geodesy, and regional studies, p. 68-82.

_____ 1961-1965... Chast' 1. Zhurnaly, trudy, biulletini. 1973. In two parts. Part 1. Opisanie izdanii (Description of publications), 535 p. Part 2. Vspomogatel'nye ukazateli (Auxiliary indexes). 271 p. 9⟨

_____ 1966-1970... Chast' 1. Zhurnaly. 1972. 131 p. Geography, p. 23-24. 9⟨

LETOPIS' periodicheskikh izdanii SSSR. Bibliograficheskii ukazatel'. Trudy, uchënye zapiski, sborniki i drugie prodolzhaiushchiesia izdaniia. (Serials of the USSR. Transactions, scientific series, collected works, and other continuing publications). Moskva, Izdatel'stvo Vsesoiuznoi Knizhnoi Palaty, or "Kniga," 1951- . Annual. Vsesoiuznaia Knizhnaia Palata. 1952- . 9⟨

LETOPIS' periodicheskikh izdanii SSSR. Novye, pereimenovannye i prekrativshiesia
zhurnaly, gazety i drugie periodicheskie izdaniia (New, renamed and discontinued
journals, newspapers and other periodicals). 1950- . Annual with cumulations
1950-1954, 1955-1960, 1961-1963, 1963-1965.   97

PERIODICHESKAIA pechat' SSSR 1917-1949. Bibliograficheskii ukazatel'. Zhurnaly,
trudy i biulleteni. (Periodicals of the USSR, 1917-1949. A bibliographical guide,
Journals, works, bulletins). (Vsesoiuznaia Knizhnaia Palata). Moskva, Izdatel'-
stvo Vsesoiuznoi Knizhnoi Palaty, 1955-1963. 10 v.   98

     v. 1, social sciences; v. 2, natural sciences and mathematics; v. 3, technology
and industry; v. 4, transportation, communications, and municipal services; v. 5,
agriculture; v. 6, culture and education; v. 7, health, medicine, physical culture,
and sport; v. 8, linguistics, literature, and art; v. 9, publishing and bibliography; v. 10, general indexes to all volumes.

     Serials in geography are listed in v. 2, chapter IV, Geological and geograph-
ical sciences, section 2, Geography, p. 49-56; general regional journals, v. 2,
p. 56-71; geology, v. 2, p. 71-98; and hydrology, meteorology, climatology, and
hydrochemistry, v. 2, p. 98-107.

     Includes periodicals published in the Soviet Union in all languages, arranged
by fields. Alphabetical index in Russian. Alphabetical index of periodicals in
other languages of the USSR or in foreign languages. Index of place of publication. Index of issuing agencies and organizations.

BIBLIOGRAFIIA periodicheskikh izdanii Rossii, 1901-1916. (Bibliography of periodicals
of Russia, 1901-1916). L. N. Beliaeva, and others, comps. V. M. Barashenkov, ed.
Leningrad, 1958-1961. 4 v. (Gosudarstvennaia Publichnaia Biblioteka imeni Saltykova-Shchedrina).   99

     9,713 titles, mainly holdings of the Leningrad Public Library. Subject index.

LISOVSKII, Nikolai Mikhailovich. Russkaia periodicheskaia pechat' 1703-1900 gg.
Bibliografiia i graficheskie tablitsy (Russian periodicals 1703-1900: bibliography
and graphic tables). Petrograd, G. A. Shumakher, 1915. 267 p.   100

     About 2,900 periodicals and newspapers listed chronologically. Alphabetical
and classified indexes. Also published as Bibliografiia russkoi periodicheskoi
pechati 1703-1900, Petrograd, Tip. Akts-obshch. tip. dela, 1915. 1067 p., small
format.

PERIODYCHNYI vydannia URSR, 1918-1950, zhurnaly: bibliohrafichnyi dovidnyk (Periodical
publications of the Ukrainian SSR. Journals: bibliographical handbook). Kharkiv,
Knyzhkova Palata URSR, 1956. 462 p. In Ukrainian.   101

     List of periodicals and other serials published in the Ukrainian SSR 1918-
1950 and the Moldavian SSR 1924-1940, arranged in alphabetical order. 25 indexes.
Geographical journals, p. 231-232 of index.
_____. 1951-1960. 1964. 184 p.

Inventories of Foreign Serials

SVODNYI katalog inostrannykh nauchnykh zhurnalov, postupivshikh v biblioteki SSSR. Estestvennye nauki. Meditsina. Sel'skoe khoziaistvo. Tekhnika (Union catalogue of foreign scientific journals received by libraries of the USSR. Natural sciences. Medicine. Agriculture. Technology). 1947- . Moskva, Izdatel'stvo "Kniga," 1950- . annual. (Vsesoiuznaia Gosudarstvennaia Biblioteka Inostrannoi Literatury). (1949-1950 ...v krupneishie biblioteki Moskvy i Leningrada. Estestvennye nauki 1947-1948 ...Moskva). 1947-1948 (1950), 181 p. 1949 (1951), 135 p.; 1950 (1952), 176 p.; 1951 (1953), 248 p.; 1952 (1955), 300 p.; 1953-1954 (1956), 566 p.; 1955 (1957), 376 p.; 1956 (1959), 526 p.; 1957 (1959), 679 p.; 1958 (1960), 702 p.; 1959 (1961), 571 p.; 1960 (1962), 753 p.; 1961 (1963), 623 p.; 1962 (1964), 742 p.; 1963 (1965), 719 p.; 1964 (1966), 776 p.; 1965 (1967), 704 p.; 1967 (1969), 896 p.; 1968 (1970), 806 p.; 1969 (1971), 855 p.; 1970 (1972), 846 p.
102

AKADEMIIA Nauk SSSR. Biblioteka Akademii Nauk SSSR. Svodnyi ukazatel' inostrannykh periodicheskikh izdanii, poluchaemykh Bibliotekoi Akademii Nauk SSSR (Union catalogue of foreign periodicals, received by the Library of the Academy of Sciences of the USSR). V. Ia. Khvatov, ed. Leningrad, Izdatel'skii otdel Biblioteki Akademii Nauk SSSR, 1969. 427 p.
103

8,594 periodicals arranged by subject. Geography, p. 94-100, entries 2411-2592 includes 182 periodicals in alphabetical order by titles.

Earth sciences as a whole, p. 70-103, entries 1778-2669, includes besides geography: geodesy and cartography, p. 71-73, entries 1793-1845; geophysics (including seismology, meteorology and climatology, and oceanography), p. 73-81, entries 1846-2073; geology, p. 81-94, entries 2074-2410; geomorphology, p. 100-101, entries 2593-2606; hydrology, p. 101-103, entries 2607-2649; glaciology and permafrost studies, p. 103, entries 2650-2654; and polar lands, p. 103, entries 2655-2669. List of closed serials, p. 313. Alphabetical index of titles, p. 314-427.

AKADEMIIA Nauk SSSR. Biblioteka. Periodicheskie izdaniia stran Azii i Afriki. Katalog fondov Biblioteki AN SSSR i Gosudarstvennoi Publichnoi Biblioteki imeni Saltykova-Shchedrina: Fiziko-matematicheskie, estestvennye i tekhnicheskie nauki. (Periodicals of the countries of Asia and Africa. Catalogue of the collections of the Library of the Academy of Sciences of the USSR and the Saltykov-Shchedrin State Public Library: physical-mathematical, natural, and technical sciences). S. S. Bulatov, ed. Leningard, 1967. 310 p.
104

Indexes of Articles

LETOPIS' zhurnal'nykh statei. Organ gosudarstvennoi bibliografii SSSR. Izdaetsia s 1926 goda (Annals of journal articles. Organ of state bibliography of the USSR. Published since 1926). Moskva, Izdatel'stvo "Kniga," 1926- . weekly (Vsesoiuznaia Knizhnaia Palata). (1926-1937 as Zhurnal'naia letopis').
105

Geography articles in Russian are listed in section 15.4.5, including those in geographical periodicals and serials and also in other related fields and in journals of general science. Quarterly author and geographical indexes.

For earlier indexes of periodical articles see Paul L. Horecky. Basic Russian publications, entries 36-40, p. 12-13, and Karol Maichel, Guide to Russian reference books, v. 1, General bibliographies and reference books, entries A290-A302, p. 64-65.

LETOPIS' gazetnykh statei organ gosudarstvennoi bibliografii SSSR (Annals of newspaper articles.Organ of state bibliography of the USSR) Moskva, Izdatel'stvo "Kniga," 1936- . monthly. (Vsesoiuznaia Knizhnaia Palata).
106

Indexes articles in 38 leading Soviet newspapers, 19 central publications, such as Pravda and Izvestiia, and 19 regional publications.

Index of Reviews

ŁETOPIS' retsenzii. Organ gosudarstvennoi bibliografii SSSR. Izdaetsia s 1935
  goda (Annals of reviews. Organ of state bibliography of the USSR. Published
  since 1935). Moskva, Izdatel'stvo "Kniga," 1- 1934 (1935-). quarterly.
  (Vsesoiuznaia Knizhnaia Palata).   107

  Arranged by 31 subject classes.

  Indexes of Soviet authors, foreign authors, and reviewers in each issue,
  cumulated annually in the fourth issue of each year. Geography is included
  within section 15, natural sciences.

DISSERTATIONS

AVTOREFERATY dissertatsii (Authors essential portions of dissertations), 1954- in
  Knizhnaia letopis', 1955- .   108

  This section at the end of each issue provides a useful listing of the
  essential portions of dissertations, usually about 20 pages, which in small
  editions of about 200 copies are distributed to selected institutions and libraries within the Soviet Union, and thus are more widely available than the typed
  dissertations themselves. Not all dissertations for which essential portions
  are distributed are accepted. Furthermore substantial time is typically involved
  between the successful defense of a dissertation and the confirmation of the
  degree by the appropriate state organs.

  From 1955, no. 1 to 1960, no. 52 this section is in the regular numbers of
  Knizhnaia letopis'. After 1960 it is part of the supplementary issues (Knizhnaia
  letopis'. Dopolnitel'nyi vypusk).

KATALOG kandidatskikh i doktorskikh dissertatsii, postupivshikh v Biblioteku imeni
  V. I. Lenina i Gosudarstvennuiu Tsentral'nuiu Nauchnuiu Meditsinskuiu Biblioteku
  (Catalogue of candidate and doctoral dissertations received by the Lenin Library
  and the State Central Medical Library). 1956- . Moskva. quarterly. 1957- .
  (Gosudarstvennaia Biblioteka SSSR imeni V. I. Lenina).   109

  Doctoral and candidate dissertations classified into 29 major subjects, some
  of which are further subdivided. Dissertations in geography are listed in chapter
  IV, earth sciences, section 4, geographical sciences. Index of authors, names,
  and institutions where defended.

  The typed dissertations, usually available only in the library of the institution where accepted and in the Lenin State Library of the USSR in Moscow, are
  often massive documents of many hundreds of pages (some run over a thousand pages)
  with manuscript maps and diagrams.

MOSKOVSKII Gosudarstvennyi Universitet imeni M. V. Lomonosova. Nauchnaia Biblioteka
  imeni A. M. Gor'kogo. Doktorskie i kandidatskie dissertatsii zashchishchënnye v
  Moskovskom Gosudarstvennom Universitete s 1934 po 1954 g. Vypusk 2. Fakul'tety:
  geologicheskii, geograficheskii, biologo-pochvennyi (Doctoral and candidate dissertations defended in Moscow State University 1934-1954. No. 2. Geological,
  geographical, and biological-soil faculties). Moskva, Izdatel'stvo Moskovskogo
  Universiteta, 1957. 217 p.   110

  Geography faculty, p. 45-84, entries 166a-352 lists dissertations by major
  fields: general geography and paleogeography; physical geography of the USSR;
  economic geography of the USSR; physical geography of foreign countries; economic
  geography of foreign countries; geodesy, cartography, and air photo surveys,
  geomorphology; hydrology and oceanography; meteorology and climatology; soil
  science; and history of geography.

PART II. REFERENCE WORKS AND ASSOCIATED BIBLIOGRAPHIES

In geography certain types of specialized reference materials are of high value and frequent use: atlases and maps for the location of places, the clarification of space relations, and the analysis of patterns of distribution; statistical compilations for data on quantities involved-- whether of people or of production-- and their characteristics in given areas; encyclopedias, handbooks, and gazetteers for information on specific cities, regions, countries, continents, or physical features, arranged alphabetically for quick consultation; technical dictionaries for meanings of scholarly or scientific terms; toponymic lexicons for the significance of place names; and biographical directories for identification of authors of geographical publications.

The geographer interested in specific regions of the Soviet Union may make heavy use of such reference materials. A scholar interested in the Ukraine, for example, is likely to find valuable materials in the general, climatic, agroclimatic, or agricultural atlases of the Ukraine, the quite substantial annual statistical yearbook of the Ukraine, or the large 17-volume Ukrainian encyclopedia with one volume devoted to a monographic treatment of the republic. Similarly a geographer engaged in research on Uzbekistan in Soviet Middle Asia will wish to know about the atlas of the Uzbek SSR, the annual statistical yearbook of the Uzbek SSR, and Uzbek Soviet encyclopedia. These materials may be in Russian, in the language of the republic, or in both.

Thus although this Guide is devoted mainly to bibliographies, it seems appropriate to present here the key specialized reference volumes of particular utility in geographic research.

The corpus of Soviet geographical reference materials is very rich in atlases, statistical compilations, and encyclopedias. Care has been taken to make this volume as complete as possible in its reporting of such reference works that fall within its scope. The listings are relatively complete for the period 1946-1973, but highly selective for earlier years.

Although it excludes frequently reprinted school atlases of pedagogic interest only, this work includes 88 Soviet atlases: 19 world atlases, 20 of the Soviet Union as a whole, 16 of major regions, 28 of smaller regions, and 5 of foreign areas. Specialized thematic atlases abound on a world-wide scale (as in the physical-geographic atlas of the world [152], or the atlas of peoples of the world [161], both unrivaled), or for the Soviet Union (as in the atlas of agriculture of the USSR [175]). On the other hand many large or small regions in the Soviet Union are covered by general atlases, and some foreign areas, as in the atlas of Antarctica [228].

The great wealth of Soviet statistical publications is sometimes not realized.
This volume includes 611 entries for statistical compilations ranging from the four
multi-volume census reports for 1897, 1926, 1959, and 1970 [239-242], the annual
statistical yearbook for the Soviet Union as a whole [243], and statistical handbooks for special fields, such as industry in the USSR [258], to yearbooks for
union republics such as the RSFSR [345] or the Ukraine [350], handbooks for individual oblasts, such as Volgograd [673], or for a city, such as Yerevan [815],
or for socialist countries [820], or Latin America [843]. The number of entries
by category is as follows: bibliographies, 6; censuses, 4; general statistical
yearbooks for the USSR as a whole, 15; handbooks for special topics 87; general
yearbooks for union republics, 92; handbooks for smaller administrative units,
such as oblasts, 343; handbooks for individual cities, 40; and handbooks for
foreign areas, 24.

Thirty-one encyclopedias in languages of the Soviet Union are here included:
14 in Russian, 4 in Ukrainian, 2 in Estonian, 3 in Latvian, 3 in Lithuanian, and
one each in Belorussian, Moldavian, Armenian, Kazakh, and Uzbek. Each has a particular geographic value.

In the following section under each type of material bibliographies are listed
first, followed by a selection of the most comprehensive reference works.

MAPS AND ATLASES: BIBLIOGRAPHIES

Bibliography of Bibliographies

GAVRILOVA, Serafima Abramovna, and RIABTSEVA, Z. G. Istochniki informatsii o
kartakh i atlasakh (Sources of information on maps and atlases), in Itogi nauki:
kartografiia 1962 [v. 1] Moskva, VINITI, 1964, p. 61-82.   111

    64 serial bibliographies of maps and atlases described, by regions. List
by countries on p. 81-82.

General Bibliographies of Maps and Atlases

KARTOGRAFICHESKAIA letopis': organ gosudarstvennoi bibliografii SSSR. (Cartographic
    annals: organ of state bibliography of the USSR) (Vsesoiuznaia Knizhnaia Palata).
    Moskva, "Kniga," 1931- . Annual. (1931-1940 as Bibliografiia kartograficheskoi
    literatury i kart, quarterly; 1941-1950 and 1951-1953 covered in single volumes). 112

    The official annual bibliography of Soviet maps and atlases. The 1971 volume
(1972, 177 p.) contains 355 entries organized under the following headings: maps
of the world (by systematic fields); of continents, the Arctic, and oceans; of
individual countries, especially the USSR (by systematic fields); historical maps;
biographical maps; astronomical maps; scientific-reference atlases; school atlases.
Index of personal names. Index of geographical names. Full bibliographical
details. Contents of atlases.

SALISHCHEV, Konstantin Alekseevich. Geograficheskie atlasy (Geographical atlases),
    Itogi nauki: kartografiia 1962 [v. 1] Moskva, VINITI, 1964. p. 49-60.   113

    84 references on atlases and their compilation.

————— · ————— · ————— . 1963-1964, v. 2. 1966, p. 86-95.

    43 references for the period 1963-1964.

————— · ————— · ————— . 1965-1967, v. 3. 1968, p. 93-101.

    57 references for the period 1965-1967.

————— · ————— · ————— . 1967-1969, v. 4. 1970, p. 88-103.

    78 references for the period 1967-1969.

————— · ————— · ————— . [1969-1971], Itogi nauki i tekhniki: seriia kartografiia,
v. 5. Moskva, VINITI, 1972, p. 99-132.

    91 references for 1969-1971.

SALISHCHEV, Konstantin Alekseevich. Integrated regional geographic atlases, Soviet
    Geography: Review and Translation, v. 5, no. 7 (September, 1964), (Materialy k
    IV s"ezdu Geografiche skogo Obshchestva SSSR, Leningrad, 1964, processed).   114

    Review of atlases of Soviet republics and oblasts published 1961-1964.

CHURKIN, Vladimir Gerasimovich. Geograficheskie atlasy (Geographical atlases).
    Moskva-Leningrad, Akademiia Nauk SSSR, 1961. 116 p. Bibliography, p. 101-116.
    (Geograficheskoe Obshchestvo SSSR. Zapiski, nsv. 21).   115

    315 geographical atlases listed and 156 references in Russian and other
languages on atlases.

ČEPIENĖ, K. Lietuvos TSR Mokslų Akademijos Centrinės Bibliotekos atlasų ir
žemelapių katalogas. (Lithuanian SSR. Academy of Sciences. Central Library.
Catalogue of atlases and maps). Vilnius, 1969. 344 p. (Lietuvos TSR Mokslų
Akademija. Centrine Biblioteka). In Russian: Akademiia Nauk Litovskoi SSR.
Tsentral'naia Biblioteka. Katalog atlasov i kart. K. Chepene, comp.   116

    2,863 atlases and maps of the 16th-20th centuries, atlases in chronological
order, maps by area. Includes historical, linguistic, geological, and transport
route atlases. Titles in original languages. Author, geographical, and subject
indexes.

LEDOVSKAIA, L. S. Spisok atlasov i kart (List of atlases and maps). In the book:
Issledovaniia po matematicheskoi kartografii (Research on mathematical geography).
Moskva, Izdatel'stvo "Nedra," 1971, p. 48-53. (Glavnoe Upravlenie Geodezii i
Kartografii pri Sovete Ministrov SSSR. Tsentral'nyi Nauchno-Issledovatel'skii
Institut Geodezii, Aeros"ëmki i Kartografii. Trudy, vypusk 189).   117

    136 entries.

KATALOG obshchegeograficheskikh i spetsial'nykh kart postupivshikh v Tsentral'nyi
kartografo-geodezicheskii fond za period s 1940 g. po 1 Okt. 1949 g. (Catalogue
of general geographical and specialized maps received in the central cartographic-
geodetic collection in the period 1940 to October 1, 1949). V. S. Shchekut'eva
and M. I. Prudnichenkova, comps. Moskva, Geodezizdat, 1950. 137 p. (Glavnoe
Upravlenie Geodezii i Kartografii).   118

    609 entries.

IVANOVICH, V. A. Katalog nomenklaturnykh kart v masshtabakh 1:200,000, 1:500,000
i 1:1,000,000 izd. Gl. Geodez. Upr-niem NKTP SSSR v 1932-1934 gg.i Gl. Upr-
niem Gos. s"emki i kartografii NKVD SSSR v. 1935-1938 gg. (Catalogue of listed
maps at scales of 1:200,000, 1:500,000, and 1:1,000,000 published by the Glavnoe
Geodezicheskoe Upravlenie, 1932-1934, and Glavnoe Upravlenie Gosudarstvennoi
S"ëmki i Kartografii, 1935-1938). Tsentral'noe Biuro Kartografo-Geodezicheskoi
Izuchennosti GUGK pri SNK SSSR. Moskva, 1939. 79 p.   119

    The catalogue describes 475 sheets on a scale of 1:200,000; 224 sheets on a
scale 1:500,000; and 57 sheets on a scale of 1:1,000,000.

Thematic Maps

KEL'NER, Iurii Georgievich. Karty prirody. (Maps of natural conditions), in Itogi
nauki, seriia geografiia: kartografiia 1963-1964, v. 2. Moskva, VINITI, 1966,
p. 61-75.   120

    127 entries on mapping of physical features.

LIPATOVA, V. V. Materialy k bibliografii po voprosam kartografirovaniia rastitel'-
nosti (Material for a bibliography on questions of mapping vegetation), in the
book: Printsipy i metody geobotanicheskogo kartografirovaniia, (Principles and
methods of geobotanical mapping). Moskva-Leningrad, Akademiia Nauk SSSR, 1962.
p. 265-295.   121

    729 references, of which 270 from the Soviet Union.

PREOBRAZHENSKII, Arkadii Ivanovich. Russkie ekonomicheskie karty i atlasy (Russian
economic maps and atlases). Moskva, Geografgiz, 1953. 326 p. Bibliography,
p. 266-320.   122

    Chronological list of the more important economic maps and atlases published
in pre-Revolutionary Russia and the Soviet Union, p. 266-320. Textual history of
Russian and Soviet economic maps and atlases. Many illustrations.

BARANSKII, Nikolai Nikolaevich, and PREOBRAZHENSKII, Arkadii Ivanovich. Ekonomicheskaia kartografiia (Economic cartography). Moskva, Geografgiz, 1962. 286 p.
Short list of economic maps and atlases, p. 265-280. Bibliography, p. 281-284. 123

    List of 218 economic maps and atlases produced in the Soviet Union, arranged
by systematic field: atlases, general economic maps, maps of manufacturing, agricultural maps, maps of transportation and trade, and population maps. 218 atlases
and maps produced in other countries listed by countries in alphabetical order.
Bibliography of 88 publications on economic maps and atlases.

NIKISHOV, Maksim Ivanovich. Ekonomicheskie karty, (Maps of economic features), in
Itogi nauki, seriia geografiia: kartografiia 1963-1964, v. 2. Moskva, VINITI,
1966, p. 76-85.
                                                                             124
    56 references on mapping of cultural and economic phenomena.

NIKISHOV, Maksim Ivanovich. Sel'skokhoziaistvennye karty i atlasy (Agricultural
maps and atlases). Moskva, Geodezizdat, 1957. 184 p.   125
    607 Russian agricultural maps and atlases. 87 references on Russian works
on agricultural geography.

EVTEEV, Oleg Aleksandrovich. Karty naseleniia v sovetskikh kartograficheskikh
izdaniiakh i geograficheskoi literature za 20 let (1940-1960). Obzor i bibliograficheskii ukazatel' (Maps of population in Soviet cartographical publications
and geographical literature in the 20 years 1940-1960. Review and bibliographical
guide), Voprosy geografii, v. 56 (1962), p. 209-228.   126
    Description of 190 maps.

Historical Maps and Atlases

BAGROV, Lev Semenovich. Istoriia geograficheskoi karty. Ocherk i ukazatel'
literatury (History of geographical maps. Essay and guide to the literature),
Vestnik arkheologii i istorii, no. 22 (1917), p. 1-136. Bibliography, p. 61-136.
                                                                             127
RUSSKIE geograficheskie atlasy. XVIII vek. Svodnyi katalog (Russian geographical
atlases of the 18th century: union catalogue). N. V. Lemus, comp. Leningrad,
1961. 243 p. (Gosudarstvennaia Publichnaia Biblioteka imeni Saltykova-Shchedrina).
                                                                             128
    93 atlases, both separately published and supplements to other works. In
chronological order of first publication. Detailed annotations, including list
of maps. Chronological table. Indexes.

KATALOG inostrannykh kart Rossii XV-XVII vv. po fondam GPB, vypusk 1. Severo-Zapad
Evropeiskoi chasti Rossii i Pribaltiki (Catalogue of foreign maps of Russia of
the 15th-17th centuries in the collection of the State Public Library, part 1.
Northwest European part of Russia and the Baltic). Comp. by E. K. Mikhailova.
Leningrad. 1965. 156 p. (Gosudarstvennaia Publichnaia Biblioteka imeni
Saltykova-Shchedrina).
                                                                             129
    Description of 119 maps in chronological order of first publication. Index
of authors, compilers, artists, engravers, and publishers. Geographical index.
Index of originals and reproductions. Index of objects of description.

NAVROT, M. I. Katalog rukopisnykh istoricheskikh kart XVIII-XX vv. khraniashchikhsia v fondakh Gosudarstvennogo Istoricheskogo Muzeia (Catalogue of manuscript historical maps of the 18th-20th centuries preserved in the collections of the State historical museum), in the book: Istoriia geograficheskikh znanii i istoricheskaia geografiia, Etnografiia, vypusk 2. Moskva, 1967, p. 17-19.   130

   Maps and catalogues arranged systematically by 12 categories: 1, general maps of Russia 18th-19th centuries; 2, maps of Siberia 18th-20th centuries; 3, maps of the Caucasus 18-19th centuries; 4, maps of individual regions of the Russian Empire. 18-19th centuries; 5, maps of seas, bays, straits, and lagoons of the Russian Empire and adjacent states; 6, maps of the course of streams, lakes, and also plans of dams and flood-control installations; 7, military maps; 8, road maps of Russia of the 18-19th centuries; 9, economic maps of individual regions of Russia 18-20th centuries; 10, historical maps and of archeological excavations; 11, maps of cities, forts, and ports of Russia, 18-19th centuries; and 12, surveying plans of individual estates, villages, and bases. Short notes on maps of each section.

NAVROT, M. I. Katalog gravirovannykh i rukopisnykh kart Sibiri XVI-XVII vv., khraniashchikhsia v sektore kartografii Istoricheskogo Muzeia. (Catalogue of engraved and manuscript maps of Siberia in the 16th-17th century preserved in the section of cartography of the Historical museum). In the collection, Istoriia geograficheskikh znanii i istoricheskaia geografiia. Etnografiia, no. 5, Moskva, 1971, p. 34-38.   131

   112 maps arranged by: engraved maps of the 16th century; engraved maps of the 17th century; engraved maps of Siberia of the 18th century compiled by Russian cartographers; engraved maps of Siberia of the 18th century by West-European cartographers; manuscript maps of Siberia in the 18th century. Biblio-graphical details.

KATALOG gravirovannykh kart Rossii XVI-XVIII vv. i reproduktsii s nikh, khraniash-chikhsia v Otdele istoricheskoi geografii i kartografii Gosudarstvennogo Istori-cheskogo muzeia (Catalogue of engraved maps of Russia of the 16th-18th centuries and reproductions with them, preserved in the Section of historical geography and cartography of the State Historical Museum), comp. by M. I. Navrot, in the collection, Istoriia geograficheskikh znanii i istoricheskaia geografiia. Etno-grafiia, v. 4. Moskva, 1970. p. 39-43.   132

   633 maps.

REESTR landkartam, chertezham i planam Rossiiskoi Imperii, nakhodiashchimsia v Geograficheskom departamente pri Imperatorskoi Akademii Nauk (List of maps, sketches and plans of the Russian Empire, found in the Geographical Department of the Imperial Academy of Sciences). St. Peterburg, Akademiia Nauk, 1748. 80 p.   133

KOLGUSHKIN, V. V. Opisanie starinnykh atlasov, kart i planov XVI, XVII, XVIII vv. i poloviny, XIX v khraniashchikhsia v arkhive Tsentral'nogo Kartograficheskogo Proizvodstva VMF (Description of ancient atlases, maps, and plans of the 16th, 17th, 18th, and first half of the 19th centuries preserved in the archives of the Central Cartographic Establishment of the Naval Institute). Leningrad. Voenno-Morskaia Flota, 1958. 270 p. (Upravlenie Nachal'nika Gidrograficheskoi Sluzhby Voenno-Marskogo Flota).   134

Foreign Maps and Atlases

KODES, Iraida Ivanovna and KULIKOVA, V. A. Katalog inostrannykh geograficheskikh atlasov, izdannykh s 1917 po 1970 gg.: fond otdela kartografii (Catalogue of foreign geographical atlases published from 1917 to 1970: collection of the section on cartography). Leningrad, 1972 [1973]. 211 p. (Gosudarstvennaia Publichnaia Biblioteka imeni M. E. Saltykova-Shchedrina).   135

GOSUDARSTVENNAIA Biblioteka SSSR imeni V. I. Lenina. Otdel Kartografii. Katalog inostrannykh kart i atlasov, postupivshikh v Biblioteku imeni V. I. Lenina v 1964 g. (Catalogue of foreign maps and atlases received by the Lenin State Library of the USSR in 1964). S. M. Gul', comp. Moskva. 1965. 48 p. 219 entries arranged by area. Geographical index. Index of maps and atlases by contents (specialization) and by name.   136

_____. _____. v 1965 g. 1966. 63 p. 285 entries. S. M. Gul' and A. V. Kozlova, comps.

_____. _____. v 1966 g. 1967. 51 p. 200 entries. S. M. Gul' and A. M. Nekhoroshikh, comps.

_____. _____. v 1967 g. 1968. 42 p. 185 entries.

_____. _____. v 1968 g. 1969. 56 p. 269 entries.

_____. _____. v 1969 g. 1970. 54 p. 190 entries.

_____. _____. v 1970 g. 1971. 85 p.

_____. _____. v 1971 g. 1972. 45 p.

_____. _____. v 1972 g. 1973. 44 p.

BABKINA, O. I.; GUL', S. M.; KOZLOVA, A. V.; MOZGOVA, G. N.; and NEKHOROSHIKH, A. M., comps. Katalog inostrannykh kart i atlasov, postupivshikh v Biblioteku imeni V. I. Lenina v 1961-1963 gg. (Catalogue of foreign maps and atlases received by the Lenin Library 1961-1963). Moskva, 1964. 138 p. (Gosudarstvennaia Biblioteka SSSR imeni V. L. Lenina. Otdel Kartografii).   137

    About 750 titles, arranged by area of origin.

KUZNETSOVA, S. S., and LAGUN, A. M. Svodnyi katalog inostrannykh geograficheskikh kart i atlasov, poluchënnykh Bibliotekoi AN SSSR, bibliotekami eë seti i bibliotekami Akademii nauk soiuznykh respublik (Union catalogue of foreign geographical maps and atlases received in the library of the Academy of Sciences of the USSR, libraries in its system and in libraries of the academies of sciences of the union republics). M. P. Petrov, ed. Leningrad, 1971. 205 p. (Akademiia Nauk SSSR. Biblioteka).   138

MEL'NIKOVA, Tat'iana Nikolaevna, STANCHUL, T. A., comps. LUPPOV, S. P., ed. Katalog inostrannykh geograficheskikh atlasov Biblioteki AN SSSR izdannykh v 1940-1963 gg. (Catalogue of foreign geographical atlases in the Library of the Academy of Sciences of the USSR, published 1940-1963). Moskva-Leningrad, Izdatel'stvo "Nauka," 1965. 164 p. (Akademiia Nauk SSSR. Biblioteka Akademii Nauk SSSR).
                                                                                  139
    Includes 524 atlases from outside the Soviet Union received by the Library of the Academy of Sciences, with annotations and tables listed by continent and country. Indexes by name, authors, editors, publishers, and agencies. English annotation, p. 5.

MEL'NIKOVA, Tat'iana Nikolaevna, and STANCHUL, T. A. <u>Katalog inostrannykh geo-
grafiacheskikh kart izdannykh v 1940-1958 gg</u>. (Catalogue of foreign geographical
maps published 1940-1958). S. P. Luppov, ed. Moskva-Leningrad, Izdatel'stvo
Akademii Nauk SSSR, 1960. 209 p. (Biblioteka Akademii Nauk SSSR. Otdel Kartografii). Title on cover: <u>Katalog inostrannykh geograficheskikh kart Biblioteki
AN SSSR izdannykh v 1940-1958 gg</u>.   140

> Bibliographical annotations on 1,327 foreign maps or map sets by countries.
> Index of authors, editors, publishers, and institutions. Index of geographical
> names. Tables with guides to general maps on scales larger than 1:200,000, on
> scales of 1:200,000 to 1:1,000,000, and on scales smaller than 1:1,000,000.
> Guide to physical-geographic maps. Guide to social-economic maps.

<u>SOVREMENNYE inostrannye geograficheskie atlasy</u>. <u>Katalog vystavki atlasov organiz.
K IV s''ezda Geograficheskogo Obshchestva SSSR i XI Soveshchaniiu direktorov
bibliotek Akademii Nauk SSSR i akademii soiuznykh respublik v Leningrade v 1970 g</u>.
(Current foreign geographical atlases. Catalogue of exhibition of atlases organized for the 4th Congress of the Geographical Society of the USSR and the 11th
conference of the directors of libraries of the Academy of Sciences of the USSR
and the academies of the union republics in Leningrad in 1970). comp. by T. N.
Mel'nikova. Leningrad, Biblioteka Akademii Nauk SSSR. Otdel kartografii, 1970.
29 p.   141

> 120 atlases grouped by three sections: general geographic, comples, and
> thematic.

<u>ZARUBEZHNYE goroda: katalog inostrannykh kart i atlasov</u> (Foreign cities: catalogue
of foreign maps and atlases). M. N. Usan and L. A. Khobotova, comps. Moskva,
1972. 103 p. (Gosudarstvennaia Biblioteka SSSR imeni V. I. Lenina).   142

Catalogues

USSR. Glavnoe Upravlenie Geodezii i Kartografii. Otdel Sbyta. <u>Karty i atlasy:
katalog</u> (Maps and atlases: catalogue). I. V. Gurevich and others, comps. Moskva,
GUGK, 1951. 127 p.   143

____. ____. <u>Karty i atlasy: katalog</u>. A. N. Voronina, and others, comps. Moskva,
GUGK, 1955. 85 p.   144

ATLASES

Note: Atlases are listed under titles as the sponsoring agencies often are
multiple and complex. Editorial roles also may be diverse: an atlas may have a
chairman of the editorial board, a senior editor, a scientific editor, and a
technical editor. As an aid to identification, the chief editors are here listed,
but without distinction as to the various roles, and sometimes merely the first
name of a long list of editors or authors presented in alphabetical order. Since
in atlases the size of the plates is a significant feature, the length of each
atlas is given in centimeters. Simple atlases for elementrary or secondary school
use are excluded.

GUGK is the abbreviation for Glavnoe Upravlenie Geodezii i Kartografii (Main
Administration of Geodesy and Cartography), the chief producer and publisher of
maps and atlases in the Soviet Union.

World

General

ATLAS mira (World atlas). A. N. Baranov, V. N. Lysiuk, S. I. Shurov, eds. Moskva,
GUGK, 1954. 283 p. 51 cm.
                                                                                                    145
Excellent world atlas covering the USSR in detail in 54 maps, ranging in
scale from 1:10,000,000 to 1:1,500,000. Other countries also well covered.
Physical maps show locations and relief. Major sections: maps of USSR, p. 9-84;
Europe, p. 85-137; Asia, p. 138-190; Africa, p. 191-208; North America, p. 209-
250; South America, p. 251-265; Australia, Oceania, and polar countries, p. 266-
283. English translation of preface, table of contents, titles, and legends, by
Vladimir G. Telberg. New York, Telberg Book Co. 1956. loose leaf. 37 cm.

2nd ed. A. N. Baranov, S. I. Shurov, G. B. Udintsev, eds. 1967. 250 p.
51 cm., though revised, has fewer maps on the Soviet Union.

____. Ukazatel' geograficheskikh nazvanii (Index of geographical names). Moskva,
1954. 572 p. 29 cm. (2nd ed., 1968. 533 p.).
                                                                                                    146
Location on maps of 205,000 geographical names.

WORLD atlas. A. N. Baranov, V. N. Lysiuk, S. I. Shurov, eds. 2nd ed., Moskva,
GUGK, 1967. 250 plates of maps. 51 cm. (Glavnoe Upravlenie Geodezii i Kartografii.
Chief Administration of Geodesy and Cartography).
                                                                                                    147
English-language version of the revised 2nd edition of Atlas Mira, with all
names in Latin letters and text in English. Maps of USSR, p. 7-45; Europe, p. 46-
101; Asia, p. 102-154; Africa, p. 155-175; North America, p. 176-217; South America,
p. 218-232; Australia, Oceania, and polar countries, p. 233-250. One of the really
fine world atlases.

The Index-gazetteer, Moskva, 1968, 1,010 p., includes all proper names found
on the maps of the atlas, and indicates map location. About 200,000 names.
Includes glossary of geographical terms and key to transliteration system for
Bulgarian, Greek, Mongolian, Russian and to Chinese phonetic alphabet.
                                                                                                    148
ATLAS mira (World atlas). M. K. Kudriavtsev, A. Ia. Marusov, eds. Moskva, Voenizdat,
1958. 460 p. 23 cm. (Voenno-Topograficheskoe Upravlenie General'nogo Shtaba.
Biblioteka Ofitsera).
                                                                                                    149
Major sections: introduction, p. 2-12; maps of the world, land areas, and the
oceans, p. 14-76; maps of the USSR, p. 78-130; maps of foreign countries, p. 132-
236; capitals of countries and largest cities of the world, p. 238-268; handbook-
statistical data, p. 271-321. Index of geographical names.

ATLAS mira (World atlas). A. I. Shurov, ed. Moskva, GUGK, 1959. 325 p. 35 cm.
p. 1-182 map plates.   150

Regional maps of countries of the world showing political boundaries. Maps
of the Soviet Union and its major regions, p. 9-50, Europe, p. 51-90, Asia, p.
91-128, Africa, p. 129-144, North America, p. 145-168, South America, p. 169-175,
Australia and Oceania, p. 176-178; the oceans, p. 179-182, the Arctic and Antarctic,
p. 7-8. Maps characterized by legibility and clarity. Index of geographical names,
p. 183-325, includes about 65,000 places. English translation and key, comp. by
Harriet Schwartz, ed. by V. G. Telberg. New York, Telberg Book Co., 1960.

MALYI atlas mira (Small atlas of the world). V. N. Salmanova, L. N. Kolosova, eds.
Moskva, GUGK, 1972. 159 p. of maps, 143 p. text and indexes. 19 cm. Earlier
editions from 1965 with printings annually. Preceded by Atlas mira (World atlas).
I. M. Itenberg, ed. Moskva, Glavnoe Upravlenie Geodezii i Kartografii, 1954.
165 p. maps, 136 p. text. 19 cm. frequently reprinted through 1962.   15

Small-format atlas published in large annual editions for wide circulation.

Thematic

FIZIKO-GEOGRAFICHESKII atlas mira (Physical-geographic atlas of the world). I. P.
Gerasimov, A. N. Baranov, F. F. Davitaia, Iu. V. Filippov, M. I. Budyko, E. M.
Lavrenko, M. V. Muratov, G. M. Senderova, and others, eds. Moskva, GUGK, 1964.
298 p. 51 cm. (Akademiia Nauk SSSR. Glavnoe Upravlenie Geodezii i Kartografii). 15

Superb atlas of distribution of principal physical features of the earth.
Both world thematic maps and a series of continental maps, including many maps of
the USSR. Includes relief, tectonics, geology, minerals, earthquakes and volcanoes,
Quaternary deposits, sediments of the oceans, geomorphology, climatic zones and
regions, annual march of climatic elements, radiation, evaporation, radiation
balance, heat loss and exchange, air temperatures, frost-free periods, circulation
of the atmosphere, pressure and winds, precipitation, characteristics of ocean
waters, runoff, soils, vegetation, zoogeography, and natural landscapes. Prepared
by scientists of the Academy of Sciences of the USSR.

An English translation of all legends and text is available as "Physical-
geographic atlas of the world, Moscow, 1964," special issue of Soviet Geography:
Review and Translation, May-June 1965, 403 p., published by the American Geographical Society of New York.

BOL'SHOI sovetskii atlas mira (Great Soviet world atlas), A. F. Gorkin, O. Iu.
Schmidt, V. E. Motylev, M. V. Nikitin, and B. M. Shaposhnikov, eds. GUGK, 1937-
1939. 2 v. v. 1, 1937, 168 plates; v. 2, 1939, 143 p., 135 plates. 54 cm.
Ukazatel' geograficheskikh nazvanii I-go toma (Index of geographic terms of volume
1). Moskva, 1940.   15

The greatest of the Soviet atlases prior to World War II. Plates 84-168 of
volume one depict for the Soviet Union as a whole the distribution of many elements of the physical and economic geography. Plates 16-133 of volume two consist
of two maps for each region of the Soviet Union, one physical and the other economic, showing land use and industrial production in 1935. English translation
of the legends of volume 1 by Andrew Perejda and Vera Washburne under the direction
of George B. Cressey, Great Soviet World Atlas, Ann Arbor, Michigan, 1940, 103 p.
28 cm.

MORSKOI atlas (Marine atlas). Moskva, Leningrad, Izdanie Morskogo General'nogo
Shtabe, 1950-1958. 3 v. 51 cm. (v. 1, Voenno-Morskoe Ministerstvo Soiuza SSR.
v. 2-3, Ministerstvo Oborony Soiuza SSR).

Tom 1. Navigatsionno-geograficheskii (v. 1. Navigational-geographic). I. S.
Isakov, V. A. Petrovskii, L. A. Dëmin, eds. 1950. 83 folded p. loose-leaf. Large
plates of regional maps of the world's oceans.   15

Tom 2. Fiziko-geograficheskii (v. 2. Physical-geographic). I. S. Isakov,
V. V. Shuleikin, L. D. Dëmin, eds. 1953. 76 folded p. loose-leaf. Large plates
depicting major ocean voyages, oceanography (bed of the ocean-relief of the ocean
floor, earthquakes and volcanism, geomorphology, and types of coasts; temperature,
salinity, density, and ice; currents, tides, swells; plant and animal life of the
oceans); climate (heat balance, air masses, circulation, climatic zones, temperature, and winds); terrestrial magnetism, cartography, and astronomy. 5 p. index,
separately numbered following table of contents, p. i-xviii.   155

Tom 3. Voenno-istoricheskii, chast' pervaia, listy 1-45 (v. 3, part 1.
Military-historical, plates 1-45). G. I. Levchenko, L. A. Dëmin, N. S. Frumkin,
eds. 1958. 45 folded bound p. Large plates of naval actions from the wars of
ancient Greece and Rome through World War I. Index, p. 1-23, following plates. 156

Tom 3. Voenno-istoricheskie karty listy 46-52 (Military-historical maps,
plates 46-52). 7 folded p. loose-leaf. 1957. Large plates showing military
actions 1917-1922 for the October Revolution, foreign intervention and the civil
war, and formation of the USSR.   157

Major massive detailed atlas of the oceans and activities on them.

Tom 1. Ukazatel geograficheskikh nazvanii (v. 1. Index to geographical
names). 1952. 543 p. 25 cm.   158

Gives map locations and latitude and longitude of about 115,000 place
names in· Russian form and alphabetical order followed by spelling in Latin letters
of all names in countries using the Latin alphabet.

GEOGRAFICHESKII atlas dlia uchitelei srednei shkoly (Geographical atlas for teachers
in secondary schools). V. S. Apenchenko and T. M. Pavlova, eds. 3rd ed. Moskva,
GUGK, 1967. 198 p. 38 cm. (1st ed., 1953, Iu. V. Filippov, A. I. Semenov, eds.,
192 p. 2nd ed., 1959, 191 p.). Reprinted 1968 and 1969.   159

A very good well-rounded atlas for teachers of geography with physical,
political, and thematic maps for the world, continents, major regions, and the
Soviet Union. Includes maps of climatic elements, geology, minerals, soils,
vegetation, manufacturing, agriculture, and population. Index of geographical
names, p. 165-198.

GROKLIMATICHESKII atlas mira (Agroclimatic atlas of the world). I. A. Gol'tsberg,
ed. Moskva, Leningrad, Glavnoe Upravlenie Geodezii i Kartografii, Gidrometeoizdat,
1972. 115 p. maps plus text and tables. 41 cm. (Glavnoe Upravlenie Gidrometeorologicheskoi Sluzhby pri Sovete Ministrov SSSR. Glavnaia Geofizicheskaia Observatoriia imeni A. I. Voeikova).   160

Includes maps of distribution of heat and moisture for the world as a whole,
of temperature and moisture conditions critical for agriculture by continents and
for the USSR, of winter conditions for crops, and of agroclimatic analogues.

ATLAS narodov mira (Atlas of peoples of the world). S. I. Bruk and V. S. Apenchenko,
eds. Moskva, Glavnoe Upravlenie Geodezii i Kartografii, Institut Etnografii im.
N. N. Miklukho-Maklaia Akademii Nauk SSSR, 1964. 184 p. 34 cm.   161

Finest atlas of the world distribution of ethnic groups. Regional maps covering all parts of the world with detailed distribution of each group. Excellent
detail on minorities in traditional and rural societies but less successful on
western and urban societies, which are more difficult to depict cartographically.
Tables of numbers in each ethnic group in 1961 (based primarily on language), by
ethnic groups and by regions. Alphabetical index of names of groups with equivalents of names in Latin letters. Telberg translation of titles and legends of the
maps, by V. G. Telberg. New York, Telberg Book Corp., 1965.

ATLAS istorii geograficheskikh otkrytii i issledovanii (Atlas of the history of geographical discoveries and exploration). A. V. Efimov, Z. F. Karabaeva, K. B. Martova, and others, eds. K. A. Salishchev,chairman, editorial board. K. B. Martova, responsible editor. Moskva, GUGK, 1959. 109 p. 35 cm.   162

Shows the routes of famous explorers with emphasis on the territory of the USSR and on Russian travelers and explorers. Index of place names.

ATLAS geograficheskikh otkrytii v Sibiri i v Severo-Zapadnoi Amerike XVII-XVIII vv. (Atlas of geographical discoveries in Siberia and in Northwestern America in the 17th and 18th centuries). A. V. Efimov, M. I. Belov, and O. M. Medushevskaia. A. V. Efimov, ed. Moskva, Izdatel'stvo "Nauka," 1964. 194 p. maps and 136 p. text. 41 cm. (Akademiia Nauk SSSR. Institut Etnografii imeni N. N. Miklukho-Maklaia). Supplementary English title, foreword, p. xv-xvi, and list of maps, p. 132-135.   163

Includes reproductions of original diagrams and maps of the explorers made on the spot; and general cartographic summaries. Extensive commentaries.

U. S. S. R. as a Whole

General

ATLAS SSSR (Atlas of the USSR). A. N. Baranov and others, eds. 2nd ed. Moskva, GUGK, 1969, 199 p. 38 cm. (1st ed., 1962. 185 p.).   164

A regional physical and economic atlas of the Soviet Union, with three main divisions: (1) hypsometric maps of regions of the country (37 maps) at scales varying from 1:1,500,000 for densely populated areas such as the Moscow region and the Donbass to 1:8,000,000 for Eastern Siberia and the Far East; (2) 23 physical-geographic maps for the country as a whole at scales mostly of 1:17,000,000 or 1:35,000,000 covering the systematic fields of physical geography; and (3) economic maps for systematic fields of economic geography for the country as a whole (22 maps) or complex economic-geographic maps for the principal regions (20) maps). Includes an index of geographical names, p. 153-199. English key to the Atlas of the USSR. Compiled by V. G. Telberg. New York, Telberg Book Corp., 1963. 115 p. 28 cm.

Excellent, handy, comprehensive, and moderately priced atlas of the USSR.

ATLAS SSSR (Atlas of the USSR). M. I. Svinarenko and O. A. Beloglazova, eds. 2nd ed. Moskva, GUGK, 1955. 147 p. 26 cm. p. 1-76 plates, p. 77-147. Index of names. (1st ed. 1950).   165

Hypsometric maps of regions of the USSR. Index of geographical names. English translation of table of contents, titles, and legends. New York, Telberg Book Co., 1956. 34 p. 27 cm.

GEOGRAFICHESKII atlas dlia srednei shkoly: Soiuz Sovetskikh Sotsialisticheskikh Respublik (Geographical atlas for the secondary school: Union of Soviet Socialist Republics). Moskva, GUGK, 1941. 41 p. 30 cm. Reissued, 1946, 1949, 1950. Also: Geograficheskii atlas dlia 7-go i 8-go klassov srednei shkoly: Soiuz Sovetskikh Sotsialisticheskikh Respublik, Moskva, GUGK, 1951. 76 p. 30 cm., reissued 1953, 1954, 1958.   166

MALYI atlas SSSR (Small atlas of the USSR). L. N. Mesiatseva and L. A. Polianskaia, eds. Moskva, GUGK, 1973. 117 p. maps. 80 p. text. 19 cm.   167

Small-format atlas of the Soviet Union by political subdivisions. Index of geographical names.

ATLAS SSSR (Atlas of the USSR). O. A. Beloglazova, ed. Moskva, GUGK, 1956. 194 p.
    15 cm.   168

        Small-format atlas of maps of administrative regions of the USSR. Index of
    geographical names.

ATLAS istorii SSSR dlia srednei shkoly (Atlas of history of the USSR for the secon-
    dary school). Moskva, GUGK, 1949-1952. 3 v. v. 1-2, 1949; v. 3, 1952. Reprinted
    1954. Parts frequently reprinted. 30 cm.   169

        v. 1 includes maps from the 1st to the 17th century.
        v. 2 includes maps from the 17th to the end of the 19th century.
        v. 3 includes maps of the 20th century.

ATLAS Soiuza Sovetskikh Sotsialisticheskikh Respublik (Atlas of the Union of Soviet
    Socialist Republics). Moskva, Izdanie TsIK SSSR, 1928. 108 p. 36 cm. Title also
    in Ukrainian, Belorussian, Armenian, Georgian, Arabic, German, French, and English
    on two added title pages. 36 colored maps, 7 with transparent overlays.   170

ATLAS Soiuza Sovetskikh Sotsialisticheskikh Respublik, primenitel'no k raionam
    ekonomicheskogo raionirovaniia Gosplana SSSR (Atlas of the Union of Soviet Socialist
    Republics, according to economic regionalization of Gosplan of the USSR). Compiled
    by A. F. Belavin with the participation of N. N. Aleksandrov, S. I. Sul'kevich,
    and S. V. Obodovskii. 2nd ed. Moskva-Leningrad, Gosudarstvennoe Izdatel'stvo,
    1928. 13 p. 80, 85 l. with 79 colored maps. 36 cm. 4 parts.   171

        Part 1. Ves' mir i SSSR (The world including the USSR).
        Part 2. SSSR v tselom (USSR as a whole).
        Part 3. Fizicheskii, politicheskii i ekonomicheskii obzory raionov SSSR
    (Physical, political, and economic surveys of the regions of the USSR).
        Part 4. National'nye ob"edineniia SSSR (National units of the USSR).

ATERIALY po istorii russkoi kartografii: karty vsei rossii i iuzhnykh eë oblastei
    do poloviny XVII veka. (Materials on the history of Russian cartography: maps of
    all Russia and its southern regions up to the middle of the 17th century). Izdanie
    Kievskoi Kommissii dlia razbora drevnikh aktov. Kiev, S. V. Kul'zhenko, 1899-1906.
    2 v. Edited by V. A. Kordt.   172

        Collection of 58 facsimile maps of Russia dating from 1474 to 1651, with
    accompanying text.

                                      Thematic

ATLAS razvitiia khoziaistva i kul'tury SSSR (Atlas of the development of the economy
    and culture of the USSR). A. N. Voznesenskii, and others, eds. Moskva, GUGK,
    1967. 172 p. 34 cm.   173

        Maps of the Soviet Union grouped by major categories: natural conditions and
    natural resources; industrialization of the USSR; development of agriculture of
    the USSR; development of transportation in the USSR; the cultural revolution and
    growth of the well-being of the population of the USSR; development of the economy
    and culture of the union republics and economic regions of the RSFSR; and peaceful
    competition of two systems. English key to the economic and cultural atlas of
    USSR by V. G. Telberg. Sag Harbor, New York, Telberg Book Corp., 1968.

ATLAS SSSR v deviatoi piatiletke (Atlas of the USSR in the ninth 5-year plan).
    M. V. Serebriakov, N. M. Terekhov, G. P. Venevtseva, eds. Moskva, GUGK, 1972.
    40 p. 29 cm.   174

        Maps of plans for the economy of the USSR under the 9th 5-year plan, 1971-
    1975.

ATLAS sel'skogo khoziaistva SSSR (Atlas of agriculture of the USSR). A. I. Tulupnikov, M. I. Nikishov, and others, eds. Moskva, GUGK, 1960. 309 p. 34 cm.   175

An exemplary atlas with large detailed maps depicting distribution of natural conditions, such as precipitation, temperature, and soils (plates 6-59), agricultural organization (60-89), crops (90-175), and livestock (176-209) for the Soviet Union as a whole, agriculture and land use for individual republics and regions (210-269), and value of production or yields (270-297). English key to the Agricultural atlas of USSR. Edited by V. G. Telberg. New York, Telberg Book Corp. ca 1962. variously paged. Part I consists of a syllabus of all 300 maps; part II consists of translations of the legends of the most important maps.

ATLAS avtomobil'nykh dorog SSSR (Atlas of automobile highways of the USSR). N. T. Markova, ed. 4th ed., Moskva, GUGK, 1972. 167 p. 26 cm. (1st ed., 1966; 2nd ed., 1969; 3rd ed., 1971. Preceded by Atlas avtomobil'nykh dorog SSSR. O. A. Beloglazova, ed., Moskva, Glavnoe Upravlenie Geodezii i Kartografii, 1959. 165 p. 27 cm., 2nd ed., 1960; 3rd ed. 1960; 4th ed., 1961).   176

Includes 18 maps of individual routes and 62 maps of regions of the USSR.

ATLAS skhem zheleznykh dorog SSSR (Atlas of diagrams of railroad lines of the USSR). O. A. Beloglazova, ed. Moskva, GUGK, 1972. 101 p. 16 cm.   177

Diagrams of railroad lines, p. 1-61; alphabetical list of railroad stations, p. 63-85; tables of train routes, p. 86-97; alphabetical list of urban settlements with names different from the railroad stations in them, p. 98-101.

ZHELEZNYE dorogi SSSR. Napravleniia i stantsii. Atlas putei soobshcheniia (Railroads of the USSR. Routes and stations. Atlas of means of transportation). O. A. Beloglazova, ed. 6th ed. Moskva, GUGK, 1971. 150 p. 16 cm. (1st ed. 1965, 2nd ed. 1966, 1967, 3rd ed. 1968, 4th ed. 1969, 5th ed. 1970. Preceded by Atlas skhem zheleznykh dorog SSSR. S. B. Baudina, ed., Moskva, GUGK, 1960. 143 p. 16 cm., and 1961, 1962, 1963, and 1964).   178

Diagrams of railroad lines, p. 5-94; alphabetical index to railroad stations, p. 96-141; list of main routes of trains, p. 142-150.

SPUTNIK passazhira: skhemy zheleznodorozhnykh marshrutov (The passenger's companion: diagrams of railroad lines). Leonid Nikolaevich Anatol'ev, author-compiler. Moskva, Izdatel'stvo "Transport," 1964. 159 p. 16 cm.   179

Diagrams of railroad lines, p. 6-128; alphabetical list of stations, p. 129-145; distances, times, and speeds, p. 146-155, prices of tickets, p. 156-159.

ATLAS energeticheskikh resursov SSSR (Atlas of energy resources of the USSR). A. V. Vinter, G. M. Krzhizhanovskii, G. I. Lomov, eds. Moskva, Gosudarstvennoe Energeticheckoe Izdatel'stvo, 1933-1935. 1 v. plates. 2 v. text. 50 cm. (Glavnoe Energeticheskoe Upravlenie).   180

23 plates depicting power resources.

ATLAS promyshlennosti SSSR na nachalo 2-i piatiletki [Geograficheskii atlas sostavlen po materialam, sobrannym Otdelom ekonomicheskogo kartografirovanniia Kartotresta] (Atlas of industry of the USSR at the beginning of the second 5-year plan: geographical atlas compiled from materials gathered by the Division of economic cartography of the Cartographic Trust). Prospekt. Moskva, 1934. 6 p. 4 colored maps. 26 cm. (Redaktsionnyi Sovet NKTP SSSR po Kartografirovaniiu Promyshlennosti   181

TLAS promyshlennosti SSSR (Atlas of industry in the USSR). Moskva, Izdatel'stvo
Prezidiuma V.S.N.Kh. SSSR,1929-1931. 5 v. 61 cm. (Glavnoe Geodezicheskoe Upravlenie, except v. 1 under earlier name: Glavnyi Geodezicheskii Komitet). v. 1 has
supplementary text volume, with the same name, 1930. 219 p.   182

    v. 1. Tsenzovaia promyshlennost' (Industry covered by census statistics).
F. G. Dubovikov, comp., 1929.
    v. 2. Melkaia i kustarno-remeslennaia promyshlennost' (Small and handicraft
industry). N. Ia. Vorob'ëv, comp., 1930.
    v. 3. Poleznye iskopaemye (Minerals). Compiled under the direction of V.
A. Mironov and S. I. Mikhailov, 1931.
    v. 4. Les, torf, energetika (Wood, peat, energy). 1931.
    v. 5. 1. Prirodnye usloviia SSSR (Natural conditions in the USSR). 1931.
          2. Trudovye resursy SSSR (Labor resources of the USSR). B. V. Avilov,
comp., 1931.

INANSOVO-STATISTICHESKII atlas Rossii... (Financial-statistical atlas of Russia).
P. A. Antropov. S.-Peterburg, Izdanie A. F. Marksa, 1898. 19 p. 29 l. of colored
maps. 34 cm.   183

                Union Republics and Larger Regions of the U. S. S. R.

       (In geographical order, Union republics first, then other regions.
                     See maps inside front and back covers.)

TLAS Ukrainskoi SSR i Moldavskoi SSR (Atlas of the Ukrainian SSR and Moldavian
SSR). V. G. Bondarchuk, M. A. Koroleva, and others, eds. Moskva, GUGK, 1962.
90 p. 35 cm.   184

    Maps of natural conditions, population, transport, manufacturing, agriculture,
culture, and regions.

LIMATICHESKII atlas Ukrainskoi SSR (Climatic atlas of the Ukrainian SSR). G. F.
Prikhot'ko, N. I. Guk, G. I. Sladkovich, eds. Leningrad, Gidrometeoizdat, 1968.
232 p. 26 cm. (Glavnoe Upravlenie Gidrometeorologicheskoi Sluzhby pri Sovete
Ministrov SSSR. Ukrainskii Nauchno-Issledovatel'skii Gidrometeorologicheskii
Institut).   185

    Maps of distribution of mean and extreme values of climatic characteristics;
climatic regionalization.

ROKLIMATICHESKII atlas Ukrainskoi SSR (Agroclimatic map of the Ukrainian SSR).
S. A. Sapozhnikova, ed. Kiev, Izdatel'stvo "Urozhai," 1964. 46 p., 10 of text
and 36 of maps. 45 cm. (Glavnoe Upravlenie Gidrometeorologicheskoi Sluzhby .
Ukrainskii Nauchno-Issledovatel'skii Gidrometeorologicheskii Institut).   186

    Part 1. Agroclimatic and climatic data, p. 1-31. Part 2. Seasonal and
annual regimes of rivers, p. 32-36.

LAS sil's'koho hospodarstva Ukraïns'koï RSR (Atlas of agriculture of the Ukrainian
SSR). I. N. Romanenko, A. S. Kharchenko, and others, eds. Kyïv, Vydavnytstvo
Kyïvs'koho Derzhavnoho Universytetu, 1958. 8 p. text. 47 p. map plates, some
folded. 40 cm. (Kyïvs'kyi Derzhavnyi Universytet im. T. H. Shevchenka.
Heohrafichnyi Fakul'tet). In Ukrainian.   187

    Maps on a scale of 1:4,000,000 on general characteristics of agriculture,
p. 1-7; natural conditions, p. 8-16; crops, p. 17-39; livestock, p. 40-44;
agricultural productivity, p. 45-47.

ATISTICHESKII atlas Latviiskoi Sovetskoi Sotsialisticheskoi Respubliki (Statistical
atlas of the Latvian SSR). F. Deglav, E. Abolin', A. Driuchin, eds. Riga,
Izdatel'stvo Akademii Nauk Latviiskoi SSR, 1960. 41 p. Latvian and Russian
parallel text. 40 p. maps. 34 cm. (Akademiia Nauk Latviiskoi SSR. Institut
Ekonomiki; Latviiskaia SSR. Tsentral'noe Statisticheskoe Upravlenie).   188

                                                                  182 - 188

Maps and cartograms on location, administrative structure, population, economic ties, manufacturing, agriculture, transport, finance and trade, health and culture.

ATLAS Belorusskoi Sovetskoi Sotsialisticheskoi Respubliki (Atlas of the Belorussian SSR). S. N. Malinin, K. I. Lukashev, I. S. Lupinovich, A. N. Avksent'ev, P. P. Rogovoi, and V. V. Urusov, eds. Minsk, Moskva, Akademiia Nauk BSSR, Glavnoe Upravlenie Geodezii i Kartografii, 1958. 140 p. 28 cm.   18

Maps on natural conditions, population, economy, culture, history, and political administrative divisions. In two editions, one in Russian, one in Belorussian. Belorussian title: Atlas Belaruskaĭ Savetskaĭ Satsyialistychnaĭ Respubliki.

ATLAS Gruzinskoi Sovetskoi Sotsialisticheskoi Respubliki (Atlas of the Georgian SSR). A. N. Dzhavakhishvili, A. F. Aslanikashvili, and others, eds. Tbilisi, Moskva, GUGK, 1964. 269 p. 30 cm. (Akademiia Nauk Gruzinskoi SSR. Institut Geografii imeni Vakhushti).   19

Major sections: introduction, p. 1-7; natural conditions, p. 8-164; population, p. 165-175; economy, p. 176-228; culture and health, p. 229-243; history and archeology, p. 244-269. Two editions, one in Russian, the other in Georgian. Georgian title: Sak'art'velos Sabtchot'a Sotsialisturi Respublikis atlasi (Sak'art' velos SSR Metsnierebat'a Akademia. Vakhushtis Sakhelobis Geographiis Instituti).

ATLAS Armianskoi Sovetskoi Sotsialisticheskoi Respubliki (Atlas of the Armenian SSR). A. B. Bagdasarian, ed. Erevan-Moskva, 1961. 111 p. 35 cm. (Akademiia Nauk Armianskoi SSR. Glavnoe Upravlenie Geodezii i Kartografii).   19

Maps of administrative areas, natural conditions, population, culture, economy, and history. Two editions, one in Russian, the other in Armenian. Armenian title: Haykakan Sovetakan Sots'ialistakan Rhespowbljkayi atlas (Haykakan SSH Gitowt'yownneri Akademia).

ATLAS Azerbaidzhanskoi Sovetskoi Sotsialisticheskoi Respubliki (Atlas of the Azerbaidzhan SSR). I. K. Abdullaev, K. K. Giul', A. I. Ibragimov, M. A. Kashkai, Iu. G. Mamedaliev, Sh. F. Mekhtiev, and others, eds. Baku, Moskva, GUGK, 1963. 213 p. 32 cm. (Akademiia Nauk Azerbaidzhanskoi SSR. Institut Geografii).   19

Maps on administrative structure, population, natural conditions, economy, culture, and history. Two editions, one in Russian, the other in Azerbaijani.

ATLAS Uzbekskoi Sovetskoi Sotsialisticheskoi Respubliki (Atlas of the Uzbek SSR). L. N. Babushkin, V. M. Volkova-Volk, S. B. Baudina, eds. Tashkent, Moskva, Akademiia Nauk Uzbekskoi SSR, Glavnoe Upravlenie Geodezii i Kartografii, 1963, 53 p. 34 cm. (Akademiia Nauk Uzbekskoi SSR; Glavnoe Upravlenie Geodezii i Kartografii).   1

Maps of natural conditions.

ATLAS Tadzhikskoi Sovetskoi Sotsialisticheskoi Respubliki (Atlas of the Tadzhik SSR). K. V. Staniukovich, ed. Dushanbe, Moskva, GUGK, 1968. 200 p. 34 cm. (Akademiia Nauk Tadzhikskoi SSR. Sovet po Izucheniiu Proizvoditel'nykh Sil).   1

Maps of natural conditions, population, economy, culture, and history.

ATLAS l'dov Baltiiskogo moria i prilegaiushchikh raionov (Ice atlas of the Baltic sea and adjacent regions). Iu. V. Preobrazhenskii, ed. Leningard, Gidrometeoizdat. (Gosudarstvennyi Okeanograficheskii Institut. Leningradskoe Otdelenie).   1

Chast' I. Baltiiskoe more, Rizhskii zaliv, Datskie prolivy i prilegaiushchai chast' Severnogo moria (Part 1. Baltic sea, Gulf of Riga, Danish straits, and adjacent part of the North Sea). M. E. Dvoskinaia, ed. 1960. 8 p. text. 64 p. maps. 9 p. tables. 45 cm.

General maps, maps of the ice regime, and maps of navigational characteristic of ice, and tables.

ATLAS Komi Avtonomnoi Sovetskoi Sotsialisticheskoi Respubliki (Atlas of the Komi
ASSR). S. V. Kalesnik, A. G. Isachenko, S. T. Riabchikov, and others, eds.
Moskva, GUGK, 1964. 112 p. 33 cm. (Leningradskii Gosudarstvennyi Universitet
imeni A. A. Zhdanova. Nauchno-Issledovatel'skii Geografo-Ekonomicheskii Institut;
Akademiia Nauk SSSR. Komi Filial).   196

    Major sections: natural conditions; population and administrative regions;
economy; culture; history.

ATLAS Tselinnogo kraia (Atlas of the Virgin-Lands District). I. N. Guseva, ed.
Moskva, GUGK, 1964. 49 p. 48 cm. (Moskovskii Gosudarstvennyi Universitet imeni
M. V. Lomonosova. Geograficheskii Fakul'tet).   197

    Sections: political-administrative maps, natural conditions, agriculture and
industry, construction, transport and economic ties, population and culture.

ATLAS Zabaikal'ia (Buriatskaia ASSR i Chitinskaia Oblast'). (Atlas of the Trans-
Baykal area: Buryat ASSR and Chita oblast). V. B. Sochava, V. M. Kartushin, V.
P. Shotskii, and others, eds. Moskva, Irkutsk, GUGK, 1967. 176 p. 34 cm.
(Akademiia Nauk SSSR. Sibirskoe Otdelenie. Institut Geografii Sibiri i Dal'nego
Vostoka).   198

    Main sections: history and location, p. 1-8; surface and subsurface, p. 10-28;
climate and waters, p. 29-52; soils and biogeography, p. 54-67; landscapes and
natural regions, p. 70-76; Lake Baykal, p. 78-95; population and culture, p. 97-
110; medical-geographic evaluation of the area and public health, p. 112-120;
industry, p. 122-136; agriculture, p. 137-164; regional economic maps, p. 165-168.
Notes on the maps, p. 169-176.

CHERTEZHNAIA kniga Sibiri, sostavlennaia Tobol'skim synom boiarskim Semenom
Remezovym v 1701 godu (Sketch book of Siberia, compiled by Tobol'sk son of the
boyar Semën Remezov). Izdano Arkheograficheskoiu Kommissieiu, izhdiveniem kor-
respondenta ego, P. I. Likhareva. S.-Peterburg, Tip. A. M. Kotomina, 1882.
23 p. maps.   199

    Photolithographic reproduction of the original map of Siberia compiled by
Remezov and his three sons in 1701, by the order of Peter the Great, from maps
previously made by Remezov, and from some other sources. Supplement. S.-Peterburg,
Tip. (byvshaia) A. M. Kotomina, 1882. Text p. 1-14. Index, p. 15-58.

Oblasts or Smaller Regions of the U. S. S. R.

(In alphabetical order. See map inside back
cover and alphabetical key.)

ATLAS Astrakhanskoi oblasti (Atlas of Astrakhan' oblast). N. M. Terekhov and N. A.
Lobzova, eds. Moskva, GUGK, 1968. 38 p. 30 cm. (Astrakhanskii Gosudarstvennyi
Pedagogicheskii Institut imeni S. M. Kirova).   200

    Maps of natural conditions, economy, population, culture, and history.

ATLAS Baikala (Atlas of Baikal). G. I. Galazii and B. F. Lut, eds. Irkutsk,
Moskva, GUGK, 1969. 30 p. 33 cm. (Akademiia Nauk SSSR. Sibirskoe Otdelenie.
Limnologicheskii Institut).   201

    Maps of natural conditions of Lake Baykal.

ATLAS Iaroslavskoi oblasti (Atlas of Yaroslavl' oblast). A. B. Ditmar, ed. Moskva,
GUGK, 1964. 28 p. 28 cm.   202

    Maps of natural conditions, population, economy, culture, and history.

ATLAS Irkutskoi oblasti (Atlas of Irkutsk oblast). I. P. Zarutskaia, V. P. Shotskii,
and others, eds. Moskva, Irkutsk, GUGK, 1962. 182 p. 35 cm. (Moskovskii
Gosudarstvennyi Universitet imeni M. V. Lomonosova. Geograficheskii Fakul'tet;
Akademiia Nauk SSSR. Sibirskoe Otdelenie. Vostochno-Sibirskii Filial).   20

Maps organized by sections: general part, natural conditions and resources,
population, agriculture, transport, culture and health, history of investigations,
history.

ATLAS Kalininskoi oblasti (Atlas of Kalinin oblast). A. V. Gaveman, ed. Moskva,
GUGK, 1964. 34 p. 28 cm. (Kalininskii Pedagogicheskii Institut imeni Kalinina).
20
Maps of natural conditions, population, economy, culture, and history.

ATLAS Karagandinskoi oblasti (Atlas of Karaganda oblast). M. I. Semenova, ed.
Moskva, GUGK, 1969. 48 p. 32 cm. (Akademiia Nauk Kazakhskoi SSR. Institut
Ekonomiki).   20

Maps of natural conditions, population, economy, culture, and history.

ATLAS Kirovskoi oblasti (Atlas of Kirov oblast). N. M. Terekhov, G. P. Fedorovskaia,
and others, eds. Moskva, GUGK, 1968. 38 p. 28 cm. (Leningradskii Universitet.
Nauchno-Issledovatel'skii Geografo-Ekonomicheskii Institut).   20

Maps of natural conditions, economy, population, culture, and history.

ATLAS Kurskoi oblasti (Atlas of Kursk oblast). N. A. Antimonov, N. M. Terekhov,
eds. Moskva, GUGK, 1968. 40 p. 27 cm. (Kurskii Gosudarstvennyi Pedagogicheskii
Institut; Geograficheskoe Obshchestvo SSSR. Kurskii Otdel).   20

Maps of natural conditions, population, economy, culture, and history.

ATLAS Kustanaiskoi oblasti (Atlas of Kustanay oblast). I. P. Zarutskaia, ed.
Moskva, GUGK, 1963. 79 p. 39 cm. (Moskovskii Gosudarstvennyi Universitet imeni
M. V. Lomonosova. Geograficheskii Fakul'tet; Akademiia Nauk Kazakhskoi SSR).   20

Maps of administrative divisions (at various dates), history, natural conditions and resources, population, agriculture, industry and transport, and
culture.

ATLAS Leningradskoi oblasti (Atlas of Leningrad oblast). A. G. Durov, ed. Moskva,
GUGK, 1967. 82 p. 29 cm. (Leningradskii Gosudarstvennyi Universitet imeni A. A.
Zhdanova. Nauchno-Issledovatel'skii Geografo-Ekonomicheskii Institut).   2

Major sections: natural conditions, p. 12-42; population and economy, p. 48-
62; educational and cultural institutions and public health, p. 63-72; history, p.
73-82.

ATLAS Leningradskoi oblasti i Karel'skoi ASSR (Atlas of Leningrad oblast' and the
Karelian ASSR). Leningradskaia Oblastnaia Planovaia Komissiia and Geografo-
Ekonomicheskii Nauchno-Issledovatel'skii Institut. Leningrad, Izdanie GENII,
1934. variously paged. 48 cm. Alphabetical index to the atlas, 1935. 80 p.
23 cm. Text volume, 1935. 341 p. 23 cm. (NKP RSFSR. Geografo-Ekonomicheskii
Nauchno-Issledovatel'skii Institut Leningradskogo Gosudarstvennogo Universiteteta
imeni A. S. Bubnova). Maps titles in Russian, Finnish, and English. Text volume
has English summaries at the end of each chapter.   2

Maps of administrative divisions, population, labor, physical features,
geology, mineral resources, soils, climate, drainage, vegetation, animal life,
industries, electric power, forests, and agriculture.

ATLAS Moskovskoi oblasti (Atlas of Moscow oblast). V. M. Sokolov. S. I. Shurov,
N. A. Lobzova, and others, eds. Moskva, GUGK, 1964. 12 p. 28 cm. (Tsentral'nyi
Nauchno-Issledovatel'skii Institut Geodezii, Aeros"ëmki i Kartografii).   2

Maps of Moscow and suburbs, administrative divisions, natural conditions,
population, and economy.

203 - 211

ΓLAS Moskovskoi oblasti (Atlas of Moscow oblast). S. E. Guberman, and others, eds.
Moskva, Izdatel'stvo Mosoblispolkoma, 1933- [1934]. 67 l. 40 cm. Text with same
title, 1934. 35 p. 25 cm. (Moskovskaia oblast'. Nauchno-Issledovatel'skii
Institut Ekonomiki).   212

ΓLAS Murmanskoi oblasti (Atlas of Murmansk oblast). A. G. Durov, ed. Moskva, GUGK,
1971. 32 p. 34 cm. (Leningradii Gosudarstvennyi Universitet imeni A. A. Zhdanova.
Nauchno-Issledovatel'skii Geografo-Ekonomicheskii Institut).   213

   Maps of natural conditions, population, economy, and culture.

ΓLAS Orenburgskoi oblasti (Atlas of Orenburg oblast). A. S. Vetrov, G. I. Guliuk,
and A. S. Svirskii, eds. Moskva, GUGK, 1969. 36 p. 33 cm.   214

   Maps of natural conditions, economy, population, culture, and history.

ΓLAS Pskovskoi oblasti (Atlas of Pskov oblast). A. G. Durov, and others, eds.
Moskva, GUGK, 1969. 44 p. 32 cm. (Leningradskii Gosudarstvennyi Universitet
imeni A. A. Zhdanova. Nauchno-Issledovatel'skii Geografo-Ekonomicheskii Institut).
                                                                                        215
   Maps of natural conditions, population, economy, culture, and history.

ΓLAS Riazanskoi oblasti (Atlas of Ryazan' oblast). S. N. Soldatov, ed. Moskva,
GUGK, 1965. 36 p. 26 cm.   216

   Maps of natural conditions, population, economy, culture, and history.

ΓLAS Sakhalinskoi Oblasti (Atlas of Sakhalin). G. V. Komsomol'skii and I. M. Siryk,
eds. Moskva, GUGK, 1967. 135 p. 34 cm. (Sakhalinskaia Oblast'. Sovet Deputatov
Trudiashchikhsia. Ispolnitel'nyi Komitet; Akademiia Nauk SSSR. Sibirskoe Otdelenie. Sakhalinskii Kompleksnyi Nauchno-Issledovatel'skii Institut).   217

   Maps of natural conditions of Sakhalin island, of Kuril islands, and of the
Okhotsk sea and Tatar straits.

ΓLAS Severo-Osetinskoi ASSR (Atlas of the North Oset ASSR). M. I. Serebriannaia,
V. A. Stankevich, L. K. Kurtskhaliia, eds. Moskva, GUGK, 1967. 31 p. 29 cm.
(Severo-Osetinskii Gosudarstvennyi Pedagogicheskii Institut imeni K. L. Khetagu-
rova).   218

   Maps of natural conditions, population, economy, culture, and history.

ΓLAS Smolenskoi oblasti (Atlas of Smolensk oblast). V. G. Vasil'ev, S. N. Soldatov,
N. A. Lobzova, eds. Moskva, GUGK, 1964. 31 p. 29 cm.   219

   Maps of natural conditions, population, economy, culture, and history.

ΓLAS Stavropol'skogo kraia (Atlas of Stavropol'kray). N. T. Dorokhin, V. G.
Gnilovskoi, V. N. Sergeev, eds. Moskva, GUGK, 1968. 40 p. 29 cm. (Geografi-
cheskoe Obshchestvo SSSR. Stavropol'skii Otdel; Stavropol'skii Gosudarstvennyi
Pedagogicheskii Institut).   220

   Includes maps on natural conditions, the economy, population, culture, and
history.

ΓLAS Tambovskoi oblasti (Atlas of Tambov oblast). A. M. Kirillov, G. I. Guliuk,
and A. S. Svirskii, eds. Moskva, GUGK, 1966. 32 p. 28 cm. (Tambovskii Gosudarstvennyi Pedagogicheskii Institut). Earlier edition, Geograficheskii atlas Tambovskoi oblasti (Geographical atlas of Tambov oblast), 1960. 16 p. 28 cm.).   221

   Maps of natural conditions, population, economy, culture, and history.

ATLAS Volgogradskoi oblasti (Atlas of Volgograd oblast). A. G. Liakhova, N. M.
Terekhov, and N. A. Lobzova, eds. Moskva, GUGK, 1967. 32 p. 30 cm. (Geo-
graficheskoe Obshchestvo SSSR. Volgogradskii Otdel; Volgogradskii Pedagogicheskii
Institut).   222

   Maps of natural conditions, population, economy, culture, and history.

ATLAS Vologodskoi oblasti (Atlas of Vologda oblast). S. N. Soldatov, ed. Moskva,
GUGK, 1965. 38 p. 33 cm.   223

   Maps of natural conditions, population, economy, culture, and history.

ATLAS Voronezhskoi oblasti (Atlas of Voronezh oblast). N. N. Smirnov, ed. Moskva,
GUGK, 1968. 32 p. 27 cm. (Voronezhskii Gosudarstvennyi Universitet).   224

   Maps of physical conditions, population, economy, culture, and history.

GEOGRAFICHESKII atlas Dagestanskoi ASSR (Geographical atlas of the Dagestan ASSR).
V. A. Gimmel'reikh, I. L. Bocharova, E. A. Shishkin, eds. Makhachkala, Dagestan-
skoe Uchebno-Pedagogicheskoe Izdatel'stvo, Dagestanskii Filial Geograficheskogo
Obshchestva SSSR, 1964. 20 p. 33 cm.   225

   A school atlas.

HEOHRAFIIA Kyïvs'koï oblasti. Atlas (Geography of Kiev oblast. Atlas). A. S.
Kharchenko, A. P. Zolovs'kyi, E. D. Desiatov, O. I. Korbush, eds. Kyïv,
Vydanytstvo Kyïvs'koho Derzhavnoho Universytetu, 1962. 98 p. 30 cm. (Kyïvs'kyi
Derzhavnyi Universytet im. T. H. Shevchenka. Heohrafichnyi Fakul'tet). In
Ukrainian. Russian title: Geografiia Kievskoi oblasti. Atlas.   226

   Maps grouped in seven sections: location and general characteristics; natural
conditions, industry, agriculture, culture and health, the 7-year plan of 1959-
1965, and progress in fulfillment of the 7-year plan.

PETERBURG-LENINGRAD. Istoriko-geograficheskii atlas (Petersburg-Leningrad.
Historical-geographical atlas). S. V. Kalesnik, and others, eds. Leningrad,
Izdatel'stvo Leningradskogo Universiteta, 1957. xiv, 56 p. 48 cm. (Leningradskii
Gosudarstvennyi Universitet imeni A. A. Zhdanova. Geografo-Ekonomicheskii Nauchno-
Issledovatel'skii Institut).   22[?]

   Series of historical maps of Leningrad.

Areas Outside the Soviet Union

ATLAS Antarktiki (Atlas of Antarctica). v. 1. V. G. Bakaev, D. I. Shcherbakov,
E. I. Tolstikov, G. A. Avsiuk, E. S. Korotkevich, M. I. Svinarenko, A. V. Nudel'man,
and others, eds. Moskva, GUGK, 1966. xxiii, 255 p. 60 cm. v. 2. E. I. Tolstikov, G. A. Avsiuk, and E. S. Korotkevich, V. G. Aver'ianov, eds. Leningrad,
Gidrometeoizdat, 1969. 598 p. 30 cm. (Sovetskaia Antarkticheskaia Ekspeditsiia).
                                                                              22[?]
   Detailed atlas of Antarctica in 225 plates, covering history of exploration
(plates 4-12), map coverage (13-15), general maps (16-24), aeronomy and physics
of the earth (25-57), geology and relief (58-72), climate (73-93), glaciation
(94-96), waters of the Southern Ocean (97-126), biology (127-132), physical-
geographical regionalization (133-135), and with regional plates of the continent
and adjacent islands (136-180) and of the Southern Ocean (181-225). Index of
about 5,000 place names, p. xiii-xxiii. An English translation of the legends
Atlas of Antarctica, vol. 1, Moscow, 1966, was published as a special issue of
Soviet geography: review and translation, v. 8, nos. 5-6 (May-June, 1967), p. 261-
507.

   v. 2, a text volume, is a monographic study of the geography, exploration,
geophysics, geology, climate, glaciation, water and ice biology, and landscapes
of Antarctica. It has a supplementary English table of contents, detailed bibliographies by specialists. Indexes of persons, and names of ships [entry 2384].

ATLAS Afriki (Atlas of Africa). S. I. Shurov, V. G. Brugger, and others, eds.
Moskva, GUGK, 1968. 118 p. 32 cm.   229

    Major sections: historical; natural conditions; regions; population and the
economy. Index of geographical names, p. 73-118. Supplement, Atlas of Africa,
English translation of legends and text, 1968. 39 p. 28 cm.

ATLAS S. Sh. A. (Atlas of the United States). V. N. Lysiuk, M. M. Mekler, and
others, eds. Moskva, GUGK, 1966. 72 p. 34 cm.   230

    Maps on natural conditions, population, industry, agriculture, major regions,
possessions, foreign economic relations, and cities. Index of geographical names,
p. 63-72. Table of contents also in English.

ATLAS Latinskoi Ameriki (Atlas of Latin America). L. N. Mesiatseva, V. V. Vol'skii,
and others, eds. Moskva, GUGK, 1968. 94 p. 32 cm.   231

    Major sections: historical; natural conditions; regions; population and the
economy. Index of geographical names, p. 63-94. Supplement, Atlas de America
Latina. Spanish translation of legends and text, 1968. 33 p. 28 cm.

NATSIONAL'NYI atlas Kuby. V oznamenovanie desiatoi godovshchiny revoliutsii
(National atlas of Cuba. On the occasion of the 10th anniversary of the revolution). Antonio Nuñez Jiménez and Pedro Cañas Abril, eds. Havana, Moskva, GUGK,
1970. 132 p. of maps plus index of geographical names. 48 cm. (Academia de
Ciencias de Cuba. Instituto de Geografía. Akademiia Nauk SSSR. Institut Geografii).   232

    Main sections: Introduction; general geographical maps; maps of natural
conditions; maps of agriculture, industry, and transport; maps of population
and culture; historical maps. Also available in edition in Spanish: Atlas
Nacional de Cuba.

STATISTICS

Note: For technical reasons the names of statistical agencies and titles
of works are given first in Russian, even for union republics other than the
Russian Soviet Federated Socialist Republic and even if the text is in another
language of the Soviet Union. Whenever possible the Russian is followed by the
rendering in the other language, particularly in Ukrainian, Estonian, Latvian,
and Lithuanian.

Bibliographies

IZDATEL'STVO "Statistika." <u>Statistika i uchët: annotirovannyi katalog izdatel'stva
"Statistika" 1966-1970 gg.</u> (Statistics and accounting: annotated catalogue of
books of the Statistika publishing house, 1966-1970). Moskva, Izdatel'stvo
"Statistika," 1971. 423 p. (Komitet po Pechati pri Sovete Ministrov SSSR).   233

_____. <u>Knigi Izdatel'stva "Statistika:" bibliograficheskii annotirovannyi ukazatel'
literatury za 1949-1965 gg.</u> (Books of the publishing house "Statistika:" bibliographic annotated guide to the literature of 1949-1965). E. A. Shashikhin and N.
I. Shatunova, comps. Moskva, Izdatel'stvo "Statistika," 1966. 247 p. (Komitet
po Pechati pri Sovete Ministrov SSSR).   234

NOPRIENKO, Georgii Karpovich. <u>Bibliograficheskii ukazatel' statei i materialov po
statistike i uchëtu. Zhurnal "Vestnik statistiki" za 50 let (1919-1968 gg.).</u>
(Bibliography of articles and materials on statistics and accounting. The journal
"Vestnik statistiki" for 50 years, 1919-1968). Moskva, Izdatel'stvo "Statistika,"
1971. 315 p.   235

STATISTICHESKIE sborniki, izdannye v 1956-1957 gg (Statistical compilations published in 1956-1957), <u>Vestnik statistiki</u>, 1958, no. 1, p. 77-78.   236

_____ v 1958 g., 1959, no. 2, p. 93-94.

_____ v 1959 g., 1959, no. 12, p. 78. Compiled by G. A. Aliutin.

_____ v 1960 g., 1960, no. 12, p. 82-83. G. A. Aliutin.

_____ v 1960-1961 gg., 1962, no. 1, p. 76-77. G. A. Aliutin.

_____ v 1962 g., 1963, no. 7, p. 78. G. A. Aliutin.

_____ v 1963 g. i v pervom polugodii 1964 g. (in 1963 and the first half of 1964),
1964, no. 10, p. 74-75. G. A. Aliutin.

OVYE statisticheskie sborniki (New statistical compilations). G. A. Aliutin, comp.
<u>Vestnik statistiki</u>, 1965, no. 7, p. 73-74; 1966, no. 7, p. 83-84; 1967, no. 7,
p. 85-86; 1968, no. 3, p. 88-89; 1969, no. 4, p. 80; 1970, no. 8, p. 77-78; 1972,
no. 2, p. 83.   237

The most comprehensive list and best summary of statistical compilations,
especially yearbooks, published in the USSR over the period 1956-1971.

OZULOV, Avdei Il'ich. <u>Perepisi naseleniia zemnogo shara; khronologicheskie tablitsy</u>
(Censuses of population of the world: chronological tables). Moskva, Statistika,
1970. 173 p.   238

Chronological lists of population censuses of each country of the world,
grouped by continents, 1790-1968, in two tables, p. 127-170, with bibliography.

Censuses of the USSR and of Russia

USSR. Tsentral'noe Statisticheskoe Upravlenie. _Itogi vsesoiuznoi perepisi naseleniia 1970 goda_ (Results of the All-Union census of population of 1970). Moskva, Izdatel'stvo "Statistika," 1972-1974. 7 v.   239

Tom I. _Chislennost' naseleniia SSSR_... (Population of the USSR). 1972. 175 p.

Tom II. _Pol, vozrast i sostoianie v brake naseleniia SSSR_... (Sex, age, and marital status of the population of the USSR). 1972. 271 p.

Tom III. _Uroven' obrazovaniia naseleniia SSSR_... (Education of the population of the USSR). 1972. 575 p.

Tom IV. _Natsional'nyi sostav naseleniia SSSR_... (Ethnic structure of the population of the USSR). 1973. 647 p.

Tom V. _Raspredelenie naseleniia SSSR... po obshchestvennym gruppam, istochnikam sredstv sushchestvovaniia i otrasliam narodnogo khoziaistva_ (Distribution of the population of the USSR by social groups, by sources of livelihood, and by branches of the economy). 1973. 301 p.

Tom VI. _Raspredelenie naseleniia SSSR i soiuznykh respublik po zaniatiam_ (Distribution of the population of the USSR and union republics by occupations). 1973. 807 p.

Tom VII. _Migratsiia naseleniia, chislo i sostav semei v SSSR_... (Migration of the population, numbers and structure of families in the USSR). 1974. 454 p.

Basic data of the 1970 census. Except for volume 6, which is by union republics only, all volumes contain data for union republics, autonomous republics, krays, and oblasts, as well as for the country as a whole.

_____. _____. _Itogi vsesoiuznoi perepisi naseleniia 1959 goda_ (Results of the all-Union census of population of 1959).   240

SSSR: Svodnyi tom (USSR: summary volume). 1962. 284 p.

RSFSR. 1963. 456 p.

Ukrainskaia SSR (Ukrainian SSR). 1963. 210 p.

Estonskaia SSR (Estonian SSR). 1962. 108 p.

Latviiskaia SSR (Latvian SSR). 1962. 106 p.

Lietuvos TSR. Litovskaia SSR (Lithuanian SSR). 1963. 179 p. In Russian and Lithuanian.

Belorusskaia SSR (Belorussian SSR). 1963. 146 p.

Moldavskaia SSR (Moldavian SSR). 1962. 104 p.

Gruzinskaia SSR (Georgian SSR). 1963. 162 p.

Armianskaia SSR (Armenian SSR). 1963. 116 p.

Azerbaidzhanskaia SSR (Azerbaidzhan SSR). 1963. 158 p.

Kazakhskaia SSR (Kazakh SSR). 1962. 202 p.

Kirgizskaia SSR (Kirgiz SSR). 1963. 150 p.

Uzbekskaia SSR (Uzbek SSR). 1962. 168 p.

Turkmenskaia SSR (Turkmen SSR). 1963. 150 p.

Tadzhikskaia SSR (Tadzhik SSR). 1963. 140 p.

Basic source for both the 1959 and the 1939 censuses, since the latter was not previously published.

___. ____. Vsesoiuznaia perepis' naseleniia 1926 goda (All-Union census of
population of 1926). Moskva, Izdanie TsSU, 1928-1933. 56 v.   241

The most detailed census taken during the Soviet period, with much greater
detail on the geographical distribution of the characteristics of the population
than in later Soviet censuses.

'SSIA. Tsentral'nyi Statisticheskii Komitet. Pervaia vseobshchaia perepis'
naseleniia Rossiiskoi Imperii 1897 goda (First general census of population of
the Russian Empire of 1897). Sankt-Peterburg, 1899-1905. 89 v.   242

Detailed census with a wealth of basic data geographically distributed. In
some respects unmatched by any later censuses.

General Statistical Yearbooks and Handbooks
for the USSR as a Whole

SR. Tsentral'noe Statisticheskoe Upravlenie. Narodnoe khoziaistvo SSSR:
statisticheskii sbornik (Economy of the USSR: statistical compilation). Moskva,
Gosstatizdat, 1956. 264 p.   243

The statistical yearbook for 1955, was the first general statistical compilation for the Soviet Union in many years and therefore of wide interest when published, though largely now replaced by later statistical yearbooks. Available in
several editions in English: The national economy of the USSR: a statistical
compilation. Moscow, State Statistical Publishing House, 1956. 270 p. Statistical handbook of the USSR. With introduction, additional tables, and annotations
by Harry Schwartz and foreword by John S. Sinclair. New York, National Industrial
Conference Board, 1957. 122 p. The USSR economy: a statistical abstract. London,
Lawrence and Wishart, 1957. 263 p. Available in German: Die UdSSR in Zahlen:
statistisches Sammelwerk. Berlin, Verlag Die Wirtschaft, 1956. 269 p. Editions
also published in Moscow in French, 1957. 230 p., and in Spanish, 1957, 235 p.

___. ____. Narodnoe khoziaistvo SSSR v 1956 godu: statisticheskii ezhegodnik
(Economy of the USSR in 1956: statistical yearbook). Moskva, Gosstatizdat, 1957.
296 p. (Also available in English: National economy of the USSR, 1956: statistical
returns. Moscow, Foreign Language Publishing House, 1957. 230 p.).   244

___. ____. ____ v 1958 godu. 1959. 959 p.
___. ____. ____ v 1959 godu. 1960. 896 p.
___. ____. ____ v 1960 godu. 1961. 943 p.
___. ____. ____ v 1961 godu. 1962. 861 p.
___. ____. ____ v 1962 godu. 1963. 735 p.
___. ____. ____ v 1963 godu. Moskva, Izdatel'stvo "Statistika," 1965. 760 p.
___. ____. ____ v 1964 godu. 1965. 887 p.
___. ____. ____ v 1965 godu. 1966. 910 p.
___. ____. ____ v 1967 g. 1968. 1008 p.
___. ____. ____ v 1968 g. 1969. 831 p.
___. ____. ____ v 1969 g. 1970. 864 p.
___. ____. ____ v 1970 g. 1971. 823 p.
___. ____. ____ 1922-1972 gg.: iubileinyi statisticheskii ezhegodnik (1922-
1972: jubilee statistical yearbook). 1972. 848 p. (To the 50th anniversary of
the formation of the USSR).
___. ____. ____ v 1972 g.: statisticheskii ezhegodnik. 1973. 824 p.

The official annual statistical summary of the Soviet Union. Data for the
USSR as a whole but many tables are broken down by the 15 union republics. Major
sections are population and area, science and technical progress, industry, agriculture, transportation and communication, capital formation, labor, material
welfare, trade, service, education and culture, public health, finance and credit,
and foreign economic relations of the USSR.

PINKHENSON, Dmitrii Moiseevich. Ekonomicheskaia geografiia v tsifrakh (Economic
geography in figures). Moskva, Izdatel'stvo "Prosveshchenie," 1970. 190 p.   24
    Collection of statistics on the areal distribution of economic activities
in the world economy, p. 5-15, the Communist countries, p. 16-22, the Soviet Union,
p. 23-128, capitalist countries, p. 129-137, with East-West comparisons, p. 138-
144. The section on the Soviet Union provides regional breakdowns for economic
activities.

USSR. Tsentral'noe Statisticheskoe Upravlenie. SSSR v tsifrakh: statisticheskii
sbornik (The USSR in figures: statistical compilation). Moskva, Gosstatizdat,
1958. 469 p.   24

_____. _____. SSSR v tsifrakh v 1959 godu: kratkii statisticheskii sbornik (The
USSR in figures in 1959: short statistical compilation). Moskva, Gosstatizdat,
1960. 302 p. (Also available in English and French editions, 1961. 265 p.).

_____. _____. _____ v 1960 godu. 1961. 381 p.
_____. _____. _____ v 1961 godu. 1962. 400 p.
_____. _____. _____ v 1962 godu. 1963. 360 p.
_____. _____. _____ v 1963 godu. Moskva, Izdatel'stvo "Statistika," 1964. 233 p.
_____. _____. _____ v 1964 godu. 1965. 160 p. (Also in English. The USSR in
figures for 1964: brief statistical returns. Moscow Progress, 1966, 152 p.).
_____. _____. _____ v 1965 godu. 1966. 168 p. Also in English: The USSR in
figures for 1965: brief statistical returns. French and Spanish editions also
published in Moscow.
_____. _____. _____ v 1966 g. 1967. 192 p. Moscow, Progress Publishers, 1967.
131 p.
_____. _____. _____ v 1967 g. 1968. 159 p.
_____. _____. _____ v 1970 godu. 1971. 235 p.
_____. _____. _____ v 1971 godu. 1972. 240 p.
_____. _____. _____ v 1972 godu. 1973. 215 p.

    Small-format booklet with selection of key statistics on the USSR for a wide
audience. Much less detailed than Narodnoe khoziaistvo SSSR: statisticheskii
ezhegodnik. Most tables give only figures for the Soviet Union as a whole, though
a few give data for constituent union republics and some have international comparisons.

_____. _____. SSSR i zarubezhnye strany posle pobedy Velikoi Oktiabr'skoi
sotsialisticheskoi revoliutsii: statisticheskii sbornik (The USSR and foreign
countries after the victory of the Great October socialist revolution: statistical
compilation). Moskva, Izdatel'stvo "Statistika," 1970. 319 p.   2

_____. _____. Strana sovetov za 50 let: sbornik statisticheskikh materialov
(The country of the Soviets after 50 years: compilation of statistical materials).
Moskva, Izdatel'stvo "Statistika, 1967. 351 p. Available in English: Soviet
Union 50 years: statistical returns. Moscow, Progress Publishers, 1969. 342 p.
Available in French: Le pays des Soviets en 50 ans: recueil de statistiques.
Moscow, "Progrès, 1968. 375 p. Available in Spanish. Moscú, Progreso, 1968.
352 p.   2

_____. _____. SSSR v tsifrakh: statisticheskii sbornik za 1913-1957 gg. (The
USSR in figures: a statistical compilation for the years 1913-1957). Moskva,
Gosstatizdat, 1958. 467 p.   2

_____. _____. Dostizheniia Sovetskoi vlasti za sorok let v tsifrakh: statistichesk
sbornik (Progress of Soviet power through forty years in figures: a statistical
handbook). Moskva, Gosstatizdat, 1957. 370 p. Available in English: Forty years
of Soviet power in facts and figures. Moscow, Foreign Language Publishing House,
1958. 319 p. Available in French: Les Progrès du pouvoir soviétique depuis 40
ans en chiffres: recueil statistique. Moscow, Éditions en langues étrangères,
1958. 335 p. Also available in Spanish. 1959. 345 p.   2

245 - 250

_____. Tsentral'noe Upravlenie narodnokhoziaistvennogo uchëta. Sotsialisticheskoe stroitel'stvo SSSR, 1933-1938 gg.: statisticheskii sbornik (Socialist construction of the USSR, 1933-1938: a statistical compilation). Moskva, Gosplanizdat, 1939. 206 p.   251

_____. Tsentral'noe Statisticheskoe Upravlenie. Sotsialisticheskoe stroitel'stvo SSSR (Socialist construction of the USSR), 1934-1936.   252

_____. _____. Statisticheskii spravochnik SSSR (Statistical handbook of the USSR). 1924-1925, 1927-1928, and 1932. Moskva.   253

Title varies: 1924-1925 as Narodnoe khoziaistvo Soiuza SSR v tsifrakh.

_____. _____. Itogi desiatiletiia Sovetskoi vlasti v tsifrakh 1917-1927. (Results of 10 years of Soviet power in figures 1917-1927). Moskva, Tsentral'noe Statisticheskoe Upravlenie SSSR, 1927. 514 p.   254

Available in English: Ten years of Soviet power in figures, 1917-1927. Moscow, Central Statistical Board, 1927. 516 p. Rearrangement of material published in Statisticheskii spravochnik SSSR 1927.

~USSIA. Tsentral'nyi Statisticheskii Komitet. Statisticheskii ezhegodnik Rossii (Statistical Yearbook of Russia). 1904-1915. Sankt-Peterburg. (Petrograd). (Title varies: 1904-1909 as Ezhegodnik Rossii).   255

Legends in Russian and French.

~ASHIN, Adol'f Grigor'evich. Naselenie Rossii za 100 let, 1811-1913 gg.: statisticheskie ocherki (The population of Russia for a hundred years, 1811-1913: statistical essays). S. G. Strumilin, ed. Moskva, Gosstatizdat, 1956. 350 p.   256

Convenient assemblage and interpretation of statistics on the population of Russia over the period 1811-1913.

~ABUZAN, Vladimir Maksimovich. Narodonaselenie Rossii v XVIII-pervoi polovine XIX v. po materialam revizii (Demography of Russia in the 18th and first half of the 19th centuries based on the materials of the revizii). Moskva, Izdatel'stvo Akademii Nauk SSSR, 1963. 228 p.   257

Statistical Handbooks for Specialized Topics in the USSR

Economic: Industry

~SSR. Tsentral'noe Statisticheskoe Upravlenie. Promyshlennost' SSSR: statisticheskii sbornik (Industry in the USSR: statistical compilation). Moskva, Gosstatizdat, 1957. 448 p. Also available in English: USSR industry: a statistical compilation. Moscow, State Statistical Publishing House, 1957. 434 p.   258

_____. _____. Moskva, Izdatel'stvo "Statistika," 1964. 495 p.

~SFSR. Tsentral'noe Statisticheskoe Upravlenie. Promyshlennost' RSFSR: statisticheskii sbornik (Industry in the RSFSR: statistical compilation). Moskva, Gosstatizdat, 1961. 344 p.   259

~ELORUSSIAN SSR. Tsentral'noe Statisticheskoe Upravlenie. Promyshlennost' Belorusskoi SSR: statisticheskii sbornik (Industry in the Belorussian SSR: statistical compilation). Minsk, Izdatel'stvo "Statistika," Belorusskoe Otdelenie, 1965. 479 p.   260

Economic Agriculture

USSR. Tsentral'noe Statisticheskoe Upravlenie. Sel'skoe khoziaistvo SSSR:
statisticheskii sbornik (Agriculture in the USSR: statistical compilation).
Moskva, Gosstatizdat, 1960. 666 p.   26

_____. _____. _____. Moskva, Izdatel'stvo "Statistika," 1971. 711 p. Covers
1913-1970.

UKRAINIAN SSR. Tsentral'noe Statisticheskoe Upravlenie (Ukrainian: Tsentral'ne
Statystychne Upravlinnia). Sel'skoe khoziaistvo Ukrainskoi SSR: statisticheskii
sbornik (Agriculture in the Ukrainian SSR: statistical compilation). Kiev,
Izdatel'stvo "Statistika," 1969 [1970]. 685 p. In Ukrainian. Ukrainian: Sil's'ke
hospodarstvo URSR: statystychnyĭ zbirnyk, Kyïv, Vydavnytstvo "Statystyka."   26

_____. _____. Kolkhozy Ukrainskoi SSR v tsifrakh (Collective farms in the
Ukrainian SSR in figures). Kiev, Izdatel'stvo "Urozhai," 1969. 182 p. In
Ukrainian. Ukrainian: Kolhospy Ukraïns'koi RSR v tsyfrakh. Kyïv, Vydavyntstvo
"Urozhai."   26

_____. _____. Valovye sbory i urozhainost' sel'skokhoziaistvennykh kul'tur po
Ukrainskoi SSR: statisticheskii sbornik (Gross harvest and productivity of
agricultural crops in the Ukrainian SSR: statistical compilation). Kiev,
Gosstatizdat, 1959. 339 p. In Ukrainian. Ukrainian: Valovi zbory i urozhaĭnist'
sil's'kohospodars'kykh kul'tur: statystychnyĭ zbirnyk. Kyïv, Derzhstatvydav.   26

LITHUANIAN SSR. Tsentral'noe Statisticheskoe Upravlenie. Sel'skoe khoziaistvo
Litovskoi SSR: kratkii sbornik statisticheskikh dannykh (Agriculture of the
Lithuanian SSR: short collection of statistical data). Vilnius, "Statistika,"
1966. 64 p. In Lithuanian.   26

_____. _____. _____: kratkii statisticheskii sbornik. 1968. 70 p. Two editions:
in Russian and in Lithuanian.

MOLDAVIAN SSR. Tsentral'noe Statisticheskoe Upravlenie. Osnovnye pokazateli
razvitiia kolkhozov Moldavskoi SSR (za 1950-1968 gody): statisticheskii sbornik
(Basic indexes of the development of collective farms in the Moldavian SSR during
1950-1968: statistical compilation). Kishinëv, 1969. 124 p.   26

ARMENIAN SSR. Tsentral'noe Statisticheskoe Upravlenie. (Armenian: Kentronakan
Vichakagrakan Varch'owt'yown). Sel'skoe khoziaistvo Armianskoi SSR: statisticheskiĭ
sbornik (Agriculture in the Armenian SSR: statistical compilation). Erevan,
Aipetrat, 1961. 482 p.   26

_____. _____. _____. Erevan, 1969. 517 p.

KAZAKH SSR. Tsentral'noe Statisticheskoe Upravlenie. Kolkhozy Kazakhstana: kratkii
statisticheskii sbornik (The collective farms of Kazakhstan: short statistical
compilation). Alma-Ata, "Kainar," 1969. 165 p.   26

USSR. Tsentral'noe Statisticheskoe Upravlenie. Posevnye ploshchadi SSSR:
statisticheskii sbornik (Sown area of the USSR: statistical compilation). Moskva,
Gosstatizdat, 1957. 2 v.   26

    Tom I. Statisticheskii sbornik (v. 1. Statistical compilation). 515 p.

    Tom II. [Tekhnicheskie kul'tury, kartofel', ovoshche-bakhchevye i kormovye
kul'tury] (v. 2. Industrial crops, potatoes, vegetables and melons, and feed
crops). 503 p.

USSR. Tsentral'noe Statisticheskoe Upravlenie. Sortovye posevy SSSR; zernovye
kul'tury i podsolnechnik: statisticheskii sbornik (Crops sown from selected seeds:
food grains and sunflowers: statistical compilation). Moskva, Gosstatizdat, 1957.
423 p.   270

SSR. Tsentral'noe Statisticheskoe Upravlenie. <u>Zhivotnovodstvo SSSR: statisticheskii</u>
<u>sbornik</u>. (Livestock raising of the USSR: statistical compilation). Moskva,
Gosstatizdat, 1959. 252 p.   271

KRAINIAN SSR. Tsentral'noe Statisticheskoe Upravlenie (Ukrainian: Tsentral'ne
Statystychne Upravlinnia). <u>Zhivotnovodstvo Ukrainskoi SSR: statisticheskii sbornik</u>
(Livestock rearing in the Ukrainian SSR: statistical compilation). Kiev,
Gosstatizdat, 1960. 251 p. In Ukrainian. <u>Ukrainian: Tvarynnytsvo Ukrains'koi</u>
<u>RSR: statystychnyi zbirnyk</u>. Kyïv. Derzhavne Statystychne Vydavnytstvo. 1960.
251 p.   272

IRGIZ SSR. Tsentral'noe Statisticheskoe Upravlenie. <u>Sostoianie zhivotnovodstva</u>
<u>v Kirgizskoi SSR: statisticheskii sbornik</u> (Status of livestock production in the
Kirgiz SSR: statistical compilation). Frunze, Gosstatizdat, Kirgizskoe otdelenie,
1960. 205 p.   273

SSR. Tsentral'noe Statisticheskoe Upravlenie. <u>Chislennost' skota v SSSR:</u>
<u>statisticheskii sbornik</u> (Numbers of livestock in the USSR: statistical compila-
tion). Moskva, Gosstatizdat, 1957. 619 p.   274

SSR. Tsentral'noe Statisticheskoe Upravlenie. <u>Chislennost' porodnogo skota v</u>
<u>kolkhozakh i sovkhozakh SSSR na 1 ianvaria 1960 g.: statisticheskii sbornik</u>.
(Numbers of breeds of livestock on collective and state frams of the USSR on
January 1, 1960: statistical compilation). Moskva, Gosstatizdat, 1961. 516 p.   275

Economic: Transport and Communications

SSR. Tsentral'noe Statisticheskoe Upravlenie. <u>Transport i sviaz' SSSR: statis-</u>
<u>ticheskii sbornik</u> (Transportation and communications of the USSR: statistical
compilation). Moskva, Gosstatizdat, 1957. 260 p. Also available in English:
<u>USSR transport and communications: a statistical compilation</u>. Moscow, State
Statistical Publishing House, 1957.   276

____. ____. ____. Moskva, Izdatel'stvo "Statistika," 1967. 332 p.

____. ____. ____. Moskva, 1972. 320 p.

RMENIAN SSR. Tsentral'noe Statisticheskoe Upravlenie. <u>Transport i sviaz' Armianskoi</u>
<u>SSR za riad let: statisticheskii sbornik</u> (Transport and communications of the
Armenian SSR over a series of years: statistical compilation). Erevan, "Statistika,"
1967. 149 p.   277

Economic: Internal Trade

SSR. Tsentral'noe Statisticheskoe Upravlenie. <u>Sovetskaia torgovlia: statisticheskii</u>
<u>sbornik</u>. (Soviet trade: statistical compilation). Moskva, Gosstatizdat, 1956.
352 p.   278

____. ____. ____. Moskva, Izdatel'stvo "Statistika," 1964. 503 p.

SFSR. Tsentral'noe Statisticheskoe Upravlenie. <u>Sovetskaia torgovlia v RSFSR:</u>
<u>statisticheskii sbornik</u> (Soviet trade in the RSFSR: statistical compilation).
Moskva, Gosstatizdat, 1958. 343 p.   279

KRAINIAN SSR. Tsentral'noe Statisticheskoe Upravlenie (Ukrainian: Tsentral'ne
Statystychne Upravlinnia). <u>Sovetskaia torgovlia v Ukrainskoi SSR: statisticheskii</u>
<u>sbornik</u> (Soviet trade in the Ukrainian SSR: statistical compilation). Kiev,
Gosstatizdat, 1963. 319 p. In Ukrainian. Ukrainian: <u>Radians'ka torhivlia v</u>
<u>Ukraïns'kii RSR: statystychnyi zbirnyk</u>. Kyïv, Derzhavne Statystychne Vydavnytstvo,
1963. 319 p.   280

UKRAINIAN SSR. Tsentral'noe Statisticheskoe Upravlenie. Razvitie obshchestvennogo pitaniia v Ukrainskoi SSR: osnovnye pokazateli (Development of public catering in the Ukrainian SSR: basic indices). Kiev, Gosstatizdat, 1960. 63 p. 281

LATVIAN SSR. Tsentral'noe Statisticheskoe Upravlenie. Razvitie sovetskoi torgovli v Latviiskoi SSR: statisticheskii sbornik (Development of Soviet trade in the Latvian SSR: statistical compilation). Riga, Latgosizdat, 1957. 144 p. In Russian and Latvian. 282

_____. _____. Razvitie torgovli v Latviiskoi SSR: statisticheskii sbornik (Development of trade in the Latvian SSR: statistical compilation). Riga, "Statistika," 1968. 232 p.

ARMENIAN SSR. Tsentral'noe Statisticheskoe Upravlenie. Sovetskaia torgovlia v Armianskoi SSR: statisticheskii sbornik (Soviet trade in the Armenian SSR: statistical compilation). Erevan, "Statistika," 1966. 212 p. 283

AZERBAIDZHAN SSR. Tsentral'noe Statisticheskoe Upravlenie. Razvitie sovetskoi torgovli v Azerbaidzhanskoi SSR: statisticheskii sbornik (Development of Soviet trade in the Azerbaidzhan SSR: statistical compilation). Baku, Gosstatizdat, 1962. 240 p. 284

Economic: Foreign Trade

USSR. Planovo-Ekonomicheskoe Upravlenie. Vneshniaia torgovlia Soiuza SSR za 1956 god: statisticheskii obzor (Foreign trade of the USSR in 1956: statistical review). Moskva, Vneshtorgizdat, 1958. 154 p. 285

_____. _____. _____. 1957 g. 1958. 157 p.
_____. _____. _____. 1958 g. 1959. 159 p.
_____. _____. _____. 1959 g. 1960. 184 p.
_____. _____. _____. 1960 g. 1961. 212 p.
_____. _____. _____. 1961 g. 1962. 232 p.
_____. _____. _____. 1962 g. 1963. 235 p.
_____. _____. _____. 1963 g. 1964. 264 p.
_____. _____. _____. 1959-1963 gody: statisticheskii sbornik. 1965. 483 p.
_____. _____. _____. 1964 g.: statisticheskii obzor. 1965. 295 p.
_____. _____. _____. 1965 g. 1966. 324 p.
_____. _____. _____. 1966 g. 1967. 334 p.
_____. _____. _____. 1967 g. 1968. 312 p.
_____. _____. _____. 1968 g. 1969. 300 p.
_____. _____. _____. 1969 g. 1970. 294 p.
_____. _____. _____. 1970 g. 1971. 297 p.
_____. _____. _____. 1971 g. 1972. 317 p.
_____. _____. _____. 1972 g. 1973. 316 p.

Economic: Labor

USSR. Tsentral'noe Statisticheskoe Upravlenie. Trud v SSSR: statisticheskii sbornik (Labor in the USSR: statistical compilation). Moskva, Izdatel'stvo "Statistika," 1968. 342 p. 286

LATVIAN SSR. Tsentral'noe Statisticheskoe Upravlenie. Trud v Latviiskoi SSR: statisticheskii sbornik (Labor in the Latvian SSR: statistical compilation). Riga, 1973. 192 p. 287

ARMENIAN SSR. Tsentral'noe Statisticheskoe Upravlenie. Trud v Armianskoi SSR: statisticheskii sbornik (Labor in the Armenian SSR: statistical compilation). Erevan, "Statistika," Armianskoe otdelenie, 1970. 205 p. 28

Economic: Capital

USSR. Tsentral'noe Statisticheskoe Upravlenie. Kapital'noe stroitel'stvo v SSSR:
statisticheskii sbornik (Capital formation in the USSR: statistical compilation).
Moskva, Gosstatizdat, 1961. 280 p.   289

ARMENIAN SSR. Tsentral'noe Statisticheskoe Upravlenie. Kapital'noe stroitel'stvo
v Armianskoi SSR (Capital formation in the Armenian SSR). Erevan, "Statistika,"
1968. 268 p.   290

Culture

USSR. Tsentral'noe Statisticheskoe Upravlenie. Kul'turnoe stroitel'stvo SSSR:
statisticheskii sbornik (Cultural development of the USSR: statistical compilation). Moskva, Gosstatizdat, 1956. 332 p.   291

     Available in English: Cultural progress in the USSR: statistical returns.
Moscow, Foreign Languages Publishing House, 1958. 325 p. Available in French:
L'édification culturelle en U.R.S.S.: recueil statistique. Moscou, Éditions en
langues étrangères, 1958. 326 p.

RSFSR. Tsentral'noe Statisticheskoe Upravlenie. Kul'turnoe stroitel'stvo RSFSR:
statisticheskii sbornik (Cultural development in the RSFSR: statistical compilation). Moskva, Gosstatizdat, 1958. 459 p.   292

USSR. Tsentral'noe Statisticheskoe Upravlenie. Narodnoe obrazovanie, nauka i
kul'tury v SSSR: statisticheskii sbornik (Education, science, and culture in the
USSR: statistical compilation). G. M. Kharat'ian and others, eds. Moskva,
Izdatel'stvo "Statistika," 1971. 403 p.   293

LATVIAN SSR. Tsentral'noe Statisticheskoe Upravlenie (Latvian: Centrālā Statistikas
Pārvalde). Pechat' i kul'turno-prosvetitel'nye uchrezhdeniia Latviiskoi SSR:
statisticheskii sbornik. (Publishing and cultural-educational establishments of
the Latvian SSR: statistical compilation). Riga, "Statistika," Latviiskoe
otdelenie, 1967. 179 p.   294

LATVIAN SSR. Tsentral'noe Statisticheskoe Upravlenie (Latvian: Centrālā Statistikas Pārvalde). Kul'turnoe stroitel'stvo Latviiskoi SSR: statisticheskii sbornik
(Cultural development in the Latvian SSR: statistical compilation). Riga,
Latgosizdat, 1957. 144 p. Text in Russian and Latvian. Latvian: Latvijas PSR
kulturas celtnieciba:statistisko datu krājuma. Rīgā, Latvijas valsts izdevniecība,
1957. 144 p.   295

LITHUANIAN SSR. Tsentral'noe Statisticheskoe Upravlenie (Lithuanian: Centrinė
Statistikos Valdyba). Prosveshchenie i kul'tura Litovskoi SSR: statisticheskii
sbornik (Education and culture in the Lithuanian SSR: statistical compilation).
Vilnius, "Statistika," 1964. 207 p.   296

ARMENIAN SSR. Tsentral'noe Statisticheskoe Upravlenie (Armenian: Kentronakan
Vichakagrakan Varch'owt'yown). Kul'turnoe stroitel'stvo Armianskoi SSR: statisticheskii sbornik (Cultural development in the Armenian SSR: statistical compilation). Erevan, Gosstatizdat, Armianskoe otdelenie, 1962. 121 p.   297

ARMENIAN SSR. Tsentral'noe Statisticheskoe Upravlenie (Armenian: Kentronakan
Vichakagrakan Varch'owt'yown). Prosveshchenie i kul'tura Armianskoi SSR: statisticheskii sbornik (Education and culture in the Armenian SSR: statistical compilation). Erevan, 1970. 237 p.   298

AZERBAIDZHAN SSR. Tsentral'noe Statisticheskoe Upravlenie. Kul'turnoe stroitel'stvo Azerbaidzhanskoi SSR: statisticheskii sbornik (Cultural development of the Azerbaidzhan SSR: statistical compilation). Baku, Gosstatizdat, 1961. 149 p.   29

KAZAKH SSR. Tsentral'noe Statisticheskoe Upravlenie. Kul'turnoe stroitel'stvo Kazakhskoi SSR: statisticheskii sbornik (Cultural development in the Kazakh SSR: statistical compilation). Alma-Ata, Gosstatizdat, 1960. 116 p.   30

TURKMEN SSR. Tsentral'noe Statisticheskoe Upravlenie. Kul'turnoe stroitel'stvo Turkmenskoi SSR: statisticheskii sbornik (Cultural development in the Turkmen SSR: statistical compilation). Ashkhabad, Turkmengosizdat, 1960. 131 p.   30

Education

USSR. Tsentral'noe Statisticheskoe Upravlenie. Vysshee obrazovanie v SSSR: statisticheskii sbornik (Higher education in the USSR: statistical compilation). Moskva, Gosstatizdat, 1961. 255 p.   30

USSR. Tsentral'noe Statisticheskoe Upravlenie. Srednee spetsial'noe obrazovanie v SSSR: statisticheskii sbornik (Specialized secondary education in the USSR: statistical compilation). Moskva, Gosstatizdat, 1962. 155 p.   30

LATVIAN SSR. Tsentral'noe Statisticheskoe Upravlenie (Latvian: Centrālā Statistikas Pārvalde). Vysshee i srednee spetsial'noe obrazovanie v Latviiskoi SSR: statisticheskii sbornik (Higher and specialized secondary education in the Latvian SSR: statistical compilation). Riga, Latgosizdat, 1964. 157 p. Text in Russian and in Latvian. Latvian: Augstākā un vidējā izglītība Latvijas PSR: statistisko datu krājuma. Rīgā, Latvijas valsts izdevniecība, 1964. 157 p.   30

Women and Children

USSR. Tsentral'noe Statisticheskoe Upravlenie. Zhenshchiny i deti v SSSR: statisticheskii sbornik (Women and children in the USSR: statistical compilation). Moskva, Gosstatizdat, 1961. 229 p.   30

　　　　Available in English: Women and children in the USSR: brief statistical returns. Moscow, Foreign Languages Publishing House, 1963. 196 p. Also published in French and Spanish.

_____. _____. _____. 2nd ed. Moskva, Gosstatizdat, 1963. 203 p.

_____. _____. _____. Moskva, Izdatel'stvo "Statistika," 1969. 208 p.

_____. _____. Women in the Soviet Union: statistical returns. Moscow, Progress Publishers, 1970. 54 p.   30

_____. _____. Zhenshchiny v SSSR: statisticheskie materialy (Women in the USSR: statistical materials). Reprints from the journal, Vestnik statistiki, 1966, no. 1, 11 p.; 1967, no. 1, 15 p.; 1968, no. 1, 16 p.; 1969, no. 1, 16 p.; 1970, no. 1, 10 p.   30

_____. _____. Zhenshchina v SSSR: kratkii statisticheskii spravochnik (Woman in the USSR: short statistical handbook). Moskva, Gosstatizdat, 1960. 102 p. Available in English: Women in the USSR. Brief statistics. Moscow, Foreign Language Publishing House, 1960. 99 p. Available in French: La femme en U.R.S.S. Moscou. Editions en langues étrangères, 1960. 98 p. A special edition in Spanish was also published in 1961.   30

MOLDAVIAN SSR. Tsentral'noe Statisticheskoe Upravlenie. Zhenshchina Moldavii: kratkii statisticheskii spravochnik (Woman of Moldavia; short statistical handbook). Kishinёv, Gosstatizdat, 1961. 70 p.   30

KIRGIZ SSR. Tsentral'noe Statisticheskoe Upravlenie. Zhenshchina v Kirgizskoi SSR: kratkii statisticheskii spravochnik (Woman in the Kirgiz SSR: short statistical handbook). Frunze, Gosstatizdat, Kirgizskoe otdelenie, 1960. 96 p.   310

TURKMEN SSR. Tsentral'noe Statisticheskoe Upravlenie. Zhenshchina v Turkmenskoi SSR: kratkii statisticheskii spravochnik (Woman in the Turkmen SSR: short statistical handbook). 84 p. Two editions: in Russian and in Turkmen.   311

TADZHIK SSR. Tsentral'noe Statisticheskoe Upravlenie. Zhenshchina v Tadzhikskoi SSR: kratkii statisticheskii spravochnik (Woman in the Tadzhik SSR: short statistical handbook). Stalinabad [Dushanbe], Gosstatizdat, 1960. 95 p.   312

Public Health

USSR. Tsentral'noe Statisticheskoe Upravlenie. Zdravookhranenie v SSSR: statisticheskii sbornik (Public health in the USSR: statistical compilation). Moskva, Gosstatizdat, 1960. 272 p.   313

____. Ministerstvo Zdravookhraneniia SSSR. Otdel Meditsinskoi Statistiki. Zdravookhranenie v SSSR: statisticheskii sbornik (Public health in the USSR: statistical compilation). Moskva, Medgiz, 1956. 130 p.   314

____. ____. ____. Zdravookhranenie v SSSR [statisticheskii sbornik]. (Public health in the USSR). Moskva, Medgiz, 1957. 179 p.

UKRAINIAN SSR. Ministerstvo Zdravookhraneniia. Otdel Meditsinskoi Statistiki. Ukrainskoe Nauchno-Issledovatel'skii Biuro Sanitarnoi Statistiki. Zdravookhranenie v USSR: statisticheskii spravochnik (Public health in the Ukrainian SSR: statistical handbook). Kiev, Gosmedizdat Ukrainskoi SSR, 1957. 140 p.   315

MOLDAVIAN SSR. Ministerstvo Zdravookhraneniia. Republikanskii Dom Sanitarnogo Prosveshcheniia. Zdravookhranenie v Moldavskoi SSR: statisticheskii spravochnik (Public health in the Moldavian SSR: statistical handbook). Kishinev, Gosizdat Moldavii, 1958. 80 p.   316

GEORGIAN SSR. Ministerstvo Zdravookhraneniia. Otdel Meditsinskoi Statistiki. Metodologicheskoe Biuro Sanitarnoi Statistiki. Zdravookhranenie v Gruzinskoi SSR: statisticheskii spravochnik (Public health in the Georgian SSR: statistical compilation). Tbilisi, "Ganatleba," 1966. 119 p.   317

ARMENIAN SSR. Ministerstvo Zdravookhraneniia. Respublikanskoe Nauchno-Metodicheskoe Biuro Sanitarnoi Statistiki. Zdravookhranenie v Armianskoi SSR: statisticheskii sbornik (Public health in the Armenian SSR: statistical compilation). Erevan, 1960. 81 p.   318

UZBEK SSR. Ministerstvo Zdravookhranenie. Otdel Meditsinskoi Statistiki. Zdravookhranenie v Uzbekskoi SSR: statisticheskii spravochnik (Public health in the Uzbek SSR: statistical handbook). Tashkent, Uzmedgiz, 1958. 180 p.   319

TURKMEN SSR. Ministerstvo Zdravookhraneniia. Otdel Meditsinskoi Statistiki. Zdravookhranenie Turkmenskoi SSR: statisticheskii spravochnik (Public health of the Turkmen SSR: statistical handbook). Ashkhabad, 1958. 63 p.   320

TADZHIK SSR. Ministerstvo Zdravookhraneniia. Nauchno-Metodicheskoe Biuro Sanitarnoi Statistiki. Zdravookhranenie Tadzhikistana: statisticheskii spravochnik (Public Public health of Tadzhikistan: statistical handbook). Stalinabad [Dushanbe], 1957. 107 p.   321

Housing, Utilities, and Services

LATVIAN SSR. Tsentral'noe Statisticheskoe Upravlenie. <u>Zhilishchno-kommunal'noe</u>
<u>khoziaistvo i bytovoe obsluzhivanie naseleniia v Latviiskoi SSR: statisticheskii</u>
<u>sbornik</u> (Housing, utilities, and services of the population in the Latvian SSR:
statistical compilation). Riga, 1969. 192 p.   322

ARMENIAN SSR. Tsentral'noe Statisticheskoe Upravlenie. <u>Zhilishchno-kommunal'noe</u>
<u>khoziaistvo Armianskoi SSR: statisticheskii sbornik</u> (Housing and utilities in
the Armenian SSR: statistical compilation). Erevan, "Statistika," 1967. 169 p.
                                                                                            32

Publishing

USSR. Vsesoiuznaia Knizhnaia Palata. <u>Pechat' SSSR v 1956 i 1957 godakh: statis-</u>
<u>ticheskie materialy</u> (Publishing in the USSR in 1956 and 1957: statistical
materials). Moskva, Izdatel'stvo Vsesoiuznoi Knizhnoi Palaty, 1958. 192 p.   32

_____. _____. _____ v 1958 g. 1959. 180 p.
_____. _____. _____ v 1959 g. 1960. 180 p.
_____. _____. _____ v 1960 g. 1961. 180 p.
_____. _____. _____ v 1961 g. 1962. 179 p.
_____. _____. _____ v 1962 g. 1963. 184 p.
_____. _____. _____ v 1963 g. Moskva, Izdatel'stvo "Kniga," 1964. 175 p.
_____. _____. _____ v 1964 g. 1965. 304 p.
_____. _____. _____ v 1965 g. 1966. 199 p.
_____. _____. _____ v 1966 g. 1967. 200 p.
_____. _____. _____ v 1967 g. 1968. 200 p.
_____. _____. _____ v 1968 g. 1969. 200 p.
_____. _____. _____ v 1969 g. 1970. 200 p.
_____. _____. _____ v 1970 g. 1971. 200 p.
_____. _____. _____ v 1971 g. 1972. 200 p.
_____. _____. _____ v 1972 g. 1973. 200 p.

_____. _____. <u>Pechat' SSSR za sorok let, 1917-1957: statisticheskie materialy</u>
(Publishing in the USSR over 40 years, 1917-1957: statistical materials). Moskva,
Izdatel'stvo Vsesoiuznoi Knizhnoi Palaty, 1957. 144 p.   32

_____. _____. <u>Sovetskaia pechat', 1951-1955: statisticheskie materialy</u> (Soviet
publishing, 1951-1955: statistical materials). Moskva, Izdatel'stvo Vsesoiuznoi
Knizhnoi Palaty, 1956. 84 p.   32

_____. _____. <u>Sovetskaia pechat' v period mezhdu XX i XXII s"ezdami KPSS:</u>
<u>statisticheskie materialy</u> (Soviet publishing in the period between the 20th
and 22nd congresses of the Communist Party of the Soviet Union: statistical
materials). Moskva, Izdatel'stvo Vsesoiuznoi Knizhnoi Palaty, 1961. 168 p.   32

_____. _____. <u>Sovetskaia pechat' k 400-letiiu russkogo knigopechataniia: statis-</u>
<u>ticheskie materialy</u> (Soviet publishing on the 400th anniversary of Russian book
printing: statistical materials). Moskva, Izdatel'stvo Vsesoiuznoi Knizhnoi
Palaty, 1964. 45 p.   3

_____. _____. <u>Pechat' SSSR v gody piatiletok: statisticheskie materialy</u> (Publish-
ing in the USSR during the years of the five-year plans). Moskva, Izdatel'stvo
"Kniga," 1971. 96 p.   3

UKRAINIAN SSR. Knizhnaia Palata. <u>Pressa Ukrainskoi SSR, 1917-1966: statisticheskii</u>
<u>spravochnik</u> (The press in the Ukrainian SSR, 1917-1966: statistical handbook).
Khar'kov, 1967. 146 p. In Ukrainian. Ukrainian: Knyzhkova Palata Ukrains'koi
RSR. <u>Presa Ukraïns'koi RSR, 1917-1966: statystychnyĭ dovidnyk</u>. Kharkiv.   3

———. ———. Pechat' Ukrainskoi SSR, 1918-1957: statisticheskie materialy.
(Publishing in the Ukrainian SSR. 1918-1957: statistical materials). Khar'kov,
Izdatel'stvo Knizhnoi Palaty Ukrainskoi SSR, 1958. 103 p. In Ukrainian.
Ukrainian: Presa Ukrains'koi RSR 1918-1957: statystychni materialy. Kharkiv,
Knyzhkova Palata Ukrains'koi RSR, 1958. 103 p.   331

STONIAN SSR. Gosudarstvennaia Knizhnaia Palata. Statistika pechati Estonskoi SSR,
1963 (Statistics of publishing in the Estonian SSR, 1963). Tallin, Estgosizdat,
1964. 64 p. In Russian and Estonian.   332

———. ———. ———, 1964. Tallin, "Eesti Raamat," 1965. 81 p. In Russian and
Estonian.

———. ———. ———, 1965. 1966. 79 p. In Russian and Estonian.
———. ———. ———, 1966. 1968. 80 p. In Russian and Estonian.
———. ———. ———, 1967. 1968. 80 p. In Russian and Estonian.
———. ———. ———, 1968. 1969. 78 p. In Russian and Estonian.
———. ———. ———, 1969. 1970. 80 p. In Russian and Estonian.
———. ———. ———, 1970. 1971. 83 p. In Russian and Estonian.
———. ———. ———, 1971. 1972. 93 p. In Russian and Estonian.
———. ———. ———, 1972. 1973. 86 p. In Russian and Estonian.

———. ———. 25 let pechati Sovetskoi Estonii, 1940-1965: sbornik statisticheskikh
materialov (25 years of publishing in Soviet Estonia, 1940-1965: collection of
statistical materials). Tallin, "Eesti Raamat," 1971. 183 p. In Russian and
Estonian.   333

ATVIAN SSR. Knizhnaia Palata. Pechat' Latviiskoi SSR, 1940-1956: statisticheskie
materialy (Publishing in the Latvian SSR, 1940-1956: statistical materials).
Riga, Latgosizdat, 1958. 124 p. In Russian and Latvian.   334

———. ———. ——— 1962. 1963. 33 p. In Russian and Latvian.
———. ———. ——— 1963. 1964. 35 p. In Russian and Latvian.
———. ———. ——— 1964. 1965. 37 p. In Russian and Latvian.

ITHUANIAN SSR. Knizhnaia Palata. Statistika pechati Litovskoi SSR, 1940-1955
(Statistics of publishing in the Lithuanian SSR, 1940-1955). Vilnius, Gosudar-
stvennoe Izdatel'stvo Politicheskoi i Nauchnoi Literatury, 1957. 116 p. In
Lithuanian.   335

———. ———. ——— 1956-1957. 31 p. In Lithuanian.
———. ———. ——— 1958. Vilnius, Gospolitizdat, 1960. 42 p. In Russian and
Lithuanian.
———. ———. ——— 1959. 1960. 16 p. In Russian and Lithuanian.
———. ———. ——— 1956-1960. 1962. 138 p. In Lithuanian.
———. ———. ——— 1961. 1962. 16 p. In Lithuanian.
———. ———. ——— 1963. 1964. 16 p. In Russian and Lithuanian.
———. ———. ——— 1964. 1965. 16 p. In Lithuanian.
———. ———. ——— 1961-1965. Vilnius, "Mintis," 1969. 179 p. In Lithuanian.
———. ———. ——— 1966. 1968. 28 p. In Lithuanian.
———. ———. ——— 1967. 1969. 30 p. In Lithuanian.
———. ———. ——— 1968. 1970. 32 p. In Lithuanian.
———. ———. ——— 1969. Vilnius, "Periodika," 1970. 32 p. In Lithuanian.
———. ———. ——— 1966-1970. Vilnius, "Mintis," 1972. 152 p. In Lithuanian.
———. ———. ——— 1971. Vilnius, "Periodika," 1972. 61 p. In Lithuanian.
Foreword and table of contents also in Russian.

LORUSSIAN SSR. Gosudarstvennaia Biblioteka BSSR imeni V. I. Lenina. Knizhnaia
Palata BSSR. Pechat' Belorusskoi SSR, 1918-1965: statisticheskie materialy
(Publishing in the Belorussian SSR, 1918-1965: statistical materials). Minsk,
1967. 92 p. On the 50th anniversary of Great October. In Belorussian.   336

———. ———. ———. 1966-1970. 1972. 128 p. In Belorussian.
———. ———. ———. Statistika pechati Belorusskoi SSR za 1957 god (Statistics
of publishing in the Belorussian SSR in 1957). Minsk, 1959. 28 p. Supplement
to Letapisu druku BSSR. In Russian and Belorussian.

MOLDAVIAN SSR. Gosudarstvennaia Knizhnaia Palata. Pechat' Moldavskoi SSR, 1925-
1960: statisticheskie materialy (Publishing in the Moldavian SSR, 1925-1960:
statistical materials). Kishinëv, "Karta Moldoveniaske," 1962. 47 p. In two
editions: in Russian and in Moldavian.   337

_____. _____. _____ v tsifrakh. 1968. 26 p. In Russian and Moldavian.

ARMENIAN SSR. Ministerstvo Kul'tury. Gosudarstvennaia Knizhnaia Palata. Kniga
i periodicheskaia pechat' v Armianskoi SSR za 40 let: statisticheskie materialy
(Book and periodical publishing in the Armenian SSR over 40 years: statistical
materials). Erevan, 1960. 79 p.   338

AZERBAIDZHAN SSR. Azerbaidzhanskaia Gosudarstvennaia Knizhnaia Palata. Pechat'
Azerbaidzhanskoi SSR (1920-1956): statisticheskie materialy (Publishing in the
Azerbaidzhan SSR, 1920-1956: statistical materials). Baku, 1958. 47 p. In
Azerbaijani.   339

KAZAKH SSR. Gosudarstvennaia Knizhnaia Palata. Pechat' Kazakhskoi SSR, 1921-1957:
statisticheskie materialy (Publishing in the Kazakh SSR, 1921-1957: statistical
materials). Alma-Alta, Knizhnaia Palata Kazakhskoi SSR, 1958. 55 p.   340

_____. _____. _____, 1920-1960: statisticheskie materialy. 1960. 39 p.
_____. _____. _____ za 50 let (1917-1967): statisticheskie materialy. Alma-Alta,
"Kazakhstan," 1969. 51 p.

.KIRGIZ SSR. Knizhnaia Palata. Pechat' Kirgizskoi SSR, 1926-1963: statisticheskie
materialy (Publishing in the Kirgiz SSR, 1926-1963: statistical materials).
1964. 75 p.   341

UZBEK SSR. Gosudarstvennaia Knizhnaia Palata. Sovetskaia pechat' Uzbekskoi SSR za
1918-1963 gody: statisticheskie materialy k 40-letiiu Sovetskoi vlasti v Uzbekskoi
SSR (Soviet publishing in the Uzbek SSR during 1918-1963: statistical materials on
the 40th anniversary of the Soviet regime in the Uzbek SSR). Tashkent, 1965. 61 p.
On the cover: Pechat' Uzbekskoi SSR za 1918-1963 gg.... Uzbekistana.   342

_____. _____. Pechat' Uzbekskoi SSR za 1964-1970 gody: statisticheskie materialy
(Publishing in the Uzbek SSR, 1964-1970: statistical materials). Tashkent, 1973.
52 p.

TURKMEN SSR. Gosudarstvennaia Knizhnaia Palata. Pechat' Turkmenskoi SSR, 1927-1956:
statisticheskie materialy (Publishing in the Turkmen SSR, 1927-1956: statistical
materials). Ashkhabad, 1958. 52 p.   34

_____. _____. _____ v 1957 i 1958 godakh: statisticheskie materialy. 1959. 60 p.
In Russian and Turkmen.
_____. _____. _____ v 1959 godu. 1962. 75 p. In Russian and Turkmen.
_____. _____. _____ v 1960 godu. 1962. 74 p. In Russian and Turkmen.
_____. _____. _____ v 1961 godu. Ashkhabad, Turkmengosizdat, 1962. 64 p. In
Russian and Turkmen.
_____. _____. _____ v 1962 i 1963 godakh. 1964. 66 p. In Russian and Turkmen.
_____. _____. _____ v 1964 i 1965 godakh. Ashkhabad, "Turkmenistan," 1968. 39 p.
In Russian and Turkmen.
_____. _____. _____ v 1967 godu. 1969. 35 p. In Russian and Turkmen.
_____. _____. _____ v 1968 godu. 1970. 37 p. In Russian and Turkmen.
_____. _____. _____ v 1969 godu. 1971. 43 p. In Russian and Turkmen.

TADZHIK SSR. Knizhnaia Palata. Pechat' Tadzhikskoi SSR, 1928-1958: statisticheskie
materialy (Publishing in the Tadzhik SSR, 1928-1958: statistical materials).
Stalinabad [Dushanbe], 1959. 74 p. In Russian and Tajik.   34

_____. _____. _____ 1959-1963. Dushanbe, 1964. 62 p.
_____. _____. _____ 1964-1968. 1971. 87 p. In Russian and Tajik.

Statistical Handbooks for Union Republics of the USSR

The principal statistical yearbook is listed first (usually with title Narodnoe khoziaistvo...) with individual volumes in chronological order, generally followed by a much smaller booklet series with summary data only (often with the title... v tsifrakh), and then by other statistical compilations with diverse titles in reverse chronological order, often published on particular anniversaries.

In successive editions of yearbooks and other similar statistical compilations, only those elements in the entry which do not repeat are recorded. Thus if the issuing agency, title and publisher remain the same, the entry may consist solely of the new year in the title, the year of publications, and the number of pages.

The RSFSR and the Ukrainian SSR are listed first, followed by other union republics arranged geographically from northwest to southeast. See map on inside front cover.

Russian Soviet Federated Socialist Republic (RSFSR)

Note: The National Union Catalogue of the United States lists publications of the RSFSR under Russia (1917- R.S.F.S.R.) and care must be taken to distinguish this heading from that for the Soviet Union as a whole: Russia (1923- U.S.S.R.)

RSFSR. Tsentral'noe Statisticheskoe Upravlenie. Narodnoe khoziaistvo RSFSR: statisticheskii sbornik (Economy of the RSFSR: statistical compilation). Moskva, Gosstatizdat, 1957. 372 p. Available also in English: The national economy of the RSFSR: a statistical compilation. Moscow, State Statistical Publishing House, 1957. 364 p.   345

_____.	_____.	_____	v 1958 godu: statisticheskii ezhegodnik. 1959. 508 p.
_____.	_____.	_____	v 1959 godu. 1960. 600 p.
_____.	_____.	_____	v 1960 godu. 1961. 572 p.
_____.	_____.	_____	v 1961 godu. 1962. 624 p.
_____.	_____.	_____	v 1962 godu. 1963. 608 p.
_____.	_____.	_____	v 1963 godu. Moskva, Izdatel'stvo "Statistika," 1965. 600 p.
_____.	_____.	_____	v 1964 godu. 1965. 576 p.
_____.	_____.	_____	v 1965 godu. 1966. 616 p.
_____.	_____.	_____	v 1967 g. 1968. 648 p.
_____.	_____.	_____	v 1968 g. 1969. 463 p.
_____.	_____.	_____	v 1969 g. 1970. 468 p.
_____.	_____.	_____	v 1970 g. 1971. 488 p.
_____.	_____.	_____	v 1971 g. 1972. 459 p.
_____.	_____.	_____	v 1972 g. 1973. 511 p.

Data are given not only for the RSFSR as a whole but frequently also for the Northwestern, Central, Volga-Vyatka, Central Black-Earth, Volga, North-Caucasus, Urals, West-Siberian, East-Siberian, and Far Eastern regions and for 73 smaller administrative units such as autonomous republics (ASSR), krays, oblasts, and even two cities (Moscow and Leningrad). Since the RSFSR covers 76 per cent of the total territory of the USSR, the distribution of data by these areal divisions is extremely important in the study of geographical patterns in the Soviet Union. Thus this yearbook along with those for Kazakhstan (12 per cent of the territory of the USSR) and Ukraine (3 per cent) need to be used in conjunction with the statistical yearbook of the USSR as a whole, which often provides data only for these republics as units.

Main sections are: administrative-territorial divisions and population; general section; science and technical progress; industry; agriculture; transport and communications; capital formation; labor; trade; service; public health; education and culture; and finance and credit.

The 1972 yearbook provides comparative data for the years 1965, 1970, 1971, and 1972, and occasionally also for 1940.

———. ———. RSFSR v 1959 godu: kratkii statisticheskii spravochnik (The RSFSR
in 1959: a short statistical handbook). Moskva, Gosstatizdat, 1950. 224 p.   346
———. ———. RSFSR v 1962 godu: kratkii statisticheskii sbornik (The RSFSR in
1962: short statistical compilation). Moskva, Gosstatizdat, 1963. 237 p.
———. ———. ——— v 1963 godu. 1964. 174 p.
———. ———. ——— v 1964 godu. Moskva, Izdatel'stvo "Statistika," 1965. 127 p.
———. ———. RSFSR v tsifrakh v 1965 godu: kratkii statisticheskii sbornik (The
RSFSR in figures in 1965: short statistical compilation). Moskva, Izdatel'stvo
"Statistika," 1966. 125 p.
———. ———. ——— v 1966 g. 1967. 160 p.
———. ———. ——— v 1967 g. 1968. 128 p.
———. ———. ——— v 1968 g. 1969. 127 p.
———. ———. ——— v 1969 g. 1970. 112 p.
———. ———. ——— v 1970 g. 1971. 112 p.
———. ———. ——— v 1971 g. 96 p.
———. ———. ——— v 1972 g. 1973. 96 p.
———. ———. RSFSR za 50 let: statisticheskii sbornik (RSFSR over 50 years:
statistical compilation). Moskva, "Statistika," 1967. 255 p.   347

———. ———. Rossiiskaia Federatsiia: statisticheskii spravochnik (The Russian
Federation: statistical handbook). Moskva, Izdatel'stvo "Sovetskaia Rossiia,"
1959. 239 p.   348

———. ———. RSFSR za 40 let: statisticheskii sobrnik (The RSFSR through 40 years:
statistical compilation). Moskva, Izdatel'stvo "Sovetskaia Rossiia," 1957, [1958].
223 p.   349

Ukrainian SSR

Note: The National Union Catalogue of the United States lists publications of
the Ukrainian SSR under Ukraine.

UKRAINIAN SSR. Tsentral'noe Statisticheskoe Upravlenie (Ukrainian: Tsentral'ne
Statystychne Upravlinnia). Narodnoe khoziaistvo Ukrainskoi SSR: statisticheskii
sbornik (Economy of the Ukrainian SSR: statistical compilation). Kiev,
Gosstatizdat, 1957. 536 p. In Ukrainian. Ukrainian: Narodne hospodarstvo
Ukraïns'koï RSR: statystychnyǐ zbirnyk. Kyïv, Derzhavne Statystychne Vydavnytstvo,
1957. 536 p.   350

———. ———. ——— 1957 g. Kiev, Statizdat, 1958. 264 p. In Ukrainian.
Ukrainian: 1957 r. Kyïv, Statvydav.
———. ———. ——— v 1959 godu: Statisticheskii ezhegodnik. Kiev, Gosstatizdat,
1960. 731 p. In Ukrainian: ... Statystychnyǐ shchorichnyk.
———. ———. ——— v 1960 godu. 1961. 556 p. In Ukrainian.
———. ———. ——— v 1961 godu. 1962. 751 p. In Ukrainian.
———. ———. ——— v 1962 godu. 1963. 675 p. In Ukrainian.
———. ———. ——— v 1963 godu. Kiev, Izdatel'stvo "Statistika," 1964. 654 p.
In Ukrainian.
———. ———. ——— v 1964 godu. 1965. 694 p. In Ukrainian.
———. ———. ——— v 1965 g. 1966. 715 p. In Ukrainian.
———. ———. ——— v 1966 g. 1967. 638 p. In Ukrainian.
———. ———. ——— v 1967 g. 1968. 579 p. In Ukrainian.
———. ———. ——— v 1968 g. 1969. 609 p. In Ukrainian.
———. ———. ——— v 1969 g. 1970. 611 p. In Ukrainian.
———. ———. ——— v 1970 g. 1971. 565 p. In Ukrainian.
———. ———. ——— v 1971 g. 1972. 547 p. In Ukrainian. Ukrainian title:
Narodne hospodarstvo Ukraïn'skoï RSR v 1971 rotsi: iubileǐnyǐ statystychnyǐ
shchorichnyk.

Data given for the Ukrainian SSR as a whole and often by the Donets-Dnepr, Southwest, and Southern regions and by 23 oblasts. Main sections: area and population; general data; industry; agriculture; transport and communications; capital formation; labor, level of material welfare; trade; service, science, and culture; public health; and finance and credit. The 1971 yearbook provides comparative data for the years 1940, 1960, 1965, 1970 and 1971.

_____. _____. Sovetskaia Ukraina v tsifrakh: statisticheskii sbornik (The Soviet Ukraine in figures: statistical compilation). Kiev, Gospolitizdat USSR, 1960. 356 p. In Ukrainian. Ukrainian: Radians'ka Ukraïna v tsyfrakh: statystychnyĭ zbirnyk. Kyïv, Derzhavne Statystychne Vydavnytstvo. 1960. 356 p.   351

_____. _____. Ukrainskaia SSR v tsifrakh v 1961 godu: kratkii statisticheskii spravochnik (The Ukrainian SSR in figures in 1961: short statistical handbook). Kiev, Gosstatizdat, 1962. 270 p. In Ukrainian. Ukrainian: Ukraïns'ka RSR v tsyfrakh v 1961 rotsi: korotkyĭ statystychnyĭ dovidnyk. Kyïv, Derzhavne Statystychne Vydavnytstvo.

_____. _____. _____ v 1962 godu: kratkii statisticheskii sbornik. 1963. 261 p. In Ukrainian.

_____. _____. _____ v 1964 godu: Kiev, Izdatel'stvo "Statistika," 1965. 182 p. In Ukrainian. Ukrainian: Ukraïns'ka RSR v tsyfrakh v 1964 rotsi; korotkyĭ statystychnyĭ dovidnyk. Vydavnytstvo "Statystyka," 1965. 182 p.

_____. _____. _____ [1966]. 1967. 192 p. In four editions: Ukrainian, Russian, English, and French. English: Ukrainian SSR in figures. Kiev, 1967. 192 p.

_____. _____. _____ v 1967 g. 1968. 167 p. In Ukrainian.

_____. _____. _____ [1968]. Kiev, Politizdat Ukrainy, 1969. 175 p. In Ukrainian and in Russian.

_____. _____. _____ v 1969 g. Kiev, Izdatel'stvo "Statistika," 1970. 199 p. In Ukrainian.

_____. _____. _____ v 1970 g. 1971. 182 p. In Ukrainian.
_____. _____. _____ v 1971 g. 1972. 143 p. In Ukrainian.
_____. _____. _____ v 1972 g. Kiev, Politizdat Ukrainy, 1973. 159 p. In Ukrainian.

_____. _____. Ukraina za piat'desiat let (1917-1967): statisticheskii spravochnik (The Ukraine over 50 years, 1917-1967: statistical handbook). Kiev, Politizdat Ukrainy, 1967. 271 p. In Ukrainian. Ukrainian: Ukraïna za p'iatdesiat rokiv (1917-1967): statystychnyĭ dovidnyk. Kyïv, Vydavnytstvo Politychnoï Literatury Ukraïny, 1967. 271 p.   352

_____. _____. Sovetskaia Ukraina [1944-1964]: statisticheskii sbornik (Soviet Ukraine, 1944-1964: statistical compilation). Kiev, Politizdat Ukrainy, 1964. 190 p. (20th anniversary of the liberation of the Soviet Ukraine from Fascist occupation). In Ukrainian. Ukrainian: Ukraïna radians'ka. Kyïv, Vydavnytstvo Polit. Lit-ry Ukraïny. 1964. 190 p.   353

_____. _____. Narodnoe khoziaistvo Ukrainskoi SSR nakanune XXI s"ezda Kommunis-ticheskoi partii Ukrainy (The economy of the Ukrainian SSR on the eve of the 21st Congress of the Communist party of the Ukraine). Kiev, Gosstatizdat, 1960. 159 p. In Ukrainian.   354

_____. _____. Dostizheniia Sovetskoi Ukrainy za 40 let: statisticheskii sbornik (Achievements of the Soviet Ukraine through 40 years: statistical compilation). Kiev, Gosstatizdat, 1957. 152 p. In Ukrainian. Ukrainian: Dosiahnennia Radians'koï Ukraïny za sorok rokiv: statystychnyĭ zbirnyk. Kyïv, Derzhavne Statystychne Vydavnytstvo,1957. 152 p.   355

Estonian SSR

ESTONIAN SSR. Tsentral'noe Statisticheskoe Upravlenie (Estonian: Statistika Keskvalitsus) Narodnoe khoziaistvo Estonskoi SSR: statisticheskii sbornik. Tallin, Estonskoe Gosudarstvennoe Izdatel'stvo, 1957. 308 p. Text in Russian and in Estonian. Estonian: Eesti NSV rahvamajandus: statistiline kogumik. Tallinn, Eesti Riiklik Kirjastus, 1957. 308 p.   356

_____. _____. v 1967 g.: statisticheskii sbornik. Tallin, "Statistika," Estonskoe otdelenie. 1968. 319 p. Text in Russian and Estonian. Estonian: Eesti NSV rahvamajandus 1967 aastal: statistiline kogumik. Tallinn, Kirjastuse "Statistika," Eesti Osakond, 1968. 319 p.

_____. _____. v 1968 g.: statisticheskii ezhegodnik. 1969. 290 p. In two editions: in Russian and in Estonian. Estonian: Eesti NSV rahvamajandus 1968, aastal: statistiline aastaraamat, 1969. 286 p.

_____. _____. _____ v 1969 g. 1970. 384 p. In two editions: in Russian and in Estonian.

_____. _____. _____ v 1970 g. 1971. 421 p. In two editions: in Russian and in Estonian.

_____. _____. _____ v 1971 g. 1972. 429 p. In two editions: in Russian and in Estonian.

_____. _____. Estonskaia SSR za gody Sovetskoi vlasti: kratkii statisticheskii sbornik (The Estonian SSR in the years of Soviet power: short statistical compilation). Tallin, "Statistika," Estonskoe otdelenie, 1967. 245 p. In two editions: in Russian and in Estonian. Estonian: Eesti NSV nõukogude võimu aastail: lühike statistiline kogumik. Tallinn, 1967. 245 p.   357

_____. _____. Razvitie narodnogo khoziaistva Estonskoi SSR: kratkii statisticheskii sbornik (Development of the economy of the Estonian SSR: short statistical compilation). Tallin, "Eesti Raamat," 1967. 111 p. In two editions: in Russian and in Estonian.   358

_____. _____. Sovetskaia Estoniia za 25 let: statisticheskii sbornik (Soviet Estonia over 25 years: statistical compilation). Tallin, "Eesti Raamat," 1965. 185 p. In two editions: in Russian and in Estonian. Estonian: 25 [Kakskümmend viis] aastat Nõukogude Eestit: statistiline kogumik. Tallinn, Kirjastus "Eesti Raamat," 1965. 175 p.   359

_____. _____. Razvitie narodnogo khoziaistva i kul'tury Estonskoi SSR v tsifrakh (Development of the economy and culture of the Estonian SSR in figures). Tallin [Estgosstatizdat], 1964. 135 p.   360

_____. _____. Dostizheniia Sovetskoi Estonii za 20 let: statisticheskii sbornik (Achievements of Soviet Estonia during 20 years: statistical compilation). Tallin. Estgosizdat, 1960. 176 p. In two editions: Russian and Estonian. Estonian: Nõukogude Eesti saavutusi 20 aasta jooksul: statistiline kogumik. Tallinn, Eesti Riiklik Kirjastus, 1960. 174 p.   361

Latvian SSR

LATVIAN SSR. Tsentral'noe Statisticheskoe Upravlenie (Latvian: Centrālā Statistikas Pārvalde). Narodnoe khoziaistvo Latviiskoi SSR: statisticheskii sbornik (Economy of the Latvian SSR: statistical compilation). Riga, Gosstatizdat, 1957. 232 p. Text in Russian and Latvian. Latvian: Latvijas PSR tautas saimniecība: statistisko datu krājums. Rīgā, Valsts Statistikas Izdevniecība, 1957. 232 p.   362

_____. _____. Narodnoe khoziaistvo Sovetskoi Latvii za 20 let: statisticheskii sbornik (Economy of Soviet Latvia over 20 years: statistical compilation). Riga, Gosstatizdat, Latviiskoe otdelenie. 1960. 311 p. Text in Russian and Latvian. Latvian: Padomju Latvijas tautas saimniecība 20 gados: statistisko datu krājums. Rīgā, Valsts Statistikas Izdevniecības Latvijas Nodaļa, 1960. 311 p.

___. ____. Razvitie narodnogo khoziaistva Latviiskoi SSR: statisticheskii
bornik (Development of the economy of the Latvian SSR: statistical compilation).
iga, Gosstatizdat, Latviiskoe otdelenie. 1962. 374 p. Text in Russian and in
atvian. Latvian: Latvijas PSR tautas saimniecības attīstība: statistiko datu
rājums. Rīgā, Valsts Statistikas Izdevniecības Latvijas Nolaļa, 1962. 374 p.

___. Narodnoe khoziaistvo Sovetskoi Latvii: statisticheskii sbornik (Economy of
oviet Latvia: statistical compilation). Riga, "Statistika," Latviiskoe otdelenie,
968. 541 p. Text in Russian and Latvian.

___. Narodnoe khoziaistvo Latviiskoi SSR v 1970 g.: statisticheskii sbornik.
972. 607 p. Text in Russian and Latvian. Latvian: Latvijas PSR tautas
aimniecība 1970 gadā: statistisko datu krājums. Rīgā, Izdevniecibas, "Statistika,"
atvijas nodaļa, 1972. 607 p.

___. ____ v 1971 g. 1972. 417 p. (Dedicated to the 50th anniversary of the
ormation of the USSR).

___. ____ v 1972. Riga, "Liesma," 1973. 463 p.

___. ____. Latviiskaia SSR v tsifrakh v 1960 godu: kratkii statisticheskii
bornik (The Latvian SSR in figures in 1960: short statistical compilation).
iga, Gosstatizdat, Latviiskoe otdelenie, 1961. 344 p. In two editions: in
ussian and in Latvian.   363

___. ____. ____ v 1961 godu. 1962. 199 p.
___. ____. ____ v 1962 godu. 1963. 212 p.
___. ____. ____ v 1968 g.: statisticheskii sbornik. Riga, "Statistika,"
atviiskoe otdelenie, 1969. 466 p.
___. ____. ____ v 1969 g. 1971. 622 p.
___. ____. ____ v 1971 g. 1972. 353 p.
___. ____. ____ v 1972 g. 1973. 416 p.

___. ____. Latviiskaia SSR v 1968.: kratkii statisticheskii sbornik (The
atvian SSR in 1968: short statistical compilation). Riga, 1969. 337 p.   364

___. ____. ____ v 1969 g. 1970. 376 p.
___. ____. ____ v 1970 g. 1971. 386 p.

___. ____. Latviia za gody Sovetskoi vlasti: statisticheskii sbornik (Latvia
n the years of Soviet power: statistical compilation). Riga, "Statistika,"
atviiskoe otdelenie, 1967. 403 p.   365

___. ____. Ekonomika i kul'tura Sovetskoi Latvii: statisticheskii sbornik
Economy and culture of Soviet Latvia: statistical compilation). Riga, "Statis-
ika," 1966. 508 p. Text in Russian and Latvian. Latvian: Padomju Latvijas
konomika un kultūra: statistiko datu krājums. Rīgā, Izdevniecības "Statistika,"
atvijas nodaļa, 1966. 508 p.   366

___. ____. Sovetskaia Latviia v tsifrakh. 1940-1963 gg.: statisticheskii
bornik (Soviet Latvia in figures, 1940-1963: statistical compilation). Riga,
Statistika," 1965. 400 p. Text in Russian and Latvian. Latvian: Padomju
atvija skaitlos: statistisko datu krājums. Rīgā, Izdevniecības "Statistika"
atvijas nodaļa, 1965. 400 p.   367

___. ____. Sovetskaia Latvia za 25 let: kratkii statisticheskii sbornik
Soviet Latvia through 25 years: short statistical compilation). Riga, "Statis-
ika," Latviiskoe otdelenie, 1965. 274 p.   368

___. ____. Tekhnicheskii progress v narodnom khoziaistve Latviiskoi SSR:
tatisticheskii sbornik (Technical progress in the economy of the Latvian
SR: statistical compilation). Riga, 1971. 207 p.   369

_____. _____. Dinamika mezhotraslevykh i mezhrespublikanskikh ekonomicheskikh
sviazei Latviiskoi SSR: sbornik statisticheskikh materialov (The dynamics of
economic ties of the Latvian SSR among branches of the economy and among the
republics). Riga, 1971. 285 p.   3

Lithuanian SSR

LITHUANIAN SSR. Tsentral'noe Statisticheskoe Upravlenie. (Lithuanian: Centrine
Statistikos Valdyba). Ekonomika i kul'tura Litovskoi SSR v 1963 godu: statisticheskii ezhegodnik (Economy and culture of the Lithuanian SSR in 1963:
statistical yearbook). Vilnius, "Statistika," 1964. 221 p. In two editions:
in Russian and in Lithuanian. Lithuanian: Lietuvos TSR ekonomika ir kultūra:
statistikos metraštis, 215 p.   3

_____. _____. _____ [1964]. 1965. 195 p. In two editions: in Russian and in
Lithuanian. (To the 15th Congress of the Communist party in Lithuania).
_____. _____. _____ [1917-1966]: statisticheskii sbornik (statistinių duomenų
rinkinys) 1967. 416 p. (Dedicated to the 50th anniversary of Great October).
In two editions: in Russian and in Lithuanian.
_____. _____. _____ 1966. 1967. 194 p. In two editions: in Russian and in
Lithuanian.
_____. _____. _____ v 1967 g.: statisticheskii sbornik. Vilnius, "Statistika,"
Litovskoe otdelenie, 1968. 365 p. In two editions: in Russian and in Lithuanian.
_____. _____. _____ v 1968 g. 1969. 408 p. In two editions: in Russian and in
Lithuanian. Lithuanian ed., 399 p.
_____. _____. _____ v 1969 g.: statisticheskii ezhegodnik. 1970. 503 p. In two
editions: in Russian and in Lithuanian. Lithuanian ed., 479 p.
_____. _____. _____ v 1970 g. 1971. 514 p. In two editions: in Russian and in
Lithuanian.
_____. _____. _____ 1971. 1972. 550 p. In two editions: in Russian and in
Lithuanian. (Dedicated to the 50th anniversary of the formation of the USSR).

_____. _____. Narodnoe khoziaistvo Litovskoi SSR: statisticheskii sbornik (Economy
of the Lithuanian SSR: statistical compilation). Vilnius, Gosstatizdat, Litovskoe
otdelenie, 1957. 224 p. In two editions: in Russian and in Lithuanian.
Lithuanian: Lietuvos TSR liaudies ūkis: statistiniu duomenų rinkinys. Vilnius,
Valstybinė Statistikos Leidykla, Lietuvos skyrius, 1957. 224 p.

_____. _____. _____ v 1960 godu: kratkii statisticheskii sbornik. Vilnius,
Gosstatizdat, Litovskoe otdelenie, 1962. 192 p. In two editions: in Russian
and in Lithuanian. Abbreviated edition only.
_____. _____. _____ v 1961 godu: statisticheskii sbornik. 1963. 228 p. In two
editions: in Russian and in Lithuanian.
_____. _____. _____ v 1965 g. 1966. 432 p. In two editions: in Russian and in
Lithuanian.

POCIUS, Jonas, comp. Present-day Lithuania in figures. Vilnius, "Gintaras," 1971
131 p.

LITHUANIAN SSR. Tsentral'noe Statisticheskoe Upravlenie (Lithuanian: Centrinė
Statistikos Valdyba). Lithuania in figures, 1965. Jonas Pocius, comp. Vilnius,
"Gintaras," 1966. 95 p.

_____. _____. 25 let Sovetskoi Litvy: statisticheskii sbornik (25 years of Soviet
Lithuania: statistical compilation). Vilnius, "Statistika," Litovskoe otdelenie,
1965. 271 p. In two editions: in Russian and in Lithuanian. Lithuanian: Tarybu
Lietuvai 25 metai: statistikos duomenų rinkinya. Vilnius, Statistika, Lietuvos
skyrius, 1965. 263 p.

370 - 375

___. ____. Litovskaia SSR v tsifrakh v 1962 godu: kratkii statisticheskii
sbornik (The Lithuanian SSR in figures in 1962: short statistical compilation).
Vilnius, Gosstatizdat, Litovskoe otdelenie, 1963. 212 p. In two editions: in
Russian and in Lithuanian. Lithuanian: Lietuvos TSR skaičiais.   376

___. ____. 20 let Sovetskoi Litvy: statisticheskii sbornik (20 years of Soviet
Lithuania: statistical compilation). Vilnius, Gosstatizdat, Litovskoe otdelenie,
1960. 352 p. In two editions: in Russian and in Lithuanian. Lithuanian: Tarybų
Lietuvos dvidešimtmetis: statistinių duomenų rinkinys. Vilnius, Valstybinė
Statistikos Leidykla, Lietuvos skyrius, 1960. 352 p.   377

___. ____. Statisticheskie dannye ob ekonomike i kul'ture Litovskoi SSR mezhdu
X i XI s"ezdami Kommunisticheskoi partii Litvy (Statistical data on the economy
and culture of the Lithuanian SSR between the 10th and 11th congresses of the
Communist party of Lithuania). Vilnius, Gosstatizdat, Litovskoe otdelenie, 1959.
96 p. In two editions: in Russian and in Lithuanian.   378

Belorussian SSR

Note: The National Union Catalogue of the United States lists publications
of the Belorussian SSR under White Russia.

LORUSSIAN SSR. Tsentral'noe Statisticheskoe Upravlenie. Narodnoe khoziaistvo
Belorusskoi SSR: statisticheskii sbornik (Economy of the Belorussian SSR:
statistical compilation). Minsk Gosstatizdat, 1957. 320 p.   379

___. ____. ____. 1963. 511 p.

___. ____. ____ v 1968 g.: statisticheskii sbornik. Minsk, Izdatel'stvo
"Statistika," Belorrusskoe otdelenie, 1969. 535 p.
___. ____. ____ v 1970 g. 1971. 416 p.
___. ____. ____ v 1971 g. Minsk, "Belarus'," 1972. 223 p.

___. ____. Belorusskaia SSR v tsifrakh: kratkii statisticheskii sbornik (The
Belorussian SSR in figures: short statistical compilation). Minsk, Gosstatizdat,
Belorusskoe otdelenie, 1962. 255 p.   380

___. ____. ____. 1963. 284 p.

___. ____. ____ za 1964 god: kratkii statisticheskii sbornik. Minsk,
Izdatel'stvo "Statistika," Belorusskoe otdelenie, 1965. 271 p.
___. ____. ____ v 1965 g. 1966. 173 p.
___. ____. ____ v 1969 g.: statisticheskii sbornik. 1970. 336 p.

___. ____. Belorusskais SSR za 50 let: statisticheskii sbornik (Belorussian
SR during 50 years: statistical compilation). Minsk, "Belarus'," 1968. 260 p. 381

___. ____. Belorusskaia SSR za gody Sovetskoi vlasti: statisticheskii sbornik
(The Belorussian SSR in the years of Soviet power: statistical compilation).
Minsk, "Belarus'," 1967. 403 p.   382

___. ____. Razvitie narodnogo khoziaistva Belorusskoi SSR za 20 let 1944-1963 gg.:
statisticheskii sbornik (Development of the economy of the Belorrusian SSR over
20 years, 1944-1963: statistical compilation). Minsk, "Belarus'," 1964. 214 p.
Title on cover: Belorusskaia SSR za 20 let.   383

___. ____. Dostizheniia Sovetskoi Belorussii za 40 let: statisticheskii sbornik
(Achievements of Soviet Belorussia through 40 years: statistical compilation).
Minsk, Gosstatizdat, 1958. 211 p.   384

BORUSSIAN SSR. Gosplan. Narodnoe khoziaistvo Belorusskoi SSR za 40 let: statistiko-
konomicheskii sbornik (Economy of the Belorussian SSR through 40 years: a statis-
ical-economic collection). Minsk, 1957. 288 p.   385

76 - 385

Moldavian SSR

MOLDAVIAN SSR. Tsentral'noe Statisticheskoe Upravlenie. <u>Narodnoe khoziaistvo Moldavskoi SSR: statisticheskii sbornik</u> (Economy of the Moldavian SSR: statistical compilation). Kishinëv, Gosstatizdat, 1957. 200 p.

_____. _____. _____. 1959. 287 p.
_____. _____. _____ v 1960 g.: statisticheskii ezhegodnik. 1961. 362 p.
_____. _____. _____ v 1962 g.: statisticheskii sbornik. 1963. 384 p.
_____. _____. _____ v 1964 g. Kishinëv, "Statistika," 1965. 426 p.
_____. _____. _____ [1968]. 1969. 421 p.
_____. _____. _____ [1969]. 1970. 358 p.
_____. _____. _____ na 1/VII 1971 g. 1971. 408 p.
_____. _____. _____ [1971]. Kishinëv, "Soiuzpechat'," 1972. 360 p.

_____. _____. Sovetskaia Moldaviia k 50-letiiu Velikogo Oktiabria: statisticheskii sbornik (Soviet Moldavia on the 50th anniversary of Great October: statistical compilation). Kishinëv, "Statistika," 1967. 287 p.

_____. _____. Sovetskaia Moldaviia za 40 let: statisticheskii sbornik (Soviet Moldavia over 40 years: statistical compilation). Kishinëv, Gosstatizdat, 1964. 197 p.

_____. _____. Moldavskaia SSR v tsifrakh v 1961 godu: kratkii statisticheskii sbornik (Moldavian SSR in figures in 1961: short statistical compilation). Kishinëv, Gosstatizdat, 1962. 365 p.

Georgian SSR

Note: The National Union Catalogue of the United States lists publications of the Georgian SSR under Georgia (Transcaucasia).

GEORGIAN SSR. Tsentral'noe Statisticheskoe Upravlenie. <u>Narodnoe khoziaistvo Gruzinskoi SSR: statisticheskii sbornik</u> (Economy of the Georgian SSR: statistical compilation). Tbilisi, Gosstatizdat, 1957. 304 p.

_____. _____. _____. Tbilisi, Statizdat, 1959. 357 p.
_____. _____. _____ v 1961 godu: statisticheskii ezhegodnik (...in 1961: statistical yearbook). Tbilisi, Gosstatizdat, Gruzinskoe otdelenie, 1963. 568 p.
_____. _____. _____ v 1962 godu. 1963. 444 p.
_____. _____. _____ v 1963 godu. Tbilisi, Izdatel'stvo "Statistika," Gruzinskoe otdelenie, 1964. 354 p.
_____. _____. _____ v 1964 godu. 1965. 458 p.
_____. _____. _____ v 1967 g. 1968. 383 p. Text in both Russian and Georgian.
_____. _____. _____ [1971]. 1972. 383 p. (On the 50th anniversary of the USSR).

_____. _____. Gruzinskaia SSR v tsifrakh v 1971 godu: kratkii statisticheskii spravochnik (The Georgian SSR in figures in 1971: short statistical handbook). Tbilisi, "Statistika," Gruzinskoe otdelenie, 1972. 243 p.

_____. _____. _____ v 1972 godu. Tbilisi, "Sabchota Sakartvelo," 1973. 147 p.

_____. _____. 50 let Sovetskoi Gruzii: statisticheskii sbornik (50 years of Soviet Georgia: statistical compilation). Tbilisi, "Statistika," Gruzinskoe otdelenie, 1971. 388 p. Text in both Russian and Georgian.

_____. _____ Sovetskaia Gruziia po leninskomu puti: statisticheskii sbornik (Soviet Georgia in the path of Lenin: statistical compilation). Tbilisi, "Statistika," Gruzinskoe otdelenie, 1970. 178 p.

___. ____. Gruzinskaia SSR v 1968 godu: kratkii statisticheskii sbornik (The
Georgian SSR in 1968: short statistical compilation). Tbilisi, "Statistika,"
Gruzinskoe otdelenie, 1969. 250 p.   394

___. ____. Sovetskaia Gruziia k 50-letiiu Velikoi Oktiabr'skoi sotsialisticheskoi
revoliutsii [Statisticheskii sbornik]. (Soviet Georgia on the 50th anniversary of
the Great October socialist revolution: statistical compilation). Tbilisi, "Statistika," Gruzinskoe otdelenie. 1967. 327 p.   395

___. ____. Sovetskaia Gruziia za 40 let: statisticheskii sbornik (Soviet
Georgia through 40 years: statistical compilation). Tbilisi, Gosstatizdat, Gruzin-
koe otdelenie, 1961. 208 p. In two editions: in Russian and in Georgian.   396

___. ____. Dostizheniia Sovetskoi Gruzii za 40 let: kratkii statisticheskii
pravochnik (Achievements of Soviet Georgia over 40 years: short statistical handbook). Tbilisi, Gosstatizdat, Gruzinskoe otdelenie, 1961. 101 p.   397

Armenian SSR

ENIAN SSR. Tsentral'noe Statisticheskoe Upravlenie (Armenian: Kentronakan
ichakagrakan Varch'owt'yown). Narodnoe khoziaistvo Armianskoi SSR: statisticheskii
bornik (Economy of the Armenian SSR: statistical compilation). Erevan, Gos-
tatizdat, 1957. 180 p.   398

___. ____. ____ v 1963 godu. 1964. 272 p.
___. ____. ____ v 1964 godu: statisticheskii ezhegodnik. Erevan, "Statistika,"
965. 322 p.
___. ____. ____ v 1965 godu. 1966. 352 p.
___. ____. ____ v 1967 g. 1968. 263 p.
___. ____. ____ v 1968 g. 1969. 255 p.
___. ____. ____ v 1970 g. 1971. 224 p.

___. ____. Sovetskaia Armeniia za 50 let: sbornik statisticheskogo materialov
Soviet Armenia during 50 years: compilation of statistical materials). Erevan,
970. 426 p.   399

___. ____. Ekonomika i kul'tura Armenii k 50-letiiu Velikogo Oktiabria: statis-
icheskii sbornik (Economy and culture of Armenia on the 50th anniversary of Great
ctober: statistical compilation). Erevan, 1967. 445 p.   400

___. ____. Sovetskaia Armeniia za 40 let: statisticheskii sbornik (Soviet
rmenia through 40 years: statistical compilation). Erevan, Aipetrat, 1960. 210 p.
                                                                            401

Azerbaidzhan SSR (Azerbaijan SSR)

    Note: The National Union Catalogue of the United States lists publications
f the Azerbaidzhan SSR under Azerbaijan.

BAIDZHAN SSR. Tsentral'noe Statisticheskoe Upravlenie. Narodnoe khoziaistvo
erbaidzhanskoi SSR: statisticheskii sbornik (Economy of the Azerbaidzhan SSR:
atistical compilation). Baku, Gosstatizdat, 1957. 525 p. Text in Russian and
erbaijani.
                                                                            402
___. ____. ____ v 1962 godu. Baku, Gosstatizdat, Azerbaidzhanskoe otdelenie,
963. 255 p.
___. ____. ____ v 1963 godu: statisticheskii ezhegodnik. Baku, "Statistika,"
erbaidzhanskoe otdelenie, 1965. 294 p.
___. ____. ____ v 1964 godu. 1965. 243 p.
___. ____. ____ v 1965 godu. 1966 [1967]. 321 p.
___. ____. ____ v 1968 godu. 1970. 501 p. Text in Russian and Azerbaijani.
___. ____. ____ v 1970 godu. Baku, Soiuzuchetizdat, Azerbaidzhanskoe otdelenie,
72. 458 p.
___. ____. ____. 1972. 282 p. (Jubilee statistical yearbook).

Azerbaidzhan v tsifrakh: kratkii statisticheskii sbornik (Azerbaidzhan in figures:
    short statistical compilation). Baku, Azerneshr, 1964. 302 p.   4

————. ————. ———— v 1964 godu. 1965. 162 p.
————. ————. ———— v 1965 godu. 1966. 382 p.
————. ————. ———— v 1967 g. Baku, "Statistika," Azerbaidzhanskoe otdelenie,
1968. 103 p.
————. ————. ———— v 1968 g. 1969. 98 p.
————. ————. ———— v 1969 g. 1970. 163 p.
————. ————. ———— v 1970 g. 1971. 295 p.
————. ————. ———— v 1971 g. 1972. 154 p.
————. ————. ———— v 1972 g. 1973. 146 p.

————. ————. Sovetskii Azerbaidzhan za 50 let: statisticheskii sbornik (Soviet
Azerbaidzhan during 50 years: statistical compilation). Baku, Azerneshr, 1970.
266 p.   4

————. ————. Azerbaidzhanskaia SSR k 50-letiiu Velikogo Oktiabria: statisticheski
sbornik (Azerbaidzhan SSR on the 50th anniversary of the Great October: statistic
compilation). Baku, "Statistika," 1967. 215 p.   4

————. ————. Razvitie narodnogo khoziaistva Azerbaidzhanskoi SSR i rost material
nogo i kul'turnogo urovnia zhizni naroda: statisticheskii sbornik (Development of
the economy of the Azerbaidzhan SSR and growth of the material and cultural level
of life of the people: statistical compilation). Baku, Azerbaidzhanskoe Gosudarstvennoe Izdatel'stvo, 1964. 258 p.   4

————. ————. Razvitie ekonomiki i kul'tury Azerbaidzhanskoi SSR (1953-1963 gg.):
kratkii statisticheskii sbornik (Development of the economy and culture of the
Azerbaidzhan SSR 1953-1963: short statistical compilation). Baku, Ob"edinennoe
Izdatel'stvo TsK KP Azerbaidzhana, 1963. 280 p.   4

————. ————. Razvitie narodnogo khoziaistva Azerbaidzhanskoi SSR i rost material'
nogo i kul'turnogo urovnia zhizni naroda: statisticheskii sbornik (Development of
the economy of the Azerbaidzhan SSR and growth of the material and cultural level
of the life of the people: statistical compilation). Baku, Azerneshr, 1961. 258

————. ————. Dostizheniia Sovetskogo Azerbaidzhana za 40 let v tsifrakh: statisticheskii sbornik (Achievements of Soviet Azerbaidzhan through 40 years in figur
statistical compilation). Baku, Azerneshr, 1960. 259 p.   l

Kazakh SSR

KAZAKH SSR. Tsentral'noe Statisticheskoe Upravlenie. Narodnoe khoziaistvo
Kazakhskoi SSR: statisticheskii sbornik (Economy of the Kazakh SSR: statistical
compilation). Alma-Ata, Kazakhskoe Gosudarstvennoe Izdatel'stvo, 1957. 384 p.

————. ————. ———— v 1960 i 1961 gg.: statisticheskii sbornik. Alma-Ata,
Gosstatizdat, 1963. 544 p.
————. ————. ———— Kazakhstana [1967]. Alma-Ata, "Kazakhstan," 1968. 416 p.
(50 years of Soviet rule).
————. ————. ———— v 1968 g. 1970. 368 p.
————. ————. ———— v 1971 g. Alma-Ata, Soiuzuchetizdat, Kazakhskoe otdelenie.
1972. [1973]. 432 p.

————. ————. Kazakhstan za 50 let: statisticheskii sbornik. Alma-Ata, "Statisti
Kazakhskoe Otdelenie, 1971. 248 p.

————. ————. Kazakhstan v tsifrakh: kratkii statisticheskii sbornik (Kazakhstan
in figures: short statistical compilation). Alma-Ata, "Kazakhstan," 1971. 147 p

___. ____. Kazakhstan za 40 let: statisticheskii sbornik (Kazakhstan through
40 years: statistical compilation). Alma-Ata, Gosstatizdat, 1960. 525 p.   414

___. ____. Narodnoe khoziaistvo i kul'tura Kazakhskoi SSR mezhdu VIII i X
s"ezdami Kommunisticheskoi partii Kazakhstana (Economy and culture of the Kazakh
SSR between the 8th and 10th congresses of the Communist party of Kazakhstan).
Alma-Ata, Gosstatizdat, 1960. 211 p.   415

Soviet Middle Asia
including the Kirgiz, Uzbek, Turkmen, and Tadzhik republics

SR. Sredneaziatskoe Statisticheskoe Upravlenie. Narodnoe khoziaistvo Srednei
Azii v 1963 godu: statisticheskii sbornik (The economy of Middle Asia in 1963:
statistical compilation). Tashkent, "Uzbekistan," 1964. 372 p.   416

Kirgiz SSR (Kirghiz SSR)

RGIZ SSR. Tsentral'noe Statisticheskoe Upravlenie. Narodnoe khoziaistvo
Kirgizskoi SSR: statisticheskii sbornik (Economy of the Kirgiz SSR: statistical
compilation). Frunze, Gosstatizdat, 1957. 208 p.   417

___. ____. ____. 1960 . 183 p.
___. ____. ____ v 1960 godu: statisticheskii ezhegodnik. Frunze, Gosstatizdat,
Kirgizskoe otdelenie, 1961. 274 p.
___. ____. ____ v 1961 godu. 1962. 232 p.
___. ____. ____ v 1963 godu. Frunze, "Statistika," 1964. 238 p.
___. ____. ____ v 1964 godu. Frunze, "Statistika," Kirgizskoe otdelenie,
1965. 87 p.
___. ____. ____ v 1967 g.: statisticheskii sbornik. 1968. 152 p.
___. ____. ____. 1973. 390 p.

___. ____. Kirgiziia v tsifrakh: statisticheskii sbornik (Kirgizia in figures:
statistical compilation). Frunze, Gosstatizdat, Kirgizskoe otdelenie, 1963.
199 p.   418

___. ____. Kirgizstan v tsifrakh: statisticheskii sbornik (Kirgizstan in
figures; statistical compilation). Frunze, "Kyrgyzstan," 1971. 280 p. (100th
anniversary of the voluntary entry of Kirgizia into Russia).

___. ____. Kirgizstan v tsifrakh i faktakh, 1960-1970 gg.: kratkii statisticheskii spravochnik (Kirgizstan in figures and facts, 1960-1970: short statistical
compilation). Frunze, 1971. 190 p.   419

___. ____. Kirgizstan za gody Sovetskoi vlasti: statisticheskii sbornik
(Kirgizstan during the years of Soviet power: statistical compilation). Frunze,
"Kyrgyzstan," 1970. 247 p.   420

___. ____. Kirgizstan za 50 let Sovetskoi vlasti: statisticheskii sbornik
(Kirgizstan during 50 years of Soviet power: statistical compilation). Frunze,
"Kyrgyzstan," 1967. 223 p.   421

___. ____. Sovetskii Kirgizstan za 40 let (1926-1966): statisticheskii sbornik
(Soviet Kirgizstan over 40 years, 1926-1966: statistical compilation). Frunze,
"Kyrgyzstan," 1966. 160 p.   422

___. ____. Razvitie ekonomiki i kul'tury Kirgizskoi SSR, 1958-1965.: kratkii
statisticheskii sbornik (Development of the economy and culture of the Kirgiz
SSR, 1958-1965: short statistical compilation). Frunze, 1966. 107 p.   423

Uzbek SSR

UZBEK SSR. Tsentral'noe Statisticheskoe Upravlenie. <u>Narodnoe khoziaistvo Uzbekskoi</u>
<u>SSR: statisticheskii sbornik</u> (Economy of the Uzbek SSR: statistical compilation).
Tashkent, Gosstatizdat, 1957. 200 p.   4.

_____. _____ v 1958 godu. Tashkent, Gosstatizdat, Uzbekskoe otdelenie, 1959. 223 p
_____. _____ v 1960 godu: kratkii statisticheskii sbornik. 1961. 95 p. Abbreviat
ed edition only.
_____. _____ v 1961 godu. 1962. 227 p. Abbreviated edition only.
_____. _____ v 1965 g.: statisticheskii ezhegodnik. Tashkent, "Uzbekistan," 1966.
431 p.
_____. _____ v 1967 g. 1968. 375 p.
_____. _____ v 1968 g.: 1970. 383 p.
_____. _____ v 1969 g. 1970. 351 p.
_____. _____ v 1970 g. 1971. 374 p.
_____. _____ v 1971 g. 1972. 364 p.
_____. _____ v 1972 g. 1973. 365 p.

_____. _____. <u>Uzbekistan za gody vos'moi piatiletki (1966-1970): kratkii statis-</u>
<u>ticheskii sbornik</u> (Uzbekistan during the years of the 8th 5-year plan, 1966-1970:
a short statistical compilation). Tashkent, Izdatel'stvo TsK KP Uzbekistana,
1971. 160 p.

_____. _____. <u>Narodnoe khoziaistvo Uzbekskoi SSR za 50 let: sbornik statisticheski</u>
<u>materialov</u> (Economy of the Uzbek SSR over 50 years: collection of statistical
materials). Tashkent, "Uzbekistan," 1967. 240 p.

_____. _____. <u>Uzbekistan za 7 let, 1959-1965 gg.: kratkii statisticheskii sbornik</u>
(Uzbekistan for 7 years, 1959-1965: short statistical compilation). Tashkent,
Ob"edinennoe Izdatel'stvo TsK KP Uzbekistana, 1966. 250 p.

_____. _____. <u>Sovetskii Uzbekistan za 40 let: statisticheskii sbornik</u> (Soviet
Uzbekistan through 40 years: statistical compilation). Tashkent, "Uzbekistan,"
1964. 379 p.

_____. _____. <u>Uzbekistan za 40 let Sovetskoi vlasti: statisticheskii sbornik</u>
(Uzbekistan through 40 years of the Soviet regime: statistical compilation).
Tashkent, Gosizdat Uzbekskoi SSR, 1958. 135 p.

Turkmen SSR

TURKMEN SSR. Tsentral'noe Statisticheskoe Upravlenie. <u>Narodnoe khoziaistvo Turk-</u>
<u>menskoi SSR: statisticheskii sbornik</u> (Economy of the Turkmen SSR: statistical
compilation. Ashkhabad. Gosstatizdat, 1957. 172 p.

_____. _____. <u>Narodnoe khoziaistvo Turkmenskoi SSR nakanune XVII s"ezda</u>
<u>Kommunisticheskoi partii Turkmenistana</u> (Economy of the Turkmen SSR on the eve
of the 17th congress of the Communist party of Turkmenistan). Ashkhabad.
Gosstatizdat, 1961. 121 p.

_____. _____. <u>Narodnoe khoziaistvo Turkmenskoi SSR: statisticheskii sbornik</u>.
1962. 253 p.

_____. _____. <u>Turkmenistan za gody Sovetskoi vlasti: statisticheskii sbornik</u>
(Turkmenistan during the years of the Soviet regime: statistical compilation).
Ashkhabad, "Turkmenistan," 1967. 143 p.

_____. _____. <u>Sovetskii Turkmenistan za 40 let [1924-1964]: statisticheskii</u>
<u>sbornik</u> (Soviet Turkmenistan over the 40 years, 1924-1964: statistical compila-
tion). Ashkhabad, Turkmenizdat, 1964. 158 p.

Tadzhik SSR (Tajik SSR)

DZHIK SSR. Tsentral'noe Statisticheskoe Upravlenie. Narodnoe khoziaistvo
Tadzhikskoi SSR: statisticheskii sbornik (Economy of the Tadzhik SSR: statistical
compilation). Stalinabad [Dushanbe], Gosstatizdat, 1957. 388 p. In two editions
in Russian and in Tajik. The Tajik edition was published by Tadzhikgosizdat in
1958. 398 p.   433

___. ___. ___ v 1959 godu: statisticheskii sbornik. Stalinabad [Dushanbe],
Gosstatizdat, Tadzhikskoe otdelenie, 1960. 307 p.
___. ___. ___ v 1960 godu. 1961. 330 p.
___. ___. ___ v 1961 godu: kratkii statisticheskii sbornik. Dushanbe, 1962.
166 p. Abbreviated edition only.
___. ___. ___ v 1962 godu: statisticheskii ezhegodnik. 1963. 394 p.
___. ___. ___ v 1964 godu. 1965. 274 p.
___. ___. ___ v 1965 g. 1966. 311 p.
___. ___. ___ v 1968 g. 1969. 287 p.
___. ___. ___ v 1969 g. 1970-1971. 308 p.
___. ___. ___ v 1971. Dushanbe, Soiuzuchetizdat, Tadzhikskoe otdelenie.
1972. 225 p. (To the 50th anniversary of the USSR).

___. ___. Tadzhikistan za gody sovetskoi vlasti: sbornik statisticheskikh
materialov (Tadzhikistan during the years of Soviet regime: collection of statistical materials). Dushanbe, "Statistika," Tadzhikskoe otdelenie, 1967. 199 p.
Two editions: in Russian and in Tajik.   434

___. ___. Tadzhikistan za 40 let: statisticheskii sbornik (Tadzhikistan over
40 years: statistical compilation). Dushanbe, "Statistika," 1964. 242 p.   435

___. ___. Statisticheskie dannye o razvitii khoziaistva i kul'tury Tadzhikskoi
SSR za 1957-1959 gody (Statistical data on the development of the economy and
culture of the Tadzhik SSR in 1957-1959). Stalinabad [Dushanbe], Gosstatizdat,
1960. 110 p.   436

Statistical Handbooks for Smaller Administrative Units:
Autonomous Republics, Krays, Oblasts, and Autonomous Oblasts

Abbreviations:

ASSR.   Avtonomnaia Sovetskaia Sotsialisticheskaia Respublika (Autonomous Soviet
         Socialist Republic).

AO.   Avtonomnaia Oblast' (Autonomous Oblast)

O.   Oblast' (Oblast)

Notes. The units are listed under the appropriate union republic and within such
republics are arranged by the English alphabetical order of the names. Within
each administrative unit, entries are in reverse chronological order. The statistical office (Statisticheskoe Upravlenie) of each oblast or comparable unit is
administratively under the Central Statistical Office (Tsentral'noe Statisticheskoe
Upravlenie) of the union republic in which located.

Since the names are highly repetitive and generally entirely clear, English
translations of titles have been omitted in this section.

For location of areas see map on inside back cover.

Russian Soviet Federated Socialist Republic

GEISKAIA AO (Adygey AO).Statisticheskoe Upravlenie. Sovetskaia Adygeia k 50-
etiiu obrazovaniia SSSR: statisticheskii sbornik. Maykop, Krasnodarskoe Knizhnoe
Izdatel'stvo, Adygeiskoe otdelenie, 1973. 175 p.   437

__. ____. Narodnoe khoziaistvo Adygeiskoi avtonomnoi oblasti: statisticheskii
sbornik. Krasnodar, Gosstatizdat, 1957. 75 p.   438

AISKII Krai (Altay Kray). Statisticheskoe Upravlenie. Narodnoe khoziaistvo
ltaiskogo kraia v 1966-1970 gg.: statisticheskii sbornik. Barnaul, 1972. 303 p.
                                                                               439
__. ____. Narodnoe khoziaistvo Altaiskogo kraia za 50 let Sovetskoi vlasti:
tatisticheskii sbornik. Barnaul, "Statistika," Altaiskoe otdelenie, 1967. 109 p.
                                                                               440
__. ____. Narodnoe khoziaistvo Altaiskogo kraia: statisticheskii sbornik.
arnaul, Altaiskoe Knizhnoe Izdatel'stvo, 1958. 299 p.   441

__. ____. Narodnoe khoziaistvo Altaiskogo kraia za 40 let Sovetskoi vlasti:
tatisticheskii sbornik. Barnaul, Altaiskoe Knizhnoe Izdatel'stvo, 1957. 111 p.
                                                                               442
RSKAIA Oblast' (Amur O.). Statisticheskoe Upravlenie. Narodnoe khoziaistvo
murskoi oblasti v 1963 godu. Khabarovsk, "Statistika," Khabarovskoe otdelenie,
965. 212 p.   443

__. ____. Narodnoe khoziaistvo Amurskoi oblasti: statisticheskii sbornik.
lagoveshchensk, Amurskoe Knizhnoe Izdatel'stvo, 1957. 112 p.   444

HANGEL'SKAIA Oblast' (Archangel O.). Statisticheskoe Upravlenie. Narodnoe
hoziaistvo Arkhangel'skoi oblasti v tsifrakh: statisticheskii sbornik.
Arkhangel'sk], Severo-Zapadnoe Knizhnoe Izdatel'stvo, 1972. 224 p.   445

__. ____. Arkhangel'skaia oblast' v tsifrakh: statisticheskii sbornik.
Arkhangel'sk], Severo-Zapadnoe Knizhnoe Izdatel'stvo, 1967. 98 p.   446

__. ____. Narodnoe khoziaistvo Arkhangel'skoi oblasti: statisticheskii sbornik.
ologda, Gosstatizdat, 1962. 159 p.   447

_____. _____. Narodnoe khoziaistvo Arkhangel'skoi oblasti: statisticheskii sbornik.
Arkhangel'sk, Knizhnoe Izdatel'stvo, 1957. 147 p.   4

ASTRAKHANSKAIA Oblast' (Astrakhan' O.). Statisticheskoe Upravlenie. Narodnoe
khoziaistvo Astrakhanskoi oblasti za 50 let: statisticheskii sbornik. Volgograd,
Nizhne-Volzhshoe Knizhnoe Izdatel'stvo, 1967. 152 p.   4

_____. _____. Narodnoe khoziaistvo Astrakhanskoi oblasti: statisticheskii sbornik.
Saratov, Gosstatizdat, Saratovskoe otdelenie, 1963. 109 p.   4

BASHKIRSKAIA ASSR (Bashkir ASSR). Statisticheskoe Upravlenie. Bashkiriia v Soiuze
SSR: statisticheskii sbornik. Ufa, "Statistika," 1972. 251 p. (To the 50th
anniversary of the formation of the USSR).   4

_____. _____. Bashkiriia za 50 let: statisticheskii sbornik. Ufa, "Statistika,"
Bashkirskoe otdelenie, 1969. 134 p.   4

_____. _____. Narodnoe khoziaistvo Bashkirskoi ASSR [1917-1967]: statisticheskii
sbornik. Ufa, "Statistika," 1967 . 278 p.   4

_____. _____. Narodnoe khoziaistvo i kul'turnoe stroitel'stvo Bashkirskoi ASSR:
statisticheskii sbornik. Ufa, "Statistika," 1964. 290 p.   4

_____. _____. Narodnoe khoziaistvo i kul'turnoe stroitel'stvo Bashkirskoi ASSR:
statisticheskii sbornik. Ufa, Gosstatizdat, 1959. 170 p.   4

BELGORODSKAIA Oblast' (Belgorod O.). Statisticheskoe Upravlenie. Narodnoe
khoziaistvo Belgorodskoi oblasti: statisticheskii sbornik. Orël, Gosstatizdat,
1959. 254 p.   4

_____. _____. Narodnoe khoziaistvo Belgorodskoi oblasti: statisticheskii sbornik.
Orël, Gosstatizdat, 1957. 166 p.   4

BRIANSKAIA Oblast' (Bryansk O.). Statisticheskoe Upravlenie. Narodnoe khoziaistvo
Brianskoi oblasti (1966-1970 gg.): statisticheskii sbornik. Bryansk, 1972. 344 p.
                                                                                    4

_____. _____. Narodnoe khoziaistvo Brianskoi oblasti: statisticheskii sbornik.
Orël, Gosstatizdat, 1962. 256 p .   4

_____. _____. Narodnoe khoziaistvo Brianskoi oblasti: statisticheskii sbornik.
Orël, Gosstatizdat, 1958. 195 p.   4

BURIATSKAIA ASSR (Buryat ASSR). Statisticheskoe Upravlenie. Narodnoe khoziaistvo
Buriatskoi ASSR: statisticheskii sbornik. Ulan-Ude, Buriatskoe Knizhnoe Izdatel'-
stvo, 1971. 256 p.   4

_____. _____. Narodnoe khoziaistvo Buriatskoi ASSR: statisticheskii sbornik. Ulan-
Ude, Buriatskoe Knizhnoe Izdatel'stvo, 1967. 214 p.   4

_____. _____. Buriatskaia ASSR za 50 let: statisticheskii sbornik. Ulan-Ude,
Buriatskoe Knizhnoe Izdatel'stvo, 1967. 95 p.   4

_____. _____. Narodnoe khoziaistvo Buriatskoi ASSR: statisticheskii sbornik.
Ulan-Ude, Buriatskoe Knizhnoe Izdatel'stvo, 1963. 240 p.   4

_____. _____. Narodnoe khoziaistvo Buriat-Mongol'skoi ASSR: statisticheskii sbornik
Ulan-Ude, Buriat-Mongol'skoe Knizhnoe Izdatel'stvo, 1957. 156 p.   4

448 - 465

CHECHENO-INGUSHSKAIA ASSR (Chechen-Ingush ASSR). Statisticheskoe Upravlenie.
50 let avtonomii Checheno-Ingushetii: statisticheskii sbornik. [Groznyy, Checheno-
Ingushskoe Knizhnoe Izdatel'stvo, 1972]. 239 p.
                                                                                466

____. ____. Narodnoe khoziaistvo Checheno-Ingushskoi ASSR za 1966-1970 gody:
statisticheskii sbornik. Groznyy, Checheno-Ingushskoe Knizhnoe Izdatel'stvo, 1971.
113 p.
                                                                                467

____. ____. Checheno-Ingushskaia ASSR za 50 let: statisticheskii sbornik.
[Groznyy, Checheno-Ingushskoe Knizhnoe Izdatel'stvo, 1967]. 184 p.
                                                                                468

____. ____. Narodnoe khoziaistvo Checheno-Ingushskoi ASSR: statisticheskii
sbornik. Groznyy, Checheno-Ingushskoe Knizhnoe Izdatel'stvo, 1963. 353 p.
                                                                                469

____. ____. Checheno-Ingushskaia ASSR za 40 let: statisticheskii sbornik.
Groznyy, Checheno-Ingushskoe Knizhnoe Izdatel'stvo, 1960. 185 p.
                                                                                470

____. ____. Narodnoe khoziaistvo Checheno-Ingushskoi ASSR: statisticheskii
sbornik. Grozny, Checheno-Ingushskoe Knizhnoe Izdatel'stvo, 1957. 132 p.
                                                                                471

CHELIABINSKAIA Oblast' (Chelyabinsk O.). Statisticheskoe Upravlenie. Narodnoe
khoziaistvo Cheliabinskoi oblasti: statisticheskii sbornik. Chelyabinsk,
"Statistika," [Cheliabinskoe otdelenie], 1971. 478 p.
                                                                                472

____. ____. Narodnoe khoziaistvo Cheliabinskoi oblasti: statisticheskii sbornik.
Chelyabinsk, "Statistika," Cheliabinskoe otdelenie, 1967. 430 p. Title on cover:
Cheliabinskaia oblast' v tsifrakh.
                                                                                473

____. ____. Narodnoe khoziaistvo Cheliabinskoi oblasti: statisticheskii sbornik.
Chelyabinsk, Gosstatizdat, 1961. 178 p.
                                                                                474

____. ____. Narodnoe khoziaistvo Cheliabinskoi oblasti i goroda Cheliabinska:
statisticheskii sbornik. Chelyabinsk, Gosstatizdat, 1957. 167 p.
                                                                                475

CHITINSKAIA Oblast' (Chita O.). Statisticheskoe Upravlenie. Narodnoe khoziaistvo
Chitinskoi oblasti: statisticheskii sbornik. Irkutsk, "Statistika," 1972. 256 p.
                                                                               476

____. ____. Narodnoe khoziaistvo Chitinskoi oblasti: statisticheskii sbornik.
[Irkutsk, "Statistika," Irkutskoe otdelenie, 1965]. 165 p.
                                                                                477

____. ____. Narodnoe khoziaistvo Chitinskoi oblasti: statisticheskii sbornik.
Irkutsk, 1960. 200 p.
                                                                                478

CHUVASHSKAIA ASSR (Chuvash ASSR). Statisticheskoe Upravlenie. Chuvashiia za 50 let:
statisticheskii sbornik. Cheboksary, Chuvashskoe Knizhnoe Izdatel'stvo, 1970.
132 p.
                                                                                479

____. ____. Chuvashiia za 50 let Sovetskoi vlasti v tsifrakh: statisticheskii
sbornik. Cheboksary, Chuvashskoe Knizhnoe Izdatel'stvo, 1967. 104 p.   480

____. ____. Sovetskaia Chuvashiia za 45 let v tsifrakh: statisticheskii sbornik.
Cheboksary, Chuvashskoe Knizhnoe Izdatel'stvo, 1965. 191 p.
                                                                                481

____. ____. Chuvashiia za 40 let v tsifrakh. Chuvashskoe Gosudarstvennoe
Izdatel'stvo, 1960. 195 p.
                                                                                482

____. ____. Narodnoe khoziaistvo Chuvashskoi ASSR: statisticheskii sbornik.
Cheboksary, Chuvashskoe Gosudarstvennoe Izdatel'stvo, 1957. 155 p.
                                                                                483

DAGESTANSKAIA ASSR (Dagestan ASSR). Statisticheskoe Upravlenie. Narodnoe
khoziaistvo Dagestanskoi ASSR k 50-letiiu obrazovaniia SSR: [iubileinyi
statisticheskii sbornik]. Makhachkala, Dagestanskoe Knizhnoe Izdatel'stvo,
1972 [1973]. 222 p.

_____. _____. Dagestan za 50 let: statisticheskii sbornik. Makhachkala, "Statis-
tika," 1967. 106 p.

    See also the city of Makhachkala.

_____. _____. Narodnoe khoziaistvo Dagestanskoi ASSR: statisticheskii sbornik.
Makhachkala, ["Statistika," Dagestanskoe otdelenie], 1965. 122 p.

_____. _____. Sovetskii Dagestan za 40 let: statisticheskii sbornik. Makhachkala,
Gosstatizdat, 1960. 158 p.

_____. _____. Narodnoe khoziaistvo Dagestanskoi ASSR: statisticheskii sbornik.
Makhachkala, Dagstatizdat--Dagknigoizdat, 1958. 120 p.

GOR'KOVSKAIA Oblast' (Gorkiy O.). Statisticheskoe Upravlenie. Narodnoe khoziaistvo
Gor'kovskoi oblasti za piatiletku (1966-1970 gg.): statisticheskii sbornik.
Gor'kiy, "Statistika," 1971. 211 p.

_____. _____. Narodnoe khoziaistvo Gor'kovskoi oblasti za 50 let: statisticheskii
sbornik. Gor'kiy, "Statistika," 1967. 141 p.

_____. _____. Narodnoe khoziaistvo Gor'kovskoi oblasti: statisticheskii sbornik.
Gor'kiy, "Statistika," 1965. 184 p.

_____. _____. Narodnoe khoziaistvo Gor'kovskoi oblasti: statisticheskii sbornik.
Gor'kiy, Gosstatizdat, 1960. 245 p.

IAKUTSKAIA ASSR (Yakut ASSR). Statisticheskoe Upravlenie. Iakutiia za 50 let v
tsifrakh: [statisticheskii sbornik]. Yakutsk, ["Statistika,"] 1967. 174 p.

_____. _____. Narodnoe khoziaistvo Iakutskoi ASSR: statisticheskii sbornik.
Yakutsk, Gosstatizdat, 1964. 179 p.

IAROSLAVSKAIA Oblast' (Yaroslavl' O.). Statisticheskoe Upravlenie. Iaroslavskaia
oblast' za 50 let v tsifrakh: statisticheskii sbornik. [Yaroslavl'], Verkhne-
Volzhskoe Knizhnoe Izdatel'stvo, 1967. 80 p.

    See also the city of Yaroslavl'.

_____. _____. Iaroslavskaia oblast': kratkii statisticheskii sbornik. Yaroslavl'
Knizhnoe Izdatel'stvo, 1957. 96 p.

IRKUTSKAIA Oblast (Irkutsk O.). Statisticheskoe Upravlenie. Narodnoe khoziaistvo
Irkutskoi oblasti: statisticheskii sbornik [1962-1966 gg.]. Irkutsk, "Statistika,"
1967. 196 p.

_____. _____. Narodnoe khoziaistvo Irkutskoi oblasti: statisticheskii sbornik.
Irkutsk, Gosstatizdat, 1962. 262 p.

_____. _____. Irkutskaia oblast': kratkii ekonomiko-statisticheskii sbornik.
Irkutsk, Knizhnoe Izdatel'stvo, 1958. 166 p.

_____. _____. Razvitie otraslei narodnogo khoziaistva Irkutskoi oblasti: statis-
ticheskii sbornik. Irkutsk, Knizhnoe Izdatel'stvo, 1957. 199 p.

ANOVSKAIA Oblast' (Ivanovo O.). Statisticheskoe Upravlenie. Ivanovskaia oblast'
v vos'moi piatiletke: [1966-1970]: statisticheskii sbornik. Ivanova ["Statistika,"
Ivanovskoe otdelenie], 1971. 307 p.   501

See also city of Ivanovo.

___. ____. Ivanovskaia oblast' za 50 let: statisticheskii sbornik. Ivanovo,
"Statistika," Ivanovskoe otdelenie, 1967. 278 p.   502

___. ____. Narodnoe khoziaistvo Ivanovskoi oblasti: statisticheskii sbornik.
Ivanovo, Gosstatizdat, Ivanovskoe otdelenie, 1962. 228 p.   503

___. ____. Narodnoe khoziaistvo Ivanovskoi oblasti: statisticheskii sbornik.
Moskva, Gosstatizdat, 1957. 171 p.   504

BARDINO-BALKARSKAIA SSR (Kabardino-Balkar or Kabardinian-Balkar ASSR).
Statisticheskoe Upravlenie. 50 let Kabardino-Balkarskoi ASSR: statisticheskii
sbornik. Nal'chik, "El'brus," 1971. 135 p.   505

___. ____. Narodnoe khoziaistvo Kabardino-Balkarskoi ASSR za gody Sovetskoi
vlasti: statisticheskii sbornik. Nal'chik, Kabardino-Balkarskoe Knizhnoe Izdatelstvo, 1967. 133 p.   506

___. ____. Narodnoe khoziaistvo Kabardino-Balkarskoi ASSR: statisticheskii
sbornik. Nal'chik, Kabardino-Balkarskoe Knizhnoe Izdatel'stvo, 1964. 211 p.   507

___. ____. Narodnoe khoziaistvo Kabardino-Balkarskoi ASSR: statisticheskii
sbornik. Nal'chik, Kabardino-Balkarskoe Knizhnoe Izdatel'stvo, 1957. 113 p.   508

ININGRADSKAIA Oblast' (Kaliningrad O.). Statisticheskoe Upravlenie.
aliningradskaia oblast' v vos'moi piatiletke: statisticheskii sbornik. Kalinin-
rad, Knizhnoe Izdatel'stvo, 1972. 255 p.   509

___. ____. Kaliningradskaia oblast' v tsifrakh. Kaliningrad, Knizhnoe Izdatel'-
tvo, 1968. 138 p.   510

___. ____. Kaliningradskaia oblast' v tsifrakh [1946-1967]. Kaliningrad,
nizhnoe Izdatel'stvo, 1967. 30 p.   511

ININSKAIA Oblast' (Kalinin O.). Statisticheskoe Upravlenie. Kalininskaia
blast' za gody vos'moi piatiletki (1966-1970 gg.) v tsifrakh: statisticheskii
bornik. Vologda, "Statistika," 1971 [1972]. 253 p.   512

___. ____. Kalininskaia oblast' za 50 let v tsifrakh: statisticheskii sbornik.
oskva, "Statistika," 1967. 159 p.   513

___. ____. Narodnoe khoziaistvo Kalininskoi oblasti v 1960 godu: statisticheskii
bornik. Moskva, Gosstatizdat, 1961. 183 p.   514

___. ____. Narodnoe khoziaistvo Kalininskoi oblasti: statisticheskii sbornik.
alinin, Knizhnoe Izdatel'stvo, 1957. 110 p.   515

MYTSKAIA ASSR (Kalmyk ASSR). Statisticheskoe Upravlenie. ·Kalmytskaia ASSR za
) let Sovetskoi vlasti: statisticheskii sbornik. Elista, Kalmytskoe Gosudar-
tvennoe Izdatel'stvo, 1967. 198 p.   516

___. ____. Narodnoe khoziaistvo Kalmytskoi ASSR: statisticheskii sbornik.
ista, Kalmizdat, 1960. 142 p.   517

___. ____. Dostizheniia Sovetskoi vlasti za sorok let (tsifry velikikh pobed
tsializma): statisticheskii sbornik. Elista, Kalmytskoe Knizhnoe Izdatel'stvo,
57. 31 p. In Kalmyk.   518

501 - 518

KALUZHSKAIA Oblast (Kaluga O.). Statisticheskoe Upravlenie. Kaluzhskaia oblast'
za 50 let: statisticheskii sbornik. Kaluga, 1967. 167 p.   51

_____. _____. Narodnoe khoziaistvo Kaluzhskoi oblasti: statisticheskii sbornik.
Moskva, ["Statistika,"] 1965. 199 p.   52

_____. _____. Narodnoe khoziaistvo Kaluzhskoi oblasti v 1959 godu: statisticheskii
sbornik. Moskva, Gosstatizdat, 1960. 192 p.   52

_____. _____. Narodnoe khoziaistvo Kaluzhskoi oblasti: statisticheskii sbornik.
Moskva, Gosstatizdat, 1957. 143 p.   52

KAMCHATSKAIA Oblast' (Kamchatka O.). Statisticheskoe Upravlenie. Narodnoe
khoziaistvo Kamchatskoi oblasti: statisticheskii sbornik. Petropavlovsk-
Kamchatskii, Dal'nevostochnoe Knizhnoe Izdatel'stvo, Kamchatskoe otdelenie.
1971. 202 p.   52

_____. _____. Narodnoe khoziaistvo Kamchatskoi oblasti: statisticheskii sbornik.
Khabarovsk, "Statistika," Khabarovskoe otdelenie, 1966. 150 p.   52

KARACHAEVO-CHERKESSKAIA AO (Karachayevo-Cherkes [Circassian] AO). Statisticheskoe
Upravlenie. Karachaevo-Cherkesiia k 50-letiiu SSSR: statisticheskii sbornik.
Cherkessk, Stavropol'skoe Knizhnoe Izdatel'stvo, Karachaevo-Cherkesskoe otdelenie,
1972. 205 p.   52

KAREL'SKAIA ASSR (Karelian ASSR). Statisticheskoe Upravlenie. Karel'skaia ASSR
za 50 let: statisticheskii sbornik. Petrozavodsk, "Statistika," Karel'skoe
otdelenie, 1967. 160 p.   52

_____. _____. 40 let Karel'skoi ASSR: statisticheskii sbornik. Petrozavodsk,
Gosstatizdat, 1960. 112 p.   52

_____. _____. Narodnoe khoziaistvo Karel'skoi ASSR: statisticheskii sbornik.
Petrozavodsk, [Gosstatizdat], 1957. 158 p.   52

KEMEROVSKAIA Oblast' (Kemerovo O.). Statisticheskoe Upravlenie. Kemerovskaia
ordenonosnaia: statisticheskii sbornik. Kemerovo, Knizhnoe Izdatel'stvo, 1968.
123 p.   52

_____. _____. Kemerovskaia oblast' v tsifrakh: statisticheskii sbornik.
Novosibirsk, "Statistika," Novosibirskoe otdelenie, [1966]. 264 p.   53

_____. _____. Kemerovskaia oblast' v tsifrakh: statisticheskii sbornik.
[Novosibirsk], 1961. 264 p.   53

_____. _____. Narodnoe khoziaistvo Kemerovskoi oblasti: statisticheskii sbornik.
Kemerovo, Knizhnoe Izdatel'stvo, 1958. 141 p.   53

KHABAROVSKII Krai (Khabarovsk Kray). Statisticheskoe Upravlenie. Narodnoe
khoziaistvo Khabarovskogo kraia: statisticheskii sbornik. [Khabarovsk], Knizhnoe
Izdatel'stvo, 1957. 128 p.   53

KIROVSKAIA Oblast' (Kirovo O.). Statisticheskoe Upravlenie. Narodnoe khoziaistvo
Kirovskoi oblasti: statisticheskii sbornik. Gor'kiy, "Statistika," 1971. 183 p.
                                                                                5
_____. _____. Kirovskaia oblast' k 50-letiiu Oktiabria: statisticheskii sbornik.
Gor'kii, "Statistika," 1967. 176 p.   5

_____. _____. Narodnoe khoziaistvo Kirovskoi oblasti: statisticheskii sbornik.
Gor'kiy, Gosstatizdat, 1960. 184 p.   5

_____. _____. Narodnoe khoziaistvo Kirovskoi oblasti: statisticheskii sbornik.
Kirov, Knizhnoe Izdatel'stvo, 1957. 136 p.   5

MI ASSR (Komi ASSR). Statisticheskoe Upravlenie. Komi ASSR za 50 let: statisticheskii sbornik. Syktyvkar, Komi Knizhnoe Izdatel'stvo, 1971. 234 p.   538

___. _____. Komi ASSR k 50-letiiu Sovetskoi vlasti: statisticheskii sbornik.
Syktyvkar, Komi Knizhnoe Izdatel'stvo, 1967. 192 p.   539

___. _____. Komi ASSR za 40 let: statisticheskii sbornik. Syktyvkar, Komi
Knizhnoe Izdatel'stvo, 1961. 200 p.   540

___. _____. Narodnoe khoziaistvo Komi ASSR: statisticheskii sbornik. Syktyvkar,
Komi Knizhnoe Izdatel'stvo, 1957. 175 p.   541

STROMSKAIA Oblast' (Kostroma O.). Statisticheskoe Upravlenie. Narodnoe
khoziaistvo Kostromskoi oblasti: statisticheskii sbornik. Ivanovo, "Statistika,"
Ivanovskoe otdelenie, 1969 [1970]. 221 p.   542

___. _____. Narodnoe khoziaistvo Kostromskoi oblasti: statisticheskii sbornik.
Kostroma, Knizhnoe Izdatel'stvo, 1956. 154 p.   543

ASNODARSKII Krai (Krasnodar Kray). Statisticheskoe Upravlenie. Krasnodarskii
krai za 50 let Sovetskoi vlasti: statisticheskii sbornik. Krasnodar, "Statistika,"
1967. 178 p.
                                                                                544
___. _____. Narodnoe khoziaistvo Krasnodarskogo kraia: statisticheskii sbornik.
Krasnodar, "Statistika," 1965. 462 p.   545

___. _____. Narodnoe khoziaistvo Krasnodarskogo kraia: statisticheskii sbornik.
Krasnodar, Gosstatizdat, 1958. 234 p.   546

ASNOIARSKII Krai (Krasnoyarsk Kray). Statisticheskoe Upravlenie. Narodnoe
khoziaistvo Krasnoiarskogo kraia: statisticheskii sbornik. Krasnoyarsk. Knizhnoe
Izdatel'stvo, 1967. 259 p.   547

___. _____. Narodnoe khoziaistvo Krasnoiarskogo kraia: statisticheskii sbornik.
Krasnoyarsk, "Krasnoiarskii Rabochii," 1958. 332 p.   548

BYSHEVSKAIA Oblast' (Kuybyshev O.). Statisticheskoe Upravlenie. Narodnoe
khoziaistvo Kuibyshevskoi oblasti za 1966-1970 gg.: statisticheskii sbornik.
Kuybyshev, Knizhnoe Izdatel'stvo, 1972. 271 p.   549

___. _____. Narodnoe khoziaistvo Kuibyshevskoi oblasti za 50 let: statisticheskii
sbornik. Kuybyshev, ["Statistika," Kuibyshevskoe otdelenie], 1967. 167 p.   550

___. _____. Narodnoe khoziaistvo Kuibyshevskoi oblasti za 1958-1959 gody:
statisticheskii sbornik. Kuybyshev, 1960. 175 p.   551

___. _____. Narodnoe khoziaistvo Kuibyshevskoi oblasti i goroda Kuibysheva:
statisticheskii sbornik. Kuybyshev, Gosstatizdat, 1957. 198 p.   552

GANSKAIA Oblast' (Kurgan O.). Statisticheskoe Upravlenie. Kurganskaia oblast'
za 50 let Sovetskoi vlasti: statisticheskii sbornik. [Chelyabinsk], Iuzhno-
Ural'skoe Knizhnoe Izdatel'stvo, 1967. 155 p.   553

___. _____. Narodnoe khoziaistvo Kurganskoi oblasti: statisticheskii sbornik.
Chelyabinsk, Gosstatizdat, Cheliabinskoe otdelenie, 1963. 270 p.   554

___. _____. Narodnoe khoziaistvo Kurganskoi oblasti: statisticheskii sbornik.
Chelyabinsk, Gosstatizdat, Cheliabinskoe otdelenie, 1957. 147 p.   555

KURSKAIA Oblast' (Kursk O.). Statisticheskoe Upravlenie. Narodnoe khoziaistvo
Kurskoi oblasti: statisticheskii sbornik. [Voronezh, "Statistika," 1968]. 289 p.
                                                                                    5

_____. _____. Narodnoe khoziaistvo Kurskoi oblasti: statisticheskii sbornik.
Orël, Gosstatizdat, 1960. 139 p.   5

_____. _____. Narodnoe khoziastvo Kurskoi oblasti: statisticheskii sbornik.
Orël, Gosstatizdat, 1958. 199 p.   5

LENINGRADSKAIA Oblast' (Leningrad O.). Statisticheskoe Upravlenie. Leningrad i
Leningradskaia oblast' v tsifrakh: statisticheskii sbornik. Leningrad, Lenizdat,
1971. 263 p.   5

_____. _____. Leningradskaia oblast' za 50 let: statisticheskii sbornik. Leningrad
Lenizdat, 1967. 184 p.   5

    See also the city of Leningrad.

_____. _____. Leningrad i Leningradskaia oblast' v tsifrakh: statisticheskii
sbornik. Leningrad, Lenizdat, 1964. 251 p.   5

_____. _____. Leningrad i Leningradskaia oblast' v tsifrakh: statisticheskii
sbornik. Leningrad. Lenizdat, 1961. 287 p.   5

_____. _____. Narodnoe khoziaistvo Leningradskoi oblasti: statisticheskii sbornik.
Moskva, Gosstatizdat, 1957. 142 p.   5

LIPETSKAIA Oblast' (Lipetsk O.). Statisticheskoe Upravlenie. Narodnoe khoziaistvo
Lipetskoi oblasti za gody Sovetskoi vlasti: statisticheskii sbornik. Voronezh,
Tsentral'no-Chernozëmnoe Knizhnoe Izdatel'stvo, 1967. 166 p.   5

_____. _____. Narodnoe khoziaistvo Lipetskoi oblasti: statisticheskii sbornik.
Lipetsk, Knizhnoe Izdatel'stvo, 1959. 183 p.   5

MAGADANSKAIA Oblast' (Magadan O.). Statisticheskoe Upravlenie. Narodnoe khoziaistvo
Magadanskoi oblasti za gody Sovetskoi vlasti: statisticheskii sbornik. Magadan,
1967. 104 p.   5

_____. _____. Narodnoe khoziaistvo Magadanskoi oblasti: statisticheskii sbornik.
Magadan, 1960. 110 p.   5

MARIISKAIA ASSR (Mari ASSR). Statisticheskoe Upravlenie. Mariiskaia ASSR za 50
let: statisticheskii sbornik. Yoshkar-Ola, Marknigoizdat, 1970. 127 p.   5

_____. _____. Mariiskaia ASSR v tsifrakh: statisticheskii sbornik. Yoshkar-Ola,
[Mariiskoe Knizhnoe Izdatel'stvo], 1967. 209 p.   5

_____. _____. Narodnoe khoziaistvo Mariiskoi ASSR: statisticheskii sbornik.
Yoshkar-Ola, 1960. 220 p.   5

MORDOVSKAIA ASSR (Mordvinian ASSR). Statisticheskoe Upravlenie. Mordovskaia ASSR
za gody Sovetskoi vlasti v tsifrakh [1917-1967]: statisticheskii sbornik. Saransk,
Mordovskoe Knizhnoe Izdatel'stvo, 1967. 195 p.   5

_____. _____. Narodnoe khoziaistvo Mordovskoi ASSR: statisticheskii sbornik.
Saransk, 1965 [1966], 167 p.   5

_____. _____. Narodnoe khoziaistvo Mordovskoi ASSR: statisticheskii sbornik.
Saransk, 1960. 119 p.   5

_____. _____. Narodnoe khoziaistvo Mordovskoi ASSR: statisticheskii sbornik.
Saransk, Mordovskoe Knizhnoe Izdatel'stvo, 1958. 143 p.   5

556 - 574

)SKOVSKAIA Oblast' (Moscow O.). Statisticheskoe Upravlenie. Narodnoe khoziaistvo
Moskovskoi oblasti (1966-1970 gg.): statisticheskii sbornik. Moskva, "Statistika,"
1972. 199 p.   575
    See also the city of Moscow

___. ___. Moskovskaia oblast' v tsifrakh: statisticheskii sbornik. Moskva,
"Statistika," 1970. 166 p.   576

___. ___. Moskvoskaia oblast' za 50 let: statisticheskii sbornik. [Moskva,
"Statistika," 1967]. 256 p.   577

___. ___. Narodnoe khoziaistvo Moskovskoi oblasti: statisticheskii sbornik.
Moskva, "Statistika," 1964. 152 p.   578

___. ___. Narodnoe khoziaistvo Moskovskoi oblasti: statisticheskii sbornik.
Moskva, "Moskovskii Rabochii," 1958. 271.   579

JRMANSKAIA Oblast'(Murmansk O.). Statisticheskoe Upravlenie. Narodnoe khoziaistvo
Murmanskoi oblasti za 50 let Sovetskoi vlasti: [statisticheskii sbornik]. Murmansk,
Knizhnoe Izdatel'stvo, 1967. 95 p.   580

___. ___. Narodnoe khoziaistvo Murmanskoi oblasti: statisticheskii sbornik.
Murmansk, [Gosstatizdat], 1957. 94 p.   581

)VGORODSKAIA Oblast' (Novgorod O.). Statisticheskoe Upravlenie. Novgorodskaia
oblast' za 50 let: statisticheskii spravochnik. Novgorod, "Novgorodskaia Pravda,"
[1967]. 103 p.   582

___. ___. Narodnoe khoziaistvo Novgorodskoi oblasti: statisticheskii sbornik.
[Leningrad], "Statistika," Leningradskoe otdelenie, 1965. 192 p.   583

___. ___. Narodnoe khoziaistvo Novgorodskoi oblasti: statisticheskii sbornik.
Moskva, Gosstatizdat, 1958. 164 p.   584

___. ___. Novgorodskaia oblast' za 40 let Sovetskoi vlasti (1917-1957):
statisticheskii sbornik. Novgorod, "Novgorodskaia Pravda," 1957. 51 p.   585

)VOSIBIRSKAIA Oblast' (Novosibirsk O.). Statisticheskoe Upravlenie. Novosibirskaia
oblast' za 50 let: statisticheskii sbornik. Novosibirsk, "Statistika," Novosibirskoe otdelenie, 1967. 181 p.   586

___. ___. Narodnoe khoziaistvo Novosibirskoi oblasti: statisticheskii sbornik.
Novosibirsk, Gosstatizdat, 1961. 334 p.   587

___. ___. Narodnoe khoziaistvo Novosibirskoi oblasti i goroda Novosibirska:
statisticheskii sbornik. Novosibirsk, Knizhnoe Izdatel'stvo, 1957. 192 p.   588

MSKAIA Oblast' (Omsk O.). Statisticheskoe Upravlenie. Narodnoe khoziaistvo Omskoi
oblasti: statisticheskii sbornik. Omsk, "Statistika," Omskoe otdelenie, 1971.   589
261 p.

___. ___. Narodnoe khoziaistvo Omskoi oblasti: statisticheskii sbornik. Omsk,
"Statistika," Omskoe otdelenie, 1969. 294 p.   590

___. ___. Narodnoe khoziaistvo Omskoi oblasti: statisticheskii sbornik [1958-
1965]. Omsk, "Statistika," Omskoe otdelenie, 1967. 275 p.   591

___. ___. Narodnoe khoziaistvo Omskoi oblasti i goroda Omska: statisticheskii
sbornik. Omsk, Gosstatizdat, 1957. 171 p.   592

ORENBURGSKAIA Oblast' (Orenburg O. [formerly Chkalov O.]). Statisticheskoe
Upravlenie. Orenburgskaia oblast' za 50 let Sovetskoi vlasti: statisticheskii
sbornik. Chelyabinsk, Iuzhno-Ural'skoe Knizhnoe Izdatel'stvo, 1967. 140 p.

_____. _____. Orenburgskaia oblast' za 25 let: statisticheskii sbornik. Orenburg,
Knizhnoe Izdatel'stvo, 1960. 203 p.

_____. _____. Narodnoe khoziaistvo Chkalovskoi oblasti: statisticheskii sbornik.
Chkalov [Orenburg], Chkalovskoe Knizhnoe Izdatel'stvo, 1957. 139 p.

ORLOVSKAIA Oblast' (Orël O.). Statisticheskoe Upravlenie. Narodnoe khoziaistvo
Orlovskoi oblasti: statisticheskii sbornik. Orël, "Statistika," Orlovskoe
otdelenie, 1972. 230 p.

_____. _____. Narodnoe khoziaistvo Orlovskoi oblasti: statisticheskii sbornik.
Orël, "Statistika," 1967. 230 p.

_____. _____. Narodnoe khoziaistvo Orlovskoi oblasti: statisticheskii sbornik.
Orël, Gosstatizdat, 1960. 282 p.

_____. _____. Narodnoe khoziaistvo Orlovskoi oblasti: statisticheskii sbornik.
Orël, Gosstatizdat, 1957. 136 p.

PENZENSKAIA Oblast' (Penza O.). Statisticheskoe Upravlenie. 50 let v edinoi
mnogonatsional'noi sem'e narodov SSSR. Penzenskaia oblast' v tsifrakh: statisticheskii sbornik. Penza, Privolzhskoe Knizhnoe Izdatel'stvo, Penzenskoe
otdelenie, 1972. 236 p.

_____. _____. Penzenskaia oblast' za 50 let Sovetskoi vlasti: statisticheskii
sbornik. Saratov-Penza, Privolzhskoe Knizhnoe Izdatel'stvo, 1967. 258 p.

_____. _____. Penzenskaia oblast' v tsifrakh: statisticheskii sbornik. Penza.
1963. 244 p.

_____. _____. Narodnoe khoziaistvo Penzenskoi oblasti: statisticheskii sbornik.
Penza. 1958. 191 p.

PERMSKAIA Oblast' (Perm' O. [formerly Molotov O.]). Statisticheskoe Upravlenie.
Permskaia oblast' v tsifrakh: statisticheskii sbornik. Sverdlovsk, "Statistika,"
Sverdlovskoe otdelenie, 1970. 246 p.

_____. _____. Narodnoe khoziaistvo Permskoi oblasti za gody Sovetskoi vlasti:
statisticheskii sbornik. Perm', Knizhnoe Izdatel'stvo, 1967. 269 p.

_____. _____. Narodnoe khoziaistvo Permskoi oblasti: statisticheskii sbornik.
Sverdlovsk, Gosstatizdat, 1961. 157 p.

_____. _____. Narodnoe khoziaistvo Molotovskoi oblasti: statisticheskoi sbornik.
Molotov [Perm'], Knizhnoe Izdatel'stvo, 1957. 201 p.

PRIMORSKII Krai (Primorskiy [Maritime] Kray). Statisticheskoe Upravlenie. Narodnoe
khoziaistvo Primorskogo kraia za gody Sovetskoi vlasti: statisticheskii sbornik.
Vladivostok, Dal'nevostochnoe Knizhnoe Izdatel'stvo, 1968. 171 p.

_____. _____. Narodnoe khoziaistvo Primorskogo kraia: statisticheskii sbornik.
Vladivostok, Primorskoe Knizhnoe Izdatel'stvo, 1958. 190 p.

PSKOVSKAIA Oblast' (Pskov O.). Statisticheskoe Upravlenie. Narodnoe khoziaistvo
Pskovskoi oblasti: statisticheskii sbornik. Leningrad, Lenizdat, 1968. 291 p.

_____. _____. Narodnoe khoziaistvo Pskovskoi oblasti: statisticheskii sbornik.
Leningrad, Gosstatizdat, 1960. 176 p.

593 - 611

AZANSKAIA Oblast' (Ryazan' O.). Statisticheskoe Upravlenie. Narodnoe khoziaistvo
Riazanskoi oblasti: statisticheskii sbornik. Moskva, "Statistika," 1967. 184 p. 612

___. ___. Narodnoe khoziaistvo Riazanskoi oblasti: statisticheskii sbornik.
Moskva, Gosstatizdat, 1958. 156 p.   613

STOVSKAIA Oblast' (Rostov O.). Statisticheskoe Upravlenie. Narodnoe khoziaistvo
Rostovskoi oblasti: statisticheskii sbornik. Rostov-na-Donu, "Statistika," Rostov-
skoe otdelenie, 1971. 214 p.   614

___. ___. Rostovskaia oblast' za 50 let: statisticheskii sbornik. Rostov-na-
Donu, "Statistika," Rostovskoe otdelenie, 1967. 195 p.   615

___. ___. Narodnoe khoziaistvo Rostovskoi oblasti: statisticheskii sbornik.
Rostov-na-Donu, "Statistika," 1964. 271 p.   616

___. ___. Narodnoe khoziaistvo Rostovskoi oblasti: statisticheskii sbornik.
Rostov-na-Donu, Gosstatizdat, Rostovskoe otdelenie, 1961. 238 p.   617

KHALINSKAIA Oblast' (Sakhalin O.). Statisticheskoe Upravlenie. Sakhalinskaia
oblast' v tsifrakh za gody vos'moi piatiletki: kratkii statisticheskii sbornik.
Yuzhno-Sakhalinsk, 1971. 96 p.   618

___. ___. Sakhalinskaia oblast' v tsifrakh za 1946-1966 gody: statisticheskii
sbornik. Yuzhno-Sakhalinsk [Dal'nevostochnoe Knizhnoe Izdatel'stvo], 1967. 136 p.
                                                                                                          619

___. ___. Narodnoe khoziaistvo Sakhalinskoi oblasti: statisticheskii sbornik.
Yuzhno-Sakhalinsk, Sakhalinskoe Knizhnoe Izdatel'stvo, 1960. 104 p.   620

RATOVSKAIA Oblast' (Saratov O.). Statisticheskoe Upravlenie. Narodnoe khoziaistvo
Saratovskoi oblasti za 50 let Sovetskoi vlasti: statisticheskii sbornik. Saratov,
"Statistika," Saratovskoe otdelenie, 1967. 454 p.   621

___. ___. Narodnoe khoziaistvo Saratovskoi oblasti v 1960 godu: statisticheskii
sbornik. Saratov, Gosstatizdat, 1962, 326 p.   622

___. ___. Narodnoe khoziaistvo Saratovskoi oblasti: statisticheskii sbornik.
[Saratov], Gosstatizdat, 1959. 205 p.   623

VERO-OSETINSKAIA ASSR (North Osset ASSR). Statisticheskoe Upravlenie. Severnaia
Osetiia v vos'moi piatiletke: statisticheskii sbornik. Ordzhonikidze, "Ir," 1972.
187 p.   624

___. ___. Severnaia Osetiia za gody Sovetskoi vlasti: statisticheskii sbornik.
Ordzhonikidze, Severo-Osetinskoe Knizhnoe Izdatel'stvo, 1967. 265 p.   625

___. ___. Narodnoe khoziaistvo k 40-letiiu avtonomii Severnoi Osetii: statis-
ticheskii sbornik. [Ordzhonikidze], "Statistika," Rostovskoe n/D otdelenie, 1964.
223 p.   626

___. ___. Narodnoe khoziaistvo Severo-Osetinskoi ASSR: statisticheskii sbornik.
Ordzhonikidze, 1958. 131 p.   627

OLENSKAIA Oblast' (Smolensk O.). Statisticheskoe Upravlenie. Narodnoe khoziaistvo
Smolenskoi oblasti v 1970 godu: statisticheskii sbornik. [Smolensk], "Statistika,"
1972. 198 p.   628

___. ___. Smolenskaia oblast' za 50 let Sovetskoi vlasti: statisticheskii
sbornik. [Moskva], "Statistika," 1967. 175 p.   629

_____. _____. Narodnoe khoziaistvo Smolenskoi oblasti: statisticheskii sbornik.
Moskva, Gosstatizdat, 1963. 238 p.   630

_____. _____. Narodnoe khoziaistvo Smolenskoi oblasti za 1957 g.: statisticheskii
sbornik. Smolensk, Knizhnoe Izdatel'stvo, 1958. 160 p.   631

_____. _____. Narodnoe khoziaistvo Smolenskoi oblasti: statisticheskii sbornik.
[Smolensk], Knizhnoe Izdatel'stvo, 1957. 107 p.   632

STAVROPOL'SKII Krai (Stavropol' Kray). Statisticheskoe Upravlenie. Narodnoe
khoziaistvo Stavropol'skogo kraia: statisticheskii sbornik. Stavropol', Knizhnoe
Izdatel'stvo, 1972. 141 p.   633

_____. _____. Stavropol'e za 50 let: (1917-1967) sbornik statisticheskikh mate-
rialov. Stavropol', Knizhnoe Izdatel'stvo, 1968. 219 p.   634

_____. _____. Narodnoe khoziaistvo Stavropol'skogo kraia: statisticheskii sbornik.
Krasnodar, Gosstatizdat, Krasnodarskoe otdelenie, 1959. 310 p.   635

SVERDLOVSKAIA Oblast' (Sverdlovsk O.). Statisticheskoe Upravlenie. Sverdlovskaia
oblast' v tsifrakh, 1966-1970 gg.: statisticheskii sbornik. Sverdlovsk, Statis-
tika," Sverdlovskoe otdelenie, 1971. 148 p.   636

_____. _____. Narodnoe khoziaistvo Sverdlovskoi oblasti: statisticheskii sbornik.
K 50-letiiu Velikoi Oktiabr'skoi sotsialisticheskoi revoliutsii. Sverdlovsk,
"Statistika," Sverdlovskoe otdelenie, 1967. 147 p.   637

_____. _____. Narodnoe khoziaistvo Sverdlovskoi oblasti: statisticheskii sbornik.
[Sverdlovsk], Gosstatizdat, 1962. 231 p.   638

_____. _____. Narodnoe khoziaistvo Sverdlovskoi oblasti i goroda Sverdlovska:
statisticheskii sbornik. Sverdlovsk, Gosstatizdat, 1956. 151 p.   639

TAMBOVSKAIA Oblast' (Tambov O.) Statisticheskoe Upravlenie. Tambovskaia oblast'
za 50 let Sovetskoi vlasti: statisticheskii sbornik. Tambov, "Tambovskaia
Pravda," [1967]. 187 p.   640

_____. _____. Narodnoe khoziaistvo Tambovskoi oblasti: statisticheskii sbornik.
Voronezh, 1961. 250 p.   641

_____. _____. Narodnoe khoziaistvo Tambovskoi oblasti: statisticheskii sbornik.
Tambov, "Tambovskaia Pravda," 1957. 188 p.   642

TATARSKAIA ASSR (Tatar ASSR). Statisticheskoe Upravlenie. Narodnoe khoziaistvo
Tatarskoi ASSR: statisticheskii sbornik. Kazan', Soiuzuchetizdat, Tatarskoe
otdelenie, 1972. 267 p.   643

_____. _____. Narodnoe khoziaistvo Tatarskoi ASSR k 50-letiiu so dnia obrazovaniia:
statisticheskii sbornik. Kazan', "Statistika," 1970. 195 p. Title on cover: 50
let Tatarskoi ASSR.   644

_____. _____. Dostizheniia Tatarskoi ASSR k 50-letiiu Sovetskoi vlasti: statis-
ticheskii sbornik. Kazan', "Statistika," 1967. 141 p. Title on cover: Tatariia
za 50 let.   645

_____. _____. Narodnoe khoziaistvo Tatarskoi ASSR: statisticheskii sbornik.
Kazan', "Statistika," 1966. 285 p.   646

_____. _____. Tatarskaia ASSR za 40 let: statisticheskii sbornik. Kazan',
Tatarskoe Knizhnoe Izdatel'stvo, 1960. 172 p.   647

_____. _____. Narodnoe khoziaistvo Tatarskoi ASSR: statisticheskii sbornik.
Kazan', Tatknigoizdat, 1957. 258 p.   648
630 - 648

IUMENSKAIA Oblast (Tyumen' O.). Statisticheskoe Upravlenie. Narodnoe khoziaistvo
Tiumenskoi oblasti za gody vos'moi piatiletki (1966-1970 gody): [statisticheskii
sbornik]. Omsk, "Statistika," 1971. 286 p.   649

____.   ____. Narodnoe khoziaistvo Tiumenskoi oblasti za 50 let Sovetskoi vlasti:
statisticheskii sbornik. Omsk, "Statistika," 1967. 301 p. Title on cover:
Tiumenskaia oblast' za 50 let.   650

____.   ____. Narodnoe khoziaistvo Tiumenskoi oblasti: statisticheskii sbornik.
Tyumen'--Omsk, "Statistika," 1964. 253 p.   651

____.   ____. Narodnoe khoziaistvo Tiumenskoi oblasti [i goroda Tiumeni: statis-
ticheskii sbornik]. Omsk, Gosstatizdat, 1958. 198 p.   652

OMSKAIA Oblast' (Tomsk O.). Statisticheskoe Upravlenie. Narodnoe khoziaistvo
Tomskoi oblasti: statisticheskii sbornik. Moskva, ["Statistika,"] 1965. 199 p. 653

____.   ____. Narodnoe khoziaistvo Tomskoi oblasti: statisticheskii sbornik.
Tomsk, 1957. 204 p.   654

UL'SKAIA Oblast' (Tula O.). Statisticheskoe Upravlenie. Narodnoe khoziaistvo
Tul'skoi oblasti: statisticheskii sbornik. Tula, Priokskoe Knizhnoe Izdatel'stvo,
1973. 304 p.   655

____.   ____. Narodnoe khoziaistvo Tul'skoi oblasti: statisticheskii sbornik
[1958-1965]. Tula, Priokskoe Knizhnoe Izdatel'stvo, 1967. 367 p.   656

____.   ____. Narodnoe khoziaistvo Tul'skoi oblasti: statisticheskii sbornik.
Tula, Knizhnoe Izdatel'stvo, 1958. 216 p.   657

UVINSKAIA ASSR (Tuvinian or Tuva ASSR). Statisticheskoe Upravlenie. Narodnoe
khoziaistvo Tuvinskoi ASSR: statisticheskii sbornik. Kyzyl, Tuvknigoizdat, 1971.
307 p.   658

____.   ____. Narodnoe khoziaistvo Tuvinskoi ASSR: statisticheskii sbornik. Kyzyl,
[Tuvinskoe Knizhnoe Izdatel'stvo], 1967. 205 p.   659

____.   ____. Narodnoe khoziaistvo Tuvinskoi ASSR: statisticheskii sbornik. Kyzyl,
Tuvknigoizdat, 1962. 260 p.   660

DMURTSKAIA ASSR (Udmurt ASSR). Statisticheskoe Upravlenie. UASSR za 50 let:
statisticheskii sbornik. [Izhevsk, "Udmurtiia," 1970]. 190 p.   661

____.   ____. Udmurtskaia ASSR za 40 let: statisticheskii sbornik. Izhevsk,
Udmurtskoe Knizhnoe Izdatel'stvo, 1960. 215 p.   662

____.   ____. Narodnoe khoziaistvo Udmurtskoi ASSR: statisticheskii sbornik.
Izhevsk, [Udmurtskoe Knizhnoe Izdatel'stvo], 1957. 135 p.   663

L'IANOVSKAIA Oblast' (Ul'yanovsk O.). Statisticheskoe Upravlenie. Rodina V. I.
Lenina: ekonomiko-statisticheskii sbornik. Moskva, "Statistika," 1970. 183 p.   664

____.   ____. Ul'ianovskaia ordena Lenina oblast' za 50 let Sovetskoi vlasti:
statisticheskii sbornik. Ul'yanovsk, Privolzhskoe Knizhnoe Izdatel'stvo
Ul'ianovskoe otdelenie, 1967. 381 p.   665

____.   ____. Narodnoe khoziaistvo Ul'ianovskoi oblasti: statisticheskii sbornik.
Ul'yanovsk, 1961. 271 p.   666

____.   ____. Narodnoe khoziaistvo Ul'ianovskoi oblasti: kratkii statisticheskii
sbornik. [Ul'yanovsk], Knizhnoe Izdatel'stvo, 1958. 200 p.   667

_____. _____. Narodnoe khoziaistvo Ul'ianovskoi oblasti: statisticheskii sbornik.
Ul'yanovsk, "Ul'ianovskaia Pravda," 1957. 273 p.   6

VELIKOLUKSKAIA Oblast' (Velikiye Luki O., --now part of Pskov O.), Statisticheskoe
Upravlenie. Narodnoe khoziaistvo Velikolukskoi oblasti: statisticheskii sbornik.
VelikiyeLuki, "Velikolukskaia Pravda," 1957. 127 p.   6

VLADIMIRSKAIA Oblast' (Vladimir O.). Statisticheskoe Upravlenie. Narodnoe khoziaistvo Vladimirskoi oblasti v vos'moi piatiletke (1966-1970 gody). Vladimir. 1972
308 p.   6

_____. _____. Vladimirskaia oblast' za 50 let: statisticheskii sbornik. Vladimir,
["Statistika,"], 1967. 144 p.   6
     See also the city of Vladimir.

_____. _____. Narodnoe khoziaistvo Vladimirskoi oblasti: statisticheskii sbornik.
Gor'kiy, Gosstatizdat, 1958. 171 p.   6

VOLGOGRADSKAIA Oblast' (Volgograd O. [formerly Stalingrad O.]). Statisticheskoe
Upravlenie. Narodnoe khoziaistvo Volgogradskoi oblasti v 1966-1971 gg.: statisticheskii sbornik. Saratov, Soiuzuchetizdat, Saratovskoe otdelenie, 1973. 289 p.
                                                                                     6

_____. _____. Narodnoe khoziaistvo Volgogradskoi oblasti za 50 let: statisticheski
sbornik. Volgograd, Nizhne-Volzhskoe Knizhnoe Izdatel'stvo, 1967. 263 p.   6

_____. _____. Narodnoe khoziaistvo Volgogradskoi oblasti: statisticheskii sbornik.
Saratov, Gosstatizdat, Saratovskoe otdelenie, 1962. 279 p.   6

_____. _____. Narodnoe khoziaistvo Stalingradskoi oblasti: statisticheskii sbornik
Saratov, Gosstatizdat, 1957. 319 p.   6

VOLOGODSKAIA Oblast' (Vologda O.). Statisticheskoe Upravlenie. Narodnoe khoziaistvo
Vologodskoi oblasti v vos'moi piatiletke: statisticheskii sbornik. [Vologda],
Severo-Zapadnoe Knizhnoe Izdatel'stvo, [Vologodskoe otdelenie], 1971. 188 p.   6

_____. _____. Narodnoe khoziaistvo Vologodskoi oblasti za gody Sovetskoi vlasti:
statisticheskii sbornik. Vologda, ["Statistika," Vologodskoe otdelenie], 1967.
168 p.   6

_____. _____. Narodnoe khoziaistvo Vologodskoi oblasti: statisticheskii sbornik.
Vologda, Gosstatizdat, Vologodskoe otdelenie, 1960 [1961]. 133 p.   6

VORONEZHSKAIA Oblast' (Voronezh O.). Statisticheskoe Upravlenie. Narodnoe
khoziaistvo Voronezhskoi oblasti: statisticheskii sbornik. Voronezh, "Statistika,"
Voronezhskoe otdelenie, 1972. 252 p.   6

_____. _____. Narodnoe khoziaistvo Voronezhskoi oblasti za 50 let Sovetskoi vlasti
statistichesk sbornik. Voronezh, "Statistika," Voronezhskoe otdelenie, 1967. 301
                                                                                     6
_____. _____. Narodnoe khoziaistvo Voronezhskoi oblasti v 1960 godu: statistichesk
sbornik. Voronezh, Gosstatizdat, 1961. 140 p.   6

_____. _____. Narodnoe khoziaistvo Voronezhskoi oblasti: statisticheskii sbornik.
Voronezh, Knizhnoe Izdatel'stvo, 1957. 144 p.   6

668 - 683

Ukrainian SSR

CHERKASSKAIA (Cherkas'ka) Oblast' (Cherkassy O.). Statisticheskoe Upravlenie.
Narodnoe khoziaistvo Cherkasskoi oblasti: statisticheskii sbornik. Cherkassy,
1970. 312 p. In Ukrainian.
684

____. ____. Narodnoe khoziaistvo Cherkasskoi oblasti: statisticheskii sbornik.
Cherkassy, 1957. 127 p. In Ukrainian.
685

CHERNIGOVSKAIA (Chernihivs'ka) Oblast' (Chernigov O.). Statisticheskoe Upravlenie.
Narodnoe khoziaistvo Chernigovskoi oblasti: statisticheskii sbornik. Kiev,
"Statistika," 1972. 274 p. In Ukrainian.
686

____. ____. Narodnoe khoziaistvo Chernigovskoi oblasti: statisticheskii sbornik.
[Chernigov, 1968]. 391 p.
687

CHERNOVITSKAIA (Chernivets'ka) Oblast' (Chernovtsy O.) Statisticheskoe Upravlenie.
Narodnoe khoziaistvo Chernovitskoi oblasti: statisticheskii sbornik. Kiev,
"Statistika," 1973. 188 p. In Ukrainian.
688

____. ____. Narodnoe khoziaistvo Chernovitskoi oblasti: statisticheskii sbornik.
Kiev, "Statistika," 1967. 335 p. In Ukrainian.
689

____. ____. Narodnoe khoziaistvo Chernovitskoi oblasti: statisticheskii sbornik.
Chernovtsy, 1959. 172 p. In Ukrainian.
690

DNEPROPETROVSKAIA (Dnipropetrovs'ka) Oblast' (Dnepropetrovsk O.). Statisticheskoe
Upravlenie. Narodnoe khoziaistvo Dnepropetrovskoi oblasti v 1970 godu: statisticheskii sbornik. Donetsk, "Statistika," 1971. 176 p. In Ukrainian.
691

____. ____. Narodnoe khoziaistvo Dnepropetrovskoi oblasti: statisticheskii
sbornik. Donetsk, "Statistika," 1966. 282 p. In Ukrainian.
692

____. ____. Narodnoe khoziaistvo Dnepropetrovskoi oblasti: statisticheskii
sbornik [Dnepropetrovsk], Knizhnoe Izdatel'stvo, 1960. 221 p. In Ukrainian.
693

DONETSKAIA (Donets'ka) Oblast' (Donets O.). Statisticheskoe Upravlenie. Narodnoe
khoziaistvo Donetskoi oblasti: statisticheskii sbornik. Donetsk, Soiuzuchetizdat,
[Donetskoe otdelenie], 1972. 246 p.
694

____. ____. Donetskaia oblast' za 50 let: statisticheskii sbornik. Donetsk,
"Statistika," Donetskoe otdelenie, 1967. 232 p.
695

____. ____. Narodnoe khoziaistvo Donetskoi oblasti: statisticheskii sbornik.
Donetsk, "Statistika," 1966. 196 p.
696

DROGOBYCHSKAIA (Drohobychs'ka) Oblast' (Drogobych O., now part of L'vov O.).
Statisticheskoe Upravlenie. Narodnoe khoziaistvo Drogobycheskoi oblasti: statisticheskii sbornik. [Drogobych, Oblastnoe Izdatel'stvo, 1958]. 159 p. In Ukrainian.
697

IVANO-FRANKOVSKAIA (Ivano-Frankiys'ka) Oblast' (Ivano-Frankovsk O., formerly Stanislav
O.) Statisticheskoe Upravlenie. Narodnoe khoziaistvo Ivano-Frankovskoi oblasti v 1968
godu: statisticheskii sbornik. L'vov, "Statistika," L'vovskoe otdelenie, 1970.
182 p. In Ukrainian.
698

____. ____. Narodnoe khoziaistvo Ivano-Frankovskoi oblasti v 1967 godu: statis-
ticheskii sbornik. L'vov, "Statistika," L'vovskoe otdelenie, 1968. 149 p. In
Ukrainian.
699.

_____ . _____ . Narodnoe khoziaistvo Stanislavskoi oblasti v 1961 godu: statisticheskii sbornik. L'vov, Gosstatizdat, L'vovskoe otdelenie, 1962. 125 p. In
Ukrainian.   700

_____ . _____ . Narodnoe khoziaistvo Stanislavskoi oblasti za 20 let Sovetskoi
vlasti: statisticheskii sbornik. L'vov, Gosstatizdat, 1960. 95 p. In Ukrainian.
701

KHAR'KOVSKAIA (Kharkivs'ka) Oblast' (Khar'kov O.). Statisticheskoe Upravlenie.
Narodnoe khoziaistvo Khar'kovskoi oblasti: statisticheskii sbornik. Khar'kov,
"Statistika," 1972. 236 p. (To the 50th anniversary of the formation of the
USSR). In Ukrainian.   702

_____ . _____ . Khar'kovshchina za 50 let: statisticheskii sbornik. Khar'kov,
"Statistika," 1967. 282 p. In Ukrainian.   703

_____ . _____ . Narodnoe khoziaistvo Khar'kovskoi oblasti: statisticheskii sbornik.
Khar'kov, "Statistika," 1965. 124 p. In Ukrainian.   704

KHERSONSKAIA (Khersons'ka) Oblast' (Kherson O.). Statisticheskoe Upravlenie.
Narodnoe khoziaistvo Khersonskoi oblasti: statisticheskii sbornik. Odessa,
"Statistika," Odesskoe otdelenie, [1969]. 202 p.   705

_____ . _____ . Narodnoe khoziaistvo Khersonskoi oblasti: statisticheskii sbornik.
Kherson, Knizhno-gazetnoe Izdatel'stvo; Odessa, Gosstatizdat, 1961 [1960].   706

KHMEL'NITSKAIA (Khmel'nyts'ka) Oblast' (Khmel'nitskiy O.). Statisticheskoe
Upravlenie. Narodnoe khoziaistvo Khmel'nitskoi oblasti: statisticheskii sbornik.
Kiev, "Statistika," 1972. 202 p. In Ukrainian.   707

_____ . _____ . Narodnoe khoziaistvo Khmel'nitskoi oblasti: statisticheskii sbornik.
Kiev, "Statistika," 1967. 286 p. In Ukrainian.   708

KIEVSKAIA (Kyïvs'ka) Oblast' (Kiev O.). Statisticheskoe Upravlenie. Narodnoe
khoziaistvo Kievskoi oblasti: statisticheskii sbornik. Kiev, "Statistika," 1967.
239 p. In Ukrainian.   709

     See also the city of Kiev.

_____ . _____ . Narodnoe khoziaistvo Kievskoi oblasti: statisticheskii sbornik.
Kiev, Gosstatizdat, 1959. 256 p.   710

KIROVOGRADSKAIA (Kirovohrads'ka) Oblast' (Kirovograd O.). Statisticheskoe Upravlenie.
Narodnoe khoziaistvo Kirovogradskoi oblasti: statisticheskii sbornik. Donetsk,
"Statistika," 1966. 276 p. In Ukrainian.   711

_____ . _____ . Narodnoe khoziaistvo Kirovogradskoi oblasti: statisticheskii sbornik.
Kirovograd, [Oblastnoe Izdatel'stvo], 1957. 196 p. In Ukrainian.   712

KRYMSKAIA (Kryms'ka) Oblast' (Crimean O.). Statisticheskoe Upravlenie. Narodnoe
khoziaistvo Krymskoi oblasti: statisticheskii sbornik. Odessa, "Statistika,"
1967. 178 p.   713

_____ . _____ . Ot s"ezda k s"ezdu. Krym v tsifrakh (1961-1965 gg.): kratkii
statisticheskii sbornik. Simferopol', "Krym," 1966. 97 p.   714

_____ . _____ . Narodnoe khoziaistvo Krymskoi oblasti: statisticheskii sbornik.
Simferopol', Krymizdat, 1957. 272 p.   71

L'VOVSKAIA (L'vivs'ka) Oblast' (L'vov O.). Statisticheskoe Upravlenie. Sovetskoe L'vovshchine-30 let: sbornik dokumentov i statisticheskikh materialov. L'vov, "Kameniar," 1970. 198 p. In Ukrainian.   716

____. ____. Narodnoe khoziaistvo L'vovskoi oblasti v 1965 godu: statisticheskii sbornik. L'vov, "Statistika," 1966. 282 p. In Ukrainian.   717

____. ____. L'vovskaia oblast' v tsifrakh v 1960 godu: kratkii sbornik. L'vov, Gosstatizdat, 1961. 179 p. In Ukrainian.   718

____. ____. Narodnoe khoziaistvo L'vovskoi oblasti: statisticheskii sbornik. L'vov, Gosstatizdat, 1958. 339 p. In Ukrainian.   719

NIKOLAEVSKAIA (Mykolaïvs'ka) Oblast' (Nikolayev O.). Statisticheskoe Upravlenie. Narodnoe khoziaistvo Nikolaevskoi oblasti: statisticheskii sbornik. Odessa, "Statistika," Odesskoe otdelenie, 1970. 137 p.   720

____. ____. Narodnoe khoziaistvo Nikolaevskoi oblasti: statisticheskii sbornik. Odessa, "Maiak," 1967. 182 p.   721

____. ____. Osnovnye pokazateli razvitiia narodnogo khoziaistva Nikolaevskoi oblasti. Nikolayev, 1966. 66 p.   722

____. ____. Narodnoe khoziaistvo Nikolaevskoi oblasti: statisticheskii sbornik. Nikolayev, Gosstatizdat, Odesskoe otdelenie, 1962. 173 p.   723

DESSKAIA (Odes'ka) Oblast' (Odessa O.). Statisticheskoe Upravlenie. Narodnoe khoziaistvo Odesskoi oblasti: statisticheskii sbornik. Odessa, Soiuzuchetizdat, [Odesskoe otdelenie], 1973. 238 p.   724

____. ____. Narodnoe khoziaistvo Odesskoi oblasti: statisticheskii sbornik. Odessa, "Statistika," Odesskoe otdelenie, 1969. 218 p.   725

____. ____. Narodnoe khoziaistvo Odesskoi oblasti: statisticheskii sbornik. Odessa, Gosstatizdat, 1960. 333 p.   726

OLTAVSKAIA (Poltavs'ka) Oblast' (Poltava O.). Statisticheskoe Upravlenie. Razvitie narodnogo khoziaistva Poltavshchiny: [statisticheskii sbornik]. Poltava, 1972. 45 p. In Ukrainian.   727

____. ____. Narodnoe khoziaistvo Poltavskoi oblasti: statisticheskii sbornik. Khar'kov, "Statistika," 1971. 197 p. In Ukrainian.   728

____. ____. Narodnoe khoziaistvo Poltavskoi oblasti: statisticheskii sbornik. Khar'kov, "Statistika," 1966. 198 p. In Ukrainian.   729

OVENSKAIA (Rovens'ka) Oblast' (Rovno O.). Statisticheskoe Upravlenie. Narodnoe khoziaistvo Rovenskoi oblasti: statisticheskii sbornik. L'vov, "Statistika," L'vovskoe otdelenie, 1970. 167 p. In Ukrainian.   730

____. ____. Narodnoe khoziaistvo Rovenskoi oblasti: statisticheskii sbornik. L'vov, Gosstatizdat, 1963. 187 p. In Ukrainian.   731

UMSKAIA (Sums'ka) Oblast' (Sumy O.) Statisticheskoe Upravlenie. Narodnoe khoziaistvo Sumskoi oblasti: statisticheskii sbornik. Khar'kov, "Statistika," Khar'kovskoe otdelenie. 1970. 508 p. In Ukrainian.   732

TERNOPOL'SKAIA (Ternopil'ska) Oblast' (Ternopol' O.). Statisticheskoe Upravlenie.
Narodnoe khoziaistvo Ternopol'skoi oblasti: statisticheskii sbornik. Kiev,
"Statistika," 1972. 222 p. In Ukrainian.   733

_____. _____. Narodnoe khoziaistvo Ternopol'skoi oblasti: statisticheskii sbornik.
Kiev, "Statistika," 1967. 204 p. In Ukrainian.   734

_____. _____. Narodnoe khoziaistvo Ternopol'skoi oblasti: statisticheskii sbornik.
L'vov, Gosstatizdat, 1962. 280 p. In Ukrainian.   735

_____. _____. Narodnoe khoziaistvo Ternopol'skoi oblasti: statisticheskii sbornik.
Ternopol',Oblastnoe Izdatel'stvo, 1957. 271 p. In Ukrainian.   736

VINNITSKAIA (Vinnyts'ka) Oblast' (Vinnitsa O). Statisticheskoe Upravlenie. Narodnoe
khoziaistvo Vinnitskoi oblasti: statisticheskii sbornik. Kiev, "Statistika," 1969.
240 p. In Ukrainian.   737

_____. _____. Vinnitskaia oblast' v tsifrakh: statisticheskii sbornik. Kiev,
"Statistika," 1964. 240 p. In Ukrainian.   738

_____. _____. Osnovnye pokazateli razvitiia narodnogo khoziaistva Vinnitskoi
oblasti: statisticheskii sbornik. Vinnitsa, 1957. 278 p. In Ukrainian.   739

VOLYNSKAIA (Volyns'ka) Oblast' (Volyn' or Volhynian O.). Statisticheskoe Upravlenie.
Volynskaia oblast' za gody Sovetskoi vlasti (1940-1966 gg.): statisticheskii
sbornik. L'vov, "Kameniar," 1969. 314 p. In Ukrainian.   740

_____. _____. Narodnoe khoziaistvo Volynskoi oblasti: statisticheskii sbornik.
L'vov, "Statistika," 1965. 236 p. In Ukrainian.   741

_____. _____. Narodnoe khoziaistvo Volynskoi oblasti: statisticheskii sbornik.
L'vov, Gosstatizdat, 1958. 212 p. In Ukrainian.   742

VOROSHILOVGRADSKAIA (Voroshilovhrads'ka) Oblast' (Voroshilovgrad, formerly Lugansk
O.). Statisticheskoe Upravlenie. Narodnoe khoziaistvo Voroshilovgradskoi oblasti:
statisticheskii sbornik. Donetsk, "Statistika," Donetskoe mezhoblastnoe otdelenie,
1971. 166 p.   743

_____. _____. Luganskaia oblast' za 50 let: sbornik statisticheskikh materialov.
Donetsk, "Statistika," 1967. 205 p.   744

_____. _____. Narodnoe khoziaistvo Luganskoi oblasti: statisticheskii sbornik.
Donetsk, Gosstatizdat, Donetskoe mezhoblastnoe otdelenie, 1963. 263 p.   745

ZAKARPATSKAIA (Zakarpats'ka) Oblast' (Trans-Carpathian O.). Statisticheskoe
Upravlenie. Narodnoe khoziaistvo Sovetskogo Zakarpat'ia: statisticheskii sbornik.
Uzhgorod, "Karpaty," 1969. 192 p. In Ukrainian.   746

_____. _____. Narodnoe khoziaistvo Zakarpatskoi oblasti v 1963 godu: kratkii
statisticheskii sbornik. L'vov, Gosstatizdat, 1964. 141 p. In Ukrainian.   747

_____. _____. Sovetskoe Zakarpat'e v tsifrakh: statisticheskii sbornik. Uzhgorod,
Zakarpatskoe Knizhno-gazetnoe Izdatel'stvo, 1960. 176 p. In Ukrainian.   748

_____. _____. Narodnoe khoziaistvo Zakarpatskoi oblasti: statisticheskii sbornik.
Uzhgorod, Zakarpatskoe Oblastnoe Izdatel'stvo, 1957. 168 p. In Ukrainian.   749

ZAPOROZHSKAIA (Zaporiz'ka) Oblast' (Zaporozh'ye O.). Statisticheskoe Upravlenie.
Narodnoe khoziaistvo Zaporozhskoi oblasti: statisticheskii sbornik. Donetsk,
Soiuzuchetizdat, [Donetskoe otdelenie], 1972. 256 p. In Ukrainian.   750

_____. _____. Narodnoe khoziaistvo Zaporozhskoi oblasti: statisticheskii sbornik.
Stalino [Donetsk], 1961. 229 p.   751

733 - 751

ZHITOMIRSKAIA (Zhitomyrs'ka) Oblast' (Zhitomir O.). Statisticheskoe Upravlenie.
<u>Narodnoe khoziaistvo Zhitomirskoi oblasti: statisticheskii sbornik</u>. Kiev,
"Statistika," 1972. 200 p. In Ukrainian.
752

_____. _____. <u>Narodnoe khoziaistvo Zhitomirskoi oblasti: statisticheskii sbornik</u>.
Kiev, "Statistika," 1968. 392 p. In Ukrainian.
753

_____. _____. <u>Narodnoe khoziaistvo Zhitomirskoi oblasti: statisticheskii sbornik</u>.
Zhitomir, Oblastnoe Izdatel'stvo, 1957. 150 p. In Ukrainian.
754

Georgian SSR
(See also the city of Tbilisi.)

ABKHAZSKAIA ASSR (Abkhaz ASSR). Statisticheskoe Upravlenie. <u>Narodnoe khoziaistvo</u>
<u>Abkhazskoi ASSR: statisticheskii sbornik</u>. Tbilisi, "Statistika," 1967. 357 p.   755

_____. _____. <u>Dostizheniia Sovetskoi Abkhazii za 40 let v tsifrakh: statisticheskii</u>
<u>sbornik</u>. Tbilisi, Gosstatizdat, 1961. 207 p.
756

_____. _____. <u>Narodnoe khoziaistvo Abkhazskoi ASSR: statisticheskii sbornik</u>.
Sukhumi, Gosstatizdat, 1960. 190 p.
757

_____. _____. <u>Narodnoe khoziaistvo Abkhazskoi ASSR: statisticheskii sbornik</u>.
Sukhumi, Abgosizdat, 1957. 117 p.
758

ADZHARSKAIA ASSR (Adzhar ASSR). Statisticheskoe Upravlenie. <u>40 let Sovetskoi</u>
<u>Adzharii: statisticheskii sbornik</u>. Batumi, Gosstatizdat, Gruzinskoe otdelenie,
1961. 164 p. In Georgian.
759

_____. _____. <u>Narodnoe khoziaistvo Adzharskoi ASSR: statisticheskii sbornik</u>.
Batumi, Gosudarstvennoe Izdatel'stvo Adzharskoi ASSR, 1958. 97 p. In Georgian. 760

IUGO-OSETINSKAIA Avtonomnaia Oblast' (South Osset AO). Statisticheskoe Upravlenie.
<u>Narodnoe khoziaistvo Iugo-Osetinskoi avtonomnoi oblasti: statisticheskii sbornik</u>.
Stalinir [Tskhinvali]--[Tbilisi], Gosstatizdat, Gruzinskoe otdelenie, 1960. 240 p.
761

Azerbaidzhan SSR
(See also the city of Baku)

NAGORNO-KARABAKHSKAIA Avtonomnaia Oblast' (Mountainous Karabakh AO). Statisticheskoe
Upravlenie. <u>Nagornyi Karabakh za gody Sovetskoi vlasti: kratkii statisticheskii</u>
<u>sbornik</u>. Stepanakert, 1968 [1969], 221 p.
762

_____. _____. <u>Dostizheniia Sovetskogo Nagornogo Karabakha za 40 let v tsifrakh:</u>
<u>statisticheskii sbornik</u>. Stepanakert, 1963. 171 p.
763

NAKHICHEVANSKAIA ASSR (Nakhichevan' ASSR). Statisticheskoe Upravlenie. <u>Narodnoe</u>
<u>khoziaistvo Nakhichevanskoi ASSR v 1970 godu: statisticheskii sbornik</u>. Baku, 1972.
137 p. In Azerbaijani.
764

_____. _____. <u>Razvitie narodnogo khoziaistva Nakhichevanskoi ASSR: statisticheskii</u>
<u>sbornik</u>. Baku, 1964. 142 p. In Azerbaijani.
765

Kazakh SSR

ALMA-ATINSKAIA Oblast' (Alma-Ata O.). Statisticheskoe Upravlenie. Narodnoe khoziaistvo Alma-Atinskoi oblasti: statisticheskii sbornik. Alma-Ata, "Statistika," 1967. 115 p.   766

KARAGANDINSKAIA Oblast' (Karaganda O.). Statisticheskoe Upravlenie. Narodnoe khoziaistvo Karagandinskoi oblasti: statisticheskii sbornik. Karaganda, 1970. 161 p.   767

____. ____. Narodnoe khoziaistvo Karagindinskoi oblasti: statisticheskii sbornik. Karaganda, 1967. 137 p.   768

SEVERO-KAZAKHSTANSKAIA Oblast' (North Kazakhstan O.). Statisticheskoe Upravlenie. Narodnoe khoziaistvo Severo-Kazakhstanskoi oblasti: statisticheskii sbornik. Petropavlovsk, 1962. 207 p.   769

VOSTOCHNO-KAZAKHSTANSKAIA Oblast' (East Kazakhstan O.). Statisticheskoe Upravlenie. Narodnoe khoziaistvo Vostochno-Kazakhstanskoi oblasti: statisticheskii sbornik. Ust'-Kamenogorsk, 1967. 264 p.   770

____. ____. Vostochnyi Kazakhstan v tsifrakh: statisticheskii sbornik. Alma-Ata, Gosstatizdat, 1962. 245 p.   771

Kirgiz SSR

FRUNZENSKAIA Oblast' (Frunze O., oblast now abolished). Statisticheskoe Upravlenie. Narodnoe khoziaistvo Frunzenskoi oblasti: statisticheskii sbornik. Frunze. 1957. 125 p. (To the 40th anniversary of the Great October socialist revolution).   772

See also the city of Frunze.

OSHSKAIA Oblast' (Osh O.). Statisticheskoe Upravlenie. Narodnoe khoziaistvo Oshskoi oblasti: statisticheskii sbornik. Osh, 1963. 196 p.   773

TIAN'-SHANSKAIA Oblast' (Tyan'-Shan' O., abolished in December, 1962; now Naryn O.). Statisticheskoe Upravlenie. Narodnoe khoziaistvo Tian'-Shanskoi oblasti: statisticheskii sbornik. Frunze, 1958. 112 p.   774

Uzbek SSR

KARAKALPAKSKAIA ASSR (Karakalpak ASSR). Statisticheskoe Upravlenie. Narodnoe khoziaistvo Karakalpakskoi ASSR: statisticheskii sbornik. Nukus, "Karakalpakiia," 1967. 134 p.   775

SAMARKANDSKAIA Oblast' (Samarkand O.). Statisticheskoe Upravlenie. Narodnoe khoziaistvo Samarkandskoi oblasti: statisticheskii sbornik. Samarkand, 1967. 358 p. Title on cover: Narodnoe khoziaistvo Samarkandskoi oblasti za 50 let.   776

____. ____. Narodnoe khoziaistvo Samarkandskoi oblasti: statisticheskii sbornik. Samarkand, 1958. 96 p.   777

TASHKENTSKAIA Oblast' (Tashkent O.). Statisticheskoe Upravlenie. Tashkentskaia oblast' za piat'desiat let Sovetskoi vlasti: statisticheskii sbornik. Tashkent, 1967. 382 p.   778

See also the city of Tashkent.

Turkmen SSR

CHARDZHOUSKAIA Oblast' (Chardzhou O.). Statisticheskoe Upravlenie. Narodnoe khoziaistvo Chardzhouskoi oblasti Turkmenskoi SSR: statisticheskii sbornik. Chardzhou, 1957. 97 p.   779

Statistical Handbooks on Individual Cities in the USSR

The cities are arranged in alphabetical order. Where there are both oblasts and cities with the same name, a distinction is made whether the statistical office issuing the handbook is that of the city or that of the oblast, but in the latter case the name of the oblast is given in the English form, which is identical with the name of the corresponding city.

ASTRAKHAN' (Oblast). Statisticheskoe Upravlenie. Osnovnye pokazateli narodnogo khoziaistva goroda Astrakhani: statisticheskii sbornik. Astrakhan', "Volga," 1958. 58 p.   780

BAKU. Statisticheskoe Upravlenie. Baku v tsifrakh: statisticheskii sbornik. Baku. 1967. 103 p.   781

_____. _____. Baku za 40 let v tsifrakh: statisticheskii sbornik. Baku, Azerbaidzhanskoe Gosudarstvennoe Izdatel'stvo, 1960. 116 p.   782

CHELYABINSK (Oblast). Statisticheskoe Upravlenie. Narodnoe khoziaistvo Cheliabinskoi oblasti i goroda Cheliabinska: statisticheskii sbornik. Chelyabinsk, 1957. 167 p.
                                                                                   783
DUSHANBE. Statisticheskoe Upravlenie. Gorod Dushanbe k 50-letiiu Velikoi Oktiabr'-skoi sotsialisticheskoi revoliutsii: [statisticheskii sbornik]. Dushanbe, "Statistika," 1967. 112 p.   784

EREVAN. See Yerevan.

FRUNZE (City). Statisticheskoe Upravlenie. Gorod Frunze v tsifrakh: kratkii statisticheskii sbornik. [Frunze], "Kyrgyzstan," 1972. 172 p.   785

IVANOVO (Oblast). Statisticheskoe Upravlenie. Gorod Ivanova za 100 let: statisticheskii sbornik. Ivanovo, 1971. 110 p.   786

KAUNAS. Statisticheskoe Upravlenie. Ekonomika i kul'tura Kaunasa: statisticheskii sbornik. Kaunas, "Shviesa," 1968. 230 p. In Lithuanian.   787

KIEV (City). Statisticheskoe Upravlenie. Kiev v tsifrakh: statisticheskii sbornik. Kiev, "Statistika," 1972. 117 p. In Ukrainian.   788

_____. _____. Kiev v tsifrakh: statisticheskii sbornik. Kiev, "Statistika," 1966. 162 p. In Ukrainian.   789

_____. _____. Narodnoe khoziaistvo goroda Kieva: statisticheskii sbornik. Kiev Gosstatizdat, 1963. 182 p.   790

_____. _____. Narodnoe khoziaistvo goroda Kieva: statisticheskii sbornik. Kiev, Gosstatizdat, 1960. 151 p. In Ukrainian.   791

KISHINËV. Statisticheskoe Upravlenie. Kishinëv: statisticheskii sbornik. Gosstatizdat, 1963. 192 p.   792

KUYBYSHEV (Oblast). Statisticheskoe Upravlenie. Narodnoe khoziaistvo Kuibyshevskoi oblasti i goroda Kuibysheva: statisticheskii sbornik. Kuybyshev, Gosstatizdat, 1957. 198 p.   793

LENINGRAD i Leningradskaia oblast' v tsifrakh: statisticheskii sbornik. I. D.
Kozlov, ed. Leningrad, Lenizdat, 1971. 263 p.   794

LENINGRAD (City). Statisticheskoe Upravlenie. Leningrad za 50 let: statisticheskii
sbornik. Leningrad, Lenizdat, 1967. 175 p. (Planovaia Komissiia Ispolkoma Len-
gorsoveta).   795

_____. _____. Leningrad i Leningradskaia oblast' v tsifrakh: statisticheskii
sbornik. Leningrad, Lenizdat, 1964. 251 p.   796

_____. _____. Leningrad i Leningradskaia oblast' v tsifrakh: statisticheskii
sbornik. Leningrad, Lenizdat, 1961. 287 p.   797

_____. _____. Narodnoe khoziaistvo goroda Leningrada: statisticheskii sbornik.
Moskva, Gosstatizdat, 1957. 163 p.   798

MAKHACHKALA za 50 let: statisticheskii sbornik. [I. P. Alékhin and A. N. Kazhdaev,
comp.]. Makhachkala, Dagestanskoe Knizhnoe Izdatel'stvo, 1968. 79 p.   799

MOSCOW (Moskva. City). Statisticheskoe Upravlenie. Moskva v tsifrakh (1966-
1970 gg.): kratkii statisticheskii sbornik. Moskva, "Statistika," 1972. 168 p. 800

_____. _____. Moskva v tsifrakh za gody Sovetskoi vlasti (1917-1967 gg.): kratkii
statisticheskii sbornik. Moskva, "Statistika," 1967. 167 p.   801

_____. _____. Moskva v tsifrakh (1959-1962 gg.): kratkii statisticheskii sbornik.
Moskva, "Statistika," 1964. 159 p.   802

NOVOSIBIRSK (Oblast). Statisticheskoe Upravlenie. Narodnoe khoziaistvo Novosibirskoi
oblasti i goroda Novosibirska: statisticheskii sbornik. Novosibirsk, Knizhnoe
Izdatel'stvo, 1957. 192 p.   803

OMSK (Oblast). Statisticheskoe Upravlenie. Narodnoe khoziaistvo Omskoi oblasti i
goroda Omska: statisticheskii sbornik. Omsk, Gosstatizdat, 1957. 171 p.   804

RIGA. Statisticheskoe Upravlenie. Riga: statisticheskii sbornik, [Riga], "Statistika," Latviiskoe otdelenie, 1968. 261 p.   805

_____. _____. Kratkii sbornik po gorodu Rige. Riga, "Statistika," 1967. 163 p. 806

_____. _____. Riga: statisticheskii sbornik. [Riga], 1963. 299 p.   807

SVERDLOVSK (Oblast). Statisticheskoe Upravlenie. Narodnoe khoziaistvo Sverdlovskoi
oblasti i goroda Sverdlovska: statisticheskii sbornik. Sverdlovsk, 1956. 151 p. 808

TASHKENT (City). Statisticheskoe Upravlenie. Narodnoe khoziaistvo goroda Tashkenta:
statisticheskii sbornik. Tashkent, Gosstatizdat, 1961. 112 p.   809

TBILISI. Statisticheskoe Upravlenie. Tbilisi. K 50-letiiu Velikoi Oktiabr'skoi
sotsialisticheskoi revoliutsii: statisticheskii sbornik. Tbilisi, 1967. 178 p.
In two editions: in Russian and in Georgian.   810

_____. _____. Tbilisi. K 40-letiiu Sovetskoi vlasti v Gruzii: statisticheskii
sbornik. Tbilisi, Gosstatizdat, Gruzinskoe otdelenie, 1961. 181 p.   811

TYUMEN' (Oblast). Statisticheskoe Upravlenie. Narodnoe khoziaistvo Tiumenskoi
oblasti i goroda Tiumeni: statisticheskii sbornik. Omsk, Gosstatizdat, 1958.
198 p.   812

VLADIMIR (Oblast). Statisticheskoe Upravlenie. Narodnoe khoziaistvo goroda Vladimira
statisticheskii sbornik. Vladimir, Knizhnoe Izdatel'stvo, 1958. 39 p.   813

 AROSLAVL' (Oblast). Statisticheskoe Upravlenie. Iaroslavl'. Razvitie khoziaistva
i kul'tura goroda: statisticheskii sbornik. Yaroslavl', [Knizhnoe Izdatel'stvo],
1961. 139 p.   814

EREVAN. Statisticheskoe Upravlenie. Erevan v tsifrakh: statisticheskii sbornik.
Yerevan, 1972. 173 p. (Dedicated to the 50th anniversary of the formation of the
USSR).   815

____. ____. Erevan v tsifrakh: statisticheskii sbornik. Yerevan, 1968. 223 p.
(Dedicated to the 2750th anniversary of Yerevan, capital of the Armenian SSR).   816

____. ____. Erevan k 50-letiiu Velikogo Oktiabria: statisticheskii sbornik.
Yerevan, 1967. 190 p.   817

____. ____. Erevan za 1958-1963 gody: statisticheskii sbornik. Yerevan, 1965.
197 p.   818

____. ____. Erevan za 40 let v tsifrakh: statisticheskii sbornik. Yerevan,
1960. 162 p.   819

Statistical Handbooks for Foreign Areas

Socialist Countries

OVET Ekonomicheskoi Vzaimopomoshchi [Council for Mutual Economic Assistance,
COMECON]. Sekretariat. Statisticheskii ezhegodnik stran-chlenov Soveta Ekonomicheskoi Vzaimopomoshchi 1970 (Statistical yearbook of the member countries of the Council for Mutual Economic Assistance 1970). Moskva, Sovet Ekonomicheskoi Vzaimopo-
moschi. 1970. 463 p.   820

____. ____. ____. 1971. 1971. 453 p.

____. ____. ____. 1972. 1972. 480 p. (Inserted title page and table of
contents in English: Council for Mutual Economic Assistance. Secretariat. Statistical yearbook of the Council for Mutual Economic Assistance 1972. 7 p.).

____. ____. ____. 1973. 1973. 511 p.

    Data for the Soviet Union, German Democratic Republic, Poland, Czechoslovakia,
Hungary, Romania, Bulgaria, and the Mongolian People's Republic. Principal sections: territory and population; general; industry; investments and construction;
agriculture and forestry; transport and communications; internal trade; external
trade; labor and wages; education, culture and arts; and public health.

    The 1973 yearbook has data for 1960, 1965, 1970, 1971, and 1972. The 1971
yearbook has data also for 1950, 1955, and for individual years 1966, 1967, 1968,
and 1969.

RODNOE khoziaistvo sotsialisticheskikh stran v 1961 godu: Soobshcheniia Statisticheskikh Upravlenii (Economy of socialist countries in 1961: reports of the
statistical offices). Moskva, Ekonomizdat, 1962. 180 p. (Akademiia Nauk SSSR.
Institut Ekonomiki Mirovoi Sotsialisticheskoi Sistemy).   821

____ v 1962 godu. 1963. 206 p.
____ v 1963 godu. "Ekonomika," 1964. 191 p.
____ v 1964 godu. ["Statistika"], 1965. 256 p.
____ v 1965 godu. "Statistika," 1966. 192 p.
____ v 1966 godu. 1967. 168 p.
____ v 1967 godu. 1968. 139 p.
____ v 1968 godu. 1969. 186 p.
____ v 1969 godu. 1970. 152 p.
____ v 1970 godu. 1971. 199 p.
____ v 1971 godu. 1972. 176 p.
____ v 1972 godu. 1973. 152 p.

Mainly textual reports by the statistical offices of the German Democratic
Republic, Poland, Czechoslovakia, Hungary, Romania, Bulgaria, the Mongolian
People's Republic, and the Soviet Union with one chapter devoted to each country
and with summary tables for each country of production of a few key items.

EKONOMIKA sotsialisticheskikh stran v tsifrakh. 1962 g.: kratkii statisticheskii
sbornik. (The economy of socialist countries in figures. 1962: short statistical
compilation). Moskva, Sotsekgiz, 1963. 262 p. (Akademiia Nauk SSSR. Institut
Ekonomiki Mirovoi Sotsialisticheskoi Sistemy).   82

_____. 1963 g. Moskva, Izdatel'stvo "Mysl'," 1964. 271 p.

_____. 1964 god. 1965. 247 p.

_____. 1965 g. 1966. 237 p.

EKONOMIKA stran sotsialisticheskogo lageria v tsifrakh 1960 g.: kratkii statisticheskii sbornik. (The economy of the countries of the socialist camp in figures of
1960: short statistical compilation). Moskva, Sotsekgiz, 1961. 239 p. (Akademiia
Nauk SSSR. Institut Ekonomiki Mirovoi Sotsialisticheskoi Sistemy).   82

MIR Sotsializma v tsifrakh i faktakh 1962 g. (The world of socialism in figures
and facts). Moskva, Politizdat, 1963. 111 p.   82.
_____ 1967. 1968. 158 p.
_____ 1968. 1969. 144 p.
_____ 1972. 1973. 175 p.

RAZVITIE ekonomiki stran narodnoi demokratii Evropy i Azii: statisticheskii sbornik
(Development of the economy of the people's democracies of Europe and Asia: statistical compilation). Moskva, Vneshtorgizdat, 1961. 471 p.   82

STRANY sotsializma i kapitalizma v tsifrakh: kratkii statisticheskii spravochnik
(The countries of socialism and capitalism in figures: short statistical handbook).
Politizdat, 1963. 208 p. 2nd ed., 1966. 223 p.   82

STRANY sotsializma i kapitalizma v tsifrakh: statisticheskie materialy dlia propa-
gandistov (The countries of socialism and capitalism in figures: statistical
materials for propagandists). Moskva, Gospolitizdat, 1957. 124 p.   82

RAZVITIE narodnogo khoziaistva Germanskoi Demokraticheskoi Respubliki: statisticheskie pokazateli (Development of the economy of the German Democratic Republic:
statistical indices). Moskva, Vneshtorgizdat, 1957. 348 p.   82

RAZVITIE narodnogo khoziaistva Chekhoslovakii: statisticheskii sbornik (Development
of the economy of Czechoslovakia: statistical compilation). Moskva, Gosplanizdat,
1959. 244 p.   82

RAZVITIE narodnogo khoziaistva Chekhoslovatskoi Respubliki: statisticheskie pokazatel
[za 1948-1956 gg.] (Development of the economy of the Czechoslovak Republic: statistical indices). Moskva, Vneshtorgizdat, 1958. 388 p.   83

RAZVITIE narodnogo khoziaistva Vengerskoi Narodnoi Respubliki: statisticheskii
pokazateli (Development of the economy of the Hungarian People's Republic: statistical indices). Moskva, Vneshtorgizdat, 1957. 172 p.   83

RAZVITIE narodnogo khoziaistva Rumynskoi Narodnoi Respubliki: statisticheskie
pokazateli [za 1951-1956 gg.] (Development of the economy of the Romanian People's
Republic: statistical indices for 1951-1956). Moskva, Vneshtorgizdat, 1958. 155 p.
                                                                                       83

AZVITIE narodnogo khoziaistva Narodnoi Respubliki Bolgarii: statisticheskie
pokazateli za 1948-1956 gg. Spravochnik (Development of the economy of the
People's Republic of Bulgaria: statistical indices for 1948-1956: handbook).
Moskva, Vneshtorgizdat, 1958. 232 p.   833

AZVITIE narodnogo khoziaistva Kitaiskoi Narodnoi Respubliki: statisticheskie
pokazateli (Development of the economy of the Chinese People's Republic: statistical indices). Moskva, Vneshtorgizdat, 1956. 51 p. (Ministerstvo Vneshnei
Torgovli SSSR. Nauchno-Issledovatel'skii Kon"iunkturnyi Institut).   834

AZVITIE narodnogo khoziaistva i kul'tury Koreiskoi Narodno-Demokraticheskoi
Respubliki v 1946-1957 gg. statisticheskii sbornik (Development of the economy
and culture of the Korean People's Democratic Republic in 1956-1957: statistical
compilation). Moskva, Gosplanizdat, 1959. 91 p.   835

JBA v tsifrakh: [statisticheskii sbornik] (Cuba in figures: statistical compilation). A. D. Bakarevich, and others, eds. Moskva, 1972. 102 p. (Akademiia
Nauk SSSR. Institut Latinskoi Ameriki).   836

Other Countries

ZIIA i Afrika, 1950-1962 gg.: statisticheskii sbornik (Asia and Africa, 1950-1962:
statistical compilation). Moskva, Izdatel'stvo "Nauka," 1964. 704 p. (Institut
Narodov Azii. Institut Afriki).   837

KONOMIKA razvivaiushchikhsia stran Azii v tsifrakh, 1960-1965 gg.: statisticheskii
sbornik (Economy of the developing countries of Asia in figures, 1960-1965:
statistical compilation). Moskva, Izdatel'stvo "Nauka," 1970. 486 p.   838

PONIIA: ekonomiko-statisticheskii spravochnik (Japan: an economic-statistical
handbook). N. K. Kutsobina, comp. Moskva, Izdatel'stvo "Mysl'," 1971. 239 p.   839

RIKA: statisticheskii sbornik (Africa: statistical compilation). G. M. Ushakova,
comp. V. V. Zhalnin and G. I. Rubinshtein, eds. Moskva, Izdatel'stvo "Nauka,"
Glavnaia Redaktsiia Vostochnoi Literatury, 1969. 278 p. (Akademiia Nauk SSSR.
Institut Afriki).   840

RIKA v tsifrakh: statisticheskii spravochnik (Africa in figures: a statistical
handbook). N. G. Kovalev, ed. Moskva, Sotsekgiz, 1963. 568 p.   841

    Section 2, p. 217-529, data on individual countries in alphabetical order.

VERNAIA Amerika. S.Sh.A. Kanada. Ekonomiko-statisticheskii spravochnik (North
America. The United States. Canada. Economic-statistical handbook). Moskva,
Izdatel'stvo "Mysl'," 1969. 303 p.   842

TINSKAIA Amerika v tsifrakh (do 1968 g.): statisticheskii sbornik. Moskva, 1971.
394 p. (Akademiia Nauk SSSR. Institut Latinskoi Ameriki).   843

ENCYCLOPEDIAS

Bibliography

UFMAN, Isaak Mikhailovich. Russkie entsiklopedii. Vypusk pervyi. Obshchie
entsiklopedii. Bibliografiia i kratkie ocherki (Russian encyclopedias. No. 1.
General encyclopedias. Bibliography and short articles). Moskva, 1960. 103 p.
(Gosudarstvennaia Biblioteka SSSR imeni Lenina).   844

Discussion of 22 general encyclopedias in Russian, 17 pre-Revolutionary and
5 post-Revolutionary, with rich bibliographical information, including the Bol'shaia
sovetskaia entsiklopediia (1st and 2nd eds.), Malaia sovetskaia
entsiklopediia (1st, 2nd, and 3rd eds.), Entsiklopedicheskii slovar', Granat, and
Entsiklopedicheskii slovar', Brokgauz i Efron, here listed individually below.

In Russian

ATKAIA geograficheskaia entsiklopediia (Short geographical encyclopedia). A. A.
Grigor'ev, chief ed. Editorial board: P. M. Alampiev, D. L. Armand, A. N. Baranov,
I. P. Gerasimov, P. I. Glushakov (v. 1-3), S. V. Kalesnik, D. M. Lebedev, E. N.
Lukashova, I. M. Maergoiz, F. N. Petrov, V. V. Pokshishevskii, K. M. Popov, G. D.
Rikhter, M. S. Rozin (v. 2-5), and V. P. Tikhomirov. Moskva, Izdatel'stvo
"Sovetskaia Entsiklopediia," 1960-1965. 5 v. 1362 illustrations, 634 maps (141
colored).   845

About 18,000 articles mostly on specific places or areas, such as countries,
provinces, regions, or cities, of physical features, such as continents, oceans,
seas, mountains, rivers, lakes, plains, or gulfs. Also contains articles on the
fields of geography and some principal geographical concepts and terms. The biographical index is the best Soviet source on Russian and Soviet geographers but
is limited to persons cited in the bibliographies in the encyclopedia. Brief
bibliographies. A section of geographical tables and a listing of peoples of the
world is in volume 5.

A major geographical reference work, it combines to some degree a gazetteer
in treating specific places, an encyclopedia in treating fields of geography, a
geographical dictionary in defining some terms, a biographical directory, and a
bibliography.

L'SHAIA sovetskaia entsiklopediia (The great Soviet encyclopedia) 1st ed. O. Iu.
Schmidt, ed. Moskva, "Sovetskaia Entsiklopediia," 1926-1947. 65 v. plus supplementary v. on the USSR, 1948.   846

About 65,000 articles, many not fully replaced by later editions. More than
1,000 maps. The article on geography, v. 15, cols. 253-281 (1929), includes sections on the nature of geography and its place among the sciences by L. S. Berg,
with added statement by V. Kamenetskii; history of the development of geographical
knowledge and geographical science by G. Shenberg; school geography by V. Kamenetskii, and chronological list of the most important geographical discoveries and
travels by M. S. Bodnarskii. The article on anthropogeography, v. 3, cols. 108-
13 (1926) is by A. A. Kruber and N. N. Baranskii; on physical geography, v. 57,
cols. 300-304 (1936), by M. A. Pervukhin; on economic geography, v. 63, cols.
241-249 (1933), by V. E. Motylëv.

The supplementary volume on the USSR (1948) includes a detailed bibliography,
p. xli-lxxx and sections on population, by E. Davydov; on natural conditions--
relief of the European USSR by B. F. Dobrynin, relief of the Asiatic USSR by S.
P. Suslov, geological structure by A. N. Mazarovich, hydrology by V. P. Semënov-
Tian-Shanskii, climate by A. P. Gal'tsov, soils by N. N. Rozov, vegetation by B.
K. Shishkin, animal life by V. G. Geptner, and mineral resources by N. A. Bykhover
and M. S. Rozin; and the economy, with extensive specialized bibliographies, p.
cli-xlvi and xix-lxxii.

The article on geography, by L. S. Berg and D. Tugarinov, cols. 1351-1362 in the section on the development of science in the USSR, summarizes the evolution of physical geography in Russia and the Soviet Union, both pre-Revolutionary and post-Revolutionary, with a short bibliography, p. lxxvi.

_____. 2nd ed. S. I. Vavilov and B. A. Vvedenskii, eds. Moskva, Izdatel'stvo "Bol'shaia Sovetskaia Entsiklopediia," 1950-1958. 51 v. plus index, 1960, 2 v.   8.

About 100,000 entries. v. 51 a supplement. The two index volumes are valuable in locating information not carried under main entries. The articles on geography, v. 10, p. 457-472 (1952), physical geography, v. 45, p. 58-61 (1956), and economic geography, v. 48, p. 375-380 (1957) are unsigned.

v. 50, Soiuz Sovetskikh Sotsialisticheskikh Respublik (Union of Soviet Socialist Republics), 1957, devoted to the USSR, is organized topically with sections on natural conditions (geographical zones by I. M. Zabelin, paleogeography of the Quaternary period by E. M. Shcherbakov, reliev by N. I. Mikhailov, geological structure by M. V. Muratov, minerals by A. A. Amiraslanov and others, climate by A. P. Gal'tsov, hydrography by V. E. Ioganson, soils by N. N. Rozov, vegetation by E. M. Lavrenko, and animal life by V. G. Geptner), population, and the economy, with short bibliographies after each topic. The short article on geography as a science, p. 480-483, is confined to physical geography. Available also in English as Robert Maxwell, Information USSR, 1962, and in German as Die UdSSR: Enzyklopädie der Union der Sozialistischen Sowjetrepubliken, 1959, q.v. under Handbooks in Part VI, Bibliographies and reference works on the Soviet Union in Western languages [entries 2551 and 2557].

_____. 3rd ed. A. M. Prokhorov, ed. Moskva, Izdatel'stvo "Sovetskaia Entsiklopediia," v. 1- (1969-). In progress. To be 30 v. with about 100,000 entries.   8.

A major international encyclopedia, particularly useful for detailed coverage of the Soviet Union, socialist countries, Soviet topics, and articles on scientific fields. Geography, v. 6, p. 271-275 (1971) is by S. V. Kalesnik, A. G. Isachenko, and V. V. Pokshishevskii. There are adjacent articles on geography of soils (M. A. Glazovskaia), geography of natural resources (I. V. Komar), geography of vegetation (A. I. Tolmachëv), geobotany (A. A. Uranov), geography of population (V. V. Pokshishevskii), political geography (B. N. Semevskii), historical geography (V. K. Iatsunskii), medical geography (I. I. Ëlkin), veterinary geography (M. G. Tarshis), geography of agriculture (A. N. Rakitnikov), geography of industry (I. M. Maergoiz), geography of transport (L. I. Vasilevskii), geography of service activities (V. V. Pokshishevskii), geography of the labor force (S. A. Kovalëv), geography of the world economy (M. S. Rozin). Articles on physical geography and economic geography will be published in later volumes, in their regular alphabetical place.

There are also articles on the geographical envelope (S. V. Kalesnik), G. pathology (A. P. Avtsyn), G. school in sociology (V. I. Korovikov), G. journals (I. I. Parkhomenko), G. institutes (L. V. Kravchenko), G. maps (K. A. Salishchev), G. congresses (V. V. Annenkov), G. names (E. M. Murzaev), G. societies (I. L. Kleopov), G. discoveries (N. G. Fradkin), G. encyclopedias (V. A. Nikolaev), G. determinism (P. M. Alampiev), G. possibilism (V. V. Pokshishevskii), G. cycle, G. education (A. I. Solov'ëv), G. location, G. division of labor (P. M. Alampiev), and geobotanical maps (T. I. Isachenko).

Individual articles are also devoted to major geographical institutions: Geographical Society of the USSR (S. V. Kalesnik), Institute of Geography of the Academy of Sciences of the USSR (M. I. Neishtadt), the Institute of Geography of Siberia and the Far East (V. V. Sochava), and the International Geographical Union (V. V. Annenkov).

An abridged English translation of the 3rd edition is being published by Macmillan, New York, 1973- , as Great Soviet encyclopedia [entry 2547].

___. __Ezhegodnik__ (Yearbook). 1957- . Annual.   849

Annual yearbook supplement with major sections devoted to the Soviet Union and its constituent republics; to foreign countries, arranged alphabetically; to international organizations and conferences; and to science and technology, with a chapter covering developments in geography during the year.

__IAIA sovetskaia entsiklopediia__ (The small Soviet encyclopedia). 3rd ed. B. A. Vvedenskii, ed. Moskva, Izdatel'stvo "Bol'shaia Sovetskaia Entsiklopediia," 1958-1961. 10 v. Index, 1 v. (1st ed., 1928-1931, 10 v. 2nd ed., 1933-1947, 11 v.).   850

Handy smaller encyclopedia with detailed index volume. About 48,000 entries. The index volume is especially detailed.

__3IRSKAIA sovetskaia entsiklopediia__ (The Siberian Soviet encyclopedia). v. 1-2. M. K. Azadovskii and others, eds. v. 3. B. Z. Shumiatskii, chief ed., Novosibirsk, Sibirskoe Kraevoe Izdatel'stvo, 1929-1933. v. 1-3. A-N only. No more published.

Encyclopedia on Siberia and adjacent areas. Bibliographical references   851
accompany many articles. Articles on bibliography of Siberia.

__ISIKLOPEDICHESKII slovar' t-va Br. A. i I. Granata i Ko.__ (Encyclopedia of the partnership Brothers A. and I. Granat and Co.). 7th ed. revised and enlarged. Edited by V. Ia Zheleznov, Iu. S. Gambarov, and others. Moskva, 1910-1948. v. 1-55, 57-58. Supplementary v. A-Bar. 1936. Incomplete, since last word covered alphabetically (end of v. 54) is Eshford. v. 1-33, 37-39, and 42, published before 1917; v. 34-36, 40-41, 43-55, and 57-58, after the Revolution. v. 36 consists of 7 sub-volumes; v. 41 of 10; and v. 45 of 3. v. 56 never published.   852

Mainly a pre-Revolutionary encyclopedia but the volumes published after the revolution contain some of the best information on the early Soviet years, particularly volume 41 in 3 parts, 1929, with biographical data on persons prominent in the years 1917-1927, and volumes 57 and 58, 1939-1940, consisting of a detailed survey of the USSR on the eve or World War II, entitled the Epoch of socialist reconstruction of the economy of the USSR.

Geography, v. 13, cols. 236-253 (1912), is a comprehensive treatment by D. N. Anuchin, with extensive international bibliography, and economic geography, v. 51, cols. 214-220 (1933), is somewhat polemical, though unsigned, with bibliography limited to Soviet works published in a single year, 1932.

__SIKLOPEDICHESKII slovar'__ (Encyclopedia). I. E. Andreevskii, K. K. Arsen'ev, F. F. Petrushevskii, and V. T. Sheviakov, eds. St. Petersburg, Leipzig, F. A. Brokgauz-I. A. Efron, 1890-1907. 43 v. in 86 parts.   853

Called the Brokgauz-Efron, this was the outstanding encyclopedia of pre-revolutionary Russia. It was an adaptation and revision of the Brockhaus, but by leading Russian scholars.

Geography, tom VIII (part 15), p. 377-390 (1892) by A. I. Voeikov, has an excellent bibliography of major works in each subdivision of geography with good representation of publications in Russian, German, French, and English.

"Rossiia," v. 54, p. 1-420 and v. 55, p. 421-874 (1899), separately paged and published also separately as __Rossiia eia nastoiashchee i proshedshee__, 1900. 889 p., with some supplementary material.

Partially updated by __Novyi entsiklopedicheskii slovar'__ (New encyclopedia). K. K. Arsen'ev, ed. St. Petersburg (Petrograd), F. A. Brokgauz-I. A. Efron, 1911-1916, v. 1-29. A-Otto only. No more published.   854

EKONOMICHESKAIA entsiklopediia: promyshlennost' i stroitel'stvo (Economic encyclopedia: industry and construction). A. N. Efimov, ed. Moskva, "Sovetskaia Entsikl
pediia," 1962-1965. 3 v.   8

EKONOMICHESKAIA entsiklopediia: politicheskaia ekonomiia (Economic encyclopedia:
political economy). A. M. Rumiantsev, chief ed. Moskva, "Sovetskaia Entsiklopedi
1972- . (v. 1. A-Indeksy, 1972). In progress. (Akademiia Nauk SSSR. Otdelenie
Ekonomiki. Entsiklopedii. Slovari. Spravochniki. Nauchno-Redaksionnyi Soviet:
A. M. Prokhorov, chairman).

SEL'SKOKHOZIAISTVENNAIA entsiklopediia (Agricultural encyclopedia). V. V. Matskevic
and P. P. Lobanov, eds. 4th ed., Moskva, "Sovetskaia Entsiklopediia," 1969-
(v. 1-4, 1969-1973. A-Pripusk). In progress. (3rd ed., Moskva, Gosudarstvennoe
Izdatel'stvo Selskokhoziaistvennoi literatury, 1949-1956. 5 v.).   8

SOVETSKAIA istoricheskaia entsiklopediia (Soviet historical encyclopedia). E. M.
Zhukov, chief ed. Moskva, Izdatel'stvo "Sovetskaia Entsiklopediia," 1961- .
(v. 1-14, A.-Feleo, 1961-1973). In progress. (Nauchno-Redaksionnyi Sovet
Izdatel'stva "Sovetskaia Entsiklopediia." Otdelenie Istorii Akademii Nauk SSSR). 8

The article on historical geography, v. 6, cols. 514-517 (1965) is by V. K.
Iatsunskii, on historical cartography, v. 6, cols. 517-523, by L. A. Gol'denberg,
on the geographical environment, v. 4, cols. 218-221 (1963), by I. S. Kon, and on
the Geographical Society of the USSR, v. 4, cols. 221-223, by N. L. Rubinshtein.
Thére are also articles on geographical exploration of the major regions of the
earth.

In Ukrainian

UKRAINS'KA radians'ka entsyklopediia (Ukrainian Soviet encyclopedia). M. P. Bazhan
ed. Kyïv, Holovna Redaktsiia Ukraïns'koï Radians'koï Entsyklopedii, 1959-1965.
17 v. 1 v. Index, 1968. (Akademiia Nauk Ukraïns'koï Radians'koï Sotsialistychn
Respubliky). (Russian: Ukrainskaia sovetskaia entsiklopediia).

Heohrafiia (geography), v. 3, p. 190-191 (1960), a general view, with biblio
graphy in Ukrainian and Russian. Ekonomichna heohrafiia (economic geography), v.
4, p. 409-410 (1961), by L. A. Ustynova and M. O. Khyliuk, treats the rise of
economic geography in the USSR, in Ukraine, and abroad, with a short bibliography in Russian and Ukrainian. Fizychna heohrafiia (physical geography), v. 15,
p. 257-258 (1964), by O. M. Marynych, is a general treatise with short bibliography in Russian.

v. 17, Ukraïns'ka Radians'ka Sotsialistychna Respublika (Ukrainian SSR), ha
extensive sections on natural conditions (relief, geological structure, minerals,
climate, waters, soils, vegetation, animal life, physical-geographic regionalization of Ukraine, the influence of the economic activity of man on natural conditions in the territory of Ukraine, and conservation), p. 9-66, population, p. 67-
69, and the economy, p. 268-423, by a large number of Ukrainian scholars and with
short bibliographies of works mainly in Ukrainian. The article on the science of
geography in this volume, p. 484-487, includes a section on physical geography by
O. M. Marynych and one on economic geography by M. M. Palamarchuk, both treating
the development of geography in Ukraine and the study of Ukraine and with a short
bibliography mainly in Ukrainian. This volume also has a detailed index of persons.

The encyclopedia is universal in scope and international in coverage and has
articles on principal countries of the world, and republics and oblasts in the
USSR.

ISYKLOPEDIIA Ukraïnoznavstva (Encyclopedia of Ukraine). Volodymyr Kubiĭovych, ed. München, New York, Paris, Vydavnytstvo "Molode Zhyttia," 1949- . 2 v. (v. 1, parts 1-3, 1949, 1230 p.; v. 2, parts 1-6, 1955-1970, A-Priashivshchnya, p. 1-2400). In progress. (Naukove Tovarystvo im. Shevchenka. Shevchenko Scientific Society, Inc.).

    v. 1. systematic arrangement with monographic articles.
    v. 2. arranged alphabetically with short articles.   860

    Physical geography, v. 1, part 1, p. 35-124; population, v. 1, part 1, p. 125-183; and the economy, v. 1, part 3, p. 1028-1130. [See entry 2560].

ISYKLOPEDIIA narodnoho hospodarstva Ukraïns'koï RSR (Encyclopedia of the economy of the Ukrainian SSR). S. M. Iampol's'kyĭ, and others, eds. Kyïv, Holovna Rekaktsiia Ukraïns'koï Radians'koï Entsyklopedii, 1969- . (v. 1-3, 1969-1971, A-Rob). In progress. In Ukrainian. (Akademiia Nauk Ukraïnskoï RSR). Russian title: Entsiklopediia narodnogo khoziaistva Ukrainskoi SSR. S. M. Iampol'skii, and others, eds. Kiev, Glavnaia Redaktsiia Ukrainskoi Sovetskoi Entsiklopedii, 1969- .   861

RAÏNS'KA sil's'kohospodars'ka entsyklopediia (Ukrainian agricultural encyclopedia). J. F. Peresypkin, and others, eds. Kyïv, Ukraïns'ka Radians'ka Entsyklopediia, 1970-1972. 3 v. (Akademiia Nauk Ukraïns'koï RSR). In Ukrainian. Russian title: Ukrainskaia sel'skokhoziaistvennaia entsiklopediia.   862

In Estonian

TI nõukogude entsüklopeedia (Estonian soviet encyclopedia). Tallinn, "Valgus," 1968- . v. 1- . (v. 1-4, 1968-1972, A-Maao). In progress. (Russian: Estonskaia sovetskaia entsiklopediia).   863

    Mostly short unsigned articles without bibliographies. Geografia (geography), r. 2, p. 407-408 (1970). Eesti Nõukogude Sotsialistlik Vabariik (Estonian SSR), r. 2, p. 49-154 (1970), has sections on natural conditions, p. 50-68, by E. Varep, . Raik, and others; on population, p. 68-73, by V. Kaufmann, and the economy of stonia, p. 87-101, by V. Tarmisto, and others.

TI entsüklopeedia (Estonian encyclopedia). R. Kleis, P. Treiberg, and J. V. Veski, ds. Tartu, K./Ü. "Loodus," 1932-1937. 8 v.   864

In Latvian

TVIJAS PSR mazā enciklopēdija (Small encyclopedia of the Latvian SSR). Vilis Samsons, ed. Rīgā, Izdevniecība "Zinātne," 1967-1970. 3 v., plus index v., 1972. (Latvijas PSR Zinātņu Akadēmija). (Russian: Malaia entsiklopediia Latviiskoi SSR).
    Ģeogrāfija (geography), v. 1, p. 613-614 (1967), by J. Alksnis, J. Orzols,   865
V. Pūriņš, with short bibliography on geography of or in Latvia.

    Latvijas Padomju Sociālistikā Republika (Latvian SSR), v. 2, p. 278-287 (1968), by A. Jaunputninš, V. Pūriņš, and others covers natural conditions, population, and the economy of Latvia, with short bibliography of works in Latvian and Russian.

TVIEŠU konversācijas vārdnica (Latvian encyclopedia). Arveds Švābe, Alexandrs Būmanis, and Kārlis Dišlers, eds. Riga, A. Gulbis, 1927-1940. v. 1-21. A-
Tjepolo only. No more published.   866

    Extensive information on Latvia and other Baltic countries.

TVJU enciklopēdija Latvian encyclopaedia Arveds Švābe, ed. Stockholm, Trīs Zvaigznes, 1950-1955. 3 v. Supplement, 1962. Lidija Švābe, ed.   867

    Devoted exclusively to Latvia and Latvians. Rich in biographies. Articles under letters A-T based partly on Latviešu konversācijas vārdnica, Riga, 1927-1940.

In Lithuanian

MAŽOJI Lietuviškoji tarybinė enciklopedija (Small Lithuanian soviet encyclopedia).
J. Matulis, ed. Vilnius, Leidykla "Mintis," 1966-1971. 3 v. (Russian: Malaia
litovskaia sovetskaia entsiklopediia).

Geogrāfija (geography), v. 1, p. 548 (1966) treats the development of geography in Lithuania.

Lietuvos Tarybų Socialistinė Respublika (Lithuanian SSR), v. 2, p. 400-428
(1968), covers natural conditions, population, and the economy of Lithuania, with
detailed bibliography of works in Lithuanian and Russian. Kazys Bieliukas was
general editor for geography and Vytautas Gudelis for physical geography.

LIETUVIŲ enciklopedija (Lithuanian encyclopedia). Boston, Massachusetts, Lietuvių
Enciklopedijos Leidykla, 1953-1966. 35 v. v. 36, supplement, 1969.

Geografija (geography), v. 7, p. 156-158 by Antanas Bendorius, has a general
bibliography on geography, mostly in German and French.

Lietuva (Lithuania), v. 15 (1968), contains sections on geography by Antanas
Bendorius and on geology, minerals, hydrography, climate, biogeography, and population, with bibliographies mainly in Lithuanian. Other articles emphasize histo
and culture. Articles under letters A-I based partly on Lietuviškoji enciklopedi
Kaunas, 1933-1944, revised and abridged.

Lithuanian subjects constitute about 40 per cent of the text.

LIETUVIŠKOJI enciklopedija (Lithuanian encyclopedia). Vaclovas Biržiška, ed.
Kaunas, Leidėjas "Spaudos Fondas," 1933-1944. v. 1-10. A-Ind. only. No more
published.

In Belorussian

BELARUSKAIA savetskaia entsyklapedyia (Belorussian soviet encyclopedia). Piatrus'
U. Broŭka, and others, eds. Minsk, 1969- . v. 1- (v. 1-9. 1969-1973, A-Sochy
In progress. (Akademiia Navuk BSSR. Galoŭnaia Rėdaktsyia Belaruskaï Savetskaï
ėntsyklapedyii). (Russian: Belorusskaia sovetskaia entsiklopediia).

In Moldavian

ENCHIKLOPEDIIA sovetikė Moldoveniaskė (Encyclopedia of soviet Moldavia). Iakim. S
Grosul, ed. Kishineu, Red. Princhipale a Enchiklopediei Sovetiche Moldovenesht
1970- . v. 1- (v. 1-3. A. Kiag). In progress. (In Russian: Moldavskaia
sovetskaia entsiklopediia).

To consist of 8 volumes.

In Armenian

ARMIANSKAIA sovetskaia entsiklopediia. Yerevan. Armianskaia Sovetskaia Entsilope
10 v. In Armenian. In preparation.

In Kazakh

*KAZAKHSKAIA sovetskaia entsiklopediia (Kazakh soviet encyclopedia). M. M. Karata
ed. Alma-Ata. 1972- . v. 1- (v. 1-2, A-Vengrler, 1972-1973). In progress.
(Akademiia Nauk Kazakhskoi SSR. Glavnaia Redaktsiia Kazakhskoi Sovetskoi Entsik
lopedii).

*Russian title from Knizhnaia letopis', 1972, no. 29330 and 1973, no. 43650
Kazakh title not verifiable at time of going to press.

868 - 874

In Uzbek

EK sovet éntsiklopediiasi (Uzbek soviet encyclopedia). I. M. Muminov, ed.
oshkent, Uzbek Sovet Entsiklopediiasi Bosh Redaktsiiasi, 1971- . v. 1- (v. 1-3,
971-1972, A-Dekhli). (Uzbekiston SSR Fanlar Akademiiasi). (Russian: Uzbekskaia
ovetskaia entsiklopediia).   875

General encyclopedia with emphasis on Uzbekistan, Soviet Middle Asia, and
djacent areas.

HANDBOOKS

USSR

SR. ADMINISTRATIVNO-territorial'noe delenie soiuznykh respublik (USSR. Adminis-
trative-territorial division of the union republics). 1-, 1938- . Moskva,
Izdatel'stvo "Izvestiia Sovetov Deputatov Trudiashchikhsia. SSSR." 1938-
(Prezidium Verkhovnogo Soveta Soiuza Sovetskikh Sotsialisticheskikh Republik).   876

     Official list of administrative divisions by republic, oblast, or kray, rayon,
city, and settlements of urban type. Alphabetical index of names and of name
changes. Issued about every two or three years. 1 (1938), 2 (1940), 3 (1941),
4 (1946), 5 (1947), 6 (1949), 7 (1951), 8 (1954), 9 (1958), 10 (1960), 11 (1962),
12 (1963), January 1965, July 1, 1967, July 1, 1971. Earlier similar handbooks
were issued by the Statistical Office of the NKVD for 1923, 1925, 1928, and 1935.
Comparable handbooks are issued for the individual republics.

IUZ Sovetskikh Sotsialisticheskikh Respublik. 1917-1967. Entsiklopedicheskii
spravochnik (Union of Soviet Socialist Republics, 1917-1967: an encyclopedic handbook), B. A. Vvedenskii, and others, eds. Moskva, Sovetskaia Entsiklopediia, 1967,
647 p. Bibliography, p. 637-647.   877

     A general handbook on the Soviet Union with text, statistics, maps, and bib-
liographical references on many aspects of the country, including population, p.
33-38; natural conditions and resources, p. 49-112, including geological structure,
relief, mineral resources, climate, permafrost, waters of the land, soils, vegetation and animal life, natural zones, natural regions, and seas; economics and
living standards, p. 207-276, including economics, manufacturing,construction,
agriculture, transport, communications, and internal and external trade; science,
p. 279-368, including geography, p. 312-313; and the union republics, p. 459-636.
Earlier similar handbooks on the Soviet Union were supplementary volumes to the
Bol'shaia sovetskaia entsikopediia, 1st ed., 1948 [846], and 2nd ed., v. 50, 1957
[847]. The Ukraine is covered by a similar handbook, which constitutes v. 17 of
the Ukrains'ka radians'ka entsyklopediia, 1965 [859].

ONOMICHESKAIA zhizn' SSSR: khronika sobytii i faktov 1917-1965 (Economic life of
the USSR: a chronicle of events and facts, 1917-1965). S. G. Strumilin, ed. 2nd
ed., Moskva, Izdatel'stvo "Sovetskaia Entsiklopediia," 1967. 2 v. v. 1 (1917-
1950), p. 1-439. v. 2 (1951-1965), p. 440-932. (1st ed., 1961. 1 v. 780 p.).   878

     A chronological day-by-day listing of major events in the economic life of
the Soviet Union such as dates of completion of major projects in industry, agriculture, and transportation: factories, dams, power projects, railroads. Sources
are cited for the dating of each event. Comprehensive alphabetical name and subject index, p. 857-932, aids in location of specific information. An invaluable
source book for concrete information, for identification,and for confirmation.

Foreign Areas

World-wide

PITALISTICHESKIE i razvivaiushchiesia strany: sotsial'no-ekonomicheskii spravochnik
(Capitalist and developing countries: a social-economic handbook). V. P. Viskovskaia,
A. A. Grechikhin, V. A. Kosova, and others. A. A. Manukiana, ed. Moskva,
Politizdat, 1973. 350 p.   879

IGER, Abram Grigor'evich. Sovremennaia karta zarubezhnogo mira: administrativno-
territorial'noe delenie zarubezhnykh stran. Spravochnik (The contemporary map
of the foreign world: administrative-territorial divisions of foreign countries.
A handbook). 4th ed. Moskva, Izdatel'stvo "Mysl'," Geografgiz, 1971. 462 p.
Bibliography, p. 452-459 (3rd ed., 1965. 455 p.).   880

A handbook of administrative divisions of the countries of the world with tables of principal administrative subdivisions giving area in thousands of square kilometers and population about 1970 and with a map showing location and boundaries, and with tables giving populations of cities of more than 100,000 and administrative subdivisions in which located. Countries are grouped by continents and listed in the Russian alphabetical order under each continent. Includes small as well as large countries: 34 in Europe, 44 in Asia, 55 in Africa, 50 in America, and 24 in Australasia and Oceania, 207 in all. Some of the maps are in a pocket. Bibliography lists 301 sources, 47 international sources covering many countries, and 254 official publications of individual countries.

MEZHDUNARODNYI ezhegodnik. Politika i ekonomika (International yearbook. Politics and economics). Moskva, Gospolitizdat, 1961- . Annual. (Akademiia Nauk SSSR. Institut Mirovoi Ekonomiki i Mezhdunarodnykh Otnoshenii).

Annual review of politics and economics country-by-country, and of international organizations. Bibliographies at end of each article. Supersedes Mezhdunarodnyi politiko-ekonomicheskii ezhegodnik, 1958-1960.

MEZHDUNARODNYI politiko-ekonomicheskii ezhegodnik (International political-economic yearbook). Moskva. Gospolitizdat, 1958-1960. 3 annual volumes. (Akademiia Nauk SSSR. Institut Mirovoi Ekonomiki i Mezhdunarodnykh Otnoshenii). 1958. 727 p. 1959. 751 p. 1960. 623 p.

Articles, reviews by countries, international organizations, bibliography, and statistical data. Superseded by Mezhdunarodnyi ezhegodnik. Politika i ekonomika, 1961- . annual.

STRANY mira. Kratkii politiko-ekonomicheskii spravochnik. (Countries of the world: a short political-economic handbook). Moskva, Politizdat. 1962, 1967, 1970-. Annual

Textual discussion of political, social, and economic characteristics of each country of the world.

The 1972 edition, with 472 pages, covered, in addition to the Soviet Union, the following number of countries in alphabetical order on each continent: 34 in Europe, 46 in Asia, 55 in Africa, 51 in America, 27 in Australasia and Oceania, a total of 214 countries. No tables, maps, or bibliography.

MIROVAIA ekonomika: kratkii spravochnik (World economy: short handbook). 3rd ed. Moskva, Ekonomika, 1967. 320 p. (Akademiia Nauk SSSR. Institut Mirovoi Ekonomiki i Mezhdunarodnykh Otnoshenii).

Text and statistics by countries.

STRANY i liudi (Countries and people) by G. V. Efimov. Leningrad, Lenizdat, 1965. 304 p.

Handbook of countries and peoples for popular use.

ZARUBEZHNYE strany: politiko-ekonomicheskii spravochnik (Foreign countries: politico-economic handbook). A. I. Denisov, D. I. Ignat'ev, and N. G. Pal'gunov, eds. Moskva, Gospolitizdat, 1957. 991 p.

A country-by-country compendium, with countries listed in Russian alphabetical order by continents. Material includes notes on area, population, general economic activities, industry, agriculture, transport, trade, finance, education, health, government, politics, the press, and a chronology of events, with maps and numerous tables.

Continents

RIKA, 1961-1965 gg.: spravochnik (Africa, 1961-1965, a handbook). N. I. Gavrilov
P. I. Manchkha, V. G. Solodovnikov, eds. Moskva, Nauka, Glavnaia Redaktsiia
Vostochnoi Literatury, 1967. 391 p. (Akademiia Nauk SSSR. Institut Afriki).   887

RIKA. Entsiklopedicheskii spravochnik (Africa. An encyclopedic handbook). I. I.
Potekhin, ed. Moskva, "Sovetskaia entsiklopediia," 1963. 2 v. (Nauchnyi sovet
Izdatel'stva "Sovetskaia entsiklopediia." Akademiia Nauk SSSR. Institut Afriki).888

Articles on the countries of Africa, on individual geographical features, on
peoples, and history. First volume contains a general survey of the continent and
its natural conditions and resources, ethnic structure, distribution of population,
and economy.

RIKA segodnia. Kratkii politiko-ekonomicheskii spravochnik (Africa today: a short
political-economic handbook). Moskva, Gospolitizdat, 1962. 328 p.   889

First part includes articles on natural resources, population, and economy of
Africa as a whole.

Second part information by countries in alphabetical order.

LTINSKAIA Amerika: politiko-ekonomicheskii spravochnik (Latin America: a political-
economic handbook). V. Ia. Bobrov, ed. Kiev, Izdatel'stvo Akademii Nauk Ukrainskoi SSR, 1963. 283 p.   890

LTINSKAIA Amerika. Kratkii politiko-ekonomicheskii spravochnik (Latin America:
short political-economic handbook). M. V. Danilevich, and others, eds. Moskva,
Gospolitizdat, 1962. 312 p.   891

Part one includes articles on natural resources, people, and the economy of
Latin America as a whole. Part two provides information on individual countries
in alphabetical order.

Individual Countries

VLOV, A. M., MALYSHEVA, L. A., KUZNETSOVA, L. S., and VOLDOV, V. G. Administrativno-
territorial'noe delenie SShA: slovar'-spravochnik. (Administrative-territorial
divisions of the United States: dictionary-handbook). Moskva, 1964.   892

GAZETTEERS

(See also encyclopedias and atlases)

Bibliographies

UFMAN, Isaak Mikhailovich. Geograficheskie slovari: bibliografiia (Geographic
dictionaries: bibliography). Moskva, Izdatel'stvo "Kniga," 1964. 78 p.

271 entries for "geographical dictionaries," mostly gazetteers, produced in
Russia or the Soviet Union, 18th-20th centuries, including universal geographical
dictionaries (19); dictionaries of individual continents and countries (8); dictionaries of the USSR (48); historical-geographical dictionaries (35); dictionaries
of Slavic lands (5); toponymic dictionaries (97); dictionaries of technical terms
in physical geography (24); dictionaries of Russian transcriptions of geographical
names (12); geographical dictionaries in languages of the peoples of the Union
Republics of the USSR (23). Bibliographical details and some annotations. Index
of authors and some titles. Supersedes earlier list by the same author in Voprosy
geografii, v. 4 (1947), p. 168-188, 102 entries. Within each category entries are
arranged in chronological order.   893

KOLAEV, Vladimir Aleksandrovich. Geograficheskie entsiklopedii (Geographical encyclopedias), Bolshaia sovetskaia entsiklopediia, 3rd ed., v. 6, p. 267-268 (1971). 894

Discusses main gazetteers produced in Russia and the Soviet Union, dictionaries of terms in geography and related fields, and some regional encyclopedias
from other parts of the world.

RAVOCHNYE izdaniia: geografiia, obshchie voprosy (Reference books: geography,
general questions) in Akademiia Nauk SSSR. Biblioteka Akademii Nauk SSSR.
Ukazatel' osnovnykh otechestvennykh bibliografii i spravochnykh izdanii po
estestvennym i fiziko-matematicheskim naukam. R. L. Baldaev, comp. A. I.
Mankevich, ed. Leningrad, BAN, 1966, p. 291-294.   895

Includes bibliographical information on specialized gazetteers or volumes of
transcriptions of geographical names for the United States (entry 1805), Latin
America (1804), British Isles (1808), France (1807), Germany (1803), Iran (1790),
Japan (1793), and East Antarctica (1796), as well as gazetteers for the Soviet
Union as a whole or its constitutent areas, such as Armenia (1802) or Azerbayzhan
(1806).

Gazetteers of the U. S. S. R.

LOSTNOVA, M. B., ed. Slovar' geograficheskikh nazvanii SSSR (Dictionary of
geographical names of the USSR). Utverzhden Glavnym upravleniem geodezii i
kartografii pri Sovete Ministrov SSSR v kachestve obiazatel'nogo posobiia. Moskva,
Izdatel'stvo "Nedra," 1968. 272 p. (Glavnoe Upravlenie Geodezii i Kartografii.
Tsentral'nyi Nauchno-Issledovatel'skii Institut Geodezii, Aeros"ëmki i Kartografii).
                                                                                      896
The most convenient and up-to-date alphabetical source on names of cities,
settlements, and physical features of the USSR on maps and in geographical textbooks. Gazetteer of geographical names in the Soviet Union approved for official
use. Prepared in the section on geographical names of the Central Scientific-
Research Institute for Geodesy, Air Photography, and Cartography of the Main
Administration of Geodesy and Cartography (GUGK). About 16,000 names including
particularly cities and towns, administrative units, and physical features such
as mountains, rivers, and lakes. For urban settlements indicates political unit
within which located, and official status. Marks stress and thus pronunciation.

Supersedes M. B. Volostnova, Slovar' russkoi transkriptsii geograficheskikh
nazvanii, chast' 1, Geograficheskie nazvaniia na territorii SSSR (Dictionary of
Russian transcription of geographical names, part 1, Geographical names on the
territory of the USSR), Moskva, Uchpedgiz, 1955. 132 p.   897

SEMÉNOV, Pëtr Petrovich. Geografichesko-statisticheskii slovar' Rossiiskoi imperii (Geographical-statistical dictionary of the Russian Empire). Commissioned by the Russian Geographical Society. With the collaboration of V. Zverinskii, N. Filippov, and R. Maak. Sankt-Peterburg, 1863-1885. 5 v. v. 1, A-G, 1863, 716 p. v. 2, D-K, 1865. 898 p. v. 3, L-O, 1867. 744 p. v. 4, P-S, 1873. 868 p. v. 5, T-Ia, 1885. 1,000 p.

Detailed articles on geographical features, regions, political units, and settlements with exact and original material. Each major article has a bibliography. The greatest of the 19th century gazetteers of the Russian Empire, with about 16,000 articles.

BARSOV, Nikolai Pavlovich. Materialy dlia istoriko-geograficheskogo slovaria Rossii. Tom 1. Geograficheskii slovar' russkoi zemli IX-XVI st. (Materials for an historical-geographic dictionary of Russia. v. 1. Geographical dictionary of the Russian land, 9th-16th centuries). Vilna, Tipografiia A. Syrkina, 1865. 220 p. Reprinted, Mouton, 1970. Slavistic printings and reprintings, no. 132.

GAGARIN, Sergei Petrovich. Vseobshchii geograficheskii i statisticheskii slovar' (General geographical and statistical dictionary). Moskva, A. Semen, 1843. 3 v.

Statistical material included in articles on geography of Russia.

SHCHEKATOV, Afanasii Mikhailovich. Slovar' geograficheskii Rossiiskogo gosudarstva. (Geographical dictionary of the Russian state). Moskva, Univ. Tip., 1801-1809. 7 v. (v. 1 by L. M. Maksimovich and A. M. Shchekatov).

Revision and enlargement of work by F. A. Polunin, 1773, and by L. M. Maksimovich, 1788-1789, to a large 7-volume gazetteer with 7,306 columns, or 3,653 pages.

MAKSIMOVICH, L. M. Novyi i polnyi geograficheskii slovar' Rossiiskogo gosudarstva... (New and full geographical dictionary of the Russian state). Moskva, Univ. Tip. u N. Novikova, 1788-1789. 6 v.

Revision and substantial expansion (from 480 pages to more than 2,000) of the work by F. A. Polunin, 1773.

POLUNIN, F. A. Geograficheskii leksikon Rossiiskogo gosudarstva... (Geographical dictionary of the Russian state). G. F. Miller, ed. Moskva, Imperatorskii Moskovskii Universitet, 1773. 480 p.

Alphabetical entries on rivers, lakes, seas, mountains, cities, fortresses, monasteries, etc. of the Russian Empire.

TATISHCHEV, Vasili Nikitich. Leksikon rossiiskoi, istoricheskoi, geograficheskoi, politicheskoi i grazhdanskoi (Russian historical, geographical, political, and civil dictionary). Sankt-Peterburg, Tip. Gornogo Uchilishcha, 1793. v. 1-3, A-Kliuchnik only. No more published.

First major gazetteer of Russia, compiled in manuscript in 1745 but not published until 1793. About 1,500 entries.

World-wide Gazetteers

ENTSIKLOPEDICHESKII slovar' geograficheskikh nazvanii (Encyclopedic dictionary of geographical names). S. V. Kalesnik, chief ed. Moskva, Izdatel'stvo "Sovetskaia entsiklopediia," 1973. 806 p. Editorial board: M. B. Vol'f, I. P. Gerasimov, R. A. Eramov, E. N. Lukashova, I. M. Maergoiz, N. P. Nikitin, V. V. Pokshishevskii, K. M. Popov, G. D. Rikhter, and M. S. Rozin.

About 11,000 entries. Stress indicated. Names in Western languages given also in Latin letters. Up-to-date information on natural conditions and economic activities of places described. Especially valuable for short accurate articles on physical and economic geography of the USSR and its regions, administrative units, physical features, and settlements. Written by specialists from leading geographic institutions in the USSR. Supersedes older gazetteer by M. S. Bodnarskii, <u>Slovar' geograficheskikh nazvanii</u>, ed. by V. P. Tikhomirov, 2nd ed., Moskva, Uchpedgiz, 1958. 391 p.   906

Gazetteers of Areas Outside the USSR

)LOSTNOVA, M. B., ed. <u>Slovar' geograficheskikh nazvanii zarubezhnykh stran</u> (Dictionary of geographical names of foreign countries). M. B. Volostnova, V. I. Savina, N. P. Danilova, and others. 2nd ed., Moskva, Izdatel'stvo "Nedra," 1970. 430 p. (Glavnoe Upravlenie Geodezii i Kartografii. Tsentral'nyi Naucho-Issledovatel'skii Institut Geodezii, Aeros"ëmki i Kartografii). (1st ed. 1965. 480 p.).   907

About 37,000 place names outside the USSR, in Russian form of transcription, approved for official use by all departments and institutions in the Soviet Union. Stress and thus pronunciation in Russian indicated and also location by country and major political subdivisions.

Supersedes M. B. Volostnova. <u>Slovar' russkoi transkriptsii geograficheskikh nazvanii. Chast' II. Geograficheskie nazvaniia na territorii zarubezhnykh stran</u> (Dictionary of Russian transcription of geographical names. Part 2. Geographical names in foreign countries). Moskva, Uchpedgiz, 1959. 167 p.   908

DICTIONARIES OF GEOGRAPHICAL TERMS

(See also encyclopedias, gazetteers, and place-name dictionaries)

Bibliographies of Dictionaries

AUFMAN, Isaak Mikhailovich. <u>Geograficheskie slovari: bibliografiia</u> (Geographic dictionaries: bibliography). Moskva, Izdatel'stvo "Kniga," 1964. 78 p.   909

Section on dictionaries of terms in physical geography, p. 63-68, entries 213-236, lists 24 works including also local terms for geographical features and oceanographic and ice terms. See full annotation under gazetteers.

AUFMAN, Isaak Mikhailovich. <u>Terminologicheskie slovari: bibliografiia</u> (Terminological dictionaries: a bibliography). Moskva, "Sovetskaia Rossiia," 1961. 419 p.
910
Arranged by subjects, including social sciences, humanities, natural sciences, and technology.

LOVARI izdannye v SSSR: bibliograficheskii ukazatel' 1918-1962 (Dictionaries published in the USSR: bibliographical guide 1918-1962). M. G. Izhevskaia, and others, comps. Moskva, Izdatel'stvo "Nauka," 1966. 232 p. (Akademiia Nauk SSSR. Institut Russkogo iazyka).   911

4,048 numbered entries arranged by language groups and individual languages, including monolingual, bilingual, and multilingual dictionaries.

Dictionaries for the national languages of the Union republics are listed on the following pages: Russian, p. 13-51; Ukrainian, p. 51-63; Belorussian, p. 64-67; Estonian, p. 119-121; Latvian, p. 69-73; Lithuanian, p. 73-75; Moldavian, p. 106-107; Georgian, p. 170-176; Armenian, p. 115-118; Azerbaijani, p. 129-136; Kazakh, p. 139-142; Kirgiz, p. 143-145; Uzbek, p. 153-157; Turkmen, p. 151-153; and Tajik, p. 8-10.

Dictionaries of Technical Terms in Geography in Russian

KALESNIK, Stanislav Vikent'evich, ed. Entsiklopedicheskii slovar' geograficheskikh terminov (Encyclopedic dictionary of geographical terms). Moskva, "Sovetskaia Entsiklopediia," 1968. 435 p. Editorial board: P. M. Alampiev, M. B. Vol'f, A. G. Voronov, I. P. Gerasimov, A. A. Grigor'ev, R. A. Eramov, A. G. Isachenko, E. N. Lukashova, I. M. Maergoiz, N. P. Nikitin, F. N. Petrov, V. V. Pokshishevskii, K. M. Popov, V. S. Preobrazhenskii, G. D. Rikhter, and M. S. Rozin.   9.

Best dictionary with definitions and explanations of technical geographical terms in Russian. About 4,200 terms defined by specialists in each field. Coverage especially good for terms in physical geography. Indicates stress thus pronunciation.

AGAPOV, Sergei Vasil'evich, SOKOLOV, Sergei Nikolaevich, and TIKHOMIROV, Dmitrii Ivanovich. Geograficheskii slovar' (Geographical dictionary). 2nd ed. Moskva, "Prosveshchenie," 1968. 245 p. (Biblioteka Shkol'nika) (1st ed., 1961. 156 p.).
                                                                                  9.
Consists of two parts: concepts in physical and economic geography found in textbooks and other geographic literature; short biographies of explorers and scholarly geographers. Marks stress and thereby indicates pronunciation. Primarily for school use.

MIL'KOV, Fëdor Nikolaevich. Slovar'-spravochnik po fizicheskoi geografii (Dictionary handbook on physical geography). 2nd ed. Moskva, "Mysl'," Geografgiz, 1970. 342 p. (1st ed. 1960. 269 p.).   91

Theoretical concepts and terms in physical geography, distribution of general and local terms of regional physical geography, basic literature of complex physical geography (203 titles). Subject index. For selected terms specific literature is cited. Also includes biographical sketches of major contributions to physical geography, geographical series, and institutions.

BARKOV, Aleksandr Sergeevich. Slovar'-spravochnik po fizicheskoi geografii: posobie dlia uchitelei geografii. (Dictionary handbook on physical geography: an aid for geography teachers). 4th ed. Moskva, Uchpedgiz, 1958. 329 p. (1st ed. 1940. 200 p. 2nd ed. 1948. 303 p. 3rd ed. 1954. 307 p.).   9

Explanatory dictionary of 2,400 terms in physical geography in Russian. For several decades the standard work, now superseded by S. V. Kalesnik, ed. Entsiklopedicheskii slovar' geograficheskikh terminov, 1968.

KRATKAIA geograficheskaia entsiklopediia (Short geographical encyclopedia). Moskva, 1960-1966. 5 v.   9

Although primarily a gazetteer and encyclopedia, it contains many geographical concepts and terms with full discussions. See entry under encyclopedias. [845]

Dictionary of Local Geographical Terms in the USSR

MURZAEV, Eduard Makarovich, and MURZAEVA, Valentina G. Slovar' mestnykh geograficheskikh terminov. (Dictionary of local geographical terms). Moskva, Geografgiz, 1959. 301 p.   9

Excellent alphabetical dictionary of about 2,600 local geographical terms from many languages of the various regions of the Soviet Union that have specific geographical meanings and have become part of Russian and often of international terminology, such as tundra from Finnish; taiga, a Siberian Turkic-Mongolian word; and barkhan, a Turkic word. Bibliography, p. 267-274, with about 150 entries. Index. Some terms illustrated by photographs.

Russian-English Geographical Dictionary

RATKII russko-angliiskii geograficheskii slovar' (Short Russian-English geographical dictionary). K. I. Volynkin, comp. Moskva, 1963. 65 p. (Universitet Druzhby Narodov im. Patrisa Lumumby. Kafedra Ekonomicheskoi i Politicheskoi Geografii). 918

Very elementary. English equivalents of about 2,000 terms in geography and general science needed by foreign students using Russian texts.

Russian-English General Dictionary

USSKO-ANGLIISKII slovar'. Russian-English dictionary (Under the direction of A. I. Smirnitskii. O. S. Akhmanova, ed. O. S. Akhmanova, Z. S. Vygodskaia, T. P. Gorbunova, N. F. Rotshtein, A. I. Smirnitskii, A. M. Taube, comps. 7th ed., Moskva, Izdatel'stvo "Sovetskaia Entsiklopediia," 1965. 766 p. (1st ed., Moskva, Gosudarstvennoe Izdatel'stvo Inostrannykh i National'nykh Slovarei, 1948. 988 p.; 2nd ed., 1952. 804 p.; 3rd ed., 1958. 952 p., stereotype reprintings, 4th ed., 1959, 5th ed., 1961, and 6th ed., 1962). Frequently reprinted. 919

The standard Russian-English general dictionary with about 50,000 words.

PLACE-NAME DICTIONARIES

(See also gazetteers and toponymy)

Bibliography

AUFMAN, Isaak Mikhailovich. Geograficheskie slovari: bibliografiia (Geographic dictionaries: bibliography). Moskva, Izdatel'stvo "Kniga," 1964. 78 p. 920

Section on toponymic dictionaries, p. 44-63, entries 116-212, lists 97 works. See full annotation under gazetteers [892].

Dictionaries

IKONOV, Vladimir Andreevich. Kratkii toponimicheskii slovar' (Short toponymic dictionary). Moskva, Izdatel'stvo "Mysl'," Geografgiz, 1966. 509 p. Bibliography, p. 503-509. 921

About 4,000 names of larger geographical objects such as countries, cities, seas, rivers, islands, and mountains, with notes on the origin of the names, including different explanations in many cases, and sources. List of linguistic terms used in the dictionary, p. 496-499, abbreviations, p. 500-502, and abbreviations used in references to the literature, p. 503-509, of which 85 abbreviations refer to works in Russian and 91 to works in other languages.

EL'KHEEV, Matvei Nikolaevich. Geograficheskie imena: toponimicheskii slovar' (Geographical names: toponymic dictionary). Moskva, Uchpedgiz, 1961. 99 p. Bibliography, p. 98-99. 922

About 1,750 geographical names of the world, with some emphasis on the Soviet Union, with notes on the origin of the place names and with references by number and page to further information in the bibliography with 43 references.

UCHKEVICH, Vadim Andreevich. Toponimika Belorussii (Place names of Belorussia). Minsk, 1968. 183 p. 923

Place-name dictionary of Belorussia with about 450 names and a bibliography with about 150 references.

ROZEN, M. F. Slovar' geograficheskikh terminov Zapadnoi Sibiri (Dictionary of
geographical terms of Western Siberia). Leningrad, 1970. 101 p. Bibliography,
p. 85-91. (Geograficheskoe Obshchestvo SSSR. Komissiia Toponimiki i Transkriptsii
Geograficheskikh Nazvanii).
924

MASLENNIKOV, Boris Georgievich. Morskaia karta rasskazyvaet: spravochnik (The sea
map reveals: a handbook). Moskva, Voenizdat, 1973. 366 p. Bibliography, p.
356-363.
92

Geographical names on marine maps, p. 19-232, includes more than 2,000 geographical features on sea charts that carry names of Russian explorers or their
ships or of foreigners in Russian service. Information includes latitude and
longitude, sea in which found, and identification of the person or ship memorialized
in the name. A biographical section, p. 233-344 provides information on the life
of about 650 persons, especially official service on geographical expeditions by
sea. A short section provides data on about 150 ships, p. 345-355. The bibliography of 130 references includes archival sources.

POPOV, Sergei Vladimirovich, and TROITSKII, Vladilen Aleksandrovich. Toponimika
morei Sovetskoi Arktiki: spravochnik (Toponymy of the seas of the Soviet Arctic:
a handbook). Leningrad, 1972. 316 p. Bibliography, p. 289-297. Index of
geographical names, p. 300-315. (Geograficheskoe Obshchestvo SSSR. Gidrografi-
cheskoe Predpriiatie MMF).
926

About 2,500 geographical names.

SAVINA, V. I. Slovar' geograficheskikh terminov i drugikh slov, formiruiushchikh
toponimiiu Irana (Dictionary of geographical terms and other words, forming the
toponymy of Iran). Moskva, Izdatel'stvo "Nauka," 1971. 346 p. Bibliography,
p. 337-345. (Glavnoe Upravlenie Geodezii i Kartografii pri Sovete Ministrov SSSR).
92

BIOGRAPHICAL REFERENCE VOLUMES

Bibliographies and Guides to Archives

ᴋAUFMAN, Isaak Mikhailovich. <u>Russkie biograficheskie i biobibliograficheskie slovari</u>
(Russian biographical and bio-bibliographical dictionaries). Moskva, Gosudarstvennoe Izdatel'stvo Kul'turno-Prosvetitel'noi Literatury, 1955. 751 p. (Earlier
ed. 1950. 332 p.).   927

     Entries 263-336, Geographical science, p. 162-184. 74 entries for geographical
monographs and series giving biographical and bio-bibliographical information on
geographers, with full bibliographical details and annotations.

BIBLIOGRAFIIA trudov otechestvennykh geografov i puteshestvennikov (Bibliography of
the works of native geographers and travelers). In the book: <u>Fizicheskaia geografiia annotirovannyi perechen' otechestvennykh bibliografii izdannykh v 1810-
1966 gg</u>. (Physical geography: annotated list of native bibliographies published
1810-1966), comp. by M. N. Morozova, E. A. Stepanova, V. V. Klevenskaia, and L.
G. Kamanin. Moskva, Izdatel'stvo "Kniga," 1968. p. 43-86. Entries 136-339.
(Gosudarstvennaia Biblioteka SSSR imeni V. I. Lenina. Institut Geografii Akademii
Nauk SSSR. Sektor Seti Spetsial'nykh Bibliotek Akademii Nauk SSSR).   928

     207 entries on published biographies and lists of publications of geographers
and travelers of Russia and the Soviet Union, as follows, in Russian alphabetical
order:
R. I. Abolin, V. V. Alëkhin, N. M. Al'bov, D. N. Anuchin, V. K. Arsen'ev,
V. I. Baranov, A. S. Barkov, A. I. Bezsonov, A. N. Beketov, F. F. Bellinsgauzen,
L. S. Berg, M. S. Bodnarskii, V. G. Bondarchuk, A. A. Borzov, S. G. Boch,
P. I. Brounov, A. I. Butakov, N. A. Bush, K. M. Ber, N. I. Vavilov, Ch. Ch.
Valikhanov, V. A. Varsanof'eva, M. I. Veniukov, G. Iu. Vereshchagin, V. I.
Vernadskii, A. I. Voeikov, A. V. Voznesenskii, E. V. Vul'f, G. N. Vysotskii,
I. P. Gerasimov, K. D. Glinka, V. G. Glushkov, V. S. Govorukhin, G. G. Grigor,
B. N. Gorodkov, A. A. Grigor'ev, A. A. Grossgeim, G. E. Grumm-Grzhimailo,
K. M. Deriugin, A. N. Dzhavakhishvili, N. A. Dimo, N. Ia. Dinnik, N. I. Dmitriev,
B. F. Dobrynin, V. S. Dokturovskii, V. V. Dokuchaev, B. I. Dybovskii, S. A.
Zakharov, S. V. Zonn, N. N. Zubov, V. F. Zuev, M. M. Il'in, A. P. Il'inskii,
V. I. Kavrishvili, S. V. Kalesnik, A. A. Kaminskii, G. S. Karelin, N. Ia. Kats,
P. I. Keppen, I. K. Kirilov, N. M. Knipovich, E. P. Kovalevskii, P. K. Kozlov,
D. M. Kolosov, V. L. Komarov, N. L. Korzhenevskii, S. I. Korzhinskii, E. P.
Korovin, A. N. Krasnov, I. M. Krasheninnikov, S. P. Krasheninnikov, P. A.
Kropotkin, A. A. Kruber, P. N. Krylov, N. I. Kuznetsov, M. V. Kul'tiasov, G. I.
Langsdorf, I. V. Larin, I. I. Lepekhin, V. I. Lipskii, B. L. Lichkov, M. V.
Lomonosov, I. A. Lopatin, R. K. Maak, S. O. Makarov, G. A. Maksimovich, K. I.
Maksimovich, N. I. Maksimovich, M. A. Menzbir, A. F. Middendorf, N. N. Miklukho-
Maklai, G. F. Morozov, S. D. Muraveiskii, E. M. Murzaev, I. V. Mushketov, G. I.
Nevel'skoi, S. S. Neustruev, A. Nikitin, S. N. Nikitin, I. V. Novopokrovskii,
V. A. Obruchev, P. N. Ovchinnikov, A. V. Ogievskii, S. I. Ognev, N. Ia. Ozeret-
skovskii, E. V. Oppokov, V. F. Oshanin, A. P. Pavlov, N. V. Pavlov, P. S. Pallas,
I. K. Pachoskii, P. K. Pakhtusov, P. I. Pashino, M. V. Pevtsov, M. P. Petrov,
B. B. Polynov, G. N. Potanin, L. I. Prasolov, N. M. Przheval'skii, A. V. Prozo-
rovskii, T. A. Rabotnov, G. I. Radde, L. G. Ramenskii, V. V. Reverdatto, A. L.
Reingard, G. D. Rikhter, V. I. Roborovskii, F. I. Ruprekht, V. A. Rusanov, P. I.
Rychkov, V. V. Sapozhnikov, K. A. Satunin, N. A. Severtsov, A. N. Sedel'nikov,
V. P. Semënov-Tian-Shanskii, P. P. Semënov-Tian-Shanskii, N. M. Sibirtsev, F. I.
Soimonov, N. I. Sokolov, N. N. Sokolov, V. B. Sochava, M. F. Spasskii, S. S.
Stankov, I. I. Stebnitskii, V. N. Sukachev, P. P. Sushkin, V. I. Taliev, G. I.
Tanfil'ev, A. A. Tillo, A. I. Tolmachev, A. Ia. Tugarinov, P. A. Tutkovskii,
A. P. Fedchenko, B. A. Fedchenko, A. N. Formozov, K. T. Khlebnikov, Iu. D.
Tsinzerling, A. L. Chekanovskii, I. D. Cherskii, P. A. Chikhachev, G. I. Shelikhov,
A. P. Shennikov, O. Iu. Shmidt, Iu. M. Shokal'skii, I. B. Shpindler, A. I. Shrenk,
L. I. Shrenk, E. A. Eversman, Ia. S. Edel'shtein, and S. A. Iakovlev.

MATVEEVA, T. P., FILONOVICH, T. S., and IARUKOVA, L. I., comps. Russkie geografy i
puteshestvenniki: fondy Arkhiva Geograficheskogo Obshchestva (Russian geographers
and travelers: resources of the Archive of the Geographical Society of the USSR).
Leningrad, Izdatel'stvo "Nauka," 1971. 176 p.   92

Biographical Directory

UKAZATEL' imën puteshestvennikov i deiatelei geograficheskikh i smezhnykh nauk,
upomianutykh v stat'iakh KGE (Index to names of travelers and scientists in geography and related disciplines, noted in articles in the Short Geographic Encyclopedia), Kratkaia geograficheskaia entsiklopediia, Moskva, Izdatel'stvo "Sovetskaia Entsiklopediia," 1966, v. 5, p. 410-544.   9

About 3,000 names of geographers, explorers, and workers in related disciplines from all periods and countries but with emphasis on Russian and Soviet
geographers, with biographical sketches, list of more important publications,
and pages in the encyclopedia where they are mentioned or listed in bibliographies.
About 500 of the contemporary Soviet geographers here listed are also listed in
Directory of Soviet geographers. (Soviet geography: review and translation,
September 1967).

Biographies of Leading Geographers

Collected works of or monographic studies on individual geographers are not
here included. Hundreds of these can be located in the biographies on each geographer listed in the volumes below or in the preceding bibliographies.

BARANSKII, Nikolai Nikolaevich, DIK, Nikolai Evgen'evich; EFREMOV, Iurii Konstantinovich; SOLOV'EV, Aleksandr Ivanovich; and SOLNTSEV, Nikolai Adol'fovich, eds.
Otechestvennye fiziko-geografy i puteshestvenniki (Native physical geographers
and explorers). Moskva, Uchpedgiz, 1959. 782 p.   9

Biographies in chronological order of Russian and Soviet physical geographers
and travelers from Nestor (11th century) to O. Iu. Schmidt (died 1956), with
appraisals of work and bibliographies. 4-14 pages on each person usually with
portrait. General bibliography of the history of Russian and Soviet physical
geography, p. 775-777 with 78 entries. Bibliographies at the end of chapters on
individual geographers, include both their writings and writings on them.

Alphabetical index of subject of sketches, p. 778-780.

The 108 biographies cover, in Russian alphabetical order: N. M. Al'bov, D.
N. Anuchin, V. K. Arsen'ev, V. T. Atlasov, Afanasii Nikitin, A. A. Baranov, A. S.
Barkov, F. F. Bellinsgauzen, L. S. Berg, V. I. Bering, M. A. Bogolepov, M. S.
Bodnarskii, A. A. Borzov, A. I. Butakov, N. A. Bush, K. M. Ber, N. I. Vavilov,
M. N. Vasil'ev (and G. S. Shishmarev), B. Vakhushti, G. Iu. Vereshchagin, V. I.
Vernadskii, V. Iu. Vize, A. I. Voeikov, F. P. Vrangel', G. N. Vysotskii, V. M.
Golovnin, B. N. Gorodkov, S. G. Grigor'ev, G. E. Grumm-Grzhimailo, S. I. Dezhnev,
B. F. Dobrynin, V. V. Dokuchaev, B. M. Zhitkov, V. G. Zuev, N. M. Knipovich, E.
P. Kovalevskii, P. K. Kozlov, V. L. Komarov, O. E. Kotsebu, A. N. Krasnov, I. M.
Krasheninnikov, S. P. Krasheninnikov, P. A. Kropotkin, A. A. Kruber, I. F.
Kruzenshtern, P. N. Krylov, N. I. Kuznetsov, M. P. Lazarev, D. Ia. Laptev, Kh.
P. Laptev, I. I. Lepekhin, Iu. F. Lisianskii, F. P. Litke, M. V. Lomonosov, R. K.
Maak, S. O. Makarov, K. I. Maksimovich, S. G. Malygin, A. F. Middendorf, N. N.
Miklukho-Maklai, G. F. Morozov, I. Iu. Moskvitin, S. D. Muraveiskii, I. V.
Mushketov, G. I. Nevel'skoi, Nestor, S. S. Neustruev, S. N. Nikitin, V. A. Obruchev
D. L. Ovtsyn, V. F. Oshanin, A. P. Pavlov, P. S. Pallas, M. V. Pevtsov, B. B.
Polynov, G. N. Potanin, V. D. Poiarkov, L. I. Prasolov, N. M. Przheval'skii, V.
Pronchishchev, G. I. Radde, S. U. Remezov, V. I. Roborovskii, K. F. Rul'e, F. I.
Ruprekht, V. A. Rusanov, V. V. Sapozhnikov, G. A. Sarychev, N. A. Severtsov,
G. Ia. Sedov, P. P. Semenov-Tian-Shanskii, M. I. Sumgin, S. P. Suslov, G. I.
Tanfil'ev, A. A. Tillo, A. P. Fedchenko, A. E. Fersman, E. P. Khabarov, A. L.
Chekanovskii, S. I. Cheliuskin, I. D. Cherskii, A. I. Chirikov, P. A. Chikhachev,
G. I. Shelikhov, O. Iu. Shmidt, Iu. M. Shokal'skii, Ia. S. Edel'shtein, V. V.
Iunker.

929 - 931

ARANSKII, Nikolai Nikolaevich; NIKITIN, Nikolai Pavlovich; POKSHISHEVSKII, Vadim Viacheslavovich, and SAUSHKIN, Iulian Glebovich, eds. Ekonomicheskaia geografiia v SSSR. Istoriia i sovremennoe razvitie (Economic geography in the USSR: history and present development). Moskva, Prosveshchenie, 1965. 663 p. Bibliography, p. 616-620, comp. by S. M. Kogan.   932

About 110 references on the history of the development of economic geography in pre-revolutionary Russia and the Soviet Union. Also lists of works of foreign geographers on Soviet economic geography (about 100 entries), p. 626-630, and lists of works by Soviet economic geographers translated into foreign languages or published abroad (about 300 entries), p. 631-652. Index of about 550 names of Russian and Soviet economic geographers mentioned in the text or bibliographies, p. 656-661.

Biographical sketches with notes on publications by them and on other studies about them are provided on the following 49 Russian and Soviet economic geographers, in Russian alphabetical order (though discussions in the book are by chronological periods): I. G. Aleksandrov, K. I. Arsen'ev, N. N. Baranskii, S. V. Bakhrushin, V. P. Bezobrazov, L. S. Berg, S. V. Bernshtein-Kogan, Ch. Ch. Valikhanov, M. I. Veniukov, K. G. Voblyi, A. I. Voeikov, K. F. German, G. N. Gekhtman, the Decembrists (as a group), V. E. Den, L. Ia. Ziman, R. M. Kabo, V. A. Kamenetskii, P. I. Këppen, I. K. Kirilov, V. S. Klupt, B. N. Knipovich, N. N. Kolosovskii, N. I. Lialikov, V. I. Lavrov, M. V. Lomonosov, G. A. Mebus, D. I. Mendeleev, L. I. Mechnikov, K. N. Mirotvortsev, N. V. Morozov, L. L. Nikitin, N. P. Ogarev, A. N. Radishchev, D. I. Rikhter, A. A. Rybnikov, P. I. Rychkov, V. P. Semenov-Tian-Shanskii, P. P. Semenov-Tian-Shanskii, L. D. Sinitskii, A. I. Skvortsov, V. N. Tatishchev, A. F. Fortunatov, M. A. Tsvetkov, A. N. Chelintsev, P. I. Chelishchev, G. N. Cherdantsev, V. M. Chetyrkin, and A. I. Chuprov, p. 243-615.

ARANSKII, Nikolai Nikolaevich, NIKITIN, Nikolai Pavlovich, and SAUSHKIN, Iulian Glebovich, eds. Otechestvennye ekonomiko-geografy XVIII-XX vv. (Native economic geographers. 18th-20th centuries). Moskva, Uchpedgiz, 1957. 328 p. Bibliography, p. 318-321.   933

Articles on the contributions of 39 Russian and Soviet economic geographers from V. N. Tatishchev, 1686-1750 to N. N. Kolosovskii, 1891-1954. Bibliography has about 80 entries. Index of personal names, p. 322-326 includes about 400 names.

Of the 39 biographies, 23 are repeated in revised form in N. N. Baranskii, and others, Ekonomicheskaia geografiia v SSSR: Istoriia i sovremennoe razvitie, 1965. The following 16 other economic geographers in Russian alphabetical order are: V. P. Androsov, N. Ia. Bichurin, D. P. Zhuravskii, A. P. Zablotskii-Desiatovskii, D. A. Klemenets, D. A. Miliutin, N. A. Miliutin, S. I. Pleshcheev, V. I. Pokrovskii, A. P. Subbotin, Ia. V. Khanykov, G. Ts. Tsybikov, V. I. Chaslavskii, M. D. Chulkov, N. K. Chupin, N. M. Iadrintsev.

LUDI russkoi nauki. Ocherki o vydaiushchikhsia deiateliakh estestvoznaniia i tekhniki. Geologiia. Geografiia (Men of Russian science: Studies on leading figures in science and technology: geology and geography). I. V. Kuznetsov, ed. Moskva, Fizmatgiz, 1962. 580 p. Geography, p. 269-575.   934

Biographies and lists of publications of the following 33 Russian physical geographers and travelers in Russian alphabetical order: D. N. Anuchin, F. F. Bellinsgauzen, L. S. Berg, V. Bering, A. I. Voeikov, V. M. Golovnin, S. I. Dezhnev, I. K. Kirilov, V. O. Kovalevskii, P. K. Kozlov, O. E. Kotsebu, S. P. Krasheninnikov, I. F. Kruzenshtern, M. P. Lazarev, D. Ia. and Kh. P. Laptev, I. I. Lepekhin, Iu. F. Lisianskii, N. N. Miklukho-Maklai, A. I. Nagaev, G. I. Nevel'skoi, A. Nikitin, G. N. Potanin, N. M. Przheval'skii, V. A. Rusanov, G. Ia. Sedov, P. P. Semënov-Tian-Shanskii, F. I. Soimonov, G. I. Tanfil'ev, V. N. Tatishchev, S. I. Cheliuskin, A. I. Chirikov, Iu. M. Shokal'skii.

ŒKHTMAN, Georgii Nikolaevich. _Vydaiushchiesia geografy i puteshestvenniki_ (Leading
geographers and explorers). Tbilisi, Izdatel'stvo Akademii Nauk Gruzinskoi SSR,
1962. 307 p. (Akademiia Nauk Gruzinskoi SSR. Geograficheskoe Obshchestvo
Gruzinskoi SSR).   93

    Bibliographical dictionary including Russian and Soviet geographers and
explorers, with special attention to ones from Georgia, Armenia, and Azerbaydzhan.
247 essays with notes on writings by and about each individual.

PART III. SYSTEMATIC FIELDS OF GEOGRAPHY

A. PHYSICAL GEOGRAPHY

GENERAL (COMPREHENSIVE) PHYSICAL GEOGRAPHY

IZICHESKAIA geografiia: annotirovannyi perechen' otechestvennykh bibliografii
izdannykh v 1810-1966 gg. (Physical geography: annotated list of native bibliog-
raphies published 1810-1966). M. N. Morozova and E. A. Stepanova, comps. V. V.
Klevenskaia, ed. L. G. Kamanin, scientific consultant. Moskva, Izdatel'stvo
"Nauka," 1968. 309 p. (Gosudarstvennaia Biblioteka SSSR imeni V. I. Lenina.
Institut Geografii Akademii Nauk SSSR. Sektor Seti Spetsial'nykh Bibliotek
Akademii Nauk SSSR).   936

Section 2, bibliographies on physical geography, p. 89-276, entries 340-1336,
about a thousand Soviet bibliographies arranged systematically: general physical
geography, systematic branches of physical geography (geomorphology, climatology,
oceanography, hydrology, glaciology, geography of soils, geography of vegetation,
geography of animal life), and regional physical geography. Author and geographical indexes. See also the individual fields of physical geography.

EOBRAZHENSKII, Vladimir Sergeevich. Obshchaia fizicheskaia geografiia (zemlevedenie
i landshaftovedenie) v 1967-1970 gg. (General physical geography [earth science
and landscape studies] in the years 1967-1970), in Itogi nauki i tekhniki,
seriia Teoreticheskie voprosy fizicheskoi i ekonomicheskoi geografii, v. 1. Moskva,
VINITI, 1972, p. 44-88. Bibliography, p. 71-86. English summary, p. 86-88.   937

315 references, 285 in Russian, and 30 in other languages, on characteristics
of general physical geography or landscape science.

RG, Lev Semenovich. Ocherki po fizicheskoi geografii. (Studies in physical
geography). Moskva-Leningrad, Izdatel'stvo Akademii Nauk SSSR, 1949. 340 p.
(Akademiia Nauk SSSR. Seriia "Itogi i Problemy Sovremennoi Nauki").   938

About 400 references at the ends of the separate studies.

GOIAVLENSKII, Georgii Pavlovich. Fizicheskaia geografiia: bibliograficheskie
posobie dlia uchitelei (Physical geography: bibliographical aid for teachers).
Moskva, Uchpedgiz, 1963. 208 p.   939

Annotated systematic guide to 280 titles in Russian 1940-1962. Index of
author and titles.

ACHENKO, Anatolii Grigor'evich. Osnovnye voprosy fizicheskoi geografii (Fundamental problems of physical geography). Leningrad, Izdatel'stvo Leningradskogo
Universiteta, 1953. 392 p. Bibliography, p. 361-385. (Leningradskii Gosudarstvennyi Universitet).   940

About 630 references.

LESNIK, Stanislav Vikent'evich. Obshchie geograficheskie zakonomernosti Zemli
(General geographical regularities of the earth). Moskva, Izdatel'stvo "Mysl',"
1970. 283 p. Bibliography, p. 261-276.   941

About 420 references. Index of concepts and terms, p. 277-280.

LESNIK, Stanislav Vikent'evich. Osnovy obshchego zemlevedeniia (Fundamentals of
general geography). 2nd ed. Moskva, Uchpedgiz, 1955. 472 p. (1st ed. 1947).   942

Bibliographies at end of each chapter, devoted to general physical geography,
and many references to the literature in footnotes.

MIL'KOV, Fedor Nikolaevich. Osnovnye problemy fizicheskoi geografii (Fundamental problems of physical geography). Moskva, Izdatel'stvo "Vysshaia Shkola," 1967. 251 p. 94.

284 references in bibliographies at ends of chapters: (1) object of physical geography, p. 20; (2) concept of the landscape and the problem of physical-geographic regionalization, p. 31; (3) basic principles and methods of physical-geographic regionalization, p. 43-44; (4) typological landscape complexes, p. 68-70; (5) the problem of zones and provinces in physical-geographical regionalization, p. 87; (6) some features of the structure of the landscape sphere and of constructing its landscape complexes, p. 97; (7) current status of study of geographical zones, p. 118-119; (8) dynamics of geographical zones, p. 142; (9) recent tectonics and the problem of the formation of present landscapes, p. 161; (10) the glacial epoch and the genesis of present landscapes, p. 176-177; (11) the problem of quantitative characteristics in physical geography, p. 196-197; (12) problems of methods of field work in landscape research, p. 209-210; (13) influence of human society on nature in geographical zones, p. 233-234; (14) physical geography in the service of the economy, p. 248-249. Footnote references identify individuals who have worked on specific elements of each problem.

MUKHINA, Lidiia Ivanovna. Printsipy i metody tekhnologicheskoi otsenki prirodnykh kompleksov (Principles and methods of the technological appraisal of natural complexes). Moskva, Izdatel'stvo "Nauka," 1973. 95 p. Bibliography, p. 87-94. (Akademiia Nauk SSSR. Institut Geografii). 94

.RIABCHIKOV, Aleksandr Maksimovich. Struktura i dinamika geosfery, ee estestvennoe razvitie i izmenenie chelovekom (Structure and dynamics of the geosphere, its natural development and modification by man). Moskva, Izdatel'stvo "Mysl'," Geografgiz, 1972. 223 p. Bibliography, p. 216-222. 94

About 200 references, arranged alphabetically by name of author, on world physical geography and the modification of the surface of the earth by man.

RIKHTER, Gavriil Dmitrievich, ed. Razvitie i preobrazovanie geograficheskoi sredy: k 80-letiiu so dnia rozhdeniia Akad. A. A. Grigor'eva (Development and transformation of the geographical environment: to the 80th birthday of Academician A. A. Grigor'ev). Moskva, Izdatel'stvo "Nauka," 1964. 240 p. Bibliography, p. 222-239 and at ends of articles. (Akademiia Nauk SSSR, Institut Geografii). 94

SHCHUKIN, Ivan Semenovich, and SHCHUKINA, Ol'ga Evseevna. Zhizn' gor; opyt analiza gornykh stran kak kompleksa poiasnykh landshaftov (Life in mountains: an attempt to analyze mountainous countries as complex landscape zones). Moskva, Geografgiz, 1959. 287 p. Bibliography, p. 280-286. 94

134 references in Russian and other languages.

SVATKOV, Nikolai Mikhailovich. O predmete issledovaniia fizicheskoi geografii: Problema vneshnikh granits i sushchnosti geograficheskoi obolochki (On the object of research in physical geography: the problem of the outer boundary and nature of the geographical envelope of the earth). Moskva, Izdatel'stvo "Mysl'," 1970. 182 p. Bibliography, p. 163-181. Summary in English. 94

About 440 references.

VIKTOROV, Sergei Vasil'evich. Ispol'zovanie indikatsionnykh geograficheskikh issledovanii v inzhenernoi geologii. (The utilization of indicator geographical research in engineering geology). Moskva, Izdatel'stvo "Nedra," 1966. 120 p. Bibliography, p. 114-120. (VNII Gidrogeologii i Inzhenernoi Geologii). 9

About 170 references on the use specific geographic features as indexes to other features in engineering geology, as the utilization of particular species of plants in an ecological community as indicators of mineralized ground water, drainage conditions, or soil types.

ABELIN, Igor' Mikhailovich. Teoriia fizicheskoi geografii (Theory of physical
geography). Moskva, Geografgiz, 1959. 303 p. Bibliography, p. 290-300.   950
    About 250 references to the Soviet literature on physical geography.

IL'KOV, Fedor Nikolaevich. Chelovek i landshafty. Ocherki antropog. landshaftovedeniia (Man and the landscape. Studies in anthropogeography of landscape
science). Moskva, Izdatel'stvo "Mysl'," 1973. 224 p. Bibliography, p. 207-223. 951

BRAMOV, Lev Solomonovich. Opisaniia prirody nashei strany. Razvitie fiziko-
geograficheskikh kharakteristik. (Descriptions of natural conditions in our
country. Development of physical-geographic characteristics). Moskva, Izdatel'-
stvo "Mysl'," 1972. 277 p. Bibliography, p. 266-275. (Akademiia Nauk SSSR.
Institut Geografii).   952

LANDSCAPE STUDIES

JNEVSKII, I. M., MIKITIN, Teodor Dmitrievich, and CHORNOMAZ, M. M. Sovetskoe
landshaftovedenie 1917-1967: bibliograficheskii ukazatel' literatury (Soviet
landscape studies 1917-1967: a bibliographical guide to the literature). K. I.
Gerenchuk, ed. L'vov, 1970. 185 p. (L'vovskii Universitet imeni Iv. Franko.
Nauchnaia Biblioteka. L'vovskii Otdel Geograficheskogo Obshchestva USSR).   953

JNEVSKII, I. M. Literatura po metodike landshaftnykh issledovanii (Literature on
methodology of landscape research), Geograficheskii sbornik, Heohrafichnyi zbirnyk
(L'vovskii Universitet. Geograficheskoe Obshchestvo SSSR, L'vovskii Otdel), v. 6
(1961), p. 152-160.   954
    About 130 references in alphabetical order on the methodology of the complex
regional physical geography of the landscape, 1934-1960.

LNDSHAFTOVEDENIE (Landscape studies). A. A. Vidina, and others, eds. Moskva,
1972. 225 p. Bibliography, 1969-1970, comp. by V. I. Riabchikova and A. A.
Vidina, p. 206-224, and at ends of chapters. (Moskovskii Gosudarstvennyi Universitet. Geograficheskii Fakul'tet). Collected volume of articles in honor of
N. A. Solntsev.   955

ABCHIKOVA, V. I., VIDINA, Alida Avgustovna, and NIKOLAEV, Vladimir Aleksandrovich.
Literatura po landshaftovedeniiu za 1966-1968 gg. (Literature on landscape
studies, 1966-1968). In the book: Landshaftnyi sbornik (Collection on landscape
studies). Moskva, Izdatel'stvo Moskovskogo Universiteta, 1970. p. 378-421.   956
    About 380 entries.

RIGOR'EV, Aleksei Alekseevich, comp. Literatura po voprosam metodiki landshaftnykh
issledovanii i kartografirovaniia landshaftov za 1964-1965 gg. (Literature on
questions of the methodology of landscape research and the cartography of landscapes, 1964-1965). In the book: Landshaftnyi sbornik (Collection of landscape
studies), Moskva, Izdatel'stvo Moskovskogo Universiteta, 1970, p. 364-377.   957
    125 references.

VAIA literatura po landshaftovedeniiu (New literature on the study of the landscape), Vestnik Moskovskogo Universiteta, seriia 5. Geografiia, 1961, no. 4,
p. 71-80.   958
    230 entries for the period 1954-1960, systematically organized, including
both Soviet (about 100) and foreign (about 130) works.

L'KOV, Fedor Nikolaevich. Landshaftnaia geografiia i voprosy praktiki (Landscape
geography and practical problems). Moskva, Izdatel'stvo "Mysl'," 1966. 256 p.
Bibliography, p. 227-240.   959
    About 440 references on applied regional physical geography.

MIL'KOV, Fedor Nikolaevich. Landshaftnaia sfera Zemli (The landscape sphere of the earth). Moskva, Izdatel'stvo "Mysl'," 1970. 207 p. Bibliography, p. 195-206. 96

About 280 references.

PREOBRAZHENSIII, Vladimir Sergeevich. Landshaftnye issledovaniia (Landscape research). Moskva, Izdatel'stvo "Nauka," 1966. 127 p. Bibliography, p. 120-127. (Akademiia Nauk. Institut Geografii). 96

186 references on regional physical geography.

ZABELIN, Igor' Mikhailovich, and IAKOVER, Mendel' Borisovich. Osnovnaia literatura po sovetskomu landshaftovedeniiu (Basic literature on Soviet landscape studies), Voprosy geografii, 16 (1949), p. 204-207. 96

Alphabetical list of about 70 books and articles published 1918-1948, useful in study of the rise of the Soviet discipline of landscape studies.

Geochemistry of the landscape

GLAZOVSKAIA, Mariia Al'fredovna. Geokhimicheskie osnovy tipologii i metodiki issledovanii prirodnykh landshaftov (Geochemical foundations of the classification and of methods of research on natural landscapes). Moskva, Izdatel'stvo Moskovskog Universiteta, 1964. 230 p. Bibliography, p. 223-228. 9

About 100 references.

PEREL'MAN, Aleksandr Il'ich. Geokhimiia landshafta (Geochemistry of the landscape). Moskva, Geografgiz, 1961. 496 p. Bibliography, p. 482-490. 9

About 200 references in alphabetical order, 1922-1961.

PEREL'MAN, Aleksandr Il'ich. Ocherki geokhimii landshafta (Studies in the geochemistry of the landscape). D. I. Shcherbakov, ed. Moskva, Geografgiz, 1955. 392 p. Bibliography, p. 375-388. 9

About 300 references systematically organized.

PHYSICAL GEOGRAPHICAL REGIONALIZATION AND MAPPING

General

JZHINSKAIA, N. G. and MIKHAILOV, Nikolai Ivanovich. Literatura po voprosam
fiziko-geograficheskogo raionirovaniia (1963-1965 gg.) [Literature on questions
of physical-geographical regionalization, 1963-1965], in Itogi nauki: Geografiia
SSSR, vypusk 4, N. I. Mikhailov, Fiziko-geograficheskoe raionirovanie. Moskva.
Vsesoiuznyi Institut Nauchnoi i Tekhnicheskoi Informatsii, 1967, p. 67-143.   966

    1,136 entries on problems of physical-geographical regionalization listed
in Referativnyi zhurnal: geografiia during the years 1963-1965 organized by
general questions and the theory of regionalization (133 entries); regional work
by systematic components, including geomorphological regionalization (59), climatic
regionalization (59), hydrological regionalization (19); soil-geographic regionalization (59); biogeographical regionalization (119); regional works by complex
physical-geographical regions, including USSR as a whole (9), European
USSR (53), Urals (14), the Caucasus (20), Soviet Middle Asia and Kazakhstan (34),
Siberia and the Far East (45), foreign countries (55); questions of methodology
(76); applied work (122); studies on the landscape and regional landscape investigations (249); translations of the work of Soviet geographers into foreign
languages (11). Author index. Location of abstract in Referativnyi zhurnal:
geografiia indicated. Textual discussion with English abstract and supplementary
table of contents.

DINA, Aleksandra Efimovna. Fiziko-geograficheskoe raionirovanie (Physical-
geographic regionalization). Moskva, Izdatel'stvo Moskovskogo Universiteta,
1965. 142 p. Bibliography, p. 137-141.   967
    About 120 references.

DINA, Aleksandra Efimovna. Fiziko-geograficheskoe raionirovanie (Physical-
geographic regionalization). N. A. Gvozdetskii, ed. Moskva, Izdatel'stvo Moskovskogo Universiteta, 1973. 196 p. Bibliography, p. 193-195.   968

ACHENKO, Anatolii Grigor'evich. Fiziko-geograficheskoe kartirovanie (Physical-
geographical mapping). Leningrad, v. 1-3, 1958-1961 (Leningradskii Gosudarstvennyi
Universitet). v. 1, 1958, bibliography, p. 225-230, v. 2, 1960, bibliography,
p. 213-229, v. 3, 1961, bibliography, p. 252-266.   969
    About 920 references; the 570 in v. 1-2 arranged systematically, the 350 in
v. 3 on landscape study and landscape mapping arranged alphabetically.

ACHENKO, Anatolii Grigor'evich. Osnovy landshaftovedeniia i fiziko-geograficheskoe
raionirovanie (Foundations of landscape study and physical-geographic regionalization). Moskva, "Vysshaia Shkola," 1965. 327 p. Available in English as: Principles of landscape science and physical-geographic regionalization. Edited by
John S. Massey and N. J. Rosengren. Translated by R. J. Zatorski. Parkville,
Victoria, Australia, Melbourne University Press, 1973. 311 p.   970

KHAILOV, Nikolai Ivanovich. Fiziko-geograficheskoe raionirovanie (Physical-
geographic regionalization), v. 2, Kompleksnoe fiziko-geograficheskoe raionirovanie
(Complex physical-geographical regionalization). Moskva, 1962. 217 p. Bibliography, p. 208-216 (Moskovskii Gosudarstvennyi Universitet. Geograficheskii
Fakul'tet. Kafedra Fizicheskoi Geografii SSSR).   971
    About 140 references arranged systematically.

OKAEV, Vasilii Ivanovich. Osnovy metodiki fiziko-geograficheskogo raionirovaniia
(Basic methods of physical-geographic regionalization). Leningrad, Izdatel'stvo
"Nauka," 1967. 167 p. Bibliography, p. 160-165 (Akademiia Nauk SSSR. Geo-
graficheskoe Obshchestvo SSSR).
    About 170 references.   972

U. S. S. R. as a Whole

KOLTUN, Mariia Isaakovna. Prirodnoe (fiziko-geograficheskoe) raionirovanie territorii
Sovetskogo Soiuza: Ukazatel' literatury izdannoi v 1917-1960 gg. (Natural, or
physical-geographic, regionalization of the territory of the Soviet Union: Guide
to the literature published 1917-1960). V. V. Klevenskaia, ed. Moskva, 1962.
379 p. (Gosudarstvennaia Biblioteka SSSR imeni V. I. Lenina. Otdel Spravochno-
Bibliograficheskoi i Informatsionnoi Raboty).   97.

About 4,000 partially annotated entries of publications, 1917-1960, in
Russian, Ukrainian, and other languages of the USSR (if provided with a Russian
summary) on physical-geographic regionalization of the Soviet Union, organized
systematically by the following fields: complex natural, or physical-geographic,
geomorphological and tectonic, karst-phenomena, seismic, geochemical and hydro-
chemical, climatic, hydrological and glaciological, hydrogeological and permafrost, soil, geobotanical, marsh-and-bog, and zoogeographical regionalizations.
Name and place index.

KOLTUN, Mariia Isaakovna. comp. Prirodnoe (fiziko-geograficheskoe) raionirovanie
territorii Sovetskogo Soiuza: spisok literatury, izdannoi na russkom i ukrainskom
iazykakh v 1961-1964 gg. Chast' 1. Kompleksnoe prirodnoe (fiziko-geograficheskoe)
raionirovanie. (Natural, physical-geographic, regionalization of the territory
of the USSR: list of the literature published in the Russian and Ukrainian languages, 1961-1964. Part 1: complex natural, physical-geographical regionalization). M. E. Ekshtein, ed. Moskva, 1965, 99 p. (Gosudarstvennaia Biblioteka
SSSR imeni Lenina. Metod. materialy v pomoshch' oblastnym, kraevedcheskim, i
respublikanskim bibliotekam).   97

348 entries, systematically arranged, partly annotated, devoted to theoretical
and methodological problems of regionalization, of physical-geographical regionalization of the territory of the USSR as a whole, and of its regions. Index of
place names.

GVOZDETSKII, Nikolai Andreevich, and MIKHAILOV, Nikolai Ivanovich, eds. Fiziko-
geograficheskoe raionirovanie SSSR: obzor opublikovannykh materialov (Physical-
geographic regionalization of the USSR: survey of published materials). Moskva,
Izdatel'stvo Moskovskogo Universiteta, 1960. 287 p.   9

About 500 references distributed at ends of articles as follows:

Solntsev, Nikolai Adol'fovich. Istoriia fiziko-geograficheskogo raionirovanii
Evropeiskoi chasti SSSR (History of physical-geographic regionalization of the
European part of the USSR), p. 6-54. 27 references.

Solov'ev, Aleksandr Ivanovich, with the participation of A. A. Makunina. Iz
istorii fiziko-geograficheskogo raionirovaniia Urala (From the history of physical-
geographic regionalization of the Urals, p. 55-76. 41 references.

Mikhailov, Nikolai Ivanovich. Melkomasshtabnoe fiziko-geograficheskoe
raionirovanie Sibiri (Small-scale physical-geographic regionalization of Siberia),
p. 77-136. 100 references.

Glazovskaia, Mariia Al'fredovna. Fiziko-geograficheskoe raionirovanie
Kazakhstana (Physical-geographic regionalization of Kazakhstan), p. 137-168.
31 references.

Gvozdetskii, Nikolai Andreevich. Prirodno-geograficheskoe raionirovanie
Srednei Azii (Natural-geographic regionalization of Middle Asia), p. 169-207.
About 100 references.

Fedina, Aleksandra Efimovna. Obzor literatury po prirodnomu i fiziko-
geograficheskomu raionirovaniiu Kavkaza (Survey of the literature on natural
and physical-geographic regionalization of the Caucasus), p. 208-245. 117
references.

Parmuzin, Iurii Pavlovich, and Kachashkina, M. V. Obzor literatury po
prirodno-geograficheskomu raionirovaniiu Dal'nego Vostoka (Survey of the literature on natural-geographical regionalization of the Soviet Far East), p. 246-
265. 101 references.

CRG, Lev Semenovich. Geograficheskie zony Sovetskogo Soiuza (Geographical zones
of the Soviet Union). Geografgiz, v. 1. 3rd ed., 1947. 397 p. v. 2, 1952, 510 p.
976
Bibliographies at the end of chapters in volume 1 with about 475 references:
landscapes and changes in landscapes, p. 29-33, tundra, p. 76-80, forest zone, p.
263-264, broad-leafed forests of the Far East, p. 283-284, and the forest-steppe,
p. 387-391.

Volume 2 lacks an assembled bibliography but has more than 1,000 bibliographical footnotes, to the literature on the steppe zone, semiarid zone, temperate
desert zone, mountains of Soviet Middle Asia, subtropical zone, Caucasus mountains,
Crimean mountains, Eastern Carpathians, Urals, Altay, Sayany, Tuva region, Baykal
region, mountains of Northeast Siberia, mountains of the Soviet Far East, Sakhalin,
Kuril Islands, Kamchatka, and mountains of the Arctic.

Also available in German as Die geographischen Zonen der Sowjetunion. Leipzig,
B. G. Teubner, v. 1, 1958. 437 p. v. 2, 1959. 604 p. [2583].

RG, Lev Semenovich. Priroda SSSR (Natural conditions of the USSR). 3rd ed.,
Moskva, Geografgiz, 1955. 496 p. Bibliography, p. 462-466.   977

About 110 references for the most important literature, maps, and atlases of
a general character. Available in English as Natural regions of the USSR. Translated from the Russian by Olga Adler Titelbaum. Edited by John A. Morrison and
C. C. Nikiforoff. New York, Macmillan, 1950. 436 p. (American Council of
Learned Societies. Russian Translation Project. Series 6). [2584].

TESTVENNOISTORICHESKOE raionirovanie SSSR. comp. by S. G. Strumilin and I. S.
Lupinovich. Moskva-Leningrad, Izdatel'stvo Akademiia Nauk SSSR, 1947. 375 p.
Bibliography, p. 358-363 (Akademiia Nauk SSSR. Sovet po Izucheniiu Proizvoditel'-
nykh Sil. Komissiia po estestvennoistoricheskomu raionirovaniiu SSSR, Trudy,
tom 1).
978
204 references.

OZDETSKII, Nikolai Andreevich, ed. Fiziko-geograficheskoe raionirovanie SSSR:
kharakteristika regional'nykh edinits (Physical-geographical regionalization of
the USSR: characteristics of regional units). Moskva, Izdatel'stvo Moskovskogo
Universiteta, 1968. 576 p. Bibliography, p. 559-575.   979

About 550 references in alphabetical order by name of author. Large folded
colored map on a scale of 1:10,000,000 showing location and extent of the 304
physical-geographic units recognized.

L'KOV, Fedor Nikolaevich. Prirodnye zony SSSR. (Natural zones of the USSR).
Moskva, Izdatel'stvo "Mysl'," 1964. 325 p. Bibliography, p. 287-305.   980

About 450 references, arranged alphabetically. Indexes of plant, animal,
and geographical names.

SHINSKII, Georgii Kazimirovich, ed. Fizicheskaia geografiia SSSR (Physical
geography of the USSR), by M. I. Davydova, A. I. Kamenskii, N. P. Nekliukova,
and G. K. Tushinskii. 2nd ed. Moskva, Izdatel'stvo "Prosveshchenie," 1966.
347 p. Bibliography, p. 802-812. (1st ed., Moskva, Uchpedgiz, 1960. 680 p.). 981

About 360 references organized by systematic fields and major regions: maps and atlases, general overview of the USSR, seas of the USSR, seas of the Arctic Ocean, seas of the Pacific Ocean, seas of the Atlantic Ocean, lakes and seas, waters of the land, the Russian plain, the Crimea, the Caucasus, the Urals, islands of the Asiatic sector of the Arctic, Siberia (general survey, West Siberian plain, Middle Siberia, Northeast Siberia, the mountains of Southern Siberia, Baykal mountains), the Far East, Soviet Middle Asia.

U.S.S.R.: European Part and Caucasus

(See also 2227-2229)

ZUBOV, Sergei Mikhailovich. Fizicheskaia geografiia SSSR. Regional'nyi obzor. Chast' 1. Fiziko-geograficheskoe raionirovanie SSSR. Evropeiskaia chast' SSSR. Kavkaz (Physical geography of the USSR. Regional survey. v. 1. Physical-geographical regionalization of the USSR. European part of the USSR. The Caucasus). Minsk, Vysshaia Shkola, 1965. 364 p. Bibliography, p. 352-360. Chast' 2. Zapadno-Sibirskaia Ravnina. Kazakhstanskii Melkosopochnik i Turgaiskoe plato. Prikaspiisko-Turanskie Ravniny. Gory Srednei Azii (v. 2. The West-Siberian plain. Kazakhstan hill country and Turgay plateau. Pricaspian-Turan plains. Mountains of Middle Asia). Minsk, "Vysheish. Shkola, 1967. 272 p. Bibliography, p. 265-268.

About 260 references on the European USSR and the Caucasus and about 120 references on Western Siberia, Kazakhstan, and Soviet Middle Asia.

MIL'KOV, Fedor Nikolaevich, and GVOZDETSKII, Nikolai Andreevich. Fizicheskaia geografiia SSSR. Tom 1. Obshchii obzor. Evropeiskaia chast' SSSR, Kavkaz (Physical geography of the USSR. v. 1. General survey. European part of the USSR. The Caucasus). 3rd ed. Moskva, Izdatel'stvo "Mysl'," 1969. 461 p. Bibliography, p. 433-437. (1st ed. Geografgiz, 1958. 351 p. 2nd ed., 1962. 475 p.

147 references organized by major regions: general for the USSR, Russian Plain, Urals, the Carpathians, the Crimea, the Caucasus, and maps and atlases. Indexes of plant names, animal names, and geographical names.

DOBRYNIN, Boris Fedorovich. Fizicheskaia geografiia SSSR: Evropeiskaia chast' i Kavkaz (Physical geography of the USSR: European part and the Caucasus). 2nd ed., Moskva, Uchpedgiz, 1948. 324 p. Bibliography, p. 320-323. (1st ed., Moskva, Uchpedgiz, 1941).

About 150 references arranged by large regions: general works on the USSR, East European plain, Urals, Crimea, and the Caucasus.

PILATOV, Pavel Nikolaevich. Stepi SSSR kak uslovie material'noi zhizni obshchestva: k probleme--priroda i chelovek (Steppes of the USSR as the base for the material life of society. To the problem: nature and man). Yaroslavl', Verkhne-Volzhskoe Knizhnoe Izdatel'stvo, 1966. 287 p. Bibliography, p. 256-286 (Iaroslavskii Gosudarstvennyi Pedagogicheskii Institut. Geograficheskoe Obshchestvo SSSR, Iaroslavskii Otdel).

561 references in alphabetical order on natural conditions in the steppe zone of the USSR.

U. S. S. R.: Asiatic Part

GVOZDETSKII, Nikolai Andreevich, and MIKHAILOV, Nikolai Ivanovich. Fizicheskaia geografiia SSSR: Aziatskaia chast'(Physical geography of the USSR: Asiatic part). 2nd ed. Moskva, Izdatel'stvo "Mysl'," Geografgiz, 1970. 543 p. Bibliography, p. 509-512. (1st ed., Moskva, Geografgiz, 1963. 571 p.)

102 basic references arranged regionally: general for the Asiatic part of
the USSR, Soviet Middle Asia and Central Kazakhstan, the West Siberian Plain,
Middle Siberia, mountains of Southern Siberia, Northeast Siberia, Soviet Far
East, and maps and atlases. Indexes of plant names, animal names, and geographical names. Folded maps in pocket.

SLOV, Sergei Petrovich. Fizicheskaia geografiia SSSR: Aziatskaia chast' (Physical
geography of the USSR; Asiatic part). 2nd ed. Moskva, Uchpedgiz, 1954. 710 p.
1st ed., 1947. 544 p.   987

525 references listed at ends of chapters organized regionally for Western
and Eastern Siberia, the Soviet Far East, and Soviet Middle Asia, p. 83-84, 137-
138, 156, 193, 223, 251, 275-276, 357-358, 398, 448, 473, 517-518, 544, 640-641,
704-705, and 706. Available in English translation as Physical geography of
Asiatic Russia. San Francisco, California, W. H. Freeman and Co., 1961. 594 p.
[2585].

ZIKO-GEOGRAFICHESKOE raionirovanie Tiumenskoi oblasti (Physical-geographical
regionalization of Tyumen' oblast'). N. A. Gvozdetskii, ed. Moskva, Izdatel'stvo
Moskovskogo Universiteta, 1973. 246 p. Bibliography, p. 239-245. (Prirodnoe i
ekonomiko-geograficheskoe raionirovanie SSSR dlia sel'skogo khoziaistva).   988

Major World Areas Outside U.S.S.R.

ABCHIKOV, Aleksandr Maksimovich, ed. Fizicheskaia geografiia chastei sveta
(Physical geography of the continents). By N. V. Aleksandrovskaia, R. A. Eramov,
G. M. Ignat'ev, E. N. Lukashova, K. K. Markov, L. A. Mikhailova, and A. M.
Riabchikov. Moskva, Gosudarstvennoe Izdatel'stvo "Vysshaia Shkola," 1963. 547 p.
Bibliography, p. 519-522.   989

About 150 references to books on physical geography of the world mostly in
Russian, or Russian translations of foreign works.

GEOLOGY

General

FERATIVNYI zhurnal: geologiia (Reference journal: geology) (Vsesoiuznyi Institut Nauchnoi i Tekhnicheskoi Informatsii). Moskva. 1954- . monthly. Annual author and subject indexes. 1954-1955 combined with geography. 990

Massive and detailed coverage of the literature in all languages with abstracts and bibliographical notes. The main sections are: A. General geology; B. Stratigraphy and paleontology; G. Quaternary period and geomorphology of the land and of the floor of the sea (This section is included also in the geography volume); V. Geochemistry, mineralogy, and petrography; Zh. Ore deposits; I. Nonmetallic deposits; K. Deposits of mineral fuels; D. Geological and geochemical methods of prospecting for minerals; methods of reconnaissance and appraisal of deposits; exploration and production geophysics; L. Engineering geological-prospecting work; E. Hydrogeology, engineering geology, and study of permafrost.

VIKOV, Energii Alekseevich. Putevoditel' po geologicheskoi literature mira (Guide to the geological literature of the world). Leningrad, Izdatel'stvo "Nedra," Leningradskoe Otdelenie, 1971. 167 p. 991

U. S. S. R.

OLOGIIA. Petrografiia. Mineralogiia. Geofizika. Geokhimiia (Geology, petrography, mineralogy, geophysics, and geochemistry), Bibliografiia sovetskoi bibliografii (Bibliography of Soviet bibliography). (Vsesoiuznaia Knizhnaia Palata). 1939, 1946- . Annual. Moskva, Izdatel'stvo "Kniga," 1941, 1948- . 992

Annual list of Soviet bibliographies for geology, petrography, mineralogy, geophysics, and geochemistry published as separate works or as parts of books or articles. The 1970 annual volume (1972), p. 73-98, contains 676 entries.

OLOGICHESKAIA literatura SSSR (Geological literature of the USSR) (Vsesoiuznaia Geologicheskaia Biblioteka). Moskva, 1934, 1937, 1951-1958. Annual volumes 1934, 1937, 1951-1955, 5 nos. a year, quarterly plus index, 1956-1958). 993

Basic source of geological literature of the USSR for the years covered.

OLOGICHESKAIA literatura SSSR. Bibliograficheskii ezhegodnik za 1961 g. (Geological literature of the USSR for 1961). M. I. Vitoshinskaia, I. F. Gekker, G. P. Osanova, and others. Moskva, Izdatel'stvo "Nedra," -Saratov, 1971. 631 p. Ministerstvo Geologii SSSR. Vsesoiuznaia Geologicheskaia Biblioteka). 994

OLOGICHESKAIA literatura SSSR. Bibliograficheskii ezhegodnik za 1963 g. v. 2-kh omakh (Geological literature of the USSR. Bibliographical annual from 1963. In volumes). M. V. Kulikov and S. P. Solov'ev, eds. Leningrad, Idatel'stvo "Nauka," Leningradskoe Otdelenie, 1973. v. 1. 304 p. v. 2. 290 p. V. V. Belkina, I. F. Gekker, A. P. Illiashevich, and others, comps. (Ministerstvo Geologii SSSR. Vsesoiuznaia Geologicheskaia Biblioteka). 995

OLOGICHESKAIA literatura SSSR. Bibliograficheskii ezhegodnik za 1967 g. (Geological literature of the USSR. Bibliographical annual for 1967). Leningrad, Izdatel'stvo "Nedra," Leningradskoe Otdelenie, 1973. 2 v. v. 1. 326 p. v. 2. 20 p. V. V. Belkina, I. F. Gekker, I. E. Isaeva, and others. (Ministerstvo Geologii SSSR. Vsesoiuznaia Geologicheskaia Biblioteka). 996

GEOLOGIIA SSSR   (Geology of the USSR). I. I. Malyshev, P. Ia. Antropov, and A. V.
Sidorenko, successively eds. Leningrad or Moskva, Gosgeolizdat, later Izdatel'stvo
"Nedra," 1941- . In progress. (Ministerstvo Geologii SSSR, in co-operation with
geological ministries of union republics. Also as Komitet po Delam Geologii).   9

    Extensive series of monographs with bibliographies devoted to specific
regional segments of the Soviet Union.

GIDROGEOLOGIIA SSSR (Hydrogeology of the USSR). A. V. Sidorenko, ed. Moskva,
Izdatel'stvo "Nedra," 1966- . 50 v. In progress. (Ministerstvo Geologii.
Vsesoiuznyi Nauchno-Issledovatel'skii Institut Gidrogeologii i Inzhenernoi
Geologii "VSEGINGEO").   9

    Detailed set of regional volumes covering the Soviet Union produced in co-
operation with regional offices. Each volume has a substantial bibliography.

OBRUCHEV, Vladimir Afans'evich. Istoriia geologicheskogo issledovaniia Sibiri.
Period 1-5 (History of the geological investigation of Siberia). Moskva-
Leningrad, Izdatel'stvo Akademii Nauk SSSR, 1931-1959. 5 v.   9

    About 12,000 references.

GEOPHYSICS

REFERATIVNYI zhurnal: geofizika (Reference journal: geophysics) (Vsesoiuznyi
Institut Nauchnoi i Teknicheskoi Informatsii. Moskva. 1954- . monthly. Annual
subject and author indexes. 1954-1956 combined with physics.   10

    Includes sections on geomagnetism and upper layers of the atmosphere; physics
of the earth; meteorology and climatology (included also in geography); oceanography, hydrology, and glaciology (included also in geography); and geological and
geochemical methods of prospecting for minerals, methods of reconnaissance and
appraisal of deposits, exploration and production geophysics (included also in
geology).

MEZHDUNARODNYI geofizicheskii god. Bibliograficheskii ukazatel' literatury na
russkom iazyke za 1954-1957 gg. (International geophysical year. Bibliographical
guide to the literature in the Russian language, 1954-1957). Comp. by R. F.
Zatrutina. Moskva, Izdatel'stvo Akademii Nauk SSSR, 1958. 44 p. (Mezhdunarodnyi
Komitet po Provedeniiu Mezhdunarodnogo Geofizicheskogo Goda pri Prezidiume Akademii
Nauk SSSR).   10

_____ za 1958 g., comp. by R. F. Zatrutina and Z.M. Mikhailova. 1959. 63 p.

_____ za 1959 g., comp. by R. F. Zatrutina and L. S. Rubina. 1960. 88 p.

_____ za 1960 g., 1961. 61 p.

USHAKOV, S. A. Stroenie i razvitie Zemli (Structure and development of the earth).
V. V. Fedynskii, ed. Itogi nauki i tekhniki. Seriia "Fizika Zemli," tom 1.
Moskva, VINITI, 1974. 269 p. Bibliography, p. 251-268.   10

GEOMORPHOLOGY

General

REFERATIVNYI zhurnal: antropogenovyi period. Geomorfologiia sushi i morskogo dna (Reference journal: Quaternary period. Geomorphology of land and the sea floor). (Vsesoiuznyi Institut Nauchnoi i Tekhnicheskii Informatsii). Moskva, 1954-monthly. Included as section G in the combined volumes both for geography and for geology and since 1962 also available separately.   1003

About 3,300 abstracts a year on the literature in all languages, divided into three main sections:

The Quaternary period with sections: general questions; stratigraphy and geochronology; history of development of the fauna and flora; paleogeography; genetic types of continental deposits and structure of matter; and paleontology and archeology.

Neotectonics includes sections: general problems, methods of research and cartography; and regional neotectoncs.

Geomorphology includes sections: general section; general problems of geomorphology; geomorphology of the land; paleogeomorphology; and geomorphology of the sea floor. Geomorphology of the land is subdivided into morphostructures formed by recent tectonic movements and volcanic and pseudovolcanic processes; morphostructures formed by dissection and preparation of primary tectonic morphostructures and ancient passive geological structures; morphosculptures (forms of relief formed by the activity of exogenous processes further subdivided by running water, weathering, gravitational processes, solution, wind, ice, periglacial phenomena, waves, and organic forms),and general questions of "climatic" and dynamic geomorphology; and regional geomorphology.

ITOGI nauki i tekhniki, seriia geografiia: geomorfologiia (Results of science and technology, geography series: geomorphology). (Vsesoiuznyi Institut Nauchnoi i Tekhicheskoi Informatsii). Moskva, VINITI, 1- (1966-). Irregular.   1004

1. A. S. Devdariani. Matematicheskie metody (Mathematical methods). 1966. 142 p. 176 references.

2. D. V. Borisevich, E. A. Liubimtseva. Geomorfologicheskoe kartirovanie (Geomorphological mapping). A. I. Spiridonov, ed. 1971. 150 p. 358 references.

3. Poverkhnosti vyravnivaniia Evropy, Azii i Afriki (Erosional surfaces in Europe, Asia, and Africa). D. V. Borisevich, ed. 1973. 233 p. 737 references.

SHCHUKIN, Ivan Semenovich. Obshchaia geomorfologiia (General geomorphology). 2nd ed. Moskva, Izdatel'stvo Moskovskogo Universiteta, v. 1, 1960. 615 p. v. 2, 1964. 564 p. (1st ed., 1938. 460 p.).   1005

More than 1600 references in Russian and other languages, in lists of literature at ends of each section and chapter.

METODICHESKOE rukovodstvo po geomorfologicheskim issledovaniiam (Methodological manual on geomorphological research). M. N. Voitsov, G. S. Ganeshin, M. I. Plotnikova, and others, comps. Leningrad, "Nedra," Leningradskoe Otdelenie, 1972. 384 p. Bibliography, p. 342-361. (Vsesoiuznyi Nauchno-Issledovatel'skii Geologicheskii Institut "VSEGEI").   1006

505 references. Subject index.

REL'EF zemli: morfostruktura i morfoskul'ptura (Relief of the earth: morphostructure and morphosculpture). Moskva, Izdatel'stvo "Nauka," 1967. 331 p. Bibliography, p. 309-323, comp. by R. S. Narskikh. (Akademiia Nauk SSSR. Institut Geografii). 1007

About 700 entries.

DEVDARIANI, Anatolii Seitovich. Matematicheskie metody [v geomorfologii] (Mathematical methods [in geomorphology]). Moskva, 1966. 142 p. Bibliography, p. 131-141. (Akademiia Nauk SSSR. Institut Nauchnoi Informatsii. Itogi Nauki. Seriia "Geografiia," Geomorfologiia. Vypusk 1). 10

176 references.

DEVDARIANI, Anatolii Seitovich. Izmerenie peremeshchenii zemnoi poverkhnosti (Measurement of displacement of the earth's surface). Moskva, Izdatel'stvo "Nauka," 1964. 247 p. Bibliography, p. 225-244 (Akademiia Nauk SSSR. Institut Geografii). 10

About 500 references, arranged alphabetically, about 340 of which are of Soviet sources.

MARKOV, Konstantin Konstantinovich. Osnovnye problemy geomorfologii (Basic problems of geomorphology). Ia. S. Edel'shtein, ed. Moskva, Geografgiz, 1948. 344 p. Bibliography, p. 320-332. (Moskovskii Gosudarstvennyi Universitet. Nauchno-Issledovatel'skii Institut Geografii). 10

More than 300 references grouped by major topics.

MESHCHERIAKOV, Iurii Aleksandrovich. Strukturnaia geomorfologiia ravninnykh stran (Structural geomorphology of plains areas). Moskva, Izdatel'stvo "Nauka," 1965. 390 p. Bibliography, p. 371-388. (Akademiia Nauk SSSR. Institut Geografii). 10

About 550 references in alphabetical order, about 450 of which are Soviet sources.

NIKOLAEV, Nikolai Ivanovich. Istoriia razvitiia osnovnykh predstavlenii v geomorfologii (History of the development of the fundamental concepts in geomorphology). Ocherki po istorii geologicheskoi znanii, no. 6 (1958). Moskva, Izdatel'stvo Akademii Nauk SSSR, 1958, p. 3-96. Bibliography, p. 88-96. (Akademiia Nauk SSSR. Institut Geologii). 10

About 230 references in Russian and other languages.

OBSHCHIE raboty. Teoreticheskie voprosy. Sborniki statei. Geomorfologicheskie zhurnaly (General works. Theoretical questions. Collections of articles. Geomorphological journals), L. M. Lenau and V. A. Sokolova, comps., Geomorfologiia 1970, no. 1, p. 92-93, no. 2, p. 92-95, no. 3, p. 111-116, no. 4, p. 93-96. 10

SIMONOV, Iurii Gavrilovich. Regional'nyi geomorfologicheskii analiz (Regional geomorphological analysis). Moskva, Izdatel'stvo Moskovskogo Universiteta, 1972. 251 p. Bibliography, p. 236-250. 10

About 430 references on analysis of regional geomorphology.

SPIRIDONOV, Aleksei Ivanovich. Osnovy obshchei metodiki polevykh geomorfologicheski issledovanii (Fundamentals of general methods of field research in geomorphology) Moskva, Izdatel'stvo Moskovskogo Universiteta. v. 1, 1956. v. 2, 1959. v. 3, part 2, 1963. (v. 1 as Moskovskii Gosudarstvennyi Universitet. Uchenye zapiski, vypusk 182). Bibliography, v. 2, p. 285-298, v. 3, part 2, p. 346-369. 10

259 references in v. 2; 380 in v. 3, part 2.

ZVONKOVA, Tat'iana Vasil'evna. Izuchenie rel'efa v prakticheskikh tseliakh (The study of relief for practical ends). Moskva, Geografgiz, 1959. 304 p. Bibliography, p. 277-303.

About 600 references systematically organized by chapters: general questions of the practical study of relief and the application of geomorphology to the economy of the USSR; geomorphological investigations in the search for minerals; geomorphological research in planning engineering structures; relief in agriculture; geomorphological research in cartographic production; applied geomorphological mapping.

IVOLUTSKII, Aleksandr Evgen'evich. Zhizn' zemnoi poverkhnosti: problemy geomorfologii (Life of the surface of the globe: problems in geomorphology). Moskva,
Izdatel'stvo "Mysl'," 1971. 407 p. Bibliography, 383-405.   1017

Special Topics

EKSANDROVA, Tat'iana Davydovna. Vnutrigornye kotloviny (Intermontane basins).
Moskva, Izdatel'stvo "Nauka," 1972. 118 p. Bibliography, p. 109-117. (Akademiia
Nauk SSSR. Institut Geografii).   1018

KSIMOV, A. A. Geograficheskoe raionirovanie rechnykh dolin (Geographical regionalization of river valleys). In the book: Geograficheskie problemy Sibiri (Geographical problems of Siberia). Nauchnye Soobshcheniia Programme XXII Mezhdunarodnogo Geograficheskogo Kongressa. V. N. Saks, ed. Novosibirsk, Izdatel'stvo
"Nauka," Sibirskoe Otdelenie, 1972. p. 45-67. Bibliography, p. 62-67. Summary
in English. (Akademiia Nauk SSSR. Sibirskoe Otdelenie. Geograficheskoe
Obshchestvo SSSR. Novosibirskii Otdel).   1019

109 references.

ZNETSOV, Vladilen Aleksandrovich. Geokhimiia alliuvial'nogo litogeneza (Geochemistry of alluvial rock formation). Minsk. Izdatel'stvo "Nauka i Tekhnika,"
1973. 278 p. Bibliography, p. 262-276. (Akademiia Nauk Belorusskoi SSR.
Institut Geokhimii i Geofiziki. K IX Mezhdunarodnomu Kongressu INKVA).   1020

SKRESENSKII, Sergei Sergeevich. Dinamicheskaia geomorfologiia. Formirovanie
sklonov (Dynamic geomorphology. Formation of slopes). Moskva, Izdatel'stvo
Moskovskogo Universiteta, 1971. 229 p. Bibliography, p. 223-228.   1021

RASIMOV, Innokentii Petrovich, GANESHIN, G. S., BORISEVICH, D. V., and others.
Primenenie geomorfologicheskikh metodov v strukturno-geologicheskikh issledovaniiakh (Utilization of geomorphological methods in research in structural
geology). I. P. Gerasimov and others, eds. Moskva, Izdatel'stvo "Nedra," 1970.
296 p. Bibliography, p. 285-294.   1022

RISEVICH, Dmitrii Vasil'evich, ed. Poverkhnosti vyravnivaniia Evropy, Azii i
Afriki (Erosional surfaces in Europe, Asia, and Africa). D. V. Borisevich, D.
A. Timofeev, I. N. Oleinikov. Itogi nauki i tekhniki, seriia geomorfologiia, v.
3. Moskva, VINITI, 1973. 233 p. Bibliography, p. 170-231. (Vsesoiuznyi
Institut Nauchnoi i Tekhnicheskoi Informatsii).   1023

737 references in two alphabetical orders, first works in the Cyrillic
alphabet, 493 entries, and then in the Latin alphabet, 244 entries.

Geomorphological Mapping

OMORFOLOGICHESKOE kartovedenie SSSR i chastei sveta (Geomorphological cartography
of the USSR and the continents). L. R. Serebriannyi, ed. L. R. Serebriannyi,
N. F. Leont'ev, A. V. Zhivago, and others, authors. Moskva, Izdatel'stvo "Nauka,"
1973. 248 p. Bibliography, p. 230-247 (Akademiia Nauk SSSR. Institut Geografii).

About 700 references   1024

RISEVICH, Dmitrii Vasil'evich, and LIUBIMTSEVA, E. A. Geomorfologicheskoe
kartirovanie 1965-1969 gg. (Geomorphological mapping, 1965-1969). Itogi Nauki,
Seriia Geografiia: Geomorfologiia, no. 2, Moskva, VINITI, 1971. 150 p. Bibliography, p. 112-149 (Vsesoiuznyi Institut Nauchnoi i Tekhnicheskoi Informatsii). 1025

358 entries on geomorphological mapping in two separate alphabetical orders,
one in Cyrillic (Russian and Ukrainian), 181 entries, the other in Latin letters
(Western languages), 177 entries. Citation number is given for abstract of each
entry in Referativnyi zhurnal: geografiia.

Reviews the literature with particular attention to the legends of general geomorphological maps with conclusions on maps of geomorphological regions, maps of type of relief, and maps of forms of relief.

LITERATURA po voprosam geomorfologicheskogo kartirovaniia (Literature on problems of geomorphological mapping). In the book: Geomorfologicheskoe kartirovanie (Geomorphological mapping). Moskva, Izdatel'stvo Akademii Nauk SSSR, 1963, p. 182-189. (Akademiia Nauk SSSR. Geograficheskoe Obshchestvo SSSR. Moskovskii Filial).   10

180 references in alphabetical order.

SPIRIDONOV, Aleksei Ivanovich. Geomorfologicheskoe kartografirovanie (Geomorphological mapping). Moskva, Geografgiz, 1952. 185 p. Bibliography, p. 181-185. 10

About 110 references.

Karst (See also 2230-2231)

GVOZDETSKII, Nikolai Andreevich. Karst. Voprosy obshchego i regional'nogo karstovedeniia (Karst. Questions of general and regional karst study). 2nd ed. Moskva, Geografgiz, 1954. 352 p.   10

About 500 references arranged systematically in bibliographies for individual chapters.

_____. Regional'noe karstovedenie. Kratkii obzor novykh issledovanii i nekotorye problemy (Regional karst study. Short review of new research and some problems). In the book: Regional'noe karstovedenie. Trudy soveshchaniia po regional'nomu karstovedeniiu. Moskva. (Moskovskoe Obshchestvo Ispytatelei Prirody. Geograficheskaia Sektsiia). 1958, 80 p. Bibliography, p. 4-14.   10

109 references published 1954-1958.

GVOZDETSKII, Nikolai Andreevich. Problemy izucheniia karsta i praktika (Problems of the study of karst and practice). Moskva, Izdatel'stvo "Mysl'," 1972. 392 p. Bibliography, p. 341-391.   10

SOKOLOV, Dmitrii Sergeevich. Osnovnye usloviia razvitiia karsta (Basic conditions for the development of karst). Moskva. Gosgeoltekhizdat, 1962. 322 p. Bibliography, p. 287-320 (Vsesoiuznyi Nauchno-Issledovatel'skii Institut Gidrogeologii i Inzhenernoi Geologii, VSEGINGEO).   10

1,200 publications, about 1,150 of them Soviet.

MAKSIMOVICH, Georgii Alekseevich. Osnovy karstovedeniia. Tom 1. Voprosy morfologii karsta, speleologii i gidrogeologii karsta (Basic karst studies. v. 1. Questions of morphology of karst, speleology and hydrogeology of karst). Perm', Knizhnoe Izdatel'stvo, 1963. 444 p. Bibliography, p. 400-440 (Permskii Gosudarstvennyi Universitet).   10

About 1,200 references organized by chapters.

CHIKISHEV, Anatolii Grigor'evich. Metody izucheniia karsta (Methods of karst study). Moskva, Izdatel'stvo Moskovskogo Universiteta, 1973. 91 p. Bibliography, p. 84-90.   10

CHIKISHEV, Anatolii Grigor'evich. Karstovye peshchery SSSR (Karst caves in the USSR) In the book: Speleologiia i karstovedenie. Materialy soveshchaniia po speleologii i karstovedeniiu 17-18 dekabria 1958 g. (Speleology and karst studies. Materialy of the conference on speology and karst studies, December 17-18, 1958). Moskva, 1959. p. 7-40. Bibliography, p. 32-40. (Moskovskoe Obshchestvo Ispytatelei Prirody. Geograficheskaia Sektsiia).   10

233 references.

CHURIN, Sergei Petrovich. Termokarst na territorii SSSR (Thermokarst on the
territory of the USSR). Moskva, Izdatel'stvo Akademii Nauk SSSR, 1961. 291 p.
Bibliography, p. 264-289.   1035

  About 330 references in alphabetical order.

DIONOV, Nikolai Vasil'evich. Karst Evropeiskoi chasti SSSR, Urala i Kavkaza
(Karst of the European patt of the USSR, Utals, and the Caucasus). Moskva,
Gosgeoltekhizdat, 1963. 175 p. Bibliography, p. 163-174. (Vsesoiuznyi Nauchno-
Issledovatel'skii Institut Gidrogeologii i Inzhenernoi Geologii, VSEGINGEO, Trudy,
n.s., no. 13).   1036

  275 references arranged alphabetically.

KNADZE, T. Z. Karst massiva Arabika. Abkhazskaia ASSR (Karst of the Arabika
massif in the Abkhaz ASSR). Tbilisi, "Metsniereba," 1972. 248 p. Bibliography,
p. 227-243. Summary in French.   1037

KUASHVILI, Levan Iosifovich. Morfologicheskii analiz karstovykh peshcher: na
primere Zapadnoi Gruzii, v sopostavlenii s drugimi karstovymi oblastiami SSSR i
zarubezhnykh stran (Morphological analysis of karst caves: the example of western
Georgia with comparisons with other karst regions of the USSR and foreign countries). In the book: Ocherki po fizicheskoi geografii Gruzii (Studies in the
physical geography of Georgia). Tbilisi, "Metsniereba," 1969. p. 5-84. Bibliography, p. 79-84. (Akademiia Nauk Gruzinskoi SSR. Institut Geografii imeni
Vakhushti).   1038

Landslides

EL'IANOVA, Evgeniia Petrovna. Ukazatel' literatury po opolzniam i ustoichivosti
sklonov, izdannoi v 1936-1960 gg. (na russkom iazyke) i kratkii obzor sostoianiia
izuchennosti opolznevoi problemy v SSSR (Guide to the literature on landslides
and the stability of slopes, published in 1936-1960 in Russian and short review
of the status of the study of the slide problem in the USSR). Moskva. 1962.
93 p. Bibliography, p. 18-79. (Vsesoiuznyi Nauchno-Issledovatel'skii Institut
Gidrogeologii i Inzhenernoi Geologii, VSEGINGEO).   1039

  855 entries in chronological order. Author index. Thematic index.

GREBOV, Nikolai Fedorovich. Ukazatel' literatury po opolzniam (Guide to the
literature on landslides). Leningrad-Moskva, ONTI, 1936. 52 p. (Tsentral'nyi
Nauchno-Issledovatel'skii Geologo-Razvedochnyi Institut).   1040

  947 references in alphabetical order, 400 in Russian, 547 in other languages.
Subject index. Geographical index.

ONSHTAM, M. G., KADKINA, E. L., and MOLOKOV, L. A. Opolzni i metod ikh izucheniia
(Landslides and methods of their study). Itogi nauki i tekhniki, Gidrogeologiia
i inzhenernaia geologiia (Hydrogeology and engineering geology, v. 2. Moskva,
VINITI, 1972., p. 69-154.   1041

  252 references.

EL'IANOVA, Evgeniia Petrovna. Osnovnye zakonomernosti opolznevykh protsessov
(Fundamental regularities in landslide processes). Moskva, Izdatel'stvo "Nedra,"
1972. 310 p. Bibliography, p. 297-307.   1042

Marine Geomorphology and Coasts

ONT'EV, Oleg Konstantinovich. Kratkii kurs morskoi geologii. (Short course on
marine geology). Moskva, Izdatel'stvo Moskovskogo Universiteta, 1963. 464 p.
Bibliography, p. 446-461.   1043

About 400 references in alphabetical order, about 280 Soviet and 120 foreign, on geology and geomorphology of the sea floor.

LEONT'EV, Oleg Konstantinovich. Osnovy geomorfologii morskikh beregov (Fundamentals of the geomorphology of sea coasts). Moskva, Izdatel'stvo Moskovskogo Universiteta 1961. 418 p. Bibliography, p. 408-417. 10/

About 250 references in Russian and other languages.

SAF'IANOV, Gennadii Aleksandrovich. Dinamika beregovoi zony morei: teksty lektsii (Dynamics of coastal zone of the sea: texts of lectures). O. K. Leont'ev, ed. Moskva, Izdatel'stvo Moskovskogo Universiteta, 1973. 175 p. Bibliography, p. 166-172. (Moskovskovskii Gosudarstvennyi Universitet imeni M. V. Lomonosova. Geograficheskii Fakul'tet). 10/

LEONT'EV, Oleg Konstantinovich, and SAF'IANOV, Gennadi Aleksandrovich. Kan'ony pod morem (Canyons under the sea). Moskva, Izdatel'stvo "Mysl'," 1973. 261 p. Bibliography, p. 254-260. 10/

UDINTSEV, Gleb Borisovich. Issledovaniia rel'efa dna morei i okeanov (Research on the relief of the sea and ocean bottom). In the book: Uspekhi v izuchenii okeanicheskikh glubin: biologiia i geologiia (Progress in the study of the ocean deeps: biology and geology). Moskva, Izdatel'stvo Akademii Nauk SSSR, 1959, p. 27-90. 10/

About 450 references, 120 Soviet and 330 foreign, in alphabetical order.

PANOV, Dmitrii Gennadievich. Morfologiia dna mirovogo okeana (Morphology of the ocean bottom). Moskva-Leningrad, Izdatel'stvo Akademii Nauk SSSR, 1963. 228 p. Bibliography, p. 212-227 (Geograficheskoe Obshchestvo SSSR. Zapiski, n.s. tom 23). 10/

More than 600 references in alphabetical order, about 300 Soviet and about 300 foreign.

ZENKOVICH, Vsevolod Pavlovich. Osnovy ucheniia o razvitii morskikh beregov (Foundations of the science of the development of sea coasts). Moskva, Izdatel'stvo Akademii Nauk SSSR, 1962. 710 p. Bibliography, p. 669-700 (Akademiia Nauk SSSR. Okeanograficheskaia Komissiia). 10/

More than 1,000 references in alphabetical order, about 800 Soviet and 200 foreign.

LEONT'EV, Oleg Konstantinovich. Geomorfologiia morskikh beregov i dna (Geomorphology of sea coasts and floors). V. P. Zenkovich and I. S. Shchukin, eds. Moskva, Izdatel'stvo Moskovskogo Universiteta, 1955. 379 p. Bibliography, p. 365-373. (K 200-letiiu Moskovskogo Universiteta, 1755-1955). 10:

About 350 references in alphabetical order.

KAPLIN, Pavel Alekseevich. Fiordlovye poberezh'ia Sovetskogo Soiuza (The fiord coasts of the Soviet Union). Moskva, Izdatel'stvo Akademii Nauk SSSR, 1962. 188 p. Bibliography, p. 176-186 (Akademiia Nauk SSSR. Institut Okeanologii). 10

About 450 references in alphabetical order, about 250 Soviet and 200 foreign.

ZENKOVICH, Vsevolod Pavlovich. Morfologiia i dinamika sovetskikh beregov Chërnogo Moria (Morphology and dynamics of the Soviet shores of the Black Sea). Moskva, Izdatel'stvo Akademii Nauk SSSR, 1958-1960. 2 v. (Akademiia Nauk SSSR. Institut Okeanologii). 10

v. 1. 1958, 187 p. Bibliography, p. 176-185. About 330 references.
v. 2. The Northwest Part. 1960, 216 p. Bibliography, p. 212-215. About 130 references.

ONT'EV, Oleg Konstantinovich, and KHALILOV, A. I. Prirodnye usloviia formirovaniia beregov Kaspiiskogo moria (Natural conditions in the formation of the coasts of the Caspian Sea). Baku, Izdatel'stvo Akademii Nauk Azerbaidzhanskoi SSR, 1965. 206 p. Bibliography, p. 197-205 (Akademiia Nauk Azerbaidzhanskoi SSR. Institut Geografii).   1053

About 260 references on the geomorphology of the Caspian Sea.

MAREV, Vasilii Iosifovich. Berega Aral'skogo Moria--vnutrennego vodoëma aridnoi zony (The coasts of the Aral Sea--an internal reservoir of the arid zone). Leningrad, Izdatel'stvo "Nauka," 1967. 252 p. Bibliography, p. 239-250. (Akademiia Nauk SSSR. Geograficheskoe Obshchestvo SSSR).   1054

About 280 references.

'SHKIN, Boris Andreevich. Dinamika beregov vodokhranilishch (Dynamics of the shores of reservoirs). 3rd ed. Kiev, Izdatel'stvo "Naukova Dumka," 1973. 413 p. Bibliography, p. 387-411 (Akademiia Nauk Ukrainskoi SSR. Institut Gidromekhaniki). 2nd ed. was entitled: Voprosy dinamiki beregov vodokhranilishch.   1055

578 references.

Periglacial

STIAEV, A. G. Bibliografiia rabot sovetskikh avtorov po voprosam perigliatsial'noi morfologii za piat' let, 1955-1959 (Bibliography of the works of Soviet authors on questions of periglacial morphology, 1955-1959), in the book: Perigliatsial'nye iavleniia na territorii SSSR: sbornik statei (Periglacial phenomena on the territory of the USSR: a collection of articles). Moskva, Izdatel'stvo Moskovskogo Universiteta, 1960, p. 276-288.   1056

About 240 references. Systematic guide; general questions; macrostructure polygonal formations; the texture of frozen ground; microstructure polygonal formations; nonstructural forms; thermokarst and related forms of relief; mountain and solifluction terraces, solifluction; loess and covering loams; other phenomena; physical and chemical processes in the periglacial zone. Within each division entries are in alphabetical order.

U. S. S. R.

General and Topical

ZICHESKAIA geografiia: annotirovannyi perechen' otechestvennykh bibliografii izdannykh v 1810-1966 gg. (Physical geography: annotated list of native bibliographies published 1810-1966). M. N. Morozova and E. A. Stepanova, comps. V. V. Klevenskaia, ed. L. G. Kamanin, scientific consultant. Moskva, Izdatel'stvo "Nauka," 1968. 309 p. (Gosudarstvennaia Biblioteka SSSR imeni V. I. Lenina. Institut Geografii Akademii Nauk SSSR. Sektor Seti Spetsial'nykh Bibliotek Akademii Nauk SSSR).   1057

Geomorphology, p. 95-116, entries 365-490, 121 Soviet bibliographies of geomorphology organized systematically: bibliographies of a general character; bibliographies on individual themes and problems (relief produced by endogenous factors, relief produced by exogenous processes, karst relief, relief produced by denudational processes, gravitational relief, permafrost and periglacial relief); bibliographies on regional geomorphology (European part of the USSR and the Caucasus, European part of the USSR, North and Northwest, Center, West, Southwest, Volga region, Urals, Caucasus; Asiatic part of the USSR, Siberia and the Far East, Western Siberia, Eastern Siberia; Middle Asia and Kazakhstan, northern plains part, southern mountainous part); bibliographies on the geomorphology of the sea floor and of coasts. Author and geographical indexes.

GERENCHUK, Kalinik Ivanovich. Tektonicheskie zakonomernosti v orografii i rechnoi
seti Russkoi Ravniny (Tectonic regularities in the orography and the stream network of the Russian plain). L'vov, Izdatel'stvo L'vovskogo Universiteta, 1960.
242 p. Bibliography, p. 220-240. (Geograficheskoe Obshchestvo SSSR. Zapiski,
n.s. tom 20. Geograficheskoe Obshchestvo SSSR. L'vovskii Otdel).   10

     About 650 references on the Russian platform in alphabetical order in
Russian, Ukrainian, and other languages.

LOPATIN, Georgii Vladimirovich. Nanosy rek SSSR: Obrazovanie i perenos (Alluvium
of the rivers of the USSR: formation and transport). Moskva, Geografgiz, 1952.
367 p. Bibliography, p. 354-364. (Geograficheskoe Obshchestvo SSSR. Zapiski,
n.s. tom 14).   10

     About 270 references.

MESHCHERIAKOV, Iurii Aleksandrovich. Rel'ef SSSR: morfostruktura i morfoskul'ptura
(Relief of the USSR: morphostructure and morphosculpture). Moskva, Izdatel'stvo
"Mysl'," Geografgiz, 1972. 519 p. Bibliography, p. 498-517.   10

     About 375 references on the geomorphology of the USSR arranged alphabetically
by name of author.

NIKOLAEV, Nikolai Ivanovich. Neotektonika i ee vyrazhenie v strukture i rel'efe
territorii SSSR: voprosy regional'noi i teoreticheskoi neotektoniki (Neotectonics
and its expression in the structure and relief of the territory of the USSR: questions of regional and theoretical neotectonics). Moskva, Gosgeoltekhizdat, 1962.
392 p. Bibliography, p. 353-377.   10

     About 750 references alphabetically arranged, of which about 650 are Soviet
sources.

European U.S.S.R.

SOBOLEV, Sergei Stepanovich. Razvitie erozionnykh protsessov na territorii
Evropeiskoi chasti SSSR i bor'ba s nimi (Development of erosional processes on
the territory of the European part of the USSR and the fight against them).
Moskva, Izdatel'stvo Akademii Nauk SSSR, 1948-1960. 2 v. (Akademiia Nauk SSSR.
Pochvennyi Institut imeni V. V. Dokuchaeva).   10

     v. 1. 1948. Bibliography, p. 287-303. About 530 references.

     v. 2. 1960. Bibliography, p. 237-246. About 330 references.

GLADTSIN, Ivan Nikolaevich. Geomorfologiia SSSR. Chast' 1. Geomorfologiia
Evropeiskoi chasti SSSR i Kavkaz (Geomorphology of the USSR. v. 1. Geomor-
phology of the European part of the USSR and the Caucasus). Leningrad, Uchpedgiz,
Leningradskoe Otdelenie, 1939. 384 p. (Leningradskii Gosudarstvennyi Universitet.
Geografo-Ekonomicheskii Nauchno-Issledovatel'skii Institut).   10

     About 540 references at ends of chapters on individual regions.

KARANDEEVA, Mariia Vissarionovna. Geomorfologiia Evropeiskoi chasti SSSR (Geomorphology of the European part of the USSR). Moskva, Izdatel'stvo Moskovskogo
Universiteta, 1957, 314 p. Bibliography, p. 291-321.   10

     About 600 references arranged by regions: general for the European part of
the USSR; Karelia-Kola region; Silurian plateau, Glint lowland, and Riga-Il'men'
hilly plain and lake country; Belorussian-Mezen' region; Pechora region; Timanskii
ridge; Severnye Uvaly; Poles'ye and the Dnepr lowland; Oka-Don lowland; lower
Volga; Volynian-Podolian and Dnepr upland; Central Russian upland; Black Sea
lowland; Caspian lowland; Crimean mountains and Kerch' peninsula; and Eastern
Carpathians, Trans-Carpathian plain, and Carpathian elevated plain.

VZ, Nikolai Sidorovich. Zakonomernosti razvitiia osnovnykh morfogeneticheskikh kompleksov platformennykh ravnin: na primere tsentral'noi chasti Russkoi ravniny (Regularities in the development of the basic morphogenetic complex platform plain: the example of the central part of the Russian plain). K. S. Ovodov, ed. Voronezh, Tsentral'no-Chernozemnoe Knizhnoe Izdatel'stvo, 1970. 192 p. Bibliography, p. 179-191. (Voronezhskii Gosudarstvennyi Pedagogicheskii Institut). 1065
        About 360 references.

DKOV, Aleksei Petrovich. Ekzogennoe rel'efoobrazovanie v Kazansko-Ul'ianovskom Privolzh'e (Exogenous relief formation in the Kazan'-Ul'yanovsk Volga region). Kazan'. Izdatel'stvo Kazanskogo Universiteta, 1970. 255 p. Bibliography, p. 240-253.   1066
        About 320 references.

EDIENTOVA, Glafira Vital'evna. Proiskhozhdenie Zhigulevskoi vozvyshennosti i razvitie ee rel'efa (Origin of the Zhiguli upland and the development of its relief). Moskva, Izdatel'stvo Akademii Nauk SSSR, 1953. 248 p. Bibliography, p. 235-246. (Akademiia Nauk SSSR. Institut Geografii, Trudy, vypusk 53. Materialy po Geomorfologii i Paleogeografii SSSR, 8).   1067
        About 330 references in alphabetical order.

ZHDESTVENSKII, A. P. Noveishaia tektonika i razvitie rel'efa Iuzhnogo Priural'ia (Recent tectonics and the development of the southern Ural region). Moskva, Izdatel'stvo "Nauka," 1971. 303 p. Bibliography, p. 285-302. (Akademiia Nauk SSSR. Bashkirskii Filial. Institut Geologii).   1068

RELOV, Sergei Kuz'mich. Morfostrukturnyi analiz neftegazonosnykh territorii: na primere iugo-vostoka Russkoi ravniny (Morphostructural analysis of oil-bearing areas: the example of the southeast Russian plain). Moskva, Izdatel'stvo "Nauka," 1972. 216 p. Bibliography, p. 207-215. (Akademiia Nauk SSSR. Institut Geografii).   1069

RANDEEVA, Mariia Vissarionova, and LEONT'EV, Oleg Konstantinovich, eds. Geomorfologiia zapadnoi chasti Prikaspiiskoi nizmennosti (Geomorphology of the western part of the Caspian lowland). Moskva, 1958. 238 p. Bibliography, p. 236-238. (Moskovskii Gosudarstvennyi Universitet imeni Lomonosova. Geograficheskii Fakul'tet. Trudy Prikaspiiskoi ekspeditsii). Rotaprint.   1070
        About 180 references.

RZHUEV, Sergei Sergeevich. Rel'ef Pripiatskogo Poles'ia: strukturnye osobennosti i osnovnye cherty razvitiia (Relief of Pripyat' Poles'ye: structural features and basic lines of development). Moskva, Izdatel'stvo Akademii Nauk SSSR, 1960. 141 p. Bibliography, p. 132-140. (Akademiia Nauk SSSR. Institut Geografii).   1071
        About 200 references.

RINICH, Aleksandr Mefod'evich. Geomorfologiia iuzhnogo Poles'ia (Geomorphology of the southern Poles'ye). Kiev, Izdatel'stvo Kievskogo Universiteta. 1963. 252 p. Bibliography, p. 234-251.   1072
        About 460 references on the geomorphology of the northwest Ukraine in Poles'ye, or the Pripet (Pinsk) marshes, in Russian, Ukrainian, and Polish.

YS', Petr Nikolaevich. Geomorfologiia USSR (Geomorphology of the Ukrainian SSR). L'vov, Vidavnitstvo L'vivskoho Universytetu, 1962. 224 p. Bibliography, 209-222. In Ukrainian.   1073
        About 400 references in alphabetical order in Russian, Ukrainian, and Polish.

PROKHODSKII, S. I., and EFREMOVA, M. S. Literatura po voprosam strukturnoi geomorfologii i neotektoniki Ukrainy (Literature on questions of structural geomorphology and neotectonics of the Ukraine). Materialy Khar'kovskogo otdela
geograficheskogo obshchestva Ukrainy, no. 6. Strukturnaia geomorfologiia i
neotektonika Ukrainy, v. 1. Moskva, 1968. p. 98-106.   107

PROKHODSKII, S. I., and GOL'DFEL'D, I. A. Raboty poslednikh let po strukturnoi
geomorfologii i neotektonike Ukrainy: bibliograficheskii ukazatel' (Works in
recent years on the structural geomorphology and neotectonics of the Ukraine: a
bibliographical guide), Materialy Khar'kovskogo otdela geograficheskogo obshchestva
Ukrainy, no. 9, Strukturnaia geomorfologii i neotektonika Ukrainy, v. 2. Moskva,
1970, p. 180-186.   10'

    134 entries, mainly in the period 1966-1968, on structural geomorphology,
paleogeomorphology, and neotectonics, of the Ukraine.

SOKOLOVS'KII, I. L. Zakonomirnostii rozvytku rel'efu Ukrainy (Regularities in the
development of the relief of Ukraine). Kyïv, "Naukova Dumka," 1973. 215 p. Bibliography, p. 199-211. In Ukrainian. Summary in Russian. (Akademiia Nauk
Ukrainskoï RSR. Sektor Heohrafii).   10'

SEREBRIANNAIA, T. A. and SEREBRIANNYI, Leonid Ruvimovich. Rel'ef i chetvertichnye
otlozheniia Pribaltiki v trudakh sovetskikh issledovatelei, 1960-1963. Bibliografii (Relief and Quaternary deposits of the Baltic region in the publications
of Soviet research workers, 1960-1963. A bibliography). Baltica (International
yearbook for the Quaternary geology and paleogeography, coastal morphology and
shore processes, marine geology and recent tectonics of the Baltic area), Vilnius,
v. 2 (1965), p. 317-360.   1(

    Title of journal in Lithuanian: Baltica: Tarptautinis metraštis baltijos
jūros kvartero geologijos ir paleogeografijos, krantų morfologijos ir dinamikos,
jūrų geologijos ir neotektonikos klausimais.

    460 references, first alphabetically in Cyrillic alphabet for works in
Russian (249), then for works in the Latin alphabet (211) in Estonian, Latvian,
Lithuanian, English, German, and other languages. Subject index by broad
categories.

Caucasus and Trans-Caucasus (See also 2232)

SAFRONOV, Ivan Nikolaevich. Geomorfologiia Severnogo Kavkaza (Geomorphology of
the North Caucasus). Rostov na Donu, Izdatel'stvo Rostovskogo Universiteta,
1969. 218 p. Bibliography, p. 206-217.   1(

    About 250 references.

SHCHUKIN, Ivan Semenovich. Ocherki geomorfologii Kavkaza. Chast' 1. Bol'shoi
Kavkaz (Studies in the geomorphology of the Caucasus. v. 1. The Great Caucasus)
Moskva, Izdatel'stvo Assotsiatsii Nauchno-Issledovatel'skikh Institutov pri
Fiziko-Matematicheskogo Fakul'teta I. Moskovskogo Universiteta, 1926. 198 p.
Bibliography, p. 186-198 (Nauchno-Issledovatel'skii Institut Geografii, Trudy,
vypusk 2).   1

    304 references.

LILIENBERG, Dmitrii Anatol'evich. Rel'ef iuzhnogo sklona vostochnoi chasti Bol'shog
Kavkaza (Relief of the southern slope of the eastern part of the Great Caucasus).
Moskva, Izdatel'stvo Akademii Nauk SSSR, 1962. 244 p. Bibliography, p. 228-242
(Akademiia Nauk SSSR. Institut Geografii).   1

    About 375 references in alphabetical order.

RUASHVILI, Levan Iosifovich, and others, eds. Geomorfologiia Gruzii. Rel'ef GSSR v aspektakh plastiki, proiskhozhdeniia, dinamiki i istorii (Geomorphology of Georgia. Relief of the Georgian SSR in its plasticity, origin, dynamics and history). Tbilisi, "Metsniereba," 1971. 609 p. Bibliography, p. 554-568. Index of persons and geographical names, p. 569-605. (Akademiia Nauk Gruzinskoi SSR. Institut Geografii imeni Vakhushti).   1080

STVOROVA, Valentina Alekseevna. Formirovanie rel'efa gor: na primere Gornoi Osetii (The formation of the relief of mountains: the example of mountainous Osetia [in the Caucasus]. Moskva, Izdatel'stvo "Nauka," 1973. Bibliography, p. 132-143. (Akademiia Nauk SSSR. Institut Fiziki Zemli imeni O. Iu. Shmidta). 1081

HAVAKHISHVILI, Aleksandr Nikolaevich. Geomorfologicheskie raiony Gruzinskoi SSR. Tipy rel'efa i raiony ikh rasprostraneniia (Geomorphology of regions of the Georgian SSR: types of relief and the regions of their distribution). Moskva-Leningrad, Izdatel'stvo Akademii Nauk SSSR, 1947. 180 p. Bibliography, p. 144-172 (Akademiia Nauk SSSR. Institut Geografii. Akademiia Nauk Gruzinskoi SSR. Institut Geografii imeni Vakhushti).   1082

About 670 references in Russian and Georgian arranged systematically. Entries in the Georgian language are given in Russian translation.

TAKHOV, Nikolai Evgen'evich. Strukturnaia geomorfologiia Gruzii (Structural geo-morphology of Georgia). Tbilisi, "Metsniereba," 1973. 224 p. Bibliography, p. 207-221. English summary. (Akademiia Nauk Gruzinskoi SSR. Institut Geografii imeni Vakhushti).   1083

GOSHVILI, Lidiia Vasil'evna. Zhivaia tektonika Gruzii i ee vozdeistvie na rel'ef: opyt strukturno-geomorfologicheskikh issledovanii (Active tectonics of Georgia and its effect on relief: an experiment in structural-geomorphological research). Tbilisi, Izdatel'stvo "Metsniereba," 1970. 219 p. Bibliography, p. 212-217. (Akademiia Nauk GSSR. Institut Geografii imeni Vakhushti).   1084

OLOGIIA Armianskoi SSR. Tom 1. Geomorfologiia (Geology of Armenia, v. 1. Geo-morphology). S. S. Mkrtchian, ed. Erevan, Izdatel'stvo Akademii Nauk Armianskoi SSR, 1962. 586 p. Bibliography, p. 538-556.   1085

About 460 rererences in alphabetical order in Russian, Armenian, and other languages.

OMORFOLOGIIA Azerbaidzhanskoi SSR (Geomorphology of the Azerbaidzhan SSR). M. A. Kashkai, ed. Baku, Izdatel'stvo Akademii Nauk Azerbaidzhanskoi SSR. 1959. 371 p. Bibliography, p. 359-369 (Akademiia Nauk Azerbaidzhanskoi SSR. Institut Geografii).   1086

About 330 references.

RINOV, Naib Shirin ogly. Geomorfologiia Apsheronskoi neftenosnoi oblasti (Geo-morphology of the Apsheron oil-bearing region). Baku, Izdatel'stvo Akademii Nauk Azerbaidzhanskoi SSR, 1965. 188 p. Bibliography, p. 178-188. (Akademiia Nauk Azerbaidzhanskoi SSR. Institut Geografii).   1087

About 200 references.

RINOV, Naib Shirin ogly. Geomorfologicheskoe stroenie Kura-Araksinskoi depres'ii (Geomorphological structure of the Kura-Araks depression). Baku, 1973. 216 p. Bibliography, p. 211-215. (Akademiia Nauk Azerbaidzhanskoi SSR. Institut Geo-rafii).   1088

AGOV, Budag Ali ogli. Geomorfologiia i noveishaia tektonika iugo-vostochnogo avkaza (Geomorphology and recent tectonics of the southeast Caucasus). Baku, Izdatel'stvo "Elm," 1973. 245 p. Bibliography, p. 235-245.   1089

About 250 references in alphabetical order.

ANTONOV, Boris Alekseevich. Geomorfologiia i voprosy noveishei tektoniki iugo-
vostochnoi chasti Malogo Kavkaza (Geomorphology and questions of recent tectonics
of the southwest part of the Small Caucasus). Baku, "Elm," 1971. 162 p. Bibliography, p. 152-161 (Akademiia Nauk SSSR. Institut Geografii).   1(

ANTONOV, Boris Alekseevich, MUSEIBOV, Museib Agababa ogly, and SHIRINOV, N. Sh.
Geomorfologicheskie issledovaniia v Azerbaidzhane. Reziume (Geomorphological
research in Azerbaydzhan. Summary). Akademiia Nauk Azerbaidzhanskoi SSR. Institut Geografii. Trudy, v. 18 (1971), p. 52-133. Bibliography, p. 86-133.
Kazakhstan and Soviet Middle Asia.   1(

ABASOV, M. A. Geomorfologiia Nakhichevanskoi ASSR (Geomorphology of Nakhichevan'
ASR). Baku, "Elm," 1970. 150 p. Bibliography, p. 144-149. (Akademiia Nauk
Azerbaidzhanskoi SSR. Institut Geografii).   1(

Siberia and the Soviet Far East

VOSKRESENSKII, Sergei Sergeevich. Geomorfologiia Sibiri (Geomorphology of Siberia).
Moskva, Izdatel'stvo Moskovskogo Universiteta, 1962. 352 p. Bibliography, p.
336-349.   1(

About 400 references, arranged in four groups: part 1, West-Siberian lowland;
part 2, plateaus and lowlands of Eastern Siberia; part 3, mountains of Southern
Siberia; part 4, conclusions. Within each group works are arranged in alphabetical order of name of author.

GORODETSKAIA, Marina Evgen'evna. Morfostruktura i morfoskul'ptura iuga Zapadno-
Sibirskoi ravniny (Morphostructure and morphosculpture of the south of the West
Siberian plain). Moskva, Izdatel'stvo "Nauka," 1972. 154 p. Bibliography, p.
144-153. (Akademiia Nauk SSSR. Institut Geografii).   1

ARKHIPOV, Stanislav Anatol'evich, VDOVIN, V. V., MIZEROV, B. V., and NIKOLAEV, V. A.
Zapadno-Sibirskaia ravnina (West Siberian Plain). Moskva, Izdatel'stvo "Nauka,"
1970. 279 p. Bibliography, p. 259-278. (Akademiia Nauk SSSR. Sibirskoe
Otdelenie. Institut Geologii i Geofiziki. Istoriia razvitiia rel'efa Sibiri i
Dal'nego Vostoka).   1

About 600 references.
FLORENSOV, Nikolai Aleksandrovich, ed. Ploskogor'ia i nizmennosti Vostochnoi
Sibiri (Plateaus and lowlands of Eastern Siberia). O. M. Adamenko, I. Iu.
Dolgushin, V. V. Ermolov, and others. Moskva, Izdatel'stvo "Nauka," 1971. 320 p.
Bibliography, p. 307-318. (Akademiia Nauk SSSR. Sibirskoe Otdelenie. Institut
Zemnoi Kory. Istoriia Razvitiia rel'efa Sibiri i Dal'nego Vostoka).   10

DUMITRASHKO, Nataliia Vladimirovna. Geomorfologiia i paleogeografiia Baikal'skoi
gornoi oblasti (Geomorphology and paleogeography of the Baykal mountain region).
Moskva, Izdatel'stvo Akademii Nauk SSSR, 1952. 191 p. Bibliography, p. 179-189
(Akademiia Nauk SSSR. Institut Geografii. Trudy, v. 55. Materialy po geomorfologii i paleogeografii SSSR, no. 9).   10

About 350 references in alphabetical order.

BORSUK, O. A. Analiz shchebnistykh otlozhenii i valechnikov pri geomorfologicheskikh
issledovaniiakh, na primere Zabaikal'ia (Analysis of detrital deposits and con-
glomorerates in geomorphological research, the example of Trans-Baykalia). Moskva,
Izdatel'stvo "Nauka," 1973. 112 p. Bibliography, p. 104-111 (Akademiia Nauk
SSSR. Komissiia po Izucheniiu Chetvertichnogo Perioda).   10

STRELKOV, Sergei Aleksandrovich. Sever Sibiri (Northern Siberia). Moskva,
Izdatel'stvo "Nauka," 1965. 336 p. Bibliography, p. 311-334 (Akademiia Nauk
SSSR. Sibirskoe Otdelenie. Institut Geologii i Geofiziki. Istoriia Razvitiia
Rel'efa Sibiri i Dal'nego Vostoka).   10

More than 700 references in alphabetical order.

RANOVA, Iu. P., and BISKE, S. F. Severo-Vostok SSSR (The Northeast USSR).
Moskva, Izdatel'stvo "Nauka," 1964. 290 p. Bibliography, p. 273-289 (Akademiia
Nauk SSSR. Sibirskoe Otdelenie. Institut Geologii i Geofiziki. Istoriia Raz-
vitiia Rel'efa Sibiri i Dal'nego Vostoka).   1100

    More than 450 references in alphabetical order on geology and geomorphology.

SANOV, Boris Sergeevich, BORODENKOVA, Z. F., GONCHAROV, V. F., and others. Geo-
morfologiia Vostochnoi Iakutii (Geomorphology of Eastern Yakutiya). Yakutsk,
Iakutknigoizdat, 1967. 376 p. Bibliography, p. 357-375. (Akademiia Nauk SSSR.
Iakutskii Filial Sibirskogo Otdeleniia. Institut Geologii).   1101

    About 550 references.

RZHUEV, Sergei Sergeevich. Geomorfologiia doliny srednei Leny i prilegaiushchikh
raionov (Geomorphology of the valley of the middle Lena river and adjacent regions).
Moskva, Izdatel'stvo Akademii Nauk SSSR, 1959. 150 p. Bibliography, p. 144-149
(Akademiia Nauk SSSR. Institut Geografii).   1102

    About 160 references.

MOFEEV, Dmitrii Andreevich. Sredniaia i Nizhniaia Olëkma. Geomorfologicheskii
analiz territorii basseina (The Middle and Lower Olekma. Geomorphological
analysis of the territory of the basin). Moskva-Leningrad, Izdatel'stvo "Nauka,"
1965. 138 p. Bibliography, p. 125-136. (Akademiia Nauk SSSR. Sibirskoe
Otdelenie. Institut Geografii Sibiri i Dal'nego Vostoka).   1103

    About 350 references in alphabetical order.

                        Kazakhstan and Soviet Middle Asia

N'KO, Elizaveta Aleksandrovna. Morfostrukturnyi analiz pri izuchenii sovremennykh
tektonicheskikh dvizhenii, na primere Vostochnogo i Iuzhnogo Kazakhstana (Morphostructural analysis in the study of contemporary tectonic movements, on the example of Eastern and Southern Kazakhstan). Moskva, Izdatel'stvo "Nauka," 1973.
95 p. Bibliography, p. 88-94 (Akademiia Nauk SSSR. Institut Geografii).   1104

EREDNICHENKO, Vladimir Pavlovich. Morfologiia eolovogo rel'efa i stroitel'stvo
truboprovodov v pustyne (Morphology of eolian relief and the construction of
pipelines in the desert). M. P. Petrov, ed. Ashkhabad, "Ylym," 1973. 130 p.
Bibliography, p. 125-129. (Akademiia Nauk Turkmenskoi SSR. Institut Pustyn').   1105

STENKO, Natal'ia Petrovna. Razvitie rel'efa gornykh stran: na primere Srednei
Azii (The development of relief in mountainous countries: the example of Soviet
Middle Asia). Moskva, Izdatel'stvo "Mysl'," 1970. 367 p. Bibliography, p. 347-
364.   1106

    About 360 references.

R'EV, A. A., and UMAROV, A. U. Geomorfologiia i noveishaia tektonika Zapadnogo
Uzbekistana (Geomorphology and recent tectonics of Western Uzbekistan). Tashkent,
"FAN," 1971. 116 p. Bibliography, p. 108-115. (Ministerstvo Geologii Uzbekskoi
SSR. Institut Geologii i Razvedki Neftianykh i Gazovykh Mestorozhdenii "IGIRNIGM").

                                                                                    1107

CLIMATOLOGY AND METEOROLOGY

Major General Bibliographies

ZICHESKAIA geografiia: annotirovannyi perechen' otechestvennykh bibliografii izdannykh v 1810-1966 gg. (Physical geography: annotated list of native bibliographies published 1810-1966). M. N. Morozova and E. A. Stepanova, comps. V. V. Klevenskaia, ed. L. G. Kamanin, scientific consultant. Moskva, Izdatel'stvo "Nauka," 1968. 309 p. (Gosudarstvennaia Biblioteka SSSR imeni V. I. Lenina. Institut Geografii Akademii Nauk SSSR. Sektor Seti Spetsial'nykh Bibliotek Akademii Nauk SSSR).   1108

Climatology, p. 116-131, entries 491-580, 90 Soviet bibliographies on climatology organized systematically: bibliographies of a general character, bibliographies of publications of scientific institutions in the USSR; bibliographies by individual themes and problems (climate-forming factors, radiation factors, circulation factors, climate-forming influence of the underlying surface; individual elements of climate, wind, temperature of the atmosphere, precipitation, evaporation; changes and oscillations of climate; aeroclimatology; microclimate; classification of climates and climatic regionalization; complex climatology); bibliographies on regional climatology (USSR as a whole; European part of the USSR, Center, West, Southwest, Volga region, Urals, Caucasus; Asiatic part of the USSR, Siberia; foreign countries). Author and geographical indexes.

FERATIVNYI zhurnal: meteorologiia i klimatologiia (Reference journal: meteorology and climatology) (Vsesoiuznyi Institut Nauchnoi i Tekhnicheskoi Informatsii). Moskva, 1954- monthly. Author index in each number. Available separately (since 1962) or as section B in combined volumes with geography or with geophysics. Annual author and subject indexes in combined geophysics volume.   1109

About 6,000 abstracts a year covering the literature in all languages, arranged by the following sections: general section; meteorological instruments; methods of observation and analysis; physical meteorology; dynamic and synoptic meteorology; climatology; and applied meteorology and climatology, each further subdivided systematically.

Climatology includes subsections on general questions; climatic formation; classification of climates and climatic regionalization; individual climatic elements; complex climatic description; climate of the free atmosphere; microclimate; climate of cities; influence of solar activity on weather and climate.

Applied meteorology and climatology includes subsections on bioclimatology and medical meteorology; agricultural meteorology and climatology; aviation meteorology; transport and industrial meteorology; the influence of man on the atmosphere, weather, and climate (artifical clouds and precipitation, air pollution; anthropogenic influence on climate).

)ROLOGIIA. Meteorologiia. Klimatologiia (Hydrology, meteorology, and climatology), Bibliografiia sovetskoi bibliografii (Bibliography of Soviet bibliography). (Vsesoiuznaia Knizhnaia Palata). 1939, 1946- . Annual. Moskva, Izdatel'-stvo "Kniga," 1941, 1948- .   1110

Annual list of Soviet bibliographies in climatology and meteorology. The 1970 annual volume (1972), p. 98-103, contains 170 entries (combined with hydrology), in a sub-section of the section on geological-geographical sciences in the chapter on natural sciences and mathematics.

ERATURA po geofizike, meteorologii, gidrologii i okeanografii (Literature on geophysics, meteorology, hydrology, and oceanography), Meteorologiia i gidrologiia, 949, no. 4, p. 131-137.   1111

190 entries for 1948. For 1945-1946 see Otechestvennaia tekushchaia literatura po geofizike, meteorologii, gidrologii i okeanografii, under hydrology. For later years see Literatura po meteorologii i gidrologii below.

1108 - 1111

LITERATURA po meteorologii i gidrologii (Literature on meteorology and hydrology), Meteorologiia i gidrologiia, 1951, no. 1, p. 69-94, no. 6, p. 61-64. 199 entries for 1950.   111

_____ . 1952, no. 2, p. 57-63, no. 7, p. 56-62. 283 entries for 1951.

_____ . 1953, no. 2, p. 60-63, no. 7, p. 54-60. 232 entries for 1952.

_____ . 1956, no. 7, p. 55-60, no. 8, p. 57-60, no. 9, p. 61-63, no. 12, p. 58-60. 357 entries.

_____ . 1957, no. 2, p. 61-64, no. 4, p. 58-62, no. 6, p. 59-61, no. 7, p. 57-60, no. 8, p. 66-69. 485 entries.

_____ . 1958, no. 2, p. 53-56, no. 3, p. 64-67, no. 7, p. 60-64, no. 8, p. 59-62, no. 9, p. 52-56, no. 10, p. 59-62, no. 11, p. 64-66, no. 12, p. 55. 663 entries.

_____ . 1959, no. 1, p. 61-65, no. 2, p. 63-66, no. 3, p. 58-62, no. 4, p. 70-73, no. 7, p. 54-57, no. 8, p. 50-54, no. 9, p. 54-57, no. 12, p. 55-58. 793 entries.

_____ . 1960, no. 4, p. 56-60, no. 5, p. 56-58, no. 6, p. 46-49, no. 7, p. 51-55, no. 8, p. 56-59, no. 9, p. 55-59, no. 10, p. 55-58, no. 11, p. 52-57, no. 12, p. 56-58. About 900 entries.

_____ . 1961, no. 1, p. 61-64, no. 2, p. 55-57, no. 6, p. 57-61, no. 7, p. 52-55, no. 8, p. 57-60, no. 9, p. 60-63, no. 10, p. 58-62, no. 11, p. 61-65, no. 12, p. 52-56. About 980 entries.

_____ . 1962, no. 1, p. 60-63, no. 2, p. 64-67, no. 3, p. 61-64, no. 5, p. 57-60, no. 8, p. 56-59, no. 9, p. 58-61, no. 10, p. 57-60, no. 11, p. 62-67, no. 12, p. 51-55. About 1,000 entries.

_____ . 1963, no. 2, p. 56-59, no. 3, p. 57-60, no. 4, p. 54-57, no. 5, p. 57-59, no. 6, p. 61-62, no. 7, p. 58-61, no. 8, p. 58-62, no. 9, p. 58-63, no. 10, p. 57-62, no. 11, p. 57-61, no. 12, p. 50-53. 1,100 entries.

_____ . 1964, no. 1, p. 57-60, no. 2, p. 58-61, no. 3, p. 57-60, no. 4, p. 59-60, no. 5, p. 58-61, no. 6, p. 56-60, no. 7, p. 53-56, no. 8, p. 51-54, no. 9, p. 51-54, no. 10, p. 48-51, no. 11, p. 55-57. 1,136 entries.

_____ . 1965, no. 1, p. 63-65, no. 2, p. 46-49, no. 3, p. 51-55, no. 4, p. 57-60, no. 5, p. 61-63, no. 6, p. 52-55, no. 7, p. 61-62, no. 8, p. 60-62, no. 9, p. 57-60. 860 entries.

_____ . 1966, no. 2, p. 53-55, no. 3, p. 59-62, no. 4, p. 59-63, no. 5, p. 59-61, no. 6, p. 61-64, no. 7, p. 61-65, no. 8, p. 55-58, no. 10, p. 55-58, no. 11, p. 61-63. 890 entries.

_____ . 1967, no. 1, p. 60-62, no. 2, p. 117-120, no. 3, p. 118-122, no. 4, p. 112-116, no. 5, p. 110-115, no. 7, p. 113-116, no. 8, p. 116-119, no. 9, p. 111-115, no. 10, p. 113-116, no. 11, p. 107-111, no. 12, p. 100-102. 1,233 entries.

_____ . 1968, no. 1, p. 108-113, no. 2, p. 114-116, no. 3, p. 110-114, no. 4, p. 111-115, no. 5, p. 111-114, no. 6, p. 112-116. 639 entries.

LAVNAIA Geofizicheskaia Observatoriia imeni A. I. Voeikova. <u>Bibliograficheskii</u>
<u>ukazatel' rabot...za period 1918-1967 gg.</u> Leningrad, 1967-1973. 3 v. (Glavnoe
Upravlenie Gidrometeorologicheskoi Sluzhby pri Sovete Ministrov SSSR. Glavnaia
Geofizicheskaia Observatoriia imeni A. I. Voeikova. Biblioteka).   1113

    vypusk 1. 1918-1945 gg. E. L. Andronikova, E. A. Slobodskaia, G. A. Tsimmer,
comps. M. E. Shvets, ed. 1967. 424 p.

    vypusk 2. 1946-1960 gg. 1969. 311 p.

    vypusk 3. 1961-1967. 1973. 290 p.

KLIMATOLOGIIA (Climatology). In the book: <u>Soobshchenie o nauchnykh rabotakh po</u>
<u>meteorologii</u> (Report on scientific work in meteorology). Moskva, Izdatel'stvo
Akademii Nauk SSSR, 1957, p. 5-21. (Akademiia Nauk SSSR. Komitet po Geodezii
i Geofizike). For the 11th General Assembly of the International Union of Geodesy
and Geophysics.   1114

    437 entries on climatology published 1951-1955 in the USSR arranged by topics:
general questions, teaching aids and courses; radiation factors in climate; circulation factors in climate; moisture circulation; theory of climate and changes
and oscillations of climate; climatic factors in natural processes; modification
of climate; research methods of climate; agricultural climatology, phenology;
climatography; methods of analysis and generalization of observational data.

LIMATOLOGIIA (Climatology), in the book: <u>Soobshchenie o nauchnykh rabotakh po</u>
<u>meteorologii i fizike atmosfery, 1957-1959 gg.</u> (Report on scientific work in
meteorology and physics of the atmosphere, 1957-1959). Moskva, Izdatel'stvo
Akademii Nauk SSSR, 1960, p. 3-74 (Akademiia Nauk SSSR. Komitet po Geodezii i
Geofizike). For the 12th General Assembly of the International Union of Geodesy
and Geophysics.   1115

    684 entries on works on climatology published 1956-1958 in the USSR.

LIMATOLOGIIA (Climatology) in the book: <u>Meteorologiia i fizika atmosfery: biblio-</u>
<u>graficheskii ukazatel', 1960-1962 gg.</u> (Meteorology and physics of the atmosphere:
a bibliographical guide, 1960-1962). Moskva, Izdatel'stvo Akademii Nauk SSSR,
1965, p. 148-176 (Akademiia Nauk SSSR. Prezidium. Mezhduvedomstvennyi Geofizicheskii Komitet).   1116

    392 entries on climatology published in the USSR, 1960-1962.

EBEDEV, Aleksei Nikolaevich, and TOKAR', F. G. <u>Issledovaniia po klimatografii</u>
(Research on climatography). Obninsk, Otdel Nauchno-Tekhnicheskoi Informatsii
Otdeleniia Gidromettsentra SSSR, 1969. 54 p. (Glavnaia Geofizicheskaia Obser-
vatoriia imeni Voeikova).   1117

    A bibliographical review with 520 references, mainly for the period 1957-1966.

EKKERMAN, Iosif Matveevich. <u>Bibliograficheskii ukazatel' inostrannykh knig po</u>
<u>meteorologii, okeanografii, gidrologii za 1965-1966 gg.</u> (Bibliographical guide
to foreign books on meteorology, oceanography, and hydrology 1965-1966). Leningrad,
Gidrometeoizdat, 1967. 56 p.   1118

    272 entries.

_____ . _____ 1964-1965. 1966. 71 p.

    340 entries.

.TERATURA po voprosam kompleksnoi klimatologii (Literature on questions of complex climatology). In the book: <u>Voprosy kompleksnoi klimatologii</u> (Problems of
complex climatology). Moskva, Izdatel'stvo Akademii Nauk SSSR, 1963. 207 p.
Bibliography, p. 197-205. (Akademiia Nauk SSSR. Institut Geografii).   1119

    About 250 references in alphabetical order.

Bibliographies in General Treatises

ALISOV, Boris Pavlovich, and POLTARAUS, Boris Vasil'evich. Klimatologiia (Climatology). Moskva, Izdatel'stvo Moskovskogo Universiteta, 1962. 228 p. Bibliography, p. 220-226.   112

154 references, by chapters.

RUBINSHTEIN, Evgeniia Samoilovna, ed. Kurs klimatologii (Course in climatology). Leningrad, Gidrometeoizdat, 1952-1954. v. 1-2. Alisov, B. P., Drozdov, O. A., and Rubinshtein, E. S. Obshchaia klimatologiia; metody klimatologicheskoi obrabotki nabliudenii (General climatology; methods of climatic analysis of observations). 1952. Bibliography, p. 478-482. v. 3. Alisov, B. P., Berlin, I. A., and Mikheev, V. M. Klimaty zemnogo shara (Climates of the world). 1954. Bibliography, p. 318-320.   112

227 references, by chapters.

Temperature and Heat Balance

BUDYKO, Mikhail Ivanovich. Klimat i zhizn' (Climate and life). Leningrad, Gidrometeorologicheskoe Izdatel'stvo, 1971. 472 p.   112

Monographic analysis of the heat balance of the earth. About 700 references in bibliographies at the ends of chapters, p. 42-46, 130-136, 242-247, 273-274, 298-299, 348-350, 376-377, 412-414, 438, and 470. Available in English as Climate and life. David H. Miller, ed. New York, Academic Press, 1974. 520 p. Author and subject indexes. (International Geophysics Series).

BUDYKO, Mikhail Ivanovich. Teplovoi balans zemnoi poverkhnosti (Heat balance of earth's surface). Leningrad, Gidrometeoizdat, 1956. 255 p. Bibliography p. 242-254. Available in English: The heat balance of the earth's surface, translated by Nina A. Stepanova, Washington, D. C., U. S. Weather Bureau, 1958. 259 p.   112

About 320 references in alphabetical order.

RAUNER, Iurii L'vovich. Teplovoi balans rastitel'nogo pokrova (Heat balance of the vegetative cover). Leningrad, Gidrometeoizdat, 1972. 210 p. Bibliography, p. 198-206. Supplementary table of contents in English.   11

206 references, 147 in Russian and 59 in Western languages, each in alphabetical order by name of author.

DORONIN. Iurii Petrovich. Teplovoe vzaimodeistvie atmosfery i gidrosfery v Arktike (Heat interaction of the atmosphere and hydrosphere in the Arctic). E. P. Borisenkov, ed. Leningrad, Gidrometeoizdat, 1969. 299 p. Bibliography, p. 284-297. (Glavnoe Upravlenie Gidrometeorologicheskoi Sluzhby pri Sovete Ministrov SSSR. Arkticheskii i Antarkticheskii Nauchno-Issledovatel'skii Institut).   11

329 references.

KHANEVSKAIA, I. V. Temperaturnyi rezhim svobodnoi atmosfery nad severnym polushariem (Temperature regime of the free atmosphere in the northern hemisphere). Leningrad, Gidrometeoizdat, 1968. 299 p. Bibliography, p. 281-288. Summary in English.   11

197 references.

Radiation

DIATSIONNYE kharakteristiki atmosfery i zemnoi poverkhnosti (Radiation characteristics of the atmosphere and the surface of the earth). V. N. Konashenok, S. D.
Andreev, L. S. Ivlev, and others. K. Ia. Kondrat'ev, ed. Leningrad, Gidrometeoizdat, 1969. 564 p. Bibliography, p. 527-554. English summary.   1127
    682 references by chapters.

RLIAND, Tamara Grigor'evna. Raspredelenie solnechnoi radiatsii na kontinentakh
(Distribution of solar radiation on continents). Leningrad, Gidrometeoizdat,
1961. 227 p. Bibliography, p. 191-206. (Glavnaia Geofizicheskaia Observatoriia imeni A. I. Voeikova).   1128
    421 references in alphabetical order, 185 Soviet and 236 in foreign languages.

DIATSIONNYI rezhim territorii SSSR (Radiation regime of the territory of the
USSR). E. P. Barashkova, V. L. Gaevskii, L. N. D'iachenko, K. M. Luchina, and
Z. I. Pivovarova. Leningrad, Gidrometeoizdat, 1961. 528 p. Bibliography, p.
152-156. (Glavnaia Geofizichicheskaia Observatoriia imeni A. I. Voeikova).   1129
    145 references.

RILLOVA, T. V. Radiatsionnyi rezhim ozër i vodokhranilishch (Radiation regime of
lakes and reservoirs). Leningrad, Gidrometeoizdat. 1970. 253 p. Bibliography,
p. 238-251.   1130

VRILOVA, Mariia Kuz'michna. Radiatsionnyi klimat Arktiki (Radiation climate of
the Arctic). M. I. Budyko, ed. Leningrad, Gidrometeoizdat, 1963. 225 p. Bibliography, p. 151-162 (Glavnaia Geofizicheskaia Observatoriia imeni A. I.
Voeikova).   1131
    296 references in alphabetical order, 179 Soviet and 117 foreign.

Moisture

OZDOV, Oleg Alekseevich, and GRIGOR'EVA, Anna Sergeevna. Mnogoletnie tsiklicheskie
kolebaniia atmosfernykh osadkov na territorii SSSR. (Cyclical fluctuations extend-
ing over many years in the atmospheric precipitation of the territory of the USSR).
Leningrad, Gidrometeoizdat, 1971. 158 p. Bibliography, p. 151-157. (Glavnoe
Upravlenie Gidrometeorologicheskoi Sluzhby pri Sovete Ministrov SSSR. Glavnaia
Geofizicheskaia Observatoriia imeni A. I. Voeikova).   1132

OZDOV, Oleg Alekseevich, and GRIGOR'EVA, Anna Sergeevna. Vlagooborot v atmosfere
(Moisture circulation in the atmosphere). Leningrad, Gidrometeoizdat, 1963.
315 p. Bibliography, p. 302-313. (Glavnaia Geofizicheskaia Observatoriia imeni
A. I. Voeikova). Available in English: The hydrologic cycle in the atmosphere.
Edited by R. Hardin and translated by I. Shechtman. Jerusalem, Israel Program
for Scientific Translations, 1965. 282 p. IPST 1376. Bibliography, p. 256-277. 1133
    About 290 references

PAT'EV, Anatolii Mikhailovich. Vlagooboroty v prirode i ikh preobrazovaniia
(The moisture cycle in nature and its transformation). Leningrad, Gidrometeorologicheskoe Izdatel'stvo, 1969. 323 p. Bibliography, p. 304-319. English summary, p. 320-321.   1134
    About 450 references arranged alphabetically first in Cyrillic then in Latin
letters.

NSTANTINOV, Aleksei Rodionovich. Isparenie v prirode (Evaporation in nature).
2nd ed. Leningrad, Gidrometeoizdat, 1968. 532 p. Bibliography, p. 447-488.
(1st ed. 1963. 590 p. Bibliography, p. 507-539). 1st ed. available in English:
Evaporation in nature. Jerusalem, Israel Program for Scientific Translations,
1966. 536 p. IPST 1529.   1135
    About 1,000 references (843 in the 1st edition).

ZHAKOV, Stepan Ivanovich. Proiskhozhdenie osadkov v tëploe vremia goda (Origin of precipitation in the warm period of the year). Leningrad, Gidrometeoizdat, 1966. 251 p. Bibliography, p. 232-242.   11

    About 240 references .

NEVESSKII, Evgeniy Nikolaevich. Protsessy osadkoobrazovaniia v pribrezhnoi zone moria (Processes in the formation of precipitation in coastal zones of the sea). Moskva. Izdatel'stvo "Nauka," 1967. 255 p. Bibliography, p. 245-253. (Akademiia Nauk SSSR. Institut Okeanologii).   11

    About 280 references.

BIBLIOGRAFIIA po sukhoveiam i zasukham i meram bor'by s nimi, 1917-1955 (Bibliography of sukhovei [dry winds] and droughts and measures against them, 1917-1955). In the book: Sukhovei, ikh proiskhozhdenie i bor'ba s nimi (Sukhovei, their origin and the struggle with them). Moskva, Izdatel'stvo Akademii Nauk SSSR, 1957, p. 354-367 (Akademiia Nauk SSSR. Institut Geografii). Available in English as Sukhoveis and drought control, B. I. Dzerdzeevskii, general editor. Jerusalem, Israel Program for Scientific Translations, 1963. 376 p. IPST 890.   11

    More than 400 references in alphabetical order.

BUCHINSKII, Ivan Evstaf'evich. Zasukhi, sukhovei, pyl'nye buri na Ukraine i bor'ba s nimi (Droughts, dessicating winds, and dust storms in the Ukraine and measures against them). Kiev, Izdatel'stvo "Urozhai," 1970. 236 p. Bibliography, p. 229-235.   11

    About 180 references.

MKHITARIAN, Artavazd Mel'konovich. Nekotorye voprosy gidrodinamiki pogranichnogo sloia atmosfery. Vodnyi i teplovoi balansy vodoëmov. (Some problems of the hydrodynamics of the boundary layer of the atmosphere. Water and heat balances of reservoirs). Erevan. "Aiastan," 1970. 323 p. Bibliography, p. 289-306. (Glavnoe Upravlenie Gidrometeorologicheskoi Sluzhby pri Sovete Ministrov SSSR. Zakavkazskii Nauchno-Issledovatel'skii Gidrometeorologicheskii Institut. Erevanskii Otdel).   11

    361 references.

Snow, Frost, and Atmospheric Ice

MIKHEL',. Vasilii Mikhailovich, RUDNEVA, Anna Vladimirovna, and LIPOVSKAIA, Vera Ivanovna. Perenosy snega pri meteliakh i snegopady na territorii SSSR (Movement of snow in snowstorms and snowfall in the USSR). Leningrad, Gidrometeoizdat, 1969. 203 p. Bibliography, p. 168-173 (Glavnaia Geofizicheskaia Observatoriia imeni A. I. Voeikova).   1]

    About 140 references.

MIAGKOV, N. Ia. Zamorozki: opyt tematicheskoi bibliografii (Frost: an attempt at a systematic bibliography). In the book: Sbornik rabot Ashkhabadskoi gidrometeorologicheskoi observatorii (Collected works of the Ashkhabad Hydrometeorological Observatory), no. 1. Ashkhabad, 1958, p. 157-174.   1]

    477 references.

MAKSIMOV, S. A. Ukazatel' literatury po voprosu o zamorozkakh (Guide to the literature on the problem of frosts), Meteorologiia i gidrologiia, 1940, no. 4, p. 129-136.   1]

    Chronological, partly annotated, list of 42 Soviet works 1857-1939 and alphabetical list of 262 foreign works, 1782-1939. Subject index.

MORSKII, Aleksandr Dmitrievich. Atmosfernyi lëd. Inei, gololëd, sneg i grad (Atmospheric ice. Hoarfrost, glaze, snow, and hail). Moskva-Leningrad, Izdatel'-stvo Akademii Nauk SSSR, 1955. 377 p. Bibliography, p. 353-365 (Akademiia Nauk SSSR. Geograficheskoe Obshchestvo SSSR).   1144

About 340 references in alphabetical order.

Winds and Circulation of the Atmosphere

GOSIAN, Khoren Petrovich. Obshchaia tsirkuliatsiia atmosfery (General circula-tion of the atmosphere). Leningrad, Gidrometeoizdat, 1959. 260 p. Bibliog-raphy, p. 246-252.   1145

164 references in alphabetical order.

TS, Abram L'vovich. Sezonnye izmeneniia obshchei tsirkuliatsii atmosfery i dolgosrochnye prognozy (Seasonal changes of the general circulation of the atmosphere and long-range forecasting). Leningrad, Gidrometeoizdat, 1960. 270 p. Bibliography, p. 211-223.   1146

320 references in alphabetical order.

TVITSKII, Georgii Nikolaevich. Tsirkuliatsiia atmosfery v tropikakh (Circula-tion of the atmosphere in the tropics). Leningrad, Gidrometeoizdat, 1971. 144 p. Bibliography, p. 138-143. Summary in English.   1147

157 references.

RMAN, Emmanuil Aronovich. Mestnye vetry (Local winds). Leningrad, Gidrometeoiz-dat, 1969. 341 p. Bibliography, p. 322-339.   1148

504 references.

LIVKIN, Dmitrii Vasil'evich. Uragany, buri i smerchi: geograficheskie osobennosti i geologicheskaia deiatel'nost' (Hurricanes, storms, and tornadoes: geographical characteristics and geological activity). Leningrad, Izdatel'stvo "Nauka," Leningradskoe Otdelenie, 1970. 487 p. Bibliography, p. 454-471. (Akademiia Nauk SSSR. Otdelenie Nauk o Zemle).   1149

683 references.

LEVAKHA, V. O., and ROMUSHKEVICH, V. I. Sukhovii na Ukraini (Sukhovey [dry winds] in the Ukraine). Kyïv, Vydavnytstvo Kyïvskoho Universytetu, 1972. 140 p. Bib-liography, p. 127-139. In Ukrainian.   1150

Seasons

LAKHOV, Nikolai Nikolaevich. Izuchenie struktury klimaticheskikh sezonov goda: opyt tipizatsii klimaticheskogo rezhima vo vremeni v predelakh umerennykh shirot SSR (Study of the structure of the climatic seasons of the year: an attempt at a typology of the climatic regime through time in the boundaries of the temperate latitudes of the USSR). Moskva, Izdatel'stvo Akademii Nauk SSSR, 1959. 183 p. Bibliography, p. 175-182. (Akademiia Nauk SSSR. Institut Geografii).   1151

About 200 references.

Climatic Change and Climates of the Past

INSHTEIN, Evgeniia Samoilovna, and POLOZOVA, Larisa Gavrilovna. Sovremennoe zmenenie klimata (Current changes of climate). Leningrad, Gidrometeoizdat, 1966. 68 p. Bibliography, p. 258-263.   1152

172 references.

1144 - 1152

OSNOVNYE voprosy klimatoobrazovaniia, izmenchivosti klimaticheskikh rezhimov i
atmosferno-solnechnykh sviazei (Basic questions of the formation of climate,
variability of climatic regimes and atmospheric-solar relations). Itogi nauki,
seriia "Geofizika," meteorologiia i klimatologiia, Moskva, VINITI, v. 1, 1971,
p. 86-120.   115

    184 references.

KURDIN, R. D. O prichinakh izmenenii i kolebanii klimata (On the causes of changes
and fluctuations of climate). Alma-Atinskaia Gidrometeorologicheskaia Observatoriia
Sbornik rabot, vypusk 6 (1971), p. 3-35.   115

    86 references.

BERG, Lev Semenovich. Vopros ob izuchennii klimata v istoricheskuiu epokhu
(Question on the change of climate in historical time). In the book: Klimat i
zhizn', 2nd ed., Moskva, 1947. 356 p.; also in the book: Berg, L. S. Izbrannye
trudy (Collected works), v. 2. Fizicheskaia geografiia. Moskva, Izdatel'stvo
Akademii Nauk SSSR, 1958. Bibliography, p. 66-75. (1st ed. in Zemlevedenie,
1911, no. 3).   115

    About 340 references in alphabetical order.

SHNITNIKOV, Arsenii Vladimirovich. Vnutrivekovaia izmenchivost' komponentov obshchei
uvlazhnennosti (Variability of the components of general humidity over a century).
Leningrad, Izdatel'stvo "Nauka," 1969. 245 p. Bibliography, p. 237-244. (Akademiia Nauk SSSR. Geograficheskoe Obshchestvo SSSR).   11

    About 220 references.

BORISOV, Anatolii Aleksandrovich. Paleoklimatologiia SSSR (Paleoclimatology of
the USSR). Kaliningrad, 1973. 304 p. Bibliography, p. 296-303. (Kaliningradskii
Gosudarstvennyi Universitet).   11

    179 references. Supersedes the author's Paleoklimaty territorii SSSR, Lenin-
grad, 1965. 111 p. Bibliography, p. 109-111, with 67 references.

BUCHINSKII, Ivan Evstaf'evich. O klimate proshlogo Russkoi ravniny. (On the climate
of the past of the Russian plain). 2nd ed., Leningrad, Gidrometeoizdat, 1957.
142 p. Bibliography, p. 133-141.   11

    275 references arranged systematically.

GIRS, Aleksandr Aleksandrovich. Mnogoletnie kolebaniia atmosfernoi tsirkuliatsii
i dolgosrochnye gidrometeorologicheskie prognozy (Long-term fluctuations in
atmospheric circulation and long-range hydrometeorological forecasts). Leningrad,
Gidrometeoizdat, 1971. 280 p. Bibliography, p. 269-277.   11

    193 references.

Microclimate

SHCHERBAN', Mikhail Il'ich. Mikroklimatologiia (Microclimatology). Kiev, Izdatel'-
stvo Kievskogo Universiteta, 1968. 212 p. Bibliography, p. 183-211.   11

    About 680 references.

KOROTKEVICH, V. N. Obzor rabot po izucheniiu mikroklimata (Survey of work on the
study of microclimate). Leningrad, 1936. 82 p. Bibliography, p. 47-82 (Glavnaia
Geofizicheskaia Observatoriia, Trudy, vypusk 6. Klimatologiia 2).   11

    1,284 references in alphabetical order, 293 Soviet and 981 in foreign lan-
guages.

KROKLIMAT SSSR (Microclimate of the USSR). I. A. Gol'tsberg, ed. Leningrad, Gidrometeoizdat, 1967. 286 p. Bibliography, p. 279-284. Available in English as: Microclimate of the USSR. Translated by D. Lederman. Edited by P. Greenberg. Jerusalem, Israel Program for Scientific Translations, 1969. 236 p. Bibliography, p. 226-236. IPST no. 5345.   1162

    About 150 references.

Agricultural Climatology

MIRKOV, Iurii Ivanovich, and SHABLEVSKAIA, V. A. Sel'skokhoziaistvennaia meteorologiia i klimatologiia (Agricultural meteorology and climatology), Itogi nauki, seriia "Geofizika," meteorologiia i klimatologiia, Moskva, VINITI, v. 1, 1971, p. 121-153.   1163

    168 references.

NITSINA, N. I., and others. Agroklimatologiia (Agricultural climatology). I. A. Gol'tsberg, ed. Leningrad, Gidrometeoizdat, 1973. 344 p. Bibliography, p. 339-342.   1164

VITAIA, Feofan Farneevich, and others, eds. Agroklimaticheskie resursy prirodnykh zon SSSR i ikh ispol'zovanie (Agricultural climatic resources of natural zones of the USSR and their utilization). Leningrad, Gidrometeoizdat, 1970. 160 p. Bibliographies at ends of individual reports. (Glavnoe Upravlenie Gidrometeorologicheskoi Sluzhby pri Sovete Ministrov SSSR. Mezhduvedomstvennyi Nauchnyi Sovet po Izucheniiu Klimaticheskikh i Agroklimaticheskikh Resursov. Institut Eksperimental'noi Meteorologii. Glavnaia Geofizicheskaia Observatoriia imeni Voeikova).   1165

VITAIA, Feofan Farneevich, ed. Agroklimaticheskie i vodnye resursy raionov osvoeniia tselinnykh i zalezhnykh zemel' (Agroclimatic and water resources of the regions of development of the virgin and idle lands). Leningrad, Gidrometeoizdat, 1955. 464 p. (Glavnoe Upravlenie Gidrometeorologicheskoi Sluzhby pri Sovete Ministrov SSSR).   1166

LOSKOV, Pavel Ivanovich. Klimaticheskii faktor sel'skogo khoziaistva i agroklimaticheskoe raionirovanie (The climatic factor in agriculture and regionalization in agricultural climatology). F. F. Davitaia, ed. Leningrad, Gidrometeoizdat, 1971. 328 p. Bibliography, p. 323-325.   1167

    52 references.

GROKLIMATICHESKIE resursy... (Agricultural climatic resources...) Leningrad, Gidrometeoizdat, or regional hydrometeorological observatories.   1168

    A series of handbooks devoted to individual oblasts, or similar regions, of the Soviet Union, with bibliographies.

ROKLIMATICHESKII spravochnik... (Agroclimatic handbook...). Leningrad, Gidrometeoizdat.   1169

    Volumes for individual oblasts, krays, or autonomous republics, with statistics, maps, and text, published during the 1950's and early 1960's. Second editions were later published of some volumes.

KLIAR, Abram Khaimovich. Klimaticheskie resursy Belorussii i ispol'zovanie ikh v sel'skom khoziaistve (Climatic resources of Belorussia and their utilization in agriculture). Minsk, "Vysheish. Shkola," 1973. 430 p. Bibliography, p. 410-416.   1170

AGROMETEOROLOGICHESKII ezhegodnik za 1965-1966-...sel'skokhoziaistvennyi god.
(Yearbook of agricultural meteorology for the 1965-1966-...agricultural year).
(Upravlenie Gidrometeorologicheskoi Sluzhby). 1972-. A large series published
in regional centers of the Hydrometeorological Survey.   11

BELAREV, Sergei Alekseevich, and BYKOV, Boris Aleksandrovich, eds. Produktivnost'
pastbishch Severnogo Priaral'ia: agrometeorologicheskie usloviia, ritmy razvitiia
i urozhainost' pastbishchnykh rastenii (Productivity of the pastures of the
Northern Aral area: agrometeorological conditions, rhythm of the development and
yield of pasture plants). S. A. Belarev, G. D. Gerasimenko, E. N. Korobova, and
others. Moskva, Gidrometeoizdat, Moskovskoe Otdelenie, 1971. 289 p. Bibliography, p. 282-289. (Glavnoe Upravlenie Gidrometeorologicheskoi Sluzhby pri Sovete
Ministrov SSSR. Kazakhskii Nauchno-Issledovatel'skii Gidrometeorologicheskii
Institut. Trudy, vypusk 45).   11

Bioclimatology

LIOPO, Tamara Nikitichna, and TSITSENKO, Galina Viktorovna. Klimaticheskie usloviia
i teplovoe sostoianie cheloveka (Climatic conditions and the thermal state of
man). Leningrad, Gidrometeoizdat, 1971. 151 p. Bibliography, p. 136-149.   11

MATIUKHIN, V. A. Bioklimatologiia cheloveka v usloviiakh mussonov (Bioclimatology
of man under monsoonal conditions). Leningrad, Izdatel'stvo "Nauka," Leningradskoe Otdelenie, 1971. 138 p. Bibliography, p. 119-136 (Akademiia Nauk SSSR.
Sibirskoe Otdelenie. Institut Fiziologii).   11

GERBURT-GEIBOVICH, Aleksei Andreevich. Otsenka klimata dlia tipovogo proektirovaniia
zhilishch (Appraisal of climate for the standard projecting for housing). M. V.
Zavarina, ed. Leningrad, Gidrometeoizdat, 1971. 195 p. Bibliography, p. 184-
194 (Tsentral'nyi Nauchno-Issledovatel'skii i Proektnyi Institut Tipovogo i
Eksperimental'nogo Proektirovaniia Zhilishcha).   11

278 references.

Climatic and Meteorological Data from Satellites

KONDRAT'EV, Kirill Iakovlevich. Sputnikovaia klimatologiia (Satellite climatology)
Leningrad, Gidrometeoizdat, 1971. 65 p. Bibliography, p. 60-64.   11

102 references.

SONECHKIN, D. M. Meteorologicheskoe deshifrirovanie kosmicheskikh snimkov Zemli:
kolichestvennye metody (Meteorological interpretation of satellite photographs
of the earth: quantitative methods). Sh. A. Musaelian, ed. Leningrad, Gidrometeoizdat, 1972. 130 p. Bibliography, p. 122-129. (Glavnoe Upravlenie Gidrometeorologicheskoi Sluzhby pri Sovete Ministrov SSSR. Gidrometeorologicheskii
Nauchno-Issledovatel'skii Tsentr SSSR. Trudy, vypusk 98).   1

175 references.

MALKEVICH, M. S. Opticheskie issledovaniia atmosfery so sputnikov (Optical
research on the atmosphere from satellites). Moskva, Izdatel'stvo "Nauka,"
1973. 303 p. Bibliography, p. 290-303 (Akademiia Nauk SSSR. Institut Fiziki
Atmosfery).   1

KONDRAT'EV, Kirill Iakovlevich, and TIMOFEEV, Iurii Mikhailovich. Termicheskoe
zondirovanie atmosfery so sputnikov (Thermal exploration of the atmosphere from
artifical earth satellites). Leningrad, Gidrometeoizdat, 1970. 410 p. Bibliography, p. 382-408. Summary in English.   1

543 references, arranged by chapters.

Other Special Topics

AGO- i teploobmen nad vodoëmami i sushei v gornykh usloviiakh (Moisture and heat
exchange over water bodies and dry land in mountainous conditions). A. M.
Mkhitarian, ed. Leningard, Gidrometeoizdat, 1969. 210 p. Bibliography, p. 198-
208. (Zakavkazskii Nauchno-Issledovatel'skii Gidrometeorologicheskii Institut.
Erevanskii Otdel, Trudy, vypusk 29 (35)).   1180
   327 references.

JMILOV, A. V., KOSAREV, A. N., and LEBEDEV, V. D. Protsessy obmena na granitse
okean-atmosfera: konspekt lektsii (The process of the exchange at the boundary
of the ocean and atmosphere: summary of lectures). A. I. Duvanin, ed. Moskva,
Izdatel'stvo Moskovskogo Universitata, 1973. 205 p. Bibliography, p. 195-202.
Moskovskii Universitet imeni M. V. Lomonosova. Geograficheskii Fakul'tet).   1181

LITINKEVICH, S. S. Dinamika pogranichnogo sloia atmosfery (Dynamics of the
boundary layer of the atmosphere). Leningrad, Gidrometeoizdat, 1970. 292 p.
Bibliography, p. 250-284. Name index. Summary in English.   1182
   About 670 references.

U. S. S. R. as a Whole

BLIOGRAFICHESKII ukazatel' rabot Glavnoi geofizicheskoi observatorii za period
1918-1967 gg. no. 1. 1918-1945 gg. no. 2. 1946-1960 gg. (Bibliographical
guide to the works of the Main Geophysical Observatory in the period 1918-1967.
no. 1. 1918-1945. no. 2. 1946-1960). E. L. Andronikova, E. A. Slobodskaia, and
G. A. Tsimmer, comps. M. E. Shvets, ed. Leningrad, no. 1. 1967. 424 p. no. 2.
1969. 311 p. (Glavnaia Geofizicheskaia Observatoriia imeni Voeikova).   1183

   Chronological guide to 3,875 entries in no. 1 and 2,436 entries in no. 2 for
climatology and meteorology. No. 2 also has supplement to no. 1, with 76 additional entries. Author indexes.

ISOV, Anatolii Aleksandrovich. Klimatografiia Sovetskogo Soiuza. (Climatology
of the Soviet Union). Leningrad, Izdatel'stvo Leningradskogo Universiteta, 1970).
310 p. Bibliography, p. 306-310. (Leningradskii Gosudarstvennyi Universitet). 1184

   About 130 references on the climates of the Soviet Union arranged alphabetically by author.

ENKO, Aleksandr Ivanovich, ed. Zasukhi v SSSR: ikh proiskhozhdenie, povtoriaemost'
i vliianie na urozhai (Droughts in the USSR: their origin, frequency and effect
on yields). Leningrad, Gidrometeorologicheskoe Izdatel'stvo, 1958. 207 p.   1185

   91 references on droughts in the USSR distributed through the book at the
end of individual articles.

Regions within the U. S. S. R.

Series covering entire country

INGRAD. Glavnaia Geofizicheskaia Observatoriia imeni A. I. Voeikova. Klimat
SSR (Climate of the USSR). Leningrad, Gidrometeoizdat, 1958-1963. 6 v.
Glavnoe Upravlenie Gidrometeorologicheskoi Sluzhby. Glavnaia Geofizicheskaia
Observatoriia imeni A. I. Voeikova).   1186

   Textual discussion, graphs, tables, maps, and bibliographies. The basic
series on the climates of the USSR.

   v. 1. Evropeiskaia territoriia SSSR (European territory of the USSR), by
leksei Nikolaevich Lebedev. 1958. 366 p. Bibliography, p. 358-367.   1187
   247 references in alphabetical order.

v. 2. Kavkaz (The Caucasus), by Anastasiia Andreevna Zanina. 1961. 290 p.
Bibliography, p. 220-222. 11

85 references. Appendix of statistical tables and maps, p. 223-288.

v. 3. Sredniaia Aziia (Soviet Middle Asia), by Ol'ga Mikhailovna
Chelpanova. 1963. 447 p. Bibliography, p. 381-389. 11

197 references. Appendix of statistical tables, p. 392-445. 16 folded maps
in pocket.

v. 4. Zapadnaia Sibir' (Western Siberia), by Valentina Vladimirovna Orlova.
1962. 360 p. Bibliography, p. 311-316. 11

136 references. Appendix of statistical tables, p. 319-359. 13 folded maps
in pocket.

v. 5. Vostochnaia Sibir' (Eastern Siberia), by Elena Iakovlevna Shcherbakova
1961. 300 p. Bibliography, p. 246-249. 11

77 references. Appendix of statistical tables, p. 252-299. 13 folded maps
in pocket.

v. 6. Dal'nii Vostok (Soviet Far East), by Anastasiia Andreevna Zanina.
1958. 167 p. Bibliography, p. 155-156. 11

47 references. Appendix of maps, p. 158-165.

SPRAVOCHNIK po klimatu SSSR (Handbook on the climate of the USSR). (Glavnoe
Upravlenie Gidrometeorologicheskoi Sluzhby pri Sovete Ministrov SSSR). Leningrad,
Gidrometeoizdat, 1965- . 34 numbered sections. Some volumes carry title: Klima-
ticheskii spravochnik. 11

Climatic data by regions of the Soviet Union with 34 numbered regional sections, each with five parts, published by regional observatories. The regional
numbering is as follows: 1, Arkhangel'sk and Vologda oblasts and the Komi ASSR;
2, Murmansk oblast; 3, Karelian ASSR and Leningrad, Novgorod, and Pskov oblasts;
4, Estonian SSR; 5, Latvian SSR; 6, Lithuanian SSR and Kaliningrad oblast of the
RSFSR; 7, Belorussian SSR; 8, Yaroslavl', Kalinin, Moscow, Vladimir, Smolensk,
Kaluga, Ryazan', and Tula oblasts; 9, Perm', Sverdlovsk, Chelyabinsk, and Kurgan
oblasts and Bashkir ASSR; 10, Ukrainian SSR; 11, Moldavian SSR; 12, Tatar ASSR
and Ul'yanovsk, Kuybyshev, Penza, Orenburg, and Saratov oblasts; 13, Volgograd,
Rostov, and Astrakhan' oblasts, Krasnodar kray, Stavropol' kray, Kalmyk ASSR,
Kabardino-Balkar ASSR, Chechen-Ingush ASSR, and North Oset ASSR; 14, Georgian SSR,
15, Dagestan ASSR, Azerbaidzhan SSR, and Nakhichevan' ASSR; 16, Armenian SSR; 17,
Omsk and Tyumen' oblasts; 18, Kazakh SSR; 19, Uzbek SSR; 20, Tomsk, Novosibirsk,
and Kemerovo oblasts and Altay kray; 21, Krasnoyarsk kray and Tuva ASSR; 22,
Irkutsk oblast and western part of Buryat ASSR; 23, Buryat ASSR and Chita oblast;
24, Yakut ASSR; 25, Khabarovsk kray and Amur oblast; 26, Primorskiy (Maritime)
kray; 27, Kamchatka oblast; 28, Tambov, Bryansk, Lipetsk, Orël, Kursk, Voronezh,
and Belgorod oblasts; 29, Ivanovo, Kostroma, Kirov, and Gor'kiy oblasts and Mary
ASSR, Udmurt ASSR, Chuvash ASSR, and Mordov ASSR; 30, Turkmen SSR; 31, Tadzhik
SSR; 32, Kirgiz SSR; 33, Magadan oblast and Chukot national okrug; and 34,
Sakhalin oblast.

METEOROLOGICHESKII ezhegodnik (Meteorological yearbook). (Glavnoe Upravlenie
Gidrometeorologicheskoi Sluzhby pri Sovete Ministrov SSSR). Leningrad and other
cities. 1968- . 11

Complex set of regional volumes published on annual basic observations of the
regional hydrometeorological stations on topics such as snow cover and soil temperature.

1188 - 1194

Arctic

ERNIGOVSKII, Nikolai Trofimovich, and MARSHUNOVA, Mariia Sergeevna. Klimat
Sovetskoi Arktiki: radiatsionni rezhim   (Climate of the Soviet Arctic: radiation regime). Leningrad, Gidrometeoizdat, 1965. 199 p. Bibliography, p. 150-
154. (Arkticheskii i Antarkticheskii Nauchno-Issledovatel'skii Institut).   1195
     134 references.

HKOV, Stepan Ivanovich. Vliianie Arktiki na klimat SSSR (Influence of the
Arctic on the climate of the USSR). Leningrad, Gidrometeoizdat, 1969. 84 p.
Bibliography, p. 80-83.   1196
     89 references.

European USSR

CHURINA, R. M., and BIBILO, Iuliia Osipovna. Klimat i vody sushi Iugo-Vostoka
Evropeiskoi chasti SSSR: bibliograficheskii ukazatel' (Climate and waters of
the Southeasteastern part of the European USSR: bibliographical guide). Saratov,
Izdatel'stvo Saratovskogo Universiteta, 1961. 267 p. Bibliography on climate,
p. 13-129 (Saratovskii Gosudarstvennyi Universitet. Nauchnaia Biblioteka.
Bibliografiia Saratovskoi Oblasti, vypusk 5).   1197

     1,540 references arranged by subjects: general; solar radiation, atmospheric
nd terrestrial emission, and radiation balance; processes of the circulation of
he atmosphere; thermal regime; moisture regime; atmospheric electricity; climatic
regionalization and description of the Southeast; changes and fluctuations of
limate; microclimate and local climate; agricultural meteorology and climatology;
odification of the climate; medical meteorology and climatology; various ques-
ions of applied climatology. Index of authors, editors, and titles.

GKOV, N. Ia. Bibliografiia klimata Moskvy i Moskovskoi oblasti (Bibliography
f the climate of Moscow and Moscow oblast). Moskovskii Gosudarstvennyi Univer-
itet, Uchehye zapiski, no. 25. 1939. Geografiia, p. 45-57.

     224 references arranged systematically, partly annotated.   1198

TRIEV, A. A., and Bessonov, N. P., eds. Klimat Moskvy: osobennosti klimata
ol'shogo goroda (The climate of Moscow: characteristics of the climate of a
arge city). Leningrad, Gidrometeoizdat, 1969. 323 p. Bibliography, p. 315-
20.
                                                                                   1199
     110 references arranged by chapters.

OBOV, Nikolai Vasil'evich. Klimat Srednego Povolzh'ia (The climate of the
iddle Volga region). Kazan'. Izdatel'stvo Kazanskogo Universiteta, 1968.
52 p. Bibliography, p. 239-250.   1200
     363 references.

NIKOVA, Natal'ia Sergeevna. Klimat Severnogo Kavkaza i prilezhashchikh stepei
Climate of the North Caucasus and adjacent steppes). O. A. Drozdov, ed.
eningrad, Gidrometeoizdat, 1959. 368 p. Bibliography, p. 326-333. (Glavnoe
pravlenie Gidrometeorologicheskoi Sluzhby. Severo-Kavkazskoe Upravlenie).   1201
     278 numbered references in alphabetical order.

MAT Ukrainy (Climate of the Ukraine). G. F. Prikhot'ko, and others, eds.
eningrad, Gidrometeoizdat, 1967. 413 p. Bibliography, p. 401-412.
Ukrainskii Nauchno-Issledovatel'skii Gidrometeorologicheskii Institut).   1202
     281 references.

KONSTANTINOV, Aleksei Rodionovich, Sakali, L. I., Goisa, N. I., and Oleinik, R. N. Teplovoi i vodnyi rezhim Ukrainy (Heat and water regime of the Ukraine). Leningrad, Gidrometeoizdat, 1966. 592 p. Bibliography, p. 464-481. (Ukrainskii Nauchno-Issledovatel'skii Gidrometeorologicheskii Institut).   1

About 440 references.

BUCHINSKII, Ivan Evstaf'evich. Klimat Ukrainy v proshlom, nastoiashchem i budushchem (Climate of the Ukraine in the past, the present, and the future). Kiev, Gossel'-khozizdat USSR, 1963. 308 p. Bibliography, p. 299-306.   1

More than 200 references in Russian and Ukrainian.

LOGVINOV, K. T., BABICHENKO, V. N., KULAKOVSKAIA, M. Iu. Opasnye iavleniia pogody na Ukraine (Hazardous phenomena of weather in the Ukraine). K. T. Logvinov, ed. Leningrad, Gidrometeoizdat, 1972. 236 p. Bibliography, p. 218-235. (Glavnoe Upravlenie Gidrometeorologicheskoi Sluzhby pri Sovete Ministrov SSSR. Ukrainskii Nauchno-Issledovatel'skii Gidrometeorologicheskii Institut. Trudy, vypusk 110). 1

464 references.

Siberia and the Soviet Far East

MEZENTSEV, Varfolomei Semenovich, and KARNATSEVICH, Igor' Vladislavovich. Uvlazhnennost' Zapadno-Sibirskoi ravniny (Moisture of the West Siberian plain). Leningrad, Gidrometeoizdat, 1969. 168 p. Bibliography, p. 141-150.   1

About 240 references.

KLIMAT i gidrologiia Sibiri i Dal'nego Vostoka: informatsionnyi-bibliograficheskii biulleten' (Climate and hydrology of Siberia and the Far East: informational-bibliographical bulletin). 1967- . Bimonthly. Novosibirsk. (GPNTB Sibirskogo Otdeleniia Akademii Nauk SSSR).   1

Systematic, partly annotated guide to the literature.

GAVRILOVA, Mariia Kuz'minichna. Klimat Tsentral'noi Iakutii (The climate of Central Yakutiya). 2nd ed., Yakutsk, Knizhnoe Izdatel'stvo, 1973. 119 p. Bibliography, p. 94-96. (Akademiia Nauk SSSR. Sibirskoe Otdelenie. Institut Merzlotovedeniia).   1

MOSHENICHENKO, I. E. Ocherki razvitiia meteorologii na Dal'nem Vostoke (Studies in the development of meteorology in the Far East). Leningrad, Gidrometeoizdat, 1970. 179 p. Bibliography, p. 162-178.   1

411 references.

ISSLEDOVANIIA v oblasti meteorologii i gidrologii sushi na Dal'nem Vostoke za 50 let Sovetskoi vlasti (Research in the field of meteorology and terrestrial hydrology in the Soviet Far East in the 50 years of Soviet power). A. A. Kalendov and others. Vladivostok, 1968 [1969], 108 p. Bibliography, p. 59-75, 103-106 (Akademiia Nauk SSSR. Sibirskoe Otdelenie. Dal'nevostochnyi Filial).   1

390 references at the ends of sections.

Trans-Caucasus

ALEKSANDRIAN, G. A.   Atmosfernye osadki v Armianskoi SSR (Atmospheric precipitation in the Armenian SSR). Erevan. Izdatel'stvo Akademii Nauk Armianskoi SSR, 1971. 180 p. Bibliography, p. 171-179.   1

218 references.

[KHLINSKII, Enver Mamedovich. Razvitie klimatologii v Azerbaidzhane za 50 let.
Reziume (Development of climatology in Azerbaidzhan over 50 years. Summary),
Akademiia Nauk Azerbaidzhanskoi SSR, Institut Geografii, Trudy, v. 18 (1971),
>. 143-188. Bibliography, p. 165-188.
                                                                        1212

MAT Azerbaidzhana (The climate of Azerbaidzhan). A. A. Madatzade, E. M.
hikhlinskii, G. G. Kavetskaia, and others. Baku, Izdatel'stvo Akademii Nauk
.zerbaidzhanskoi SSR, 1968. 343 p. Bibliography, p. 334-341. (Akademiia Nauk
.zerbaidzhanskoi SSR. Institut Geografii. Upravlenie Gidrometeorologicheskoi
luzhby Azerbaidzhanskoi SSR).
                                                                        1213
    About 200 references.

KHLINSKII, Enver Mamaievich. Teplovoi balans Azerbaidzhanskoi SSR (Heat balance
f the Azerbaidzhan SSR). Baku, 1969. 200 p. Bibliography, p. 197-200. (Akademiia
auk Azerbaidzhanskoi SSR. Institut Geografii).
                                                                        1214
    86 references.

                              Soviet Middle Asia

.GKOV, N. Ia. Bibliograficheskii ukazatel' literatury po klimatu Turkmenii
Bibliographical guide to the literature on the climate of Turkmenia). Ashkhabad,
957. 105 p. (Upravlenie Gidrometsluzhby Turkmenskoi SSR . Ashkhabadskaia
idrometeorologicheskaia Observatoriia).
                                                                        1215
    1,429 entries, arranged systematically.

ANYSHEVA, Svetlana Georgievna. Mestnye vetry Srednei Azii (Local winds of
Soviet Middle Asia). P. A. Vorontsov, ed. Leningrad, Gidrometeoizdat, 1966.
120 p. Bibliography, p. 113-120 (Glavnoe Upravlenie Gidrometeorologicheskoi
Sluzhby pri Sovete Ministrov SSSR. Sredneaziatskii Nauchno-Issledovatel'skii
Gidrometeorologicheskii Institut).
                                                                        1216
    186 references in alphabetical order.

BBOTINA, O. I. Vliianie orografii na temperaturnyi rezhim v gorakh Srednei Azii
(Influence of orography on the temperature regime in the mountains of Soviet
Middle Asia). S. G. Chernysheva, ed. Leningrad, Gidrometeoizdat, 1971. 123 p.
Bibliography, p. 118-122 (Glavnoe Upravlenie Gidrometeorologicheskoi Sluzhby
ori Sovete Ministrov SSSR. Sredneaziatskii Nauchno-Issledovatel'skii Gidro-
neteorologicheskii Institut. Trudy, vypusk 59 (74).
                                                                        1217
    121 references.

SAMOV, A. V. Klimat Tashkenta (The climate of Tashkent). Tashkent, "FAN,"
966. 176 p. Bibliography, p. 162-166. (Tashkentskii Gosudarstvennyi
edagogicheskii Institut imeni Nizami).
                                                                        1218
    108 references.

HMUDOV, Khamidzhan. Klimat Dushanbe (The climate of Dushanbe). V. A.
zhordzhino, ed. Dushanbe. 1965. 212 p. Bibliography, p. 206-210. (Akademiia
auk Tadzhikskoi SSR. Fiziko-tekhnicheskii Institut imeni S. U. Umarova).
                                                                        1219
    113 references.

                          Areas Outside the U. S. S. R.

OV, Boris Pavlovich. Klimaticheskie oblasti zarubezhnykh stran (Climatic
egions of foreign countries). Moskva, Geografgiz, 1950. 352 p. Bibliography,
. 346-350 (Moskovskii Gosudarstvennyi Universitet. Nauchno-Issledovatel'skii
istitut Geografii).
                                                                        1220
    164 selected major references in alphabetical order, 58 Soviet and 106
oreign, covering general climatology and climatology of individual countries.

                                                          1212 - 1220

IVANOV, Nikolai Nikolaevich. Landshaftno-klimaticheskie zony zemnogo shara
(Landscape-climatic zones of the earth). Moskva-Leningrad. Izdatel'stvo
Akademii Nauk SSSR, 1948. 224 p. Bibliography, p. 219-223 (Geograficheskoe
Obshchestvo SSSR, Zapiski, n. s. tom 1).   1

    168 references in alphabetical order on climates and climatic regions,
especially of the USSR.

VITVITSKII, Georgii Nikolaevich. Klimaty Severnoi Ameriki (Climates of North
America). Moskva, Geografgiz, 1953. 288 p. Bibliography, p. 272-285.   1

    291 references in alphabetical order on the formation of the climates of
the northern hemisphere and on the climates of North America, 43 Soviet and 248
in foreign languages.

LEBEDEV, Aleksei Nikolaevich, and SOROCHAN, Ol'ga Georgievna, eds. Klimaty Afriki
(Climates of Africa). Leningrad, Gidrometeoizdat, 1967. 488 p. Bibliography,
p. 483-486 (Glavnaia Geofizicheskaia Observatoriia imeni Voeikova).   1

    About 160 references.

VITVITSKII, Georgii Nikolaevich. Klimaty zarubezhnoi Azii (Climates of Asia
excluding the Soviet Union). Moskva, Geografgiz, 1960. 398 p. Bibliography,
p. 381-396.   1

    325 references in alphabetical order, mostly foreign, on climate-forming
factors and climates of individual regions of Asia.

KLIMATICHESKII spravochnik zarubezhnoi Azii (Climatic handbook of Asia outside the
Soviet Union). A. N. Lebedev and O. D. Kodrau, eds. Leningrad, Gidrometeoizdat,
1974- . 2 v. (Glavnoe Upravlenie Gidrometeorologochicheskoi Sluzhby pri Sovete
Ministrov SSSR. Glavnaia Geofizicheskaia Observatoriia imeni A. I. Voeikova).   1

    Chast' 1. Kontinental'nye raiony (Mainland regions). 1974. 540 p. Bib-
liography, by I. S. Borushko and F. G. Tokar', p. 26-28.

    83 references.

LEBEDEV, Aleksei Nikolaevich, and LASHKOV, V. N., eds. Parametry tropicheskogo
klimata dlia tekhnicheskikh tselei (Parameters of the tropical climate for
technical uses). Leningrad, Gidrometeoizdat, 1973. 515 p. Bibliography, p. 505-
511. (Glavnoe Upravlenie Gidrometeorologicheskoi Sluzhby. Glavnaia Geofizicheska
Observatoriia imeni A. I. Voeikova).

    238 references.

GAIGEROV, Semën Semenovich. Aerologiia poliarnykh raionov (Aerology of polar
regions). Moskva. Gidrometeoizdat, 1964. 304 p. Bibliography, p. 233-250
(Arkticheskii i Antarkticheskii Nauchno-Issledovatel'skii Institut. Tsentral'naia
Aerologicheskaia Observatoriia).

    418 references in alphabetical order, 249 Soviet and 169 in foreign languages

SOLOPOV, Andrei Vasil'evich. Oazisy v Antarktide (Oases in Antarctica). Moskva,
Izdatel'stvo "Nauka," 1967. 143 p. Bibliography, p. 136-141. (Akademiia Nauk
SSSR. Mezhduvedomstvennyi Geofizicheskii Komitet. Rezul'taty issledovanii po
mezhdunarodnym geofizicheskim proektam. Meteorologiia, no. 14). Summary in
English.

    173 references.

GLACIOLOGY, SNOW COVER, AVALANCHES, ICE, AND PERMAFROST

Glaciers and Glaciation

'CHESKAIA geografiia: annotirovannyi perechen' otechestvennykh bibliografii zdannykh v 1810-1966 gg. (Physical geography: annotated list of native bibliographies published 1810-1966). M. N. Morozova and E. A. Stepanova, comps. V. V. levenskaia, ed. L. G. Kamanin, scientific consultant. Moskva, Izdatel'stvo Nauka," 1968. 309 p. (Gosudarstvennaia Biblioteka SSSR imeni V. I. Lenina. nstitut Geografii Akademii Nauk SSSR. Sektor Seti Spetsial'nykh Bibliotek kademii Nauk SSSR).   1229

Glaciology, p. 154-159, entries 719-750, 32 Soviet bibliographies on glaciol-gy, arranged systematically: bibliographies of a general character; bibliographies n individual themes and problems (Snow cover, avalanches; natural ice and gla-iers); bibliographies on glaciers of individual areas. Author and geographical ndexes.

LIAKOV, Vladimir Mikhailovich. Annotirovannyi spisok sovetskoi literatury po **liatsiologii,** 1956-1960 gg. (Annotated list of Soviet literature on glaciology, ?56-1960), in Materialy gliatsiologicheskikh issledovanii. Khronika. Obsuzhdeniia,). 3 (1961), p. 41-133 (Mezhdunarodnyi Geofizicheskii God. 1957-1959. Akademiia uk SSSR. Institut Geografii).   1230

661 entries. Systematic guide: Organization of research, meetings, con-erences; methods of research; general questions of glaciology; Antarctic; Arctic d Polar Urals; Caucasus; Middle Asia and Kazakhstan; Siberia and the Far East; ralanches; study of snow cover, seasonal snow cover outside areas of current .aciation. Indexes of names, subjects, and places.

____ . ____ 1960-1962, in no. 8 (1963), p. 181-299. 983 entries.

____ . ____ and LOSEVA, I. A. ____ 1962-1964, in no. 13 (1967), p. 187-316. *7 entries.

____ . ____ 1965-1966, in no. 15, p. 216-321.

____ and CHERNOVA, L. P. ____ 1967-1968, in no. 18 (1971), p. 219-331. 'O entries.

____ . LAPINA, I. A., and CHERNOVA, L. P. ____ 1969, in no. 21 (1973), p. 2-286. 444 entries.

____ . ____ 1970, in no. 22 (1973), p. 240-319. 655 entries. Name index.

iHCHENIE o nauchnykh rabotakh po gliatsiologii, 1960-1962 gg. Predstavliaetsia Mezhdunarodnoi Assotsiatsii Nauchnoi Gidrologii i Mezhdunarodnoi Komissii Snega L'da k XIII General'noi Assamblee Mezhdunarodnogo Geodezicheskogo i Geofiziches-go Soiuza. (Report on scientific work on glaciology, 1960-1962. Presented to e International Association of Scientific Hydrology and the International Com-ssion on Snow and Ice to the 13th General Assembly of the International Union Geodesy and Geophysics). Moskva, 1963. 110 p. Bibliography, p. 30-110. kademiia Nauk SSSR. Sovetskii Geofizicheskii Komitet). Table of contents also English.   1231

About 1,000 references arranged systematically under the following topics: aciers; snow cover; sea, river, and lake ice; underground ice; periglacial.

HATEL'NYI katalog po gliatsiologii za period MGG/MGS (Final catalogue on ciology in the period of the International Geophysical Year) Moskva, 1963. p. (Mirovoi Tsentr Dannykh MGG-Bl. World Data Center IGY-Bl).   1232

1,404 references. Includes data on original observations. Supplemented by alog publikatsii postupivshikh v MTsD-B s ianvaria po iiun' 1964 g. vypusk 11 talogue of publications received by World Data Center-B, January-June 1964, 11). Moskva, 1964. 24 p.

KALESNIK, Stanislav Vikent'evich. Ocherki gliatsiologii (Studies in glaciology).
Moskva, Geografgiz, 1963. 551 p. Bibliography, p. 510-541.

About 750 references, in alphabetical order, about 400 Soviet and about 350
in foreign languages, on the physical characteristics of glaciers, conditions of
their origin, activity, and evolution, and on the geographical distribution of
glaciers.   1

KRENKE, Aleksandr Nikolaevich. Klimaticheskie usloviia sushchestvovaniia lednikov
i formirovanie lednikovykh klimatov (Climatic conditions for the existence of
glaciers and the formation of glacial climates), Itogi nauki, seriia Geofizika:
Meteorologiia i klimatologiia, Moskva, VINITI, v. 1, 1971, p. 168-207.   1

   203 references.

NAZATOV, G. N. Oledeneniia i geologicheskoe razvitie Zemli (Glaciation and the geological evolution of the earth). Moskva, Izdatel'stvo "Nedra," 1971. 152 p.
Bibliography, p. 146-151.   1

   165 references.

Glaciers and Glaciation of the U.S.S.R.

RESURSY poverkhnostnykh vod SSSR: Katalog lednikov SSSR (Resources of surface waters
of the USSR: Catalogue of glaciers of the USSR). Leningrad, Gidrometeoizdat,
1966- . In process. (Glavnoe Upravlenie Gidrometeorologicheskoi Sluzhby pri
Sovete Ministrov SSSR).   1

   Numerous volumes, each covering a specific region of the USSR, and each with
an extensive bibliography.

AVSIUK, Grigorii Aleksandrovich, and SVATKOV, N. M. Sovremennoe oledenenie na
territorii SSSR (Contemporary glaciation in the USSR), Moskovskii Oblastnoi
Pedagogicheskii Institut imeni N. K. Krupskoi, Uchёnye zapiski, v. 287, no. 2
(1971). p. 5-43. Bibliography, p. 38-43.

TUSHINSKII, Georgii Kazimirovich. Ledniki, snezhniki, laviny Sovetskogo Soiuza
(Glaciers, névé fields, and avalanches of the Soviet Union). Moskva, Geografgiz,
1963. 311 p. Bibliography, p. 291-298. (Komponenty Prirody Sovetskogo Soiuza).

   About 190 references in alphabetical order.

GROSVAL'D, Mikhail Grigor'evich, KRENKE, A. N., VINOGRADOV, O. N., MARKIN, V. A.,
PSAREVA, T. V., RAZUMEIKO, N. G., SUKHODROVSKII, V. L. Oledenenie zemli Frantsa-
Iosifa. Rezul'taty issledovanii po programme Mezhdunarodnogo Geofizicheskogo
Goda. (Glaciers of Franz Josef Land. Results of research under the program of
the International Geophysical Year). G. A. Avsiuk and V. M. Kotliakov, eds.
Moskva, Izdatel'stvo "Nauka," 1973. 352 p. Bibliography, p. 332-344. (Akademiia
Nauk SSSR. Mezhduvedomstvennyi Geofizicheskii Komitet).

   400 references arranged alphabetically.

PANOV, Vasilii Danilovich. Ledniki basseina r. Tereka (Glaciers of the basin of
the river Terek). Leningrad, Gidrometeoizdat, 1971. 296 p. Bibliography, p.
287-295. (Glavnoe Upravlenie Gidrometeorologicheskoi Sluzhby pri Sovete Ministrov
SSSR. Severo-Kavkazskoe Upravlenie Gidrometeorologicheskoi Sluzhby. Rostovskaia
Gidrometeorologicheskaia Observatoriia).

   232 references.

KALESNIK, Stanislaw Vikent'evich. Gornye lednikovye raiony SSSR (Mountain glacier
regions of the USSR). Leningrad-Moskva, Gidrometeoizdat, 1937. 181 p. Bibliography, p. 166-178 (Itogi rabot lednikovykh ekspeditsii 2-go MPG, 1932-1933 gg.,
vypusk 3). Supplementary title page in English.

   369 references in alphabetical order, 328 Soviet and 41 in foreign languages.

EV, Khristofor Iakovlevich. Ocherki po oledeneniiu Bol'shogo Kavkaza (Studies
  glaciation of the Great Caucasus). Rostov-na-Donu, Izdatel'stvo Rostovskogo
  iversiteta, 1965. 192 p. Bibliography, p. 176-191 (Rostovskii Pedagogicheskii
  stitut).
                                                                              1242
    About 320 references in alphabetical order on current and ancient glaciation
  the Great Caucasus.

HERBAKOVA, Elizabeta Mikhailovna. Drevnee oledenenie Bol'shogo Kavkaza (Ancient
  laciation of the Great Caucasus). Moskva, Izdatel'stvo Moskovskogo Universiteta,
  973). 271 p. Bibliography, p. 259-269.
                                                                              1243

)VALOV, Vladimir Georgievich. Abliatsiia lednikov Srednei Azii (Ablation of
  aciers in Soviet Middle Asia). V. G. Khodakov, ed. Leningrad, Gidrometeoizdat,
  972. 158 p. Bibliography, p. 151-157. (Glavnoe Upravlenie Gidrometeorologiche-
  koi Sluzhby pri Sovete Ministrov SSSR. Sredneaziatskii Regional'nyi Nauchno-
  ssledovatel'skii Gidrometeorologicheskii Institut. Trudy, vypusk 8 [89]).   1244
    150 references.

  .OVA, O. V. Ledniki Srednei Azii: bibliografiia (The glaciers of Soviet Middle
  sia: a bibliography). Akademiia Nauk Uzbekskoi SSR. Institut Geologii. Trudy,
  rpusk 8 (1952), p. 119-160.
                                                                              1245
    297 annotated references, arranged chronologically. Author index.

  ROV, Rashit Dzhamalievich. Oledenenie Pamira (Glaciation of the Pamir).
  . K. Tushinskii, ed. Moskva, Geografgiz, 1955. 372 p. Bibliography, p. 360-
  70.
                                                                              1246
    About 300 references in alphabetical order.

Glaciers of Antarctica

.OV, Nartsiss Irinarkhovich. Shel'fovye ledniki Antarktidy (Shelf glaciers of
  tarctica). E. S. Korotkevich, ed. Leningrad, Gidrometeoizdat, 1971. 226 p.
  bliography, p. 211-225. (Glavnoe Upravlenie Gidrometeorologicheskoi Sluzhby
  i Sovete Ministrov SSSR. Arkticheskii i Antarkticheskii Nauchno-Issledovatel'-
  ii Institut. Sovetskaia Antarkticheskaia Ekspeditsiia).
                                                                              1247
    378 references.

Snow Cover

OROLOGICHESKII ezhegodnik: Nabliudeniia gidrometeorologicheskikh stantsii i
  stov nad snezhnym pokrovom snegos"ëmki (Meteorological yearbook: Observations
  hydrometeorological stations and posts on snow cover of the snow survey).
  Glavnoe Upravlenie Gidrometeorologicheskoi Sluzhby pri Sovete Ministrov SSSR).
  68- .
                                                                              1248
    Complex set of regional volumes published by regional stations annually.

  OVA, L. A. Literatura po snegu i snezhnomu pokrovu, 1955-1958 (Literature on
  ow and snow cover), in the book: Geografiia snezhnogo pokrova (Geography of
  ow cover). Moskva, Izdatel'stvo Akademii Nauk SSSR, 1960, p. 204-222 (Akademiia
  uk SSSR. Institut Geografii).
                                                                              1249
    About 450 publications. Systematic guide, continuing for 1955-1958, the
  bliography of V. M. Kotliakov for 1950-1954.

  IAKOV, Vladimir Mikhailovich. Literatura po snegu i snezhnomu pokrovu 1950-
  54 gg. (Literature on snow and snow cover, 1950-1954), in the book Sneg i
  lye vody, ikh izuchenie i ispol'zovanie (Snow and melt water, their study and
  ilization). Moskva, Izdatel'stvo Akademii Nauk SSSR, 1956, p. 251-271 (Akademiia
  uk SSSR. Institut Geografii).
                                                                              1250

About 250 references. Systematic annotated guide: the study of snow (physical qualities, influence of snow on components of the natural environment, methodology of study); study of snow melt and the formation of flow of melt waters; utilization of snow in agriculture; struggle with snow drifts and avalanches.

KOPANEV, Ivan Dmitrievich. Metody izucheniia snezhnogo pokrova (Methods of studying snow cover). Leningrad, Gidrometeoizdat, 1971. 226 p. Bibliography, p. 208-224. (Glavnoe Upravlenie Gidrometeorologicheskoi Sluzhby pri Sovete Ministrov SSSR. Glavnaia Geofizicheskaia Observatoriia imeni A. I. Voeikova).　　　　1

465 references.

KUZ'MIN, Prokofii Pavlovich. Formirovanie snezhnogo pokrova i metody opredeleniia snegozapasov (The formation of snow cover and methods of determination of snow supply). Leningrad, Gidrometeoizdat, 1960. 171 p. Bibliography, p. 162-168 (Gosudarstvennyi Gidrologicheskii Institut).　　　　1

About 150 references in alphabetical order.

KUZ'MIN, Prokofii Pavlovich. Fizicheskie svoistva snezhnogo pokrova (Physical properties of the snow cover). Leningrad, Gidrometeoizdat, 1957. 179 p. Bibliography, p. 169-174 (Gosudarstvennyi Gidrologicheskii Institut).

About 150 references in alphabetical order on density of snow and on its aqueous, thermal, radiational, electrical, radioactive, acoustical, and mechanical properties.

RIKHTER, Gavriil Dmitrievich. Rol' snezhnogo pokrova v fiziko-geograficheskom protsesse (The role of snow cover in the physical-geographic process) Moskva-Leningrad, Izdatel'stvo Akademii Nauk SSSR, 1948. 172 p. Bibliography, p. 161-169 (Institut Geografii, Trudy, vypusk 40).

About 270 references in alphabetical order.

KUZ'MIN, Prokofii Pavlovich. Protsess taianiia snezhnogo pokrova (The process of melting of the snow cover). Leningrad, Gidrometeoizdat, 1961. 345 p. Bibliography, p. 297-307 (Gosudarstvennyi Gidrologicheskii Institut). Available in English as: Melting of Snow cover, Jerusalem, Israel Program for Scientific Translations, 1971. 296 p. IPST 6023.

About 260 references in alphabetical order on methods of determination of heat and water balance, speed of warming up and of freezing of layers of snow, ways of measuring the intensity of snow melting on plains and in mountains, methods of calculating snow melting, organization of observations and methods of carrying out experimental research.

KOLOMYTS, Erland Georgievich. Struktura i rezhim snezhnoi tolshchi Zapadnosibirsko taigi (Structure and regime of depth of snow of the Western Siberian tayga). Leningrad, Izdatel'stvo "Nauka," Leningradskoe Otdelenie. 1971. 174 p. Bibliography, p. 166-172. (Akademiia Nauk SSSR. Sibirskoe Otdelenie. Institut Geograf Sibiri i Dal'nego Vostoka).

OSOKIN, Ivan Mikhailovich. Geografiia snezhnogo pokrova vostoka Zabaikal'ia (Geography of the snow cover of the Eastern Trans-Baikal region). Chita, Izdatel'stvo Zabaikal'skogo Filiala Geograficheskogo Obshchestva SSSR, 1969. 192 p. Bibliography, p. 182-191 (Geograficheskoe Obshchestvo SSSR. Zabaikal'skii Filial. Zapiski, vypusk 33. Otdelenie Fizicheskoi Geografii, no. 1).

About 190 references.

1251 - 1257

Avalanches

ICH, A. V., ed. Rasprostranenie i rezhim lavin na territorii SSSR: kratkii
atalog izvestnykh svedenii (Distribution and regime of avalanches on the ter-
itory of the USSR: a short catalogue of known information). Leningrad, Gidro-
eteoizdat, 1970. 92 p. (Glavnaia Upravlenie Gidrometeorologicheskoi Sluzhby
ri Sovete Ministrov SSSR. Sredneaziatskii Nauchno-Issledovatel'skii Gidro-
eteorologicheskii Institut).   1258

HINSKII, Georgii Kazimirovich, ed. Lavinoopasnye raiony Sovetskogo Soiuza.
Avalanche hazardous regions of the Soviet Union). Moskva, Izdatel'stvo
oskovskogo Universiteta, 1970. 198 p. Bibliography, p. 191-198.   1259

EV, Kim Semenovich. Laviny SSSR: rasprostranenie, raionirovanie, vozmozhnosti
rognoza (Avalanches in the USSR: propagation, regionalization, possibilities of
redicting). V. L. Shul'ts and N. F. Drozdovskaia. Leningrad, Gidrometeoizdat,
966. 131 p. Bibliography, p. 124-130 (Sredneaziatskii Nauchno-Issledovatel'skii
idrometeorologicheskii Institut).   1260

   About 150 references in alphabetical order.

KALEV, Iurii Diomidovich. Vozniknovenie i dvizhenie lavin (Origin and movement
f avalanches). P. M. Mashukov, ed. Leningrad, Gidrometeoizdat, 1966. 152 p.
ibliography, p. 147-152 (Glavnoe Upravlenie Gidrometeorologicheskoi Sluzhby pri
ovete Ministrov. Sreneaziatskii Nauchno-Issledovatel'skii Gidrometeorologicheskii
nstitut).   1261

   145 references in alphabetical order, 102 Soviet and 43 in foreign languages,
specially on methods of fighting avalanches.

HINSKII, Georgii Kazimirovich. Laviny, vozniknovenie i zashchita ot nikh (Ava-
anches, their origin and protection against them). K. K. Markov, ed. Moskva,
eografgiz, 1949. 215 p. Bibliography, p. 207-211 (Moskovskii Gosudarstvennyi
niversitet. Nauchno-Issledovatel'skii Institut Geografii).   1262

   135 references in alphabetical order.

ROSY lavinnogo morfolitogeneza (Questions of avalanche morpholithogenesis).
. I. Sizikov, ed. Chita, 1971. 113 p. Bibliography, p. 102-112. (Geografiche-
koe Obshchestvo SSSR. Zabaikal'skii Filial. Zapiski, vypusk 63. Otdelenie
izicheskoi Geografii, Trudy, no. 3).   1263

IKHANOV, M. Ch. Snezhno-lavinnyi rezhim i perspektivy osvoeniia gor Kabardino-
alkarii (Snow avalanche regime and prospects for the utilization of the mountains
f Kabardino-Balkariya). Nal'chik, "El'brus," 1971. 191 p. Bibliography, p. 185-
90.   1264

Ice

LIAKOVSKII, Lev Gertsovich. Poiavlenie l'da i nachalo ledostava na rekakh,
zerakh i vodokhranilishchakh: raschёty dlia tselei prognozov (The appearance
f ice and the beginning of the period of stable ice on rivers, lakes, and reser-
oirs: computations with the aim of forecasts). Moskva, Gidrometeoizdat, 1960.
16 p. Bibliography, p. 209-216. (Glavnoe Upravlenie Gidrometeorologicheskoi
luzhby pri Sovete Ministrov SSSR. Tsentral'nyi Institut Prognozov).   1265

   About 170 references in alphabetical order on the thermal and ice regimes of
ivers, lakes, and reservoirs and methods of forecasting the freezing and break-
ng up of ice in rivers and reservoirs.

MAKSIMOVICH, Georgii Alekseevich. Peshchernye l'dy (Cave ice). Geograficheskoe
Obshchestvo SSSR, Izvestiia, v. 79, no. 5 (1947), Bibliography, p. 547-549.   12•

143 references in alphabetical order, 71 Soviet and 72 in foreign languages,
on the structure and origin of cave ice and the distribution of ice caves over
the earth.

BYKOV, A. F. Ukazatel' noveishei literatury po ledovym obrazovaniiam, 1930-1934 gg.
(Guide to recent literature on ice formation, 1930-1934), Geograficheskoe Obshchestvo SSSR, Izvestiia, v. 67, no. 5 (1935), p. 636-653.   1:

463 references arranged systematically: ice and its qualities: the Glacial
Period; ice caves; glaciers; glacial deposits; permafrost; ice cover on rivers
and lakes; ice of seas; ice engineering; ice roads; general problems.

BUINITSKII, Viktor Khalampievich. Morskie l'dy i aisbergi Antarktiki (Sea ice and
iceberg of Antarctica). Leningrad, Izdatel'stvo Leningradskogo Universiteta, 1973.
255 p. Bibliography, p. 241-254. (Leningradskii Gosudarstvennyi Universitet imeni
A. A. Zhdanova).   12

BOGORODSKII, V. V., GUSEV, A. V., KHOKHLOV, G. P. Fizika presnovodnogo l'da
(Physics of fresh-water ice). V. V. Bogorodskii, ed. Leningrad, Gidrometeoizdat,
1971. 227 p. Bibliography, p. 215-223. Summary in English.   1:

BULATOV, Sergei Nikolaevich. Raschët prochnosti taiushchego ledianogo pokrova i
nachala vetrovogo dreifa l'da (Calculation of durability of the melting ice cover
and the beginning of the wind drift of the ice). Leningrad, Gidrometeoizdat,
1970. 118 p. Bibliography, p. 112-117. (Glavnoe Upravlenie Gidrometeorologicheskoi Sluzhby pri Sovete Ministrov SSSR. Gidrometeorologicheskii Nauchno-
Issledovatel'skii Tsentr SSSR. Trudy, vypusk 74).   12

144 references.

Permafrost

ERSHOVA, O. M. Bibliografiia po merzlotovedeniiu za 1955 i 1956 gg. (Bibliography
on permafrost 1955-1956), in the book: Materialy k osnovam ucheniia o merzlykh
zonakh zemnoi kory, vypusk 5. Moskva, Akademiia Nauk SSSR, 1960, p. 203-218.
(Institut Merzlotovedeniia imeni Obrucheva).   12

About 250 references. Supplements list published in Materialy..., no. 4,
1958.

MATERIALY k osnovam ucheniia o merzlykh zonakh zemnoi kory (Materials as a basis
for the study of frozen zones of earth's crust). P. D. Bondarev, ed., no. 4.
Moskva, 1958. 251 p. (Institut Merzlotovedeniia imeni Obrucheva).   12

Bibliographies at ends of articles.

VEL'MINA, N. A. Osobennosti gidrogeologii merzloi zony litosfery: Kriogidrogeologiia
(Features of the hydrogeology of the frozen zone of the lithosphere: Cryohydro-
geology). Moskva, Izdatel'stvo "Nedra," 1970. 326 p. Bibliography, p. 312-324.12

BARANOV, Ivan Iakovlevich, ed. Geokriologicheskie issledovaniia pri inzhenernykh
izyskaniia (Geocryological investigations in engineering research). Moskva,
1970. 284 p. Bibliography, p. 268-278. (Proizvod. i Nauchno-Issledovatel'skii
Institut po Inzhenernym Izyskaniiam v Stroitel'stve Gosstroia SSSR. Trudy, tom
2).   12

OVOI, Petr Nikolaevich. Osobennosti geokriologicheskikh uslovii gornykh stran
Features of geocryological conditions of mountainous areas). Moskva, Izdatel'-
tvo "Nauka," 1970. 135 p. Bibliography, p. 129-124. (Gosstroi SSSR. Proiz-
oditel'nyi i Nauchno-Issledovatel'skii Institut po Inzhenernym Izyskaniiam v
troitel'stve).   1275

     174 references.

BUNOV, Aldar Petrovich. Merzlotnye iavleniia Tian'-Shania (Permafrost phenomena
f the Tyan'-Shan'). L. G. Bondarev, 3d. Moskva, Gidrometeoizdat, Moskovskoe
tedlenie, 1970. 266 p. Bibliography, p. 257-264. (Glavnoe Upravlenie Gidro-
eteorologicheskoi Sluzhby pri Sovete Ministrov SSSR. Kazakhskii Nauchno-
ssledovatel'skii Gidrometeorologicheskii Institut. Trudy, vypusk 39).   1275a

QUATERNARY STUDIES AND PALEOGEOGRAPHY

Quaternary Studies (See also 2234)

CHINSKAIA, E. A. Osnovnye zakonomernosti razvitiia prirody territorii SSSR v
hetvertichnom periode (lednikovom periode--antropogene), v. 2. Bibliografiia
1940-1960 gg.), nos. 1-2 (Basic regularities in the development of natural con-
itions on the territory of the USSR in the Quaternary period, the glacial period,
r the Anthropogenic period, v. 2, bibliography 1940-1960, nos. 1-2). G. I.
azukov, ed. With the participation of K. V. Zhukovskaia. Moskva, 1962. no. 1,
51 p. no. 2, 272 p. (Moskovskii Gosudarstvennyi Universitet. Geograficheskii
akul'tet. Nauchnaia Biblioteka imeni A. M. Gor'kogo. Otdel Geografii).   1276

    No. 1 with about 3,600 entries covers the USSR as a whole, the European part,
nd the Urals. No. 2, with about 3,840 entries covers Siberia, the Caucasus,
oviet Middle Asia and Kazakhstan, and the seas and islands of the USSR. List of
ibliographical sources at end of no. 2. Author indexes in each number.

ASIMOV, Innokentii Petrovich, and MARKOV, Konstantin Konstantinovich. Lednikovyi
eriod na territorii SSSR. Fiziko-geograficheskie usloviia lednikovogo perioda
The glacial period on the territory of the USSR. Physical-geographic conditions
f the glacial period). Moskva-Leningrad, Izdatel'stvo Akademii Nauk SSSR, 1939.
62 p. (Akademiia Nauk SSSR. Institut Geografii. Trudy, vypusk 33). English   1277
ummary, p. 443-462.

    About 1,250 references in the following sections: ancient glaciation of the
lain (about 450), ancient glaciation of high mountain regions of the USSR (200),
ome general problems of the paleogeography of the nonglaciated regions of the
SSR (500), and the southern seas of the USSR during the glacial period (100).

HDUNARODNAIA ASSOTSIATSIIA PO IZUCHENIIU CHETVERTICHNOGO PERIODA, KONGRESS.
-i. PARIZH. 1969. (International Association for the Study of the Quaternary,
NQUA. 8th Congress. Paris. 1969). Itogi i materialy (Results and materials).
oskva, Izdatel'stvo "Nauka," 220 p. Bibliography, compiled by M. I. Neishtadt,
. 195-217 and bibliographies at ends of reports. (Akademiia Nauk SSSR. Komissiia
o Izucheniiu Chetvertichnogo Perioda).   1278

IKIE oledeneniia.Paleogeografiia Evropy v pozdnem pleistotsene. Rekonstruktsii
modeli. Opyt. maket atlasa-monografii (The great glaciation. Paleogeography
f Europe in the late Pleistocene. Reconstruction and model. An experimental
ock-up of an atlas-monograph). Authors and compilers of the text and summary
aps: A. A. Assev, V. V. Berdnikov, A. A. Velichko, and others. Moskva, 1973.
59 p. Bibliography, p. 241-259. (Akademiia Nauk SSSR. Institut Geografii.
IX kongressu MezhdunarodnoiAssotsiatsii po Izucheniiu Chetvertichnogo Perioda,
ovaia Zelandiia, XII, 1973).   1279

HANKOV, Iurii Mikhailovich. Geologicheskaia s"ëmka chetvertichnykh otlozhenii i
eomorfologicheskie issledovaniia (Geological surveys of Quaternary deposits and
eomorphological research). Leningrad, Izdatel'stvo "Nedra," Leningradskoe Otde-
enie, 1973. 239 p. Bibliography, p. 220-238. (Ministerstvo Geologii SSSR.
sesoiuznyi Nauchno-Issledovatel'skii Geologicheskii Institut "VSEGEI." Meto-
ologichesk e ukazaniia po geologicheskoi s"ëmke masshtaba 1:50,000. vypusk 6). 1280
    422 references.

OPLICHKO, Ivan Grigor'evich. O lednikovom periode (On the glacial period). Kiev,
946-1956. 4 v.   1281

    v. 1. Vozniknovenie i razvitie ucheniia o lednikovom periode (Rise and
evelopment of study of the glacial period). Kiev, Izdatel'stvo Kievskogo
osudarstvennogo Universiteta, 1946. Bibliography, p. 156-168.

    About 330 references.

v. 2. Biologicheskie i geograficheskie osobennosti evropeiskikh predstavitele chetvertichnoi fauny (Biological and geographical characteristics of European specimens of the Quaternary period). Kiev, Izdatel'stvo Akademii Nauk Ukrainskoi SSR, 1951. Bibliography, p. 242-262.

About 680 references.

v. 3. Istoriia chetvertichnoi fauny Evropeiskoi chasti SSSR (History of the Quaternary fauna of the European part of the USSR). Kiev, Izdatel'stvo Akademii Nauk Ukrainskoi SSR, 1954. Bibliography, p. 202-218.

About 600 references.

v. 4. Proiskhozhdenie valunnykh otlozhenii (Origin of the boulder deposits). Kiev, Izdatel'stvo Akademii Nauk Ukrainskoi SSR, 1956. Bibliography, p. 317-335.

About 650 references.

VELICHKO, Andrei Alekseevich. Prirodnyi protsess v pleistotsene. K IX kongressu INQUA. Novaia Zelandiia. 1973 (Natural processes in the Pleistocene. To the 9th INQUA Congress, New Zealand, 1973). Moskva, Izdatel'stvo "Nauka," 1973. 256 p. Bibliography, p. 242-254. (Akademiia Nauk SSSR. Institut Geografii. Materialy nauchnykh s"ezdov i konferentsii).   1

LUKASHEV, Konstantin Ignat'evich. Osnovnye voprosy geologii i paleogeografii antropogena (Basic questions of geology and paleogeography of the Quaternary). Minsk, Izdatel'stvo Akademii Nauk Belorusskoi SSR, 1959. 287 p. Bibliography, p. 269-285.   12

About 480 references, of which about 150 in foreign languages.

SEREBRIANNYI, Leonid Ruvimovich. Radiouglerodnyi metod i ego primenenie dlia izucheniia paleogeografii chetvertichnogo perioda. K VI kongressu INQUA v Varshave 1961 g. (The radiocarbon method and its application for the study of the paleo- geography of the Quaternary period. For the VI Congress of INQUA in Warsaw in 1961). Moskva, Izdatel'stvo Akademii Nauk SSSR, 1961. 226 p. Bibliography, p. 51-225. (Akademiia Nauk SSSR. Institut Geografii. Sovetskaia Sektsiia Mezhdunarodnoi Assotsiatsii po Izucheniiu Chetvertichnogo Perioda).   12

About 1,000 entries for the period 1946-1961. Subject index. Geographical index. Index of the most important archeological sites. List of serial publica- tions utilized.

_____. _____. K VII kongressu INQUA, S.Sh.A, 1965. (For the VII INQUA Congress in th United States in 1965). Moskva, Izdatel'stvo "Nauka," 1965. 271 p. Bibliography, p. 126-267. (Akademiia Nauk SSSR. Institut Geografii. Sovetskaia Sektsiia Mezhdunarodnoi Assotsiatsii po Izucheniiu Chetvertichnogo Perioda).   12

899 annotated entries, mostly for the period 1961-1964, both Soviet and foreign (677). Subject index. Geographical index. Index of the most important archeological sites.

KOSTIAEV, A. G. Osnovnye problemy pleistotsenovoi geologii i paleogeografii Russkoi ravniny i sopredel'nykh gliatsial'no-perigliatsial'nykh territorii v svete osoben- nostei ritmyki pleistotsenovykh otlozhenii (Basic problems of Pleistocene geology and paleogeography of the Russian plain and the contiguous glacial-periglacial territory in the light of the characteristics of rhythmics of Pleistocene deposits) Voprosy geografii, sbornik 79 (1970), p. 82-104. Bibliography, p. 99-104.   12

About 100 references.

SEREBRIANNYI, Leonid Ruvimovich, and RAUKAS, A. V. Novye puti i metody izucheniia lednikovoi istorii russkoi ravniny v verkhnem pleistotsene (New trends and methods of studying the glacial history of the Russian plain in the Upper Pleistocene), Geograficheskii sbornik (Vsesoiuznyi Institut Nauchnoi i Tekhnicheskoi Informatsii v. 4. Moskva, VINITI, 1970, p. 117-137. Bibliography, p. 133-137.   12

80 references, 57 in Russian, 23 in other languages.

'EF i stratigrafiia chetvertichnykh otlozhenii severo-zapada Russkoi ravniny.
VI kongressu INQUA v Varshave, 1961. (Relief and stratigraphy of Quaternary
eposits of the Northwest part of the Russian plain. To the 6th INQUA Congress
n Warsaw, 1961). Moskva, Izdatel'stvo Akademii Nauk SSSR, 1961. 252 p. Bibliog-
aphy, p. 233-245. Summary in English. (Akademiia Nauk SSSR. Sovetskaia
ektsiia Mezhdunarodnoi Assotsiatsii po Izucheniia Chetvertichnogo Perioda. INQUA).
    About 340 references in alphabetical order on the geology, paleogeography,   1288
nd geomorphology.

ELIS, Vitautas (Vytautas) K. Rel'ef i chetvertichnye otlozheniia Pribaltiki
Relief and Quaternary deposits of the East Baltic region). Vilnius, "Mintis,"
973. 264 p. Supplementary title and table of contents in Lithuanian and English.
ibliography, p. 245-261.   1289
    About 500 references (about 350 in Russian and 150 in Western languages) in
lphabetical order.

ILANS, Igor' Ianovich. Chetvertichnye otlozheniia Latvii (Quaternary deposits
f Latvia). Riga, "Zinatne," 1973. 312 p. Bibliography, p. 296-309. (Minis-
erstvo Geologii SSSR. Vsesoiuznyi Nauchno-Issledovatel'skii Institut Morksoi
eologii i Geofiziki).   1290

DER, Ekaterina Pavlovna. Antropogenovye otlozheniia i razvitie rel'efa Belorussii:
IX kongressu INQUA (Quaternary deposits and the development of the relief of
elorussia: for the 9th Congress of INQUA). Minsk, Izdatel'stvo "Nauka i Tekhnika,"
973. 124 p. Bibliography, p. 115-123. (Akademiia Nauk Belorusskoi SSR. In-
titut Geokhimii i Geofiziki).   1291

DARCHUK, Volodymyr Havrylovych, ed. Antropogen: Chetvertynni vidklady (Quater-
ary deposits). Kyiv, "Naukova Dumka," 1969. 326 p. Bibliography, p. 319-325.
Akademiia Nauk URSR. Instytut Heolohichnykh Nauk. Stratyhrafiia URSR tom 11).
n Ukrainian with summaries in Russian and English. Russian: Bondarchuk, Vladimir
avrilovich, ed. Antropogen: Chetvertichnye osadki (Akademiia Nauk Ukrainskoi
SR. Institut Geologicheskikh Nauk. Stratigrafiia USSR, tom 11).   1292
    About 200 references.

ADAEV-NIKONOV, K. N., and IANOVSKII, P. V. Chetvertichnye otlozheniia Moldavskoi
SR (Quaternary deposits of Moldavia). Kishinev, Izdatel'stvo "Kartia Moldovenia-
ke," 1969. 91 p. Bibliography, p. 85-89. (Akademiia Nauk Moldavskoi SSR. Otdel
aleontologii i Stratigrafii).   1293
    About 120 references.

ETSKII, G. I. Alliuvial'naia letopis' velikogo Pra-Dnepra (Alluvial yearbook
f the great proto-Dnepr). Moskva, Izdatel'stvo "Nauka," 1970. 491 p. Bibliog-
aphy, p. 472-488. (Akademiia Nauk SSSR. Komissiia po Izucheniiu Chetvertichnogo
erioda).   1294

AKOV, Vsevolod Alekseevich. Paleogeografiia Zapadno-Sibirskoi nizmennosti v
leistotsene i pozdnem pliotsene. K IX kongressu INQUA, Novaia Zelandiia. 1973
Paleogeography of the West-Siberian lowland in the Pleistocene and late Pliocene.
or the 9th INQUA Congress in New Zealand, 1973). Leningrad, Izdatel'stvo "Nauka,"
eningradskoe Otdelenie, 1972. 200 p. Bibliography, p. 190-196 (Akademiia Nauk
SSR. Geograficheskoe Obshchestvo SSSR). English summary.   1295

HIPOV, Stanislav Anatol'evich. Chetvertichnyi period v Zapadnoi Sibiri (The
uaternary period in Western Siberia). V. N. Saks, ed. Novosibirsk, Izdatel'stvo
Nauka," Sibirskoe Otdelenie, 1971. 331 p. Bibliography, p. 310-329 (Akademiia
auk SSSR. Sibirskoe Otdelenie. Institut Geologii i Geofiziki).   1296

                                     1288 - 1296

NEKRASOV, Igor' Aleksandrovich, MAKSIMOV, Evgenii Vladislavovich, and KLIMOVSKII, Igor' Vladimirovich. Poslednee oledenenie i kriolitozona Iuzhnogo Verkhoian'ia (The last glaciation and the zone of frozen ground in the Verkhoyanskii Khrebet) Yakutsk, Knizhnoe Izdatel'stvo, 1973. 151 p. Bibliography, p. 144-150. (Akademiia Nauk SSSR. Sibirskoe Otdelenie. Institut Merzlotovedeniia). 1

RAVSKII, Edmund Iosifovich. Osadkonakoplenie i klimaty Vnutrennei Azii v antropogene (Sedimentation and climate of Inner Asia in the Quaternary). Moskva, Izdatel'stvo "Nauka," 1972. 336 p. Bibliography, p. 324-335 (Akademiia Nauk SSSR. Geologicheskii Institut. Komissiia po Izucheniiu Chetvertichnogo Perioda). 1

MOSKVITIN, Aleksandr Ivanovich. Stratigrafiia pleistotsena Tsentral'noi i Zapadnoi Evropy (Stratigraphy of the Pleistocene in Central and Western Europe). Moskva, Izdatel'stvo "Nauka," 1970. 287 p. Bibliography, p. 257-276. (Akademiia Nauk SSSR. Geologicheskii Institut. Trudy, vypusk 193). Summary in German. 1

    610 references.

Paleogeography

MARKOV, Konstantin Konstantinovich. Paleogeografiia: istoricheskoe zemlevedenie (Paleogeography: historical earth science). S. Iu. Geller, ed. 2nd ed. Moskva, Izdatel'stvo Moskovskogo Universiteta, 1960. 268 p. (1st ed. Moskva, Geografgiz, 1951). 1

    200 references by chapters: object and history of paleogeography (13 references), cosmogonic bases of paleogeography (13), development of basic forms of the surface of the earth (47), development of weathering of the crust (20), development of the hydrosphere (27), development of the atmosphere, changes in climate of the earth (25), development of living matter in the biosphere (44), development of natural conditions in the Quaternary period, the Anthropogenic period (11)

MARKHOVSKII, Nikolai Iosifovich. Paleogeograficheskie osnovy poiskov nefti i gaza (Paleogeographical foundations of exploration for oil and gas). Moskva, Izdatel'stvo "Nedra," 1973. 302 p. Bibliography, p. 293-300. 1

PRONICHEVA, M. V. Paleogeormorfologiia v neftianoi geologii. Metody i opyt primeneniia (Paleogeomorphology in oil geology. Methods and an experiment in utilization). Moskva, Izdatel'stvo "Nauka," 1973. 174 p. Bibliography, p. 159-173. (Akademiia Nauk SSSR. Geomorfologicheskaia Komissiia). 1

TERRESTRIAL HYDROLOGY

(See also the section Climatology and Meteorology)

Major General Bibliographies

[ZICHESKAIA geografiia: annotirovannyi perechen' otechestvennykh bibliografii izdannykh v 1810-1966 gg. (Physical geography: annotated list of native bibliographies published 1810-1966). M. N. Morozova and E. A. Stepanova, comps. V. V. Klevenskaia, ed. L. G. Kamanin, scientific consultant. Moskva, Izdatel'stvo "Nauka," 1968. 309 p. (Gosudarstvennaia Biblioteka SSSR imeni V. I. Lenina. Institut Geografii Akademii Nauk SSSR. Sektor Seti Spetsial'nykh Bibliotek Akademii Nauk SSSR).   1303

Hydrology of the land, p. 131-132, 139-154, entries 581-584 and 628-718, 95 Soviet bibliographies on hydrology organized systematically: hydrology; hydrology of the land: bibliographies of a general character, bibliographies of publications of scientific institutions of the USSR; bibliographies on individual themes and problems (composition of terrestrial waters; terrestrial water balance, flow; thermal conditions of waters; hydrological regionalization; hydrological forecasting); hydrology of streams (stream flow; water regime of rivers; fluvial processes and the load of sediment of a stream; mouths of rivers); hydrology of lakes and bogs; hydrology of mudflows; bibliographies of regional hydrology of the land (USSR as a whole; European part of the USSR, North and Northwest, Center, West, Southwest, Volga region, Urals, the Caucasus; Asiatic part of the USSR, Siberia, Middle Asia and Kazakhstan). Author and geographical indexes.

DROLOGIIA. Meteorologiia. Klimatologiia (Hydrology, meteorology, and climatology), Bibliografiia sovetskoi bibliografii (Bibliography of Soviet bibliography). (Vsesoiuznaia Knizhnaia Palata). 1939, 1946- . Annual. Moskva, Izdatel'stvo "Kniga," 1941, 1948- .   1304

Annual list of Soviet bibliographies on hydrology (both of the land and the ocean), meteorology, and climatology, a subsection of the section on geological-geographical sciences in the chapter on natural sciences and mathematics. The 1970 annual volume (1972), p. 98-103, contains 170 entries (combined with meteorology and climatology).

'OGI nauki i tekhniki: gidrologiia (Results of science and technology: hydrology). (Vsesoiuznyi Institut Nauchnoi i Tekhnicheskoi Informatsii). Moskva, VINITI, 1- (1966-). Irregular.   1305

1. Gidrologiia sushi (Terrestrial hydrology) 1963-1964. 1966. 147 p. 933 references.

2. Gidrologiia sushi. Gliatsiologiia (Hydrology and glaciology). 1968. 132 p. 177 references.

3. Gidrologiia sushi. Gliatsiologiia (Hydrology and glaciology). 1969. 172 p.

)DNYE resursy, ikh izuchenie, ispol'zovanie i okhrana. Bibliograficheskii ukazatel' literatury za 1970 g. (Water resources, their study, utilization, and conservation: bibliographical guide to the literature for 1970). I. E. Kuksin, I. G. Pikovskaia, G. A. Levashova, N. V. Iurevich, Ivanovskii, comps. Minsk, Nauchno-Issledovatel'skii Institut Vodnykh Problem Minvodkhoza SSSR. Fundamental'naia Biblioteka Akademii Nauk Belorusskoi SSR, 1971. 607 p.   1306

_____ . 1971 g. I. E. Kuksin, I. G. Pikovskaia, G. A. Levashova, and others. 1972. 557 p.

_____ . 1972 g. I. E. Kuksin, I. G. Pikovskaia, G. A. Levashova, and others. 1973. 654 p.

SOOBSHCHENIIA o nauchnykh rabotakh po gidrologii. Predstavliaetsia v Mezhdunarodnoi
    Assotsiatsii Nauchnoi Gidrologii k XI general'noi Assamblee Mezhdunarodnogo Geo-
    dezicheskogo i Geofizicheskogo Soiuza (Report on scientific work in hydrology.
    Offered in the International Association of Scientific Hydrology to the 11th
    General Assembly of the International Union of Geodesy and Geophysics). Moskva,
    1957. 32 p. (Akademiia Nauk SSSR. Komitet po Geodezii i Geofizike).   130

        552 publications in the Soviet Union during the years 1950-1956 on research
    in hydrology: surface waters (including erosion), ground water, and glaciology.

    _____ , 1957-1959. (12th General Assembly). Moskva, 1960. 106 p.

        649 publications, 1957-1959.

    _____ , 1960-1962. (13th General Assembly). Moskva, 1963. 61 p.

        669 publications, 1960-1962.

ITOGI piatiletnikh rabot v SSSR po programme Mezhdunarodnogo gidrologicheskogo
    desiatiletiia, 1965-1969 (Results of 5 years of work in the USSR on the program
    of the International Hydrological Decade, 1965-1969). Leningrad, Gidrometeoizdat,
    1969. 75 p. Bibliography, p. 68-75. (IUNESKO. Mezhduvedomstvennyi Komitet SSSR
    po Mezhdunarodnomu Gidrologicheskomu Desiatiletiiu).   130

        169 references on basic hydrological work.

KORNILOVA, M. A. Bibliograficheskii ukazatel' izdanii Gosudarstvennogo gidrologi-
    cheskogo instituta za 1919-1956 gg. (Bibliographical guide to publications of
    the State Hydrological Institute 1919-1956). Leningrad, Gidrometeoizdat, 1957. 130
    93 p. (Gosudarstvennyi Gidrologicheskii Institut).

        725 entries organized by sections: periodicals and serials; separate pub-
    lications; instructions, directions, manuals, methodological publications; material
    of the water registry and hydrological yearbooks. Contents analyzed for collected
    works, periodicals, and serials. Systematic and author indexes.

LENINGRAD. Gosudarstvennyi Gidrologicheskii Institut. Ukazatel' sovetskoi litera-
    tury po gidrologii sushi. 1945-1950 gg. (Guide to Soviet literature on terrestrial
    hydrology, 1945-1950). M. A. Kornilova. O. A. Spengler, ed. Leningrad, Gidro-
    meteoizdat, 1960. 210 p.   131

        3,166 entries systematically organized: general section; hydrometry, hydro-
    mechanics and hydraulics; hydrophysics; hydrochemistry; rivers and canals; lakes
    and reservoirs; bogs; underground water; glaciers; snow cover; flow and water
    economy and hydromechanics. Author index.

    _____ . _____ 1951-1955 gg. 1962. 340 p.

        5,123 entries.

SHTEPANEK, S. I. Annotirovannyi ukazatel' opublikovannykh rabot Instituta Gidro-
    logii i Gidrotekhniki AN USSR za 1926-1956 gg. (Annotated guide to published
    works of the Institute of Hydrology and Hydraulic Engineering of the Academy of
    Sciences of the Ukrainian SSR 1926-1956), Akademiia Nauk Ukrainskoi SSR. Institut
    Gidrologii i Gidrotekhniki. Izvestiia, v. 15 (22), 1959, p. 110-119.   131

        64 entries grouped by types of publications. Collected works are analyzed.
    The Institute until 1938 was entitled Nauchno-Issledovatel'skii Institut Vodnogo
    Khoziaistva Ukrainy (Scientific-Research Institute of the Water Economy of the
    Ukraine).

BIBLIOGRAFIIA po gidrologii za 1937 g. Soiuz Sovetskikh Sotsialisticheskikh
    Respublik (Bibliography on hydrology in 1937. Union of Soviet Socialist
    Republics). Leningrad-Moskva. Gidrometeoizdat, 1938. 286 p. (Gosudarstvennyi
    Gidrologicheskii Institut. Mezhdunarodnaia Assotsiatsiia Gidrologii).   131

        402 entries in systematic annotated guide of publications in hydrology in
    the USSR. Author and subject indexes. Parallel text in Russian and French.

TECHESTVENNAIA literatura po gidrologii sushi za vremia voiny, 1941-1945 (Native literature on terrestrial hydrology during the war, 1941-1945), Meteorologiia i gidrologiia, 1946, no. 1, p. 100-104, no. 2, p. 86-90, no. 3, p. 96-99, no. 4, p. 99-103, no. 5, p. 106-109, no. 6, p. 93-99.   1313

    723 entries for the years 1941-1945. Earlier bibliographies cover the period 1937-March 1941. For 1945-1946 see Otechestvennaia tekushchaia literatura po geofizike, meteorologii, gidrologii i okeanografii.

TECHESTVENNAIA tekushchaia literatura po geofizike, meteorologii, gidrologii i okeanografii (Native current literature on geophysics, meteorology, hydrology, and oceanography), Meteorologiia i gidrologiia, 1947, no. 2, p. 98-102, no. 3, p. 86-89, no. 4, p. 101-104.   1314

    302 entries for 1945-1946. For 1941-1945 under the title, Otechestvennaia literatura po gidrologii sushi za vremia voiny, 1941-1945. For later years see Literatura po geofizike, meteorologii, gidrologii i okeanografii, or Literatura po meteorologii i gidrologii, both in the section on climatology.

ALININ, Gennadii Pavlovich. Problemy global'noi gidrologii. (Problems of global hydrology). Leningrad, Gidrometeoizdat, 1968. 377 p. Bibliography, p. 365-375. 1315
    260 references.

PPOKOV, Evgenii Vladimirovich. Gidrologiia kak nauka i kratkii spisok glavneishei literatury po gidrologii (Hydrology as a science and a short list of the most important literature on hydrology), Visti Naukovo-doslydchogo in-tu vodnogo gos-podarstva Ukrainy, 1929, tom 2 (1927-1928), no. 2, p. 171-230.   1316

    1,036 entries with sections: general hydrology and potamology; hydrology of ground water; oceanography; limnology; bog studies.

DOSEEV, Ivan Andreevich. Razvitie gidrologii sushi v Rossii (Development of hydrology in Russia). Moskva, Izdatel'stvo Akademii Nauk SSSR, 1960. 301 p. Bibliography, p. 289-301 (Akademiia Nauk SSSR. Institut Istorii Estestvoznaniia i Tekhniki).   1317

    330 references on the development of hydrology in Russia up to 1917.

DROGEOLOGIIA. Inzhenernaia geologiia (Hydrogeology. Engineering geology). Moskva, VINITI, 1970, 125 p. (Itogi nauki, seriia geologiia, 1969).   1318

    440 references at end of articles on methods of constructing hydrogeological maps, p. 33-47, and on landslides and on methods of their study, p. 116-125.

IITROV, I. N. Kratkii ukazatel' literatury po gidrogeologii (Short guide to the literature on hydrogeology). V. M. Maksimov and N. G. Pauker, eds. Leningrad, Gostoptekhizdat, 1959. 139 p. (Leningradskii Gornyi Institut imeni G. V. Plekhanova).   1319
    More than 4,000 entries systematically arranged.

BLIOGRAFICHESKII ukazatel' literatury po voprosam gidrotekhniki i melioratsii, izdannoi v SSSR v 1960-1962 gg. (Bibliographical guide to the literature on questions of hydraulic engineering and development, published in the USSR, 1960-1962). R. P. Pavlova, and V. G. Reznikov, comps. Moskva, 1964. 4 nos. (Goszemvodkhoz SSSR. Giprovodkhoz).   1320

    Systematic, partly annotated, guide to books, articles, dissertation abstracts, collected works, and serials in Russian and other languages of the USSR.

    no. 1. Obshchie voprosy gidrotekhniki i melioratsii. Stroitel'stvo i mekhanizatsiia gidromeliorativnykh rabot (General problems of hydraulic engineering and development. Construction and mechanization of work in hydraulic reclamation). 187 p. About 1,900 references.

no. 2. Oroshenie (Irrigation). 251 p. About 2,500 references.

no. 3. Osushenie. Pochvovedenie. Eroziia pochv i bor'ba s nei. Sel'skokhoziaistvennoe osvoenie meliorir. zemel'. Zaderzhanie i sokhranenie vlagi.
Rybokhoz. gidrotekhnika (Drainage. Soil study. Erosion of soils and measures
against it. Agricultural development of reclaimed land. Retention and preservation of moisture. Fish-producing hydraulic engineering). 171 p. About 1,600
references.

no. 4. Obvodnenie i s.-kh. vodosnabzhenie (Irrigation and agricultural
water supply). 80 p. About 720 references.

ISKUSSTVENNOE vospolnenie zapasov podzemnykh vod. Bibliograficheskii ukazatel'
literatury (Artifical filling up of the reserve of underground water. Bibliographical guide to the literature). V. S. Usenko, A. Kh. Al'tshul', I. G.
Pikovskaia, V. K. Shevchuk, comps. with the participation of Z. A. Kobetskaia
and A. A. Pisarik. Minsk, 1969. 155 p. (Nauchno-Issledovatel'skii Institut
Vodnykh Problem Ministerstva Melioratsii i Vodnogo Khoziaistva SSSR. Fundamental'naia Biblioteka imeni Kolasa AN Belorusskoi SSR).   132

1,086 entries, systematically arranged, including both Soviet and foreign
works. Author and title index.

SHEBEKO, Vassa Fedorovich. Gidrologicheskii rezhim osushaemykh territorii (Hydrological regime of drained lands). Minsk, "Urozhai," 1970. 299 p. Bibliography,
p. 293-298. (Ministerstvo Melioratsii i Vodnogo Khoziaistva SSSR. Belorusskii
Nauchno-Issledovatel'skii Institut Melioratsii i Vodnogo Khoziaistva).   132

143 references.

MECHITOV, Ivan Ivanovich, and GERSHIKOVICH, M. I. Vodokhoziaistvennye balansy: uchët
i raspredelenie vodnykh resursov (Balances in the water economy: calculation and
distribution of water resources). Tbilisi, Izdatel'stvo "Sabchota Sakartvelo,"
1970. 100 p. Bibliography, p. 101-107.   132

PLASHCHEV, Aleksandr Vasil'evich, and CHEKMAREV, Viktor Aleksandrovich.
Gidrografiia SSSR (Hydrography of the USSR). A. A. Sokolov, ed. Leningrad,
Gidrometeoizdat, 1967. 287 p. Bibliography, p. 286-287.   13.

Interrelationships of Man, Vegetation, and Water

L'VOVICH, Mark Isaakovich. Chelovek i vody: preobrazovanie vodnogo balansa i
rechnogo stoka (Man and water: the transformation of water balance and of stream
flow). Moskva, Geografgiz, 1963. 568 p. Bibliography, p. 542-566.   132

About 590 references in alphabetical order, about 540 Soviet and 52 in
foreign languages on the influence of man on the regime of rivers.

RAKHMANOV, Viktor Vasil'evich. Rechnoi stok i agrotekhnika (Stream flow and agricultural practices). A. A. Kharshan, ed. Leningrad, Gidrometeoizdat, 1973.
200 p. Bibliography, p. 190-198. (Glavnoe Upravlenie Gidrometeorologicheskoi
Sluzhby pri Sovete Ministrov SSSR. Gidrometeorologicheskii Nauchno-Issledovatel'-
skii Tsentr SSSR. Trudy, vypusk 114).   132

231 references.

INZHENERNO-GEOGRAFICHESKIE problemy proektirovaniia i ekspluatatsii krupnykh
ravninnykh vodokhranilishch (Engineering-geographical problems of construction
and exploitation of large reservoirs on plains). S. L. Vendrov, eds. S. L.
Vendrov, I. S. Glukh, I. Iu. Dolgushin, and others, authors. Moskva, Izdatel'-
stvo "Nauka," 1972. 240 p. Bibliography, p. 230-238. (Akademiia Nauk SSR.
Institut Geografii).   132

VIN, A. P. Vodnyi faktor v razmeshchenii promyshlennogo proizvodstva (The water factor in the distribution of industrial production). Moskva, Stroiizdat, 1973. 167 p. Bibliography, p. 164-166 (Akademiia Nauk SSSR. Institut Vodnykh Problem AN SSSR). 1328

NDROV, Semen Leonidovich. Problemy preobrazovaniia rechnykh sistem (Problems in the transformation of river systems). Leningrad, Gidrometeoizdat, 1970. 236 p. Bibliography, p. 225-234. 1329

207 references.

AZATEL' literatury po gidrologii i smezhnym distsiplinam raionov velikikh stroek kommunizma (Guide to the literature on hydrology and related disciplines of the regions of great constructions of communism). Leningrad, Gidrometeoizdat, 1951. (Godudarstvennyi Gidrologicheskii Institut). 1330

vypusk 1. Raiony stroitel'stva Kuibyshevskoi i Stalingradskoi [Volgograd-skoi] GES i obvoditel'nykh i orositel'nykh sistem Povolzh'ia, Zavolzh'ia i Prikas-piiskoi nizmennosti (Regions of projects of the Kuybyshev and Volgograd hydro-electric stations and irrigation systems of the Volga, Trans-Volga and Pri-Caspian lowland). V. A. Kozhina. 62 p.

About 900 entries, organized systematically. Author index.

vypusk 2. Raion stroitel'stva Glavnogo Turkmenskogo Kanala Amu-Dar'ia-Krasnovodsk, obvodnitel'nykh i orositel'nykh sistem (The region of the Great Turkmen Canal from the Amu-Dar'ya to Krasnovodsk, and the irriguation system). M. A. Kornilova. 42 p.

About 930 entries. Author index.

vypusk 3. Raion stroitel'stva Kakhovskoi gidroelektrostantsii na reke Dnepre, Iuzhno-Ukrainskogo i Severo-Krymskogo Kanalov (The region of the con-struction of the Kakhovka hydroelectric station on the Dnepr River and the South Ukraine-Crimean canals). M. A. Kornilova. 44 p.

About 800 entries in Russian and Ukrainian. Author index.

vypusk 4. Raion stroitel'stva Volgo-Donskogo kanala orositel'nykh i obvoditel'nykh sistem Rostovskoi i Stalingradskoi [Volgogradskoi] oblasti (The region of the construction of the Volga-Don canal and irrigation system of Rostov and Volgograd oblasts). M. A. Kornilova. 42 p.

About 450 entries. Author index.

HPAK, Ivan Semenovich. Vliianie lesa na vodnyi balans vodosborov (Influence of forests on the water balance of water basins). Kiev, "Naukova Dumka," 1968. 283 p. Bibliography, p. 232-242. (AN Ukrainskoi SSR. Sektor Geografii Soveta po Izucheniiu Proizvoditel'nykh Sil USSR). 1331

About 400 references.

AKHMANOV, Viktor Vasil'evich. Vliianie lesov na vodnost' rek v basseine Verkhnei Volgi (Influence of forests on the water content of rivers in the basins of the Upper Volga). A. A. Kharshan, ed. Leningrad, Gidrometeoizdat, 1971. 175 p. Bibliography, p. 168-174 (Glavnoe Upravlenie Gidrometeorologicheskoi Sluzhby pri Sovete Ministrov SSR. Gidrometeorologicheskii Nauchno-Issledovatel'skii Tsentr SSSR. Trudy, vypusk 88). 1332

179 references.

VANOV, Konstantin Evgen'evich. Osnovy gidrologii bolot lesnoi zony i raschëty vodnogo rezhima bolotnykh massivov (Basic hydrology of bogs of the forest zone and estimates of the water regime of boggy tracts). Leningrad, Gidrometeoizdat, 1957. 499 p. Bibliography, p. 398-405. (Glavnoe Upravlenie Gidrometeoro-logicheskoi Sluzhby pri Sovete Ministrov SSSR. Gosudarstvennyi Gidrologicheskii Institut). 1333

245 references in alphabetical order.

GARMONOV, I. V., EFIMOV, A. I., TOLSTOI, M. P., and others. <u>Vliianie proizvod-</u>
<u>stvennoi deiatel'nosti cheloveka na gidrogeologicheskie i inzhenerno-geologicheskie</u>
<u>usloviia</u> (The influence of the productive activity of man on hydrogeological and
engineering geological conditions). I. V. Garmonov, ed. Moskva, Izdatel'stvo
"Nedra," 1973. 278 p. Bibliography, p. 270-278. (Ministerstvo Geologii SSSR.
Vsesoiuznyi Nauchno-Issledovatel'skii Institut Gidrogeologii i Inzhenernoi Geologii).   133

DAVITAIA, Feofan Farneevich, and MEL'NIK, Iurii Sergeevich. <u>Problema prognoza</u>
<u>isparaemosti i orositel'nykh norm</u> (The problem of evaporation prediction and
irrigation rates). Leningrad, Gidrometeoizdat, 1970. 72 p.   133

Stream Flow

SOKOLOVSKII, Daniil L'vovich. <u>Rechnoi stok: osnovy teorii i praktiki raschétov</u>
(Stream flow: basic theory and practice of measurements). 2nd ed. Leningrad,
Gidrometeoizdat, 1959. 527 p. Bibliography, p. 495-508.   133

About 400 references by chapters: water balance of river and lake basins,
mean annual flow, fluctuations of annual flow.

ANDREIANOV, Vladimir Georgievich. <u>Vnutrigodovoe raspredelenie rechnogo stoka:</u>
<u>osnovnye zakonomernosti i ikh ispol'zovanie v gidrologicheskikh i vodokhoziaist-</u>
<u>vennykh raschétakh</u> (Annual regime of the flow of streams: basic regularities
and their utilization in hydrological and water-economy estimates). D. L.
Sokolovskii, ed. Leningrad, Gidrometeoizdat, 1960. 327 p. Bibliography, p. 320-
327. (Gosudarstvennyi Gidrologicheskii Institut).   133

About 260 references in alphabetical order.

CHEBOTAREV, Nikolai Petrovich. <u>Uchenie o stoke</u> (Science of stream flows).
Moskva, Izdatel'stvo Moskovskogo Universiteta, 1962. 406 p. Bibliography,
p. 395-406.   133

353 references in alphabetical order. Available in English as **Theory of**
stream runoff. Translated by Adolf Wald. Jerusalem, Israel. Program for
Scientific Translations. Springfield, Virginia, U. S. Department of Commerce.
Clearinghouse for Federal Scientific and Technical Information, 1966. 464 p.
Bibliography, p. 439-462.

KUCHMENT, L. S. <u>Matematicheskoe modelirovanie rechnogo stoka</u> (Mathematical modelling of river flow). Leningrad. Gidrometeoizdat, 1972. 191 p. Bibliography,
p. 184-189.   133

140 references.

KURDOV, A. G. <u>Minimal'nyi stok rek: osnovnye zakonomernosti formirovaniia i metody</u>
<u>raschéta</u> (Minimum flow of rivers: fundamental regularities of formation and
methods of calculation). Voronezh, Izdatel'stvo Voronezhskogo Universiteta, 1970.
251 p. Bibliography, p. 228-237.   133

233 references.

VLADIMIROV, Anatolii Mikhailovich. <u>Minimal'nyi stok rek SSSR</u> (Minimum flow of rivers
of the USSR). Leningrad, Gidrometeoizdat, 1970. 214 p. Bibliography, p. 201-
207.   134

157 references.

<u>METODICHESKIE rekomendatsii po izucheniiu rezhima poverkhnostnykh i podzemnykh vod</u>
<u>v karstovykh raionakh</u> (Methodological recommendations in the study of the regime
of surface and underground waters in karst regions). P. V. Molitvin, N. V.
Rodionov, I. V. Garmonov, and others. Leningrad, Gidrometeoizdat, 1969. 150 p.
Bibliography, p. 144-150. (Mezhduvedomstvennaia Komissiia SSSR po Mezhdunarodnomu
Gidrologicheskomu Desiatiletiiu).   134

1334 - 1341

TODICHESKIE rekomendatsii po izucheniiu rezhima poverkhnostnykh i podzemnykh vod
v karstovykh raionakh SSSR (Methodological recommendations in the study of the
regime of surface and subsurface waters in karst regions). Leningrad. 1967.
127 p. Bibliography, p. 113-120. (Vsesoiuznyi Nauchno-Issledovatel'skii Institut
Gidrologii i Inzhenernoi Geologii. Gosudarstvennyi Gidrologicheskii Institut).   1342
    114 references.

BBOTIN, A. I. Stok talykh i dozhdevykh vod po eksperimental'nym dannym (The flow
of melt and rain water on the basis of experimental data). Moskva, Gidrometeoizdat,
1966. 376 p. Bibliography, p. 312-319.   1343

    About 250 references in alphabetical order, primarily of studies in the
forest zone of the European part of the USSR.

LADIMIROV, Lev Aleksandrovich. K istorii issledovaniia zakonomernostei stoka v
gornykh oblastiakh (On the history of research on regularities in run-off in
mountain regions). Tbilisi, 1960. 146 p. Bibliography, p. 122-146 (Akademiia
Nauk Gruzinskoi SSR. Institut Geografii imeni Vakhushti).   1344

    About 500 references organized by characteristics of stream flow (mean flow,
distribution of flow throughout the year, maximum and minimum flow) in chronological order. Covers especially the mountainous regions of the USSR, particularly
the Caucasus.

ATEMATICHESKOE modelirovanie protsessa stoka gornykh rek (Mathematical modelling
of the process of flow of mountain streams). V. L. Shul'ts, ed. L. N. Borovikova,
Iu. M. Denisov, E. B. Trofimova, I. D. Shentsis. Leningrad, Gidrometeoizdat,
1972. 152 p. Bibliography, p. 146-151. (Glavnoe Upravlenie Gidrometeorologicheskoi Sluzhby pri Sovete Ministrov SSSR. Sredneaziatskii Nauchno-Issledovatel'skii
Gidrometeorologicheskii Institut. Trudy, vypusk 61 [76]).   1345

JZIN, Pavel Sergeevich. Tsiklicheskie kolebaniia stoka rek severnogo polushariia
(Cyclical fluctuations of the flow of streams in the northern hemisphere),
Leningrad, Gidrometeoizdat, 1970. 179 p. Bibliography, p. 151-155. (Glavnoe
Upravlenie Gidrometeorologicheskoi Sluzhby pri Sovete Ministrov SSSR. Mezhduve-
domstvennyi Komitet SSSR po Mezhdunarodnomu Gidrologicheskomu Desiatiletiiu.
Gosudarstvennyi Gidrologicheskii Institut).   1346

    110 references.

ARUKHANIAN, Eduard Iosifovich, and SMIRNOV, Nikolai Pavlovich. Mnogoletnie
kolebaniia stoka Volgi. Opyt geofizicheskogo analiza (Fluctuations in the flow
of the Volga over the years. An experiment in geophysical analysis). I. V.
Maksimov, ed. Leningrad, Gidrometeoizdat, 1971. 166 p. Bibliography, p. 158-
165. Summary in English.   1347

    205 references.

KAZATEL' otechestvennoi literatury po gidrometeorologicheskim prognozam za period
1957-1966 gg. vypusk 3. Rechnye prognozy (Guide to Soviet literature on hydro-
meteorological forecasts for the period 1957-1966, no. 3, river forecasts). A.
I. Afanas'ev, I. F. Babak, V. V. Klimova, and V. V. Nekrasova, comps. A. I.
Afanas'ev, ed. Moskva-Obninsk, 1967. 183 p. (Gidrometeorologicheskii Nauchno-
Issledovatel'skii Tsentr SSSR. NTB).   1348

    2,122 entries, systematically organized. Author index. List of sources
reviewed.

HARSHAN, Aleksandr Abramovich. Dolgosrochnye prognozy stoka rek Sibiri (Long-
range forecasts of the flow of the rivers of Siberia). A. I. Afanas'ev, ed.
Leningrad, Gidrometeoizdat, 1970. 210 p. Bibliography, p. 200-208. (Glavnoe
Upravlenie Gidrometeorologicheskoi Sluzhby pri Sovete Ministrov SSSR. Gidrometeorologicheskii Nauchno-Issledovatel'skii Tsentr SSSR . Trudy, vypusk 65).   1349

POPOV, Evgenii Grigor'evich. Voprosy teorii i praktiki prognozov rechnogo stoka
(Questions of theory and practice of forecasting stream flow). Moskva, Gidro-
meteoizdat, 1963. 395 p. Bibliography, p. 379-392 (Tsentral'nyi Institut
Prognozov).   135

     More than 320 references in alphabetical order.

RUKOVODSTVO po gidrologicheskim prognozam, vypusk 3. Prognozy stoka gornykh rek
(Handbook on hydrological forecasting, no. 3. Forecasting the flow of mountain
streams). A. N. Vazhnov, N. G. Dmitrieva, Sh. A. Kharshan, T. S. Abal'ian.
Moskva, Gidrometeoizdat, 1963. 294 p. Bibliography, p. 280-290 (Tsentral'nyi
Institut Prognozov).   135

     About 280 references by chapters: formation of the flow of mountain streams
in different physical-geographic regions, initial data and their processing and
analysis, methods of forecasting the flow of mountain streams. Special attention
to stream flow in regions of permafrost.

Floods

ZASHCHITA ot navodnenii: bibliograficheskii ukazatel.Otechestvennaia i zarubezhnaia
literatura za 1969-1973 gg. (Protection against floods: bibliography of native
and foreign literature for 1969-1973). Leningrad, VNIIG, 1973. 84 p. (Ministerstvo Energetiki i Elektrifikatsii SSSR. Glavniiproekt. Vsesoiuznyi Nauchno-
Issledovatel'skii Institut Gidrotekhniki imeni B. E. Vedeneeva).   1352

SHATILINA, M. K. Ukazatel' literatury po pavodkam i ikh raschëtam (Guide to the
literature on floods and their calculation). A. A. Sokolov, ed. Leningrad,
Gidrometeoizdat, 1967. 56 p. (Gosudarstvennyi Gidrologicheskii Institut).   135

     About 550 entries, systematically organized, mainly for the period 1945-
1966 on the theory of the formation of floods, their calculation, and the regulation of stream flow. Two parts, Russian and English.

NEZHIKHOVSKII, R. A.   Ruslovaia set' basseina i protsess formirovaniia stoka vody:
metodologicheskie osnovy i praktika prognozov pavodochnogo stoka rek (Channel
network of a basin and the process of the formation of the flow of water: methodological foundations and practice of forecasting floods of rivers). Leningrad,
Gidrometeoizdat, 1971. 476 p. Bibliography, p. 446-453. (Glavnoe Upravlenie
Gidrometeorologicheskoi Sluzhby pri Sovete Ministrov SSSR. Gosudarstvennyi
Gidrologicheskii Institut).   135

GARTSMAN, Il'ia Naumovich, LYLO, V. M., CHERNENKO, V. G. Pavodochnyi stok rek
Dal'nego Vostoka (Floods on rivers of the Soviet Far East). V. G. Fedorei, ed.
Leningrad, Gidrometeoizdat, 1971. 264 p. Bibliography, p. 253-262. (Glavnoe
Upravlenie Gidrometeorologicheskoi Sluzhby pri Sovete Ministrov SSSR. Dal'-
nevostochnyi Nauchno-Issledovatel'skii Gidrometeorologicheskii Institut. Trudy,
vypusk 34).   135

     218 references.

Erosion

MIRTSKHULAVA, Tsotne Evgen'evich. Inzhenernye metody raschëta i prognoza vodnoi
erozii (Engineering methods of calculation and prediction of water erosion).
Moskva, Izdatel'stvo "Kolos," 1970. 240 p. Bibliography, p. 230-238.   135

     195 references.

KUZNIK, I. A. Agrolesomeliorativnye meropriatiia, vesennii stok i eroziia pochv
(Measures for agricultural and forest reclamation, spring flows of streams, and
soil erosion). Leningrad, Gidrometeoizdat, 1962. 220 p. Bibliography, p. 212-
219.   135

199 references in alphabetical order on the effects of hydraulic engineering and agricultural and forest reclamation on stream flow and on soil erosion.

)PATIN, Georgii Vladimirovich. Nanosy rek SSSR: obrazovanie i perenos (Alluvium of rivers of the USSR: formation and transport). Moskva, Geografgiz, 1952. 367 p. Bibliography, p. 354-364 (Geograficheskoe Obshchestvo SSSR. Zapiski, n.s. tom 14).   1358

About 260 references in alphabetical order.

4CHEGLOVA, Ol'ga Petrovna. Formirovanie stoka vzveshennykh nanosov i smyv s gornoi chasti Srednei Azii (Formation of the flow of suspended alluvium and washout from the mountainous part of Soviet Middle Asia). M. I. Iveronova, ed. Leningrad, Gidrometeoizdat, 1972. 228 p. Bibliography, p. 220-227. (Glavnoe Upravlenie Gidrometeorologicheskoi Sluzhby pri Sovete Ministrov SSSR. Sredneaziatskii Nauchno-Issledovatel'skii Gidrometeorologicheskii Institut. Trudy, vypusk 60 [75]).   1359

175 references.

SLOVOI protsess (The channel process). N. E. Kondrat'ev, ed. Leningrad, Gidro-meteoizdat, 1959. 371 p. Bibliography, p. 362-368 (Gosudarstvennyi Gidro-logicheskii Institut).   1360

About 220 references in alphabetical order.

RASEV, Iosif Filippovich. Ruslovye protsessy pri perebroske stoka (The channel process with transport by the stream flow). Leningrad, Gidrometeoizdat, 1970. 267 p. Bibliography, p. 258-265.   1361

182 references.

KAVEEV, Nikolai Ivanovich. Ruslo reki i eroziia v ee basseine (Riverbeds and erosion in river basins). Moskva, Izdatel'stvo Akademii Nauk SSSR, 1955. 347 p. Bibliography, p. 327-344 (Akademiia Nauk SSSR. Institut Geografii).   1362

About 600 references in alphabetical order.

VROVSKAIA, Ol'ga Leonidovna. Problemy bor'by s vodnoi eroziei pochvy (Problems of the struggle with water erosion of soils). Moskva, [VNITEISKh], 1973. 56 p. Bibliography, p. 48-55 (Ministerstvo Sel'skogo Khoziaistva SSSR. Vsesoiuznyi Nauchno-Issledovatel'skii Institut Informatsii i Tekhn.-Ekonomicheskikh Issledovanii po Sel'skomu Khoziaistvu. Obzornaia Informatsiia).   1363

182 references.

Mudflows and Mud-laden Streams

4SOV, Andrei Iur'evich, and KRASHENINNIKOVA, Nina Vladimirovna. Selevye avleniia na territorii SSSR i mery bor'ba s nimi: ukazatel' literatury, izd. J 1850-1967 gg. (Mudflow phenomena in the USSR and measures to combat them: guide to the literature published 1850-1967). S. M. Fleishman, ed. Moskva, Izdatel'stvo Moskovskogo Universiteta, 1969. 216 p. (Moskovskii Gosudarstvennyi Universitet imeni Lomonosova. Geograficheskii Fakul'tet. Problemnaia Laboratoriia nezhnykh Lavin i Selei).   1364

1,706 references, systematically arranged, partly annotated, in Russian, krainian, and other languages of the Soviet Union. Includes books, articles, nd abstracts of dissertations. List of sources. Geographical and author indexes.

RYNDINA, V. R. Bibliografiia po selevym potokam (Bibliography on mud-laden tor-
rents). In the book: Voprosy prikladnoi gliatsiologii. Gliatsial'nye seli (Questions of applied glaciology. Glacial mud streams). Moskva, 1966, p. 107-161.
(Mezhdunarodnyi Geofizicheskii God 1957-1959. Informatsionnyi sbornik o rabotakh
po Mezhdunarodnomu geofizicheskomu godu, no. 13).   13

     860 references in alphabetical order.

SELI SSSR i mery bor'by s nimi (Mudstreams of the USSR and measures against them).
Moskva, Izdatel'stvo "Nauka," 1964. 282 p. Bibliography, p. 226-237 (Akademiia
Nauk SSSR. Institut Geografii. Komissiia po izucheniiu selevykh potokov).   13

     About 350 references in alphabetical order.

DUISENOV, Esen Duisenovich. Selevye potoki v Zailiiskom Alatau (Mudlfow streams
in the Zailiyskiy Ala-Tau). Alma-Ata, "Kazakhstan," 1971. 191 p. Bibliography,
p. 179-186.   1

GAGOSHIDZE, M. S. Selevye iavleniia i bor'ba s nimi (Mudflow phenomena and the
struggle against them). Tbilisi, "Sabchota Sakartvelo," 1970. 385 p. Bibliography, p. 378-383 (Ministerstvo Melioratsii i Vodnogo Khoziaistva SSSR.
Gruzinskii Nauchno-Issledovatel'skii Institut Gidrotekhniki i Melioratsii).   1

     156 references.

Chemical Composition of Waters

MAKSIMOVICH, Georgii Alekseevich. Khimicheskaia geografiia vod sushi (Chemical
geography of land waters). Moskva, Geografgiz, 1955. 328 p. Bibliography,
p. 280-328.   1'

     About 1,100 references systematically arranged by topics of chapters. Covers
the geographical distribution of natural waters with various chemical compositions
and regularities in the formation of the chemical make-up of land waters.

ALEKIN, Oleg Aleksandrovich. Osnovy gidrokhimii (Fundamentals of hydrochemistry).
Leningrad, Gidrometeoizdat, 1970. 444 p. Bibliography, p. 437-442. Subject
index, p. 433-436.   1'

ROSSOLIMO, Leonid Leonidovich. Ozërnoe nakoplenie kremniia i ego tipologicheskoe
znachenie (Lake accumulation of silicon and its typological significance).
Moskva, Izdatel'stvo "Nauka," 1971. 103 p. Bibliography, p. 96-102 (Akademiia
Nauk SSSR, Institut Geografii).   13

River Mouths

MIKHAILOV, V. N., and MAKAROVA, T. A. Bibliografiia po gidrologii morskikh ust'ev
rek (Bibliography on the hydrology of the mouths of rivers in seas). S. S. Baidin,
ed. Moskva, 1969. 88 p. (Gosudarstvennyi Okeanograficheskii Institut).   13

     690 entries in Russian, arranged alphabetically, mostly 1952-1968. Subject
and geographical indexes.

SAMOILOV, Ivan Vasil'evich. Ust'ia rek (Mouths of rivers). Moskva, Geografgiz,
1952. 527 p. Bibliography, p. 515-523.   13

     284 references, 230 Soviet and 54 in foreign languages.

EGOROVA, N. P., and IVANOV, V. V. Vodnyi rezhim ust'ev rek. Bibliograficheskii
ukazatel' (Water regime of river mouths. Bibliographic guide). Leningrad, 1971.
44 p. (Glavnoe Upravlenie Gidrometeorologicheskoi Sluzhby pri Sovete Ministrov
SSSR. Arkticheskii i Antarkticheskii Nauchno-Issledovatel'skii Institut).   13

KHAILOV, V. N. Dinamika potoka i rusla v neprelivnykh ust'iakh rek (Dynamics
of flow and bed of nontidal mouths of rivers). Moskva, Gidrometeoizdat, Moskovskoe
Otdelenie, 1971. 259 p. Bibliography, p. 245-253. (Glavnoe Upravlenie Gidrometeorologicheskoi Sluzhby pri Sovete Ministrov SSSR. Gosudarstvennyi Okeanograficheskii Institut. Trudy, vypusk 102).   1375

TEINMAN, B. S., and TSITSAREV, A. N. Gidrologiia ust'evoi oblasti Kury (Hydrology
of the region of the mouth of the Kura river). I. P. Beliaev, ed. Leningrad,
Gidrometeoizdat, 1971. 323 p. Bibliography, p. 314-321. (Glavnoe Upravlenie
Gidrometeorologicheskoi Sluzhby pri Sovete Ministrov SSSR. Upravlenie Gidrometeorologicheskoi Sluzhby Azerbaidzhanskoi SSR. Bakinskaia Gidrometeorologicheskaia Observatoriia).   1376

Ground Water

SLOV, B. S. Rezhim gruntovykh vod pereuvlazhnennykh zemel' i ego regulirovanie
(The regime of ground water in water-logged land and its regulation). Moskva,
Izdatel'stvo "Kolos," 1970. 232 p. Bibliography, p. 219-230.   1377

   About 260 references.

Hydrology of the U. S. S. R. as a Whole

DROLOGICHESKII ezhegodnik. (Hydrological yearbook) (Glavnoe Upravlenie Gidrometeorologicheskoi Sluzhby pri Sovete Ministrov SSSR). Leningrad, 1957- .
Gidrometeoizdat, 1962- .   1378

   A complex set of publications compiled with the cooperation of regional
stations and organized by region or subregion and year. The over-all regional
organization is 0. Basins of the White and Barents seas; 1. Basin of the Baltic
Sea; 2. Basin of the Black and Azov seas; 3. Basin of the rivers of the Caucasus;
4. Basin of the Caspian Sea; 5. Basins of the rivers of Soviet Middle Asia; 6.
Basin of the Kara Sea (western part); 7. Basin of the Kara Sea (eastern part);
8. Basin of the Eastern Siberian and Chukotsk seas; 9. Basins of the Pacific
Ocean. A typical volume provides data for a one year for a specific river basin
within one of the larger basins listed above.

URSY poverkhnostnykh vod SSSR: Osnovnye gidrologicheskie kharakteristiki
Resources of surface waters of the USSR: basic hydrological characteristics).
Leningrad, Gidrometeoizdat.   1379

   Numerous volumes each covering a specific region of the Union. A typical
volume includes 100-500 references in the bibliography.

ROGEOLOGIIA SSSR. (Hydrogeology of the USSR) A. V. Sidorenko, ed. Moskva,
"Nedra," 1966- . 50 v. In process. (Vsesoiuznyi Nauchno-Issledovatel'skii
nstitut Gidrogeologii i Inzhenernoi Geologii "VSEGINGEO).   1380

   Detailed volumes each covering a specific region of the Soviet Union.

'TSEVA, T. V., and ERMAKOVA, L. N. Ob istochnikakh informatsii po okhrane vod
 SSSR (On sources of information on the preservation of the waters of the USSR),
eograficheskii sbornik (Vsesoiuznyi Institut Nauchnoi i Tekhnicheskoi Infor-
atsii), v. 3. Moskva, VINITI, 1969, p. 91-102.   1381

   Lists 35 journals, 22 collected volumes of conferences, 19 collected volumes
n institutional series, 55 collected volumes of other types, a total of 131
ntries.

VODNYI balans SSSR i ego preobrazovanie. (The water balance of the USSR and its
transformation). M. I. L'vovich, N. N. Dreier, A. M. Grin, and others. Moskva,
Izdatel'stvo "Nauka," 1969. 338 p. Bibliography, p. 323-336. (Akademiia Nauk
SSSR. Institut Geografii. Mezhdunarodnoe Gidrologicheskoe Desiatiletie).
Summary in English.   1²

    438 references.

SOKOLOV, Aleksei Aleksandrovich. Gidrografiia SSSR: vody sushi (Hydrography of the
USSR: waters of the land). Leningrad, Gidrometeorologicheskoe Izdatel'stvo,
1964. 535 p. Bibliography, p. 530-535.   1²

    About 150 references arranged alphabetically.

DAVYDOV, Lev Konstantinovich. Gidrografiia SSSR: vody sushi (Hydrography of the
USSR: waters of the land). Leningrad, Izdatel'stvo Leningradskogo Universiteta.
1953-1955. 2 v. (Leningradskii Gosudarstvennyi Universitet imeni Zhdanova).
v. 1. Obshchaia kharakteristika vod. (General characteristics of waters). 1953.
181 p. v. 2. Gidrografiia raionov. (Hydrographic regions). 1955. 600 p.   1²

    References at the ends of chapters, 14 lists in each of the two volumes.

L'VOVICH, Mark Isaakovich. Reki SSSR. (Rivers of the USSR). Moskva, Izdatel'stvo
"Mysl'," Geografgiz, 1971. 348 p. Bibliography, p. 336-347.

    About 275 references arranged alphabetically by name of author.

KUZIN, Pavel Sergeevich. Klassifikatsiia rek i gidrologicheskoe raionirovanie SSSR
(Classification of the rivers and hydrological regionalization of the USSR).
Leningrad, Gidrometeoizdat, 1960. 455 p. Bibliography, p. 383-392 (Gosudarstven-
nyi Gidrologicheskii Institut).   1

    About 250 references in alphabetical order, mainly for the period 1930-1959.
Appendix of hydrological data for rivers of the USSR, p. 394-453.

AKADEMIIA Nauk SSSR. Institut Geografii. Ocherki po gidrografii rek SSSR (Studies
on the hydrology of rivers of the USSR). M. I. L'vovich, ed. Moskva, "Izdatel'-
stvo Akademii Nauk SSSR, 1953. 324 p. Bibliography, p. 309-313.   1

    About 100 references arranged alphabetically.

ROSSOLIMO, Leonid Leonidovich. Ocherki po geografii vnutrennikh vod SSSR. Reki i
ozëra (Studies in the geography of internal waters of the USSR. Rivers and
lakes). Moskva, Uchpedgiz, 1952. 304 p.

    Bibliographies at the end of 12 sections.

BARANOV, Ivan Vasil'evich. Limnologicheskie tipy ozër SSSR (Limnological types of
lakes of the USSR). Leningrad, Gidrometeoizdat, 1962. 276 p. Bibliography,
p. 266-273.

    178 references in order of citation in text. Includes alphabetical guide to
lakes.

DOLGOPOLOV, Konstantin Vasil'evich, and FEDOROVA, Evgeniia Fedorovna. Voda- -
national'noe dostoianie: geograficheskie problemy ispol'zovaniia vodnykh resursov
(Water as a national property: geographical problems in the utilization of water
resources). Moskva, Izdatel'stvo "Mysl'," 1973. 255 p. Bibliography, p. 248-
255. (Akademiia Nauk SSSR. Institut Geografii).

    174 references in Russian in alphabetical order.

Hydrology of Regions within the U. S. S. R.

European U.S.S.R.

ŞAROV, Valentin Dmitrievich. <u>Vesennii stok ravninnykh rek Evropeiskoi chasti SSSR,</u>
<u>sloviia ego formirovaniia i metody prognozov</u> (Spring flow of rivers of the plains
f the European part of the USSR, conditions of its formation and methods of fore-
asting). Moskva, Gidrometeoizdat, 1959. 295 p. Bibliography, p. 272-283.
Glavnoe Upravlenie Gidrometeorologicheskoi Sluzhby pri Sovete Ministrov SSSR.
sentral'nyi Institut Prognozov).   1391

About 380 references in alphabetical order.

ROLOGIIA ust'evoi oblasti Nevy (Hydrology of the mouth areas of the Neva).
. A. Skriptunov, K. I. Ermak, Ia. Kh. Ioselev, and others. S. S. Baidin, ed.
oskva, Gidrometeoizdat, 1965. 383 p. Bibliography, p. 325-335. (Gosudarstvennyi
keanograficheskii Institut. Nevskaia Ust'evaia Gidrometeorologicheskaia Stantsiia).
355 references.   1392

ESNIK, Stanislav Vikent'evich. ed. <u>Ozëra Karel'skogo peresheika. Limnologiche-</u>
<u>kie tsikly ozera Krasnogo</u> (Lakes of the Karelian isthmus. Limnological cycles
f Krasnoe Ozero [Red Lake]). Leningrad, Izdatel'stvo "Nauka," Leningradskoe
tdelenie, 1971. 531 p. Bibliography, p. 505-513, and at ends of articles.
Akademiia Nauk SSSR. Sovetskii Natsional'nyi Komitet po Mezhdunarodnoi Bio-
ogicheskoi Programme. Laboratoriia Ozerovedeniia).   1393

ESNIK, Stanislav Vikent'evich. <u>Ladozhskoe ozero</u> (Lake Ladoga). Leningrad,
idrometeoizdat, 1968. 159 p. Bibliography, p. 152-157.   1394
84 references.

EMAN, Aleksandr Vasil'evich. <u>Moskovskoe more</u> (Moscow sea). 2nd ed. Kalinin,
nizhnoe izdatel'stvo, 1955. 140 p. Bibliography, p. 135-140.   1395
121 references.

TANICH, Vadim Sergeevich. <u>Rybinskoe vodokhranilishche: annotirovannyi ukazatel'</u>
iteratury (The Rybinsk reservoir: annotated guide to the literature). Yaroslavl',
961. 80 p. (Iaroslavskaia Oblastnaia Biblioteka).   1396

467 references in alphabetical order, 1937-1960, on the wave, hydrological,
nd thermal regimes and on the influence of the reservoir on adjacent areas.
ıbject index.

TERUK, F. Ia., and KORCHAGIN, A. K. <u>Reki zapadnykh oblastei USSR i BSSR: biblio-</u>
raficheskii ukazatel' otechestvennoi i inostrannoi literatury za period 1890-
939 gg. s pril. karty (The rivers of the western regions of the Ukrainian and
elorussian republics: bibliographic guide to native and foreign literature for
he period 1890-1939 with map supplement) Moskva-Leningrad, Gosenergoizdat, 1941.
20 p. (Gidroenergoproekt).   1397

501 references in Russian, Ukrainian, Belorussian and foreign, mainly Polish,
anguages.

ETS, Grigorii Ivanovich. <u>Vydaiushchiesia gidrologicheskie iavleniia na iugo-</u>
apade SSSR (Leading hydrological phenomena in the southwest of the USSR).
eningrad, Gidrometeoizdat, 1972. 244 p. Bibliography, p. 234-243.   1398
308 references.

AKCHURINA, R. M., and BIBILO, Iu. O. <u>Klimat i vody sushi Iugo-Vostoka Evropeiskoi</u>
<u>chasti SSSR: bibliograficheskii ukazatel'</u> (Climate and waters of the land of
the Southeastern part of the European USSR: bibliographical guide). Saratov,
Izdatel'stvo Saratovskogo Universiteta, 1961. 267 p. Bibliography on waters of
the land, p. 133-246 (Saratovskii Gosudarstvennyi Universitet. Nauchnaia Biblioteka. <u>Bibliografiia Saratovskoi Oblasti</u>, vypusk 5).   1

    1,514 references arranged by subjects: general section; reservoirs, ponds,
and coastal lagoons (especially the Caspian Sea and lakes El'ton and Baskunchak);
rivers; hydrological measurements; hydrological forecasting; research on individual hydrological problems; agricultural hydrology; the water regime and the
physical-aqueous qualities of soils; hydrological information by river and lake
basins; water utilization.

BUTORIN, N. V. <u>Gidrologicheskie protsessy i dinamika vodnykh mass v vodokhranilish-</u>
<u>chakh Volzhskogo kaskada</u> (Hydrological processes and dynamics of water masses in
the reservoirs of the Volga cascade). Leningrad, Izdatel'stvo "Nauka," Leningradskoe Otdelenie, 1969. 322 p. Bibliography, p. 309-320. (Akademiia Nauk SSSR.
Institut Biologii Vnutrennykh Vod).   1

    333 references.

SMETANICH, Vadim Sergeevich. <u>Vodokhranilishcha Permskoi oblasti: bibliograficheskii</u>
<u>ukazatel' literatury</u> (Reservoirs of Perm oblast: a bibliographical guide to the
literature). Iu. M. Matarzin, ed. Perm'. 1965. 58 p. (Zapadno-Ural'skii
Sovnarkhoz. Tsentral'naia Nauchno-Tekhnicheskaia Biblioteka.Laboratoriia Vodo-
khoziaistvennykh problem Permskogo Universiteta).   1

    525 references, partly annotated, in alphabetical order on the wave, wind,
and hydrologic regime and the influence of reservoirs on adjacent areas. Indexes
of reservoirs, authors, and subjects.

BALKOV, Vladimir Aleksandrovich. <u>Vliianie karsta na stok rek Evropeiskoi territorii</u>
<u>SSSR</u> (The influence of karst on the flow of rivers in the European part of the
USSR). Leningrad, Gidrometeoizdat, 1970. 216 p. Bibliography, p. 205-215.   1

    260 references.

Siberia

SMETANICH, Vadim Sergeevich. <u>Vodokhranilishcha Sibiri: annotirovannyi ukazatel'</u>
<u>literatury</u> (Reservoirs of Siberia: an annotated guide to the literature). Kras-
noyarsk, 1962. 41 p. (Krasnoiarskaia Kraevedcheskaia Biblioteka. Bibliografiche
skii Otdel).   1

    247 references in alphabetical order on the wind, wave, hydrological, and
thermal regime and on the influence of reservoirs on adjacent areas.

POPOLZKIN, Aleksandr Grigor'evich. <u>Ozera iuga Ob'-Irtyshskogo basseina</u> (Lakes
of the south of the Ob'-Irtysh basin). Novosibirsk, Zapadno-Sibirskoe Knizhnoe
Izdatel'stvo, 1967. 350 p. Bibliography, p. 335-350. (Novosibirskii Gosudarstvennyi Pedagogicheskii Institut).   1

    About 320 references.

CHERKASOV, A. E. <u>Vodnye resursy rek Angaro-Eniseiskogo basseina</u> (Water resources
of rivers of the Angara-Yenisey basin). B. V. Zonov, ed. Irkutsk, 1969. 198 p.
Bibliography. (Irkutskii Gosudarstvennyi Universitet imeni A. A. Zhdanov.
Kafedra Gidrologii Sushi).   ▌

    132 references.

)NOV, Boris Vasil'evich, and SHUL'GIN, M. F. Gidrologiia rek basseina Bratskogo vodokhranilishcha (Hydrology of the rivers in the basin of the Bratsk reservoir). Moskva, Izdatel'stvo "Nauka," 1966. 168 p. Bibliography, p. 129-134, 168. (Akademiia Nauk SSSR. Sibirskoe Otdelenie. Limnologicheskii Institut).   1406
    About 200 references.

ZHOV, Mikhail Mikhailovich. Ocherki po baikalovedeniiu (Studies on Lake Baykal). Irkutsk, Vostochno-Sibirskoe Knizhnoe Izdatel'stvo, 1972. 254 p. Bibliography, p. 239-252. (Irkutskii Gosudarstvennyi Universitet).   1407

'ANAS'EV, Aleksandr Nikitich. Kolebaniia gidrometeorologicheskogo rezhima na territorii SSSR, v osobennosti v basseine Baikala (Variations in the hydro-meteorological regimen on the territory of the USSR, especially in the Baykal basin). Moskva, Izdatel'stvo "Nauka," 1967. 231 p. Bibliography, p. 223-229. (Akademiia Nauk SSSR. Sibirskoe Otdelenie. Institut Zemnoi Kory).   1408
    About 230 references.

RBOLOV, Vladimir Il'ich, SOKOL'NIKOV, Vladimir Mikhailovich, and SHIMARAEV, Mikhail Nikolaevich. Gidrometeorologicheskii rezhim i teplovoi balans ozera Baikal (Hydrometeorological regime and heat balance of Lake Baykal). Moskva-Leningrad, Izdatel'stvo "Nauka," 1965. 373 p. Bibliography, p. 362-371 (Akademiia Nauk SSSR. Sibirskoe Otdelenie. Limnologicheskii Institut).   1409
    About 330 references in alphabetical order.

EPANOV, V. M. Gidrogeologiia Zabaikal'ia: bibliografiia (Hydrogeology of Trans-Baykal region: a bibliography). Chast' 1. Opublikovannaia literatura (Published literature). Irkutsk, 1962. 32 p. (Akademiia Nauk SSSR. Sibirskoe Otdelenie. Vostocho-Sibirskii Geologicheskii Institut).   1410
    429 entries.

The Caspian Sea and the Trans-Caucasus

L', Kasum Kiazimovich, LAPPALAINEN, T. N., and POLUSHKIN, V. A. Kaspiiskoe lore. Referativnyi sbornik. (The Caspian Sea: a reference handbook). Moskva, 1970. 36 p. (Vsesoiuznyi Institut Nauchnoi i Tekhnicheskoi Informatsii. Akademii auk SSSR. Institut Geografii. Akademiia Nauk Azerbaidzhanskoi SSR).   1411
    1,850 entries on Soviet studies of the Caspian Sea, arranged alphabetically. eographical and subject indexes.

L', Kasum Kiazim ogly, and others. Bibliograficheskii annotirovannyi spravochnik o Kaspiiskomu moriu (Annotated bibliographic handbook on the Caspian Sea). Baku, zdatel'stvo "Elm," 1970. 217 p. (Akademiia Nauk Azerbaidzhanskoi SSR. Institut eografii. Fundamental'naia Biblioteka).   1412

TAMOV, Salekh Gadzhievich. Razvitie gidrologicheskikh issledovanii v Azerbaid-hane za 50 let. Reziume (Development of hydrological research in Azerbaydzhan ver 50 years. Summary). Akademiia Nauk Azerbaidzhanskoi SSR. Institut Geo-rafii. Trudy, v. 18 (1971), p. 213-245. Bibliography, 230-247.   1413

ROVOL'SKII, Aleksei Dmitrievich, KOSAREV, Aleksei N., and LEONT'EV, Oleg onstantinovich, eds. Kaspiiskoe More (Caspian Sea). Moskva, Izdatel'stvo oskovskogo Universiteta, 1969. 264 p. Bibliography, p. 256-263.   1414
    About 200 references on the hydrology, hydrobiology, and geomorphology of he Caspian Sea.

_', Kasum Kiazimovich. Kaspiiskoe More (Caspian Sea). Baku, Aznefteizdat,)56. 327 p. Bibliography, p. 318-325.   1415
    About 200 references in alphabetical order.

SMIRNOVA, Klavdiia Ivanovna Vodnyi balans i dolgosrochnyi prognoz urovnia Kaspii-
skogo moria (Water balance and the long-range forecast of the level of the
Caspian Sea). K. P. Vasil'ev, ed. Leningrad, Gidrometeoizdat, 1972. 123 p.
Bibliography, p. 117-122 (Glavnoe Upravlenie Gidrometeorologicheskoi Sluzhby
pri Sovete Ministrov SSSR. Gidrometeorologicheskii Nauchno-Issledovatel'skii
Tsentr SSSR. Trudy, vypusk 94).   1

     135 references.

GIUL', Kasum Kiazimovich, ABAKAROV, M. M., MEKHRALIEV, E. K., and others. Issledovanie azerbaidzhanskikh uchёnykh v reshenii problem Kaspiiskogo moria. Reziume
(Research by Azerbaydzhan scholars toward the solution of the problem of the
Caspian Sea). Akademiia Nauk Azerbaidzhanskoi SSR. Institut Geografii. Trudy,
v. 18 (1971), p. 352-405. Bibliography, p. 375-405.   1

FIZICHESKIE protsessy v Kaspiiskom more v sviazi s kolebaniiami ego urovnia:
vliianie izmeneniia urovnia Kaspiia na ego gidrologiiu i narodnoe khoziaistvo
(Physical processes in the Caspian Sea and its relations with fluctuations in
its level: influence of changes in the level of the Caspian on its hydrology
and economy). K. K. Giul', M. I. Abakarov, T. I. Furman, R. L. Reifman. Baku,
"Elm," 1971. 223 p. Bibliography, p. 217-221. (Akademiia Nauk Azerbaidzhanskoi
SSR. Institut Geografii. Sektor Problem Kaspiiskogo Moria).   1

KASHKAI, R. M. Vodnyi balans Bol'shogo Kavkaza--v predelakh Azerbaidzhanskoi SSR
(Water balance of the Great Caucasus within the borders of the Azerbaidzhan SSR).
Baku, "Elm," 1973. 84 p. Bibliography, p. 78-83. (Akademiia Nauk Azerbaidzhan-
skoi SSR. Institut Geografii).   1

ZAMANOV, Khalil Dzhalal ogly, and TARVERDIEV, Ramazan Bakhshaly ogly. Gidrologicheskie osobennosti ozёr i vodokhranilishch Bol'shogo Kavkaza (Hydrological characteristics of lakes and reservoirs of the Great Caucasus). Baku, Izdatel'stvo
Akademii Nauk Azerbaidzhanskoi SSR, 1965. 138 p. Bibliography, p. 132-136.
(Akademiia Nauk Azerbaidzhanskoi SSR. Institut Geografii).   1

     About 120 references.

ZAMANOV, Khalil Dzhalal ogly. Vodnyi balans ozёr i vodokhranilishch Malogo Kavkaza
(Water blance and reservoirs of the Lesser Caucasus). Baku, "Elm," 1969. 154 p.
Bibliography, p. 142-148. (Akademiia Nauk Azerbaidzhanskoi SSR. Institut Geo-
grafii). (In Azerbaijani).   1

VAZHNOV, Aleksandr Nikolaevich. Analiz i prognozy stoka rek Kavkaza (Analysis
and forecasting of the flow of rivers of the Caucasus). Moskva, Gidrometeoizdat,
1966. 274 p. Bibliography, p. 250-257 (Tsentral'nyi Institut Prognozov).

     About 250 references in alphabetical order.

Kazakhstan and Soviet Middle Asia

VOZNESENSKAIA, Elena Aleksandrovna, and RABINERSON, Aleksandr Ignat'evich. Ukazate
literatury po gidrologii Sredne-Aziatskikh respublik i Kazakhstana. (Guide to
the literature on the hydrology of Soviet Middle Asia and Kazakhstan). Leningrad
Izdatel'stvo Akademii Nauk SSSR, 1928. 115 p. (Akademiia Nauk SSSR. Komissiia
po Izucheniiu Estestvennykh Proizvoditel'nykh Sil Soiuza).

     1,113 annotated references arranged systematically: general section; seas
(Caspian and Aral seas); lakes (El'ton and Baskunchak); rivers; underground water
and mineral springs.

DVEDEVA, S. G., and TEMIROVA, K. T. Vodnye resursy Kazakhstana. Bibliograficheskii
ukazatel' literatury, 1917-1967 (Water resources of Kazakhstan: bibliographical
guide to the literature, 1917-1967). Alma-Ata, 1971. 312 p. (Akademiia Nauk
Kazakhskoi SSR. Tsentral'naia Nauchnaia Biblioteka. Institut Gidrogeologii i
Gidrofiziki).   1424

HMEDSAFIN, U. M., and others, eds. Formirovanie podzemnogo stoka na territorii
Kazakhstana (The formation of ground-water flow in Kazakhstan). U. M. Akhmedsafin,
M. Kh. Dzhabasov, S. Zh. Zhaparkhanov, and others. Alma-Ata, Izdatel'stvo "Nauka,"
1970. 147 p. Bibliography, p. 136-146. (Akademiia Nauk Kazakhskoi SSR. Institut
Gidrogeologii i Gidrofiziki).   1425

     280 references.

ROVIN, Vasilii Ivanovich. Vliianie gidrometeorologicheskikh uslovii na stok rek
i uroven' ozër basseinov rek verkhnego Irtysha i Balkhash-Alakol'skoi vpadiny
(Influence of hydrometeorological conditions on the run-off of rivers and level
of lakes in the basins of the rivers of the upper Irtysh and the Balkhash-Alakol'
depression). Leningrad, Gidrometeoizdat, 1966. 378 p. Bibliography, p. 355-
376. (Glavnoe Upravlenie Gidrometeorologicheskoi Sluzhby pri Sovete Ministrov
SSSR. Kazakhskii Nauchno-Issledovatel'skii Gidrometeorologicheskii Institut).   1426

     About 530 references.

SOKHOV, E. V. Solianye ozëra Kazakhstana (Salt lakes of Kakazhstan). Moskva,
Izdatel'stvo Akademii Nauk SSSR, 1955. 187 p. Bibliography, p. 180-186 (Akademiia
Nauk SSSR. Gidrokhimicheskii Institut).   1427

     About 230 references in alphabetical order.

NOGRADOV, Iu. B. Voprosy gidrologii dozhdevykh pavodkov na malykh vodosborakh
srednei Azii i Iuzhnogo Kazakhstana (Problems of the hydrology of rain floods
in small water basins of Middle Asia and southern Kazakhstan). Leningrad, Gidrometeoizdat, 1967. 262 p. Bibliography, p. 177-183 (Kazakhskii Nauchno-Issledovatel'skii Gidrometeorologicheskii Institut, Trudy, vypusk 28).   1428

     202 references.

ROSOV, Vasilii Nikiforovich. Ozero Balkhash (Lake Balkhash). Leningrad,
Izdatel'stvo "Nauka," Leningradskoe Otdelenie. 1973. 180 p. Bibliography,
p. 165-178. (Akademiia Nauk SSSR. Geograficheskoe Obshchestvo SSSR).   1429

UL'TS, Viktor L'vovich. Reki Srednei Azii (Rivers of Soviet Middle Asia).
Leningrad. Gidrometeoizdat, 1965. 2 v. 691 p. Bibliography, v. 1, p. 294-
301, v. 2, p. 683-684 (Sredneaziatskii Nauchno-Issledovatel'skii Gidrometeorologicheskii Institut).   1430

     262 references in alphabetical order.

KINEEVA, D. Kh. Aral'skoe more: ukazatel' literatury (Aral Sea: guide to the
literature). K. A. Akhmetov, ed. Kzyl-Orda, 1964, 40 p. (Kzyl-Ordinskaia
Oblastnaia Biblioteka imeni M. Gor'kogo. Spravochno-Bibliograficheskii Otdel).   1431

     248 references in systematic order: general works, history of exploration,
animals, hydrology and hydrotechnology, geography and geology, fish breeding and
fishing, and Barsa-Kel'mes reservation.

IUKANOVA, Inna Alekseevna. Vzveshennye nanosy Amurdar'i i ikh irrigatsionnoe
znachenie (Suspended alluvium of the Amu-Dar'ya and its significance for
irrigation). Moskva, Izdatel'stvo "Nauka," 1971. 112 p. Bibliography, p. 108-
111. (Akademiia Nauk SSSR. Institut Geografii).   1432

KUZNETSOV, Nikolai Timofeevich. Vody Tsentral'noi Azii (Waters of Central Asia).
Moskva, Izdatel'stvo "Nauka," 1968. 272 p. Bibliography, p. 260-271. (Akademiia
Nauk SSSR. Institut Geografii).   1/

About 300 references.

Hydrology of Europe

LAZARENKO, Nikolai Nikolaevich, and MAEVSKII, A., eds. Gidrometeorologicheskii
rezhim Vislinskogo zaliva (Hydrometeorological regime of the Vistula bay in the
Baltic sea). I. Baushis, M. P. Beliaeva, D. Vel'bin'skaia, and others. Leningrad, Gidrometeoizdat, 1971. 279 p. Bibliography, p. 270-277. (Glavnoe
Upravlenie Gidrometeorologicheskoi Sluzhby pri Sovete Ministrov SSSR. Gosudarstvennyi Gidrometeorologicheskii Institut Pol'skoi Narodnoi Republiki. Upravlenie Gidrometeorologicheskoi Sluzhby Litovskoi SSR. Morskoi Filial GGMI v
Gdyne).   14

CHERNOGLAEVA, G. M. Vodnyi balans Evropy (Water balance of Europe). M. I.
L'vovich, ed. Moskva, 1971. 140 p. Bibliography, p. 115-126 (Akademii Nauk
SSSR. Institut Geografii. Sovetskii Geofizicheskoi Komitet. Vodnyi Balans
Materikov Zemnogo Shara). Summary in English.   14

DUKICH, Dušan, and L'VOVICH, Mark Isaakovich. Vodnye resursy Evropy i puti ikh
sovremennogo ispol'zovaniia [Po programe Evropeiskoi Regional'noi Konferentsii
Mezhdunarodnogo Geofizicheskoi Soiuza] (Water resources of Europe and lines of
their contemporary utilization according to the program of the European Regional
Conference of the International Geophysical Union). Moskva, 1971. 70 p.
Bibliography, p. 66-70.   14

Hydrology of Africa

DMITRIEVSKII, Iurii Dmitrievich. Vnutrennie vody Afriki i ikh ispol'zovanie (The
internal waters of Africa and their utilization). Leningrad, Gidrometeoizdat,
1967. 382 p. Bibliography, p. 361-381.   14

About 640 references.

KARASIK, G. Ia. Vodnyi balans Afriki (Water balance of Africa). M. I. L'vovich,
ed. Moskva, 1970. 206 p. Bibliography, p. 172-189. (Akademiia Nauk SSSR.
Institut Geografii. Sovetskii Geofizicheskii Komitet). Summary in English. 14

OCEANOGRAPHY

Major General Bibliographies

ZICHESKAIA geografiia: annotirovannyi perechen' otechestvennykh bibliigrafii izdannykh v 1810-1966 gg. (Physical geography: annotated list of native bibliographies published 1810-1966). M. N. Morozova and E. A. Stepanova, comps. V. V. Klevenskaia, ed. L. G. Kamanin, scientific consultant. Moskva, Izdatel'stvo 'Nauka," 1968. 309 p. (Gosudarstvennaia Biblioteka SSSR imeni V. I. Lenina. Institut Geografii Akademii Nauk SSSR. Sektor Seti Spetsial'nykh Bibliotek Akademii Nauk SSSR).
1439

Oceanography, p. 132-139, entries 585-627, 43 Soviet bibliographies on oceanography, organized systematically: bibliographies of a general character, indexes to the contents of journals and collected volumes, bibliographies of publications of scientific institutions of the USSR; bibliographies on individual themes and problems (dynamics of the sea, sea ice and the ice regime); bibliographies of regional oceanography (Arctic Ocean, Atlantic Ocean, Pacific Ocean, Indian Ocean, internal seas). Author and geographical indexes.

TERATURA po meteorologii, gidrologii i okeanografii za 1958...god (Literature on meteorology, hydrology, and oceanography in 1958...) Meteorologiia i gidrologiia, 1959- .
1440

Classified bibliography that appears in several issues during each year.

MANOVICH, Z. S. Kratkii ukazatel' literatury po okeanografii (Short guide to the literature of oceanography). Leningrad, 1962. 63 p. (Arkticheskii i Antarkicheskii Nauchno-Issledovatel'skii Institut).
1441

Systematic guide to the literature with 559 entries. Main sections: general information on oceanography and its individual problems; theoretical and methodological works; regional works.

DEMIIA Nauk SSSR. Komitet po Geodezii i Geofizike. Soobshchenie o nauchnykh abotakh po fizicheskoi okeanografii. Predstavliaetsia v Mezhdunarodnuiu Assotiatsiiu Fizicheskoi Okeanografii k XI General'noi Assamblee Mezhdunarodnogo eodezicheskogo i Geofizicheskogo Soiuza (Report on scientific work in physical ceanography, presented in the International Association of Physical Oceanography o the 11th General Assembly of the International Union of Geodesy and Geophysics). oskva, Izdatel'stvo Akademii Nauk SSSR, 1957. 35 p.
1442

Review of work of major oceanographic institutions of the USSR for the period 950-1956. About 200 references.

DEMIIA Nauk SSSR. Komitet po Geodezii i Geofiziki. Soobshchenie o nauchnykh abotakh po fizicheskoi okeanografii, 1957-1959 gg. Predstavliaetsia v Mezhunarodnuiu Assotsiatsiiu Fizicheskoi Okeanografii k XII General'noi Assamblee ezhdunarodnogo Geodezicheskogo i Geofizicheskoi Soiuza (Report on scientific ork on physical oceanography, 1957-1959 presented in the International Association of Physical Oceanography to the 12th General Assembly of the International nion of Geodesy and Geophysics). Moskva, 1960. Bibliography, p. 14-102.
1443

Arranged systematically by the following divisions: physical oceanography; hemical oceanography; marine geology; marine biology; methods and instruments. he section on physical oceanography has about 120 references.

ANOVA, G. D., and GREVTSOVA, T. N. Ukazatel' k izdaniiam Instituta Okeanologii N SSSR, 1946-1962 (Guide to publications of the Institute of Oceanology of the cademy of Sciences of the USSR, 1946-1962), Akademiia Nauk SSSR. Institut keanologii. Trudy, v. 61 (1962), p. 155-213.
1444

Text in parallel Russian and English languages. First part is a guide to . 1-60 of the Trudy of the Institute; the second part, to 20 other publications f the Institute. Author index.

PLAKHOTNIK, Aleksandr Filipovich. Kratkaia istoriia ekspeditsionnykh issledovanii po fizicheskoi okeanologii v SSSR (Short history of expeditionary research on physical oceanology in the USSR). In the book: Voprosy istorii fizicheskoi geografii v SSSR (Questions of the history of physical geography in the USSR). Moskva, Izdatel'stvo "Nauka," 1970. p. 72-155. Bibliography, p. 147-155. (Institut Istorii Estestvoznaniia i Tekhniki).   14

PLAKHOTNIK, Aleksandr Filipovich. Istochniki i printsipy periodizatsii istorii okeanografii (Sources and principles of the division into periods of the history of oceanography), Akademiia Nauk SSSR. Institut Istorii Estestvoznaniia i Tekhniki Trudy, v. 42 (1962), p. 103-129.   14

    183 references, 142 Soviet and 41 in foreign languages.

Topical Bibliographies

SHULEIKIN, Vasilii Vladimirovich. Fizika moria (Physics of the sea). 4th ed. Moskva, Izdatel'stvo "Nauka," 1968. 1,083 p. Bibliography, p. 1,056-1,068. (Akademiia Nauk SSSR). (1st ed. 1941. 833 p.).   14

    476 references arranged by chapters.

GALERKIN, L. I. Neperiodicheskie kolebaniia urovnia moria: ukazatel' literatury na russkom i inostrannykh iazykakh (Nonperiodical fluctuations in sea level: guide to the literature in Russian and foreign languages). Moskva, Izdatel'stvo Akademii Nauk SSSR, 1962. 67 p. (Akademiia Nauk SSSR. Mezhduvedomstvennyi Geofizicheskii Komitet).   14

    1,161 entries in alphabetical order, including 259 Soviet and 902 in foreign languages. Geographical and subject indexes.

GRIGORASH, Z. K. Tsunami: annotirovannaia bibliografiia na russkom i inostrannykh iazykakh za 1726-1962 gg. (Tsunami: annotated bibliography in Russian and foreign languages, 1726-1962). I. S. Brovikov, ed. Moskva, Izdatel'stvo "Nauka," 1964. 111 p. (Akademiia Nauk SSSR. Mezhduvedomstvennyi Geofizicheskii Komitet).   14

    More than 700 entries arranged chronologically, about 200 Soviet and 500 in foreign languages). Author index.

Regional: Atlantic Ocean

BULATOV, Rudol'f Pavlovich. Tsirkuliatsiia vod Atlanticheskogo okeana i prilegaiush-chikh morei: bibliograficheskii ukazatel', 1638-1962 (Circulation of the waters of the Atlantic Ocean and adjacent seas: a bibliographical guide, 1638-1962). Moskva, 1964. 115 p. (Akademiia Nauk SSSR. Mezduvedomstvennyi Geofizicheskii Komitet).   1

    3,786 entries in alphabetical order, including 745 Soviet and 2,841 in foreig languages. Subject and geographical indexes. Tables of distribution of publica-tions by year.

ISSLEDOVANIE tsirkuliatsii i perenosa vod Atlanticheskogo okeana (Research on the circulation and movement of water of the Atlantic ocean). E. I. Baranov, V. A. Bubnov, R. P. Bulatov, I. V. Privalova. Moskva, Izdatel'stvo "Nauka," 1971. 290 p. Bibliography, p. 279-286. (Akademiia Nauk SSSR. Mezhduvedomstvennyi Geofizicheskii Komitet. Rezul'taty issledovanii po Mezhdunarodnym Geofizicheskim Proektam. Okeanologicheskie Issledovaniia, no. 22).   1

    241 references.

TI, Iu. Iu., and MARTINSEN, G. V. Problemy formirovaniia i ispol'zovaniia bio-
ogicheskoi produktsii Atlanticheskogo okeana (Problems of the formation and
utilization of biological products of the Atlantic ocean). Moskva, Izdatel'stvo
Pishchevaia Promyshlennost'," 1969. 267 p. Bibliography, p. 256-266.   1452
    About 350 references.

                                  Pacific Ocean

DEMIIA Nauk SSSR. Institut Okeanologii. Tikhii Okean (The Pacific Ocean).
. G. Kort, ed. Moskva, Izdatel'stvo "Nauka," 1966- 8 v. (Institut Okeanologii
meni Shirshova). Title and table of contents also in English. (English trans-
ation: Alexandria, Virginia, Clearinghouse for Federal Scientific and Technical
nformation, 1966- AD 651 498).   1453

    v. 1. Meteorologicheskie usloviia nad Tikhim Okeanom (Meteorological con-
itions in the Pacific Ocean). V. S. Samoilenko, ed. 1966. 390 p. Bibliography,
. 311-316.
                                                                              1454
    About 160 references, about 90 in Russian and about 70 in foreign languages.

    v. 2. Gidrologiia Tikhogo Okeana (Hydrology of the Pacific Ocean). A. D.
obrovol'skii, ed., 1968. 524 p. Bibliography, p. 501-517.   1455

    About 550 references, about 290 in Russian and about 260 in foreign languages.

    v. 3. Khimiia Tikhogo Okeana (Chemistry of the Pacific Ocean). 1966. 360 p.
ibliography, p. 342-358.
                                                                              1456
    About 500 references.

    v. 4. Berega Tikhogo Okeana (Coasts of the Pacific Ocean). V. P. Zenkovich,
d., 1967. 373 p. Bibliography, p. 358-373.   1457

    550 references, about 225 in Russian and about 325 in foreign languages.

    v. 5. Geomorfologiia i tektonika dna Tikhogo Okeana (Geomorphology and
ectonics of the floor of the Pacific Ocean).   1458

    v. 6. Osadkoobrazovanie v Tikhom Okeane (Formation of sediments in the
acific Ocean). A. P. Lisitsyn and others. 1970. 2 parts. Bibliography, part
, p. 406-423. Bibliography, part 2, p. 340-355.   1459
    About 1,100 references.

    v. 7. Biologiia Tikhogo Okeana (Biology of the Pacific Ocean). 3 parts.
art 1. Plankton (Plankton). 1967. Bibliography, p. 234-266.   1460

    About 1,000 references.

art 2. Glubokovodnaia donnaia fauna Pleiston (Deep-water benthonic fauna of
ree-floating plants). 1969. 355 p. Bibliographies at ends of sections.   1461

art 3. Ryby otkrytykh vod (Fishes of the open waters). 1967. Bibliography,
. 247-273.
                                                                              1461a
    About 900 references.

    v. 8. Mikroflora i mikrofauna v sovremennykh osadkakh Tikhogo Okeana (Micro-
lora and microfauna in current deposits of the Pacific Ocean). 1969. 203 p.
ibliography, p. 194-201.   1462

    About 240 references.

    v. 9. Geofizika dna Tikhogo Okeana. (Geophysics of the floor of the Pacific
ean). G. A. Semenov, A. G. Rainanov, V. V. Zdorovenin, and others. G. B.
intsev and V. F. Kanaev, eds. 1974. 192 p. Bibliography, p. 176-188.   1463

Arctic and Antarctic Oceans

ZUBOV, Nikolai Nikolaevich. L'dy Arktiki (Ice of the Arctic). Moskva, Izdatel'stv
Glavsevmorputi, 1945. 360 p. Bibliography, p. 353-357.   1

179 references in alphabetical order, 143 Soviet and 36 in foreign languages.

TRĔSHNIKOV. Aleksei Fedorovich. Osobennosti ledovogo rezhima Iuzhnogo Ledovitogo
okeana (Characteristics of the ice regime of the Southern ice ocean). P. A.
Gordienko, ed. Leningrad, "Morskoi Transport," 1963. 237 p. Bibliography,
p. 58-62 (Sovetskaia Kompleksnaia Antarkticheskaia Ekspeditsiia. Trudy, tom 21)
                                                                                     1
128 references in alphabetical order on the Antarctic waters, 104 Soviet and
24 in foreign languages.

NAZAROV, Vasilii Stratonovich. L'dy antarkticheskikh vod (Ice of Antarctic waters
Moskva, Izdatel'stvo Akademii Nauk SSSR, 1962. 72 p. Bibliography, p. 58-62
(Rezul'taty issledovanii po programme Mezhdunarodnogo Geofizicheskogo Goda.
Okeanologiia. X razdel programmy MGG no. 6). English summary.   1

172 references in alphabetical order, of which 121 Soviet and 51 in foreign
languages.

Seas of the U. S. S. R.

TANFIL'EV, Gavriil Ivanovich. Moria Kaspiiskoe, Chërnoe, Baltiiskoe, Ledovitoe,
Sibirskoe i Vostochnyi Okean. Istoriia issledovaniia, morfometriia, gidrologiia,
biologiia (The Caspian, Black, Baltic, Arctic, and Siberian seas and the Pacific
Ocean: history of exploration, morphometry, hydrology, and biology). Moskva-
Leningrad, GONTI, 1931. 246 p.   1

1,565 references mainly in Russian: 191 on the Caspian Sea, 633 on the Black
Sea, 209 on the Baltic Sea, 231 on the European part of the Arctic Ocean, 154 on
the Siberian part of the Arctic Ocean, and 147 on the Pacific Ocean.

DOBROVOL'SKII, Aleksei Dmitrievich, and ZALOGIN, Boris Semĕnovich. Moria SSSR:
priroda, khoziaistvo (Seas of the USSR: natural conditions and economic utilization). Moskva, Izdatel'stvo "Mysl'," 1965. 351 p. Bibliography, p. 347-349. 1

GIDROMETEOROLOGICHESKII spravochnik Azovskogo moria (Hydrometeorological handbook
of the Azov Sea). A. A. Aksenov, ed. Leningrad, Gidrometeoizdat, 1962. 856 p.
Bibliography, p. 848-853 (Upravlenie Gidrometeorologicheskoi Sluzhby Ukrainskoi
SSR. Gidrometeorologicheskaia Observatoriia Chĕrnogo i Azovskogo Morei).   1

About 170 references in alphabetical order on the hydrology and climate of
the Azov Sea, and adjacent areas.

SOIL GEOGRAPHY

Major General Bibliographies

LIOGRAFIIA (Bibliography), Pochvovedenie (Akademiia Nauk SSSR), Moskva, 1928- .

Current literature on soils, grouped by themes. Geography of soils forms a eparate section.   1470

ERATIVNYI zhurnal. Otdel'nyi vypusk 57: pochvovedenie i agrokhimiia (Reference urnal. Separate issue 57: soil science and agricultural chemistry) (Vsesoiuznyi stitut Nauchnoi i Tekhnicheskoi Informatsii). Moskva. 1963- . monthly. nual author and subject indexes. Subtitle varies.   1471

Coverage with abstracts of the literature in all languages. The literature 1 soil science was included as a section in Referativnyi zhurnal: biologiia, 954-1962.

)BIOLOGIIA. Obshchee rastenievodstvo. Pochvovedenie (Agricultural biology, eneral crop cultivation, and soil science), Bibliografiia sovetskoi bibliografii 3ibliography of Soviet bibliography). (Vsesoiuznaia Knizhnaia Palata). 1939,)46- . Annual. Moskva, Izdatel'stvo "Kniga," 1941, 1948- .   1472

The annual list of Soviet bibliographies for soil science and general agri- iltural biology and crop cultivation. The 1970 annual volume (1972), p. 197-)4, contained 194 entries.

CHESKAIA geografiia: annotirovannyi perechen' otechestvennykh bibliografii :dannykh v 1810-1966 gg. (Physical geography: annotated list of native bibliog- aphies published 1810-1966). M. N. Morozova and E. A. Stepanova, comps. V. V. Levenskaia, ed. L. G. Kamanin, scientific consultant. Moskva, Izdatel'stvo lauka," 1968. 309 p. (Gosudarstvennaia Biblioteka SSSR imeni V. I. Lenina. stitut Geografii Akademii Nauk SSSR. Sektor Seti Spetsial'nykh Bibliotek cademii Nauk SSSR).   1473

Geography of soils, p. 159-172, entries 751-831, 81 Soviet bibliographies 1 the geography of soil, arranged systematically: bibliographies of a general aracter; bibliographies on individual themes and problems (the influence of ysical-geographic factors on soils; classification of soils; types of soils, iline soils; zonal soils); bibliographies on soils of individual areas (USSR as whole; European part of the USSR, North and Northwest, Center, West, Southwest,)lga region, Urals, the Caucasus, the North Caucasus, the Trans-Caucasus; Asiatic urt of the USSR, Siberia and the Far East, Siberia, the Far East, Middle Asia id Kakakhstan, Middle Asia, Kazakhstan). Author and geographical indexes.

Bibliographies in General Treatises

)A, Viktor Abramovich. Osnovy ucheniia o pochvakh: obshchaia teoriia pochvoo- azovatel'nogo protsessa (The principles of pedology: general theory of soil rmation). Moskva, Izdatel'stvo "Nauka," 1973. 2 v. 447 p. 468 p. Bibliog- phy, v. 2, p. 429-454.   1474

About 1,000 references in alphabetical order, about 800 in Russian and 190 1 Western languages. Textual discussion of object and history of soil science, rcles of elements and the energy of soil formation, factors in soil formation, jor components of soils, the water regime and soil formation, geochemistry and il formation, basic forms of soil-forming processes, and systematization and assification of soils of the world. Appendices: list of soils and legend of e soil map of the world; correlation of names of soils. Subject index.

ZOL'NIKOV, V. G. Pochvy i prirodnye zony Zemli. Teoreticheskii analiz nekotorykh problem pochvovedeniia i geografii (Soils and natural zones of the earth. Theoretical analysis of some problems of soil science and geography). Leningrad, Izdatel'stvo "Nauka," Leningradskoe Otdelenie, 1970. 338 p. Bibliography, p. 330-336. (Akademiia Nauk SSSR. Nauchnyi Sovet po Kompleksnoi Probleme "Teoreticheskie Osnovy Pochvovedeniia").   14

GLAZOVSKAIA, Mariia Al'fredovna. Pochvy mira (Soils of the world). Moskva, Izdatel'stvo Moskovskogo Universiteta, 1972-1973. 2 v.   14

    v. 1. 1972. 231 p. Osnovnye semeistva i tipy pochv (Basic families and types of soils). Bibliography, p. 227-229.

    v. 2. 1973. 427 p. Geografiia pochv (Geography of soils). Bibliography, p. 424-426.

FRIDLAND, Vladimir Markovich. Struktura pochvennogo pokrova (Structure of soil cover). Moskva, Izdatel'stvo "Mysl'," Geografgiz, 1972. 423 p. Bibliography, p. 394-410 (Akademiia Nauk SSSR. Institut Geografii. Vsesoiuznaia Akademiia Sel'skokhoziaistvennykh Nauk imeni Lenina. Pochvennyi Institut imeni V. V. Dokuchaeva). Supplementary table of contents in English, p. 418-423.   14

    About 380 references in alphabetical order, 310 in Russian and 70 in Western languages.

VILENSKII, Dmitrii Germogenovich. Geografiia pochv (Geography of soils). Moskva, Gosudarstvennoe Izdatel'stvo "Vysshaia Shkola," 1961. 343 p. Bibliography, p. 334-339.   1

    About 150 references arranged by sections and chapters.

GERASIMOV, Innokentii Petrovich, and GLAZOVSKAIA, Mariia Al'fredovna. Osnovy pochvovedeniia i geografiia pochv (Fundamentals of soil science and the geography of soils). Moskva, Geografgiz, 1960. 490 p. Bibliography, p. 473-575. Available in English as Fundamentals of soil science and soil geography. Jerusalem, Israel Program for Scientific Translations, 1965. 2nd impression, 1970. 392 p. IPST 1355.   1

    72 references arranged by three sections: physical, chemical, and biological foundations of soil science; major problems of general theory of soil science; and the soils of the USSR and foreign countries. Subject index. 2 folded maps in pocket.

AKADEMIIA Nauk SSSR. Pochvennyi Institut imeni V. V. Dokuchaeva. Geografiia i klassifikatsiia pochv Azii. (Geography and classification of the soils of Asia). N. V. Kimberg, V. A. Kovda, E. V. Lobova, A. M. Mamytov, M. U. Umarov, editorial board. V. A. Kovda and E. V. Lobova, eds. Moskva, Izdatel'stvo "Nauka," 1965. 251 p. Available in English as: Geography and classification of the soils of Asia. Editors V. A. Kovda and E. V. Lobova. Jerusalem, Israel Program for Scientific Translations, 1968. 280 p. IPST 5270.   1

    25 studies of soils in the warm temperate, subtropical, and tropical areas of Asia presented at a UNESCO/FAO symposium in Tashkent in 1962, each with separat bibliography. The total number of references is about 275.

BYSTRITSKAIA, T. L., and TIURIUKANOV, A. N. Chernye slitye pochvy Evrazii (Black compact soils of Eurasia). Moskva, Izdatel'stvo "Nauka," 1971. 256 p. Bibliography, p. 243-255. (Akademiia Nauk SSSR. Nauchnyi Sovet po Problemam Pochvovedeniia i Melioratsii Pochv).

GLINKA, Konstantin Dmitrievich. Pochvovedenie (Soil science). 6th ed. (posthumous) Moskva, Sel'knozgiz, 1935. 631 p. (1st ed. 1908).

Bibliographies at the ends of chapters. In Section 3, Characteristics of
oil Types and the Geography of Soils, p. 297-627 bibliographies follow the
ollowing chapters: soil classification; characteristics of soil types and
arieties; the influence of the chemical nature of parent material on soil for-
ation; fossil soils and ancient soils; short characteristics of the soil zones
f the USSR and adjacent states; mountain areas of the USSR.

STRUEV, Sergei Semĕnovich. Elementy geografii pochv (Elements of the geography
f soils). L. I. Prasolov, ed. 2nd ed. Moskva-Leningrad, Sel'khozgiz, 1931.
ibliography, p. 202-212 (1st ed. 1930).   1483
    226 references.

EV, Gasan (Hassan), Alirza ogly. Pochvennye resursy i ikh issledovaniia.
eziume (Soil resources and their investigation. Summary). Baku, 1971. p. 307-
51. Bibliography, p. 321-351. (Akademiia Nauk Azerbaidzhanskoi SSR. Institut
eografii).   1484

OBUEV, Vladimir Rodionovich. Sistema pochv mira (The system of soils of the
orld). Baku, "Elm," 1973. 308 p. Bibliography, p. 297-307.   1485

                              Special Topics

OBUEV, Vladimir Rodionovich. Pochvy i klimat (Soils and climate). Baku,
zdatel'stvo Akademii Nauk Azerbaidzhanskoi SSR, 1953. 320 p. Bibliography,
. 305-317 (Akademiia Nauk Azerbaidzhanskoi SSR. Institut Pochvovedeniia i
grokhimii).   1486
    425 references in alphabetical order on the geography of soils and the
nfluence of climate on the development of the soil cover.

JL'GIN, Aleksandr Mikhailovich. Klimat pochvy i ego regulirovanie (The climate
f the soil and its regulation). 2nd ed. Leningrad, Gidrometeoizdat, 1972.
341 p. Bibliography, p. 326-339.   1487
    379 references.

OBUEV, Vladimir Rodionovich. Ekologiia pochv: ocherki (Ecology of soils:
tudies). Baku, Izdatel'stvo Akademii Nauk Azerbaidzhanskoi SSR, 1963. 260 p.
ibliography, p. 245-259 (Akademiia Nauk SSSR. Institut Pochvovedeniia i
grokhimii).   1488
    About 500 references in alphabetical order, on the geography of soils and
he interaction between soils and other elements of the geographical environment.

AGOVSKII, Anatolii Ivanovich. Isparenie pochvennoi vlagi (Evaporation of soil
oisture). Moskva, Izdatel'stvo "Nauka," 1964. 244 p. Bibliography, p. 234-
43 (Akademiia Nauk SSSR. Institut Geografii).   1489
    About 220 references in alphabetical order.

LYI, Abram Mendelevich. Vodnyi rezhim v sevooborote na chernozemnykh pochvakh
ugo-Vostoka (Water regime in crop rotation on the black-earth soils of the
outheast European USSR). Leningrad, Gidrometeoizdat, 1971. 232 p. Bibliog-
aphy, p. 207-231.   1490

DA, Viktor Abramovich. Proiskhozhdenie i rezhim zasolennykh pochv. (Origin
nd processes of saline soils). v. 2. Moskva-Leningrad, Izdatel'stvo Akademii
auk SSSR, 1947. 375 p. Bibliography, p. 358-372 (Akademiia Nauk SSSR.
ochvennyi Institut, imeni V. V. Dokuchaeva).   1491
    About 400 references in alphabetical order.

                                                              1483 - 1491

GIDROLOGICHESKIE i geokhimicheskie svoistva erodirovannykh landshaftov (Hydrological and geochemical properties of eroded landscapes). N. Eitmanavichene, G. Pauliukiavichius, K. Gikite, and others. G. Pauliukiavichius, ed. Vil'nius, 1970. 309 p. Bibliography, p. 299-309. (Akademiia Nauk Litovskoi SSR. Otdel Geografii). Summaries in Lithuanian and German.   1

NAZAROV, Georgii Vasil'evich. Zonal'nye osobennosti vodopronitsaemosti pochv SSSR (Zonal characteristics of water permeability of soils in the USSR). Leningrad, Izdatel'stvo Leningradskogo Universiteta, 1970. 184 p. Bibliography, p. 173-183. (Leningradskii Gosudarstvennyi Universitet imeni A. A. Zhdanova).   1

VILENSKII, Dmitrii Germogenovich. Russkaia pochvenno-kartograficheskaia shkola i ee vliianie na razvitie mirovoi kartografii pochv (The Russian soil-cartographic school and its influence on the development of world cartography of soils). Moskva-Leningrad, Izdatel'stvo Akademii Nauk SSSR, 1945. 143 p. Bibliography, p. 134-141 (Akademiia Nauk SSSR. Sovet po Izucheniiu Proizvoditel'nykh Sil).   1

187 references in alphabetical order, 117 Soviet and 70 in foreign languages.

U. S. S. R. as a Whole

AKADEMIIA Nauk SSSR. Sovet po Izucheniiu Proizvoditel'nykh Sil. Pochvennyi Institut imeni V. V. Dokuchaeva. Pochvenno-geograficheskoe raionirovanie SSSR-- v sviazi s sel'skokhoziaistvennym ispol'zovaniem zemel' (Soil-geographic regionalization of the USSR in relation to agricultural utilization of the land). Moskva, Izdatel'stvo Akademii Nauk SSSR, 1962. 422 p. Bibliography, p. 402-417. (Issledovaniia po voprosam razvitiia i razmeshcheniia selskokhoziaistva).

Available in English: Soil-geographical zoning of the USSR. Jerusalem, Israel Program for Scientific Translations, 1963. 494 p. IPST 2055.   1

About 500 references arranged alphabetically. Text organized by soil types.

POCHVY SSSR (Soils of the USSR). L. I. Prasolov, ed. Evropeiskaia chast' SSSR. (European part of the USSR). v. 1-3. Moskva-Leningrad, Izdatel'stvo Akademii Nauk SSSR, 1939.   1

v. 1. Usloviia pochvoobrazovaniia i kharakteristika glavneishikh tipov pochv (Conditions of soil formation and characteristics of the major types of soil). 1939. 404 p.

v. 2. Pochvy lesnykh oblastei (Soils of forest regions). 1939. 288 p.

v. 3. Pochvy lesostepnykh i stepnykh oblastei (Soils of the wooded steppe and the steppe regions). 1939. 375 p.

Bibliographies at ends of sections devoted to types of soil formation and descriptions of soils of individual regions.

GLINKA, Konstantin Dmitrievich. Pochvy Rossii i prilegaiushchikh stran (Soils of Russia and adjacent countries). Moskva-Petrograd, Gosizdat, 1923. 348 p. Bibliography, p. 334-348.   1

458 references arranged by horizontal and vertical soil zones.

VILENSKII, Dmitrii Germogenovich. Solonchaki i solontsy SSSR. Saline and alkaline soils of the Union of Socialist Soviet Republics, Pochvovedenie, 1930, no. 4. Bibliography, p. 65-85 in English translation.   1

_____. Bibliography of salty soils of the USSR. In the book: Classification, geography and cartography of soils in the USSR. Moscow, Soviet section, 1935, p. 168-190. (Trudy 5 Komissii Mezhdunarodnoi Assotsiatsii Pochvovedov, Tom. A, I). In Russian with headings in English.   1

776 references on solonchak, solonets, and other saline soils of the USSR up to 1933.

U. S. S. R. Regions

European U. S. S. R.

ЭROVOL'SKII, Gleb Vsevolodovich, and URUSEVSKAIA, I. S., eds. Geografiia pochv
i pochvennoe raionirovanie Tsentral'nogo Ekonomicheskogo Raiona SSSR (Geography
of soils and soil regionalization of the Central Economic Region of the USSR).
Moskva, Izdatel'stvo Moskovskogo Universiteta, 1972. 469 p. Bibliography, p.
454-468. (Prirodnoe i ekonomiko-geograficheskoe raionirovanie SSSR dlia tselei
sel'skogo khoziaistva). 1500

JRIUKANOV, Anatolii Nikiforovich, and BYSTRITSKAIA, Tat'iana L'vovna. Opol'ia
Tsentral'noi Rossii i ikh pochvy (Opol'ya of Central Russia and their soils).
Moskva, Izdatel'stvo "Nauka," 1971. 239 p. Bibliography, p. 230-238. (Akademiia
Nauk SSSR. Nauchnyi Sovet po Problemam Pochvovedeniia). 1501

CHKOVA, Elizaveta Khristianovna, and BIBILO, Iuliia Osipovna. Pochvy Iugo-Vostoka
Evropeiskoi chasti SSSR: bibliograficheskii ukazatel' (The soils of the Southeast
European part of the USSR: a bibliographical guide). A. P. Malianov, ed. Saratov,
Izdatel'stvo Saratovskogo Gosudarstvennogo Universiteta, 1959. 253 p. (Saratovskii
Gosudarstvennyi Universitet. Nauchnaia Biblioteka. Bibliografiia Saratovskoi
Oblasti, vypusk 4). 1502

480 references arranged systematically and partly annotated.

Trans-Caucasus

IEV, Gasan (Hassan) Alirza ogly, ALEKPEROV, Kiazim Abdul Manaf ogly, VOLOBUEV,
Vladimir Rodionovich, and others. Pochvy Azerbaidzhanskoi SSR (Soils of the
Azerbaidzhan SSR). Baku, Izdatel'stvo Akademii Nauk Azerbaidzhanskoi SSR, 1953.
451 p. Bibliography, p. 437-445. (Akademiia Nauk Azerbaidzhanskoi SSR. Institut Pochvovedeniia i Agrokhimii). 1503

EKPEROV, Kiazim Abdul Manaf ogly. Eroziia pochv v Azerbaidzhane i bor'ba s nei
(Soil erosion in Azerbaidzhan and its prevention). Baku, Izdatel'stvo Akademii
Nauk Azerbaidzhanskoi SSR, 1961. 290 p. Bibliography, p. 210-219. (Pochvenno-
Erozionnaia Stantsiia). In Azerbaijani with summary in Russian. 1503a

About 240 references in Russian.

Siberia and the Soviet Far East

CHVY Sibiri i Dal'nego Vostoka: biulleten' otechestvennoi literatury (Soils of
Siberia and the Far East: bulletin of native literature). Novosibirsk, 1964- .
quarterly. (Akademiia Nauk SSSR. Sibirskoe Otdelenie. Gosudarstvennaia Publichnaia Nauchno-Tekhnicheskaia Biblioteka). 1504

Guide to books and articles systematically organized and partly annotated.

LOV, Anatolii Dmitrievich. Vodnaia eroziia pochv Novosibirskogo Priob'ia (Water
erosion of soils of the Novosibirsk Ob' region). R. V. Kovalev, ed. Novosibirsk,
Izdatel'stvo "Nauka," Sibirskoe Otdelenie, 1971. 175 p. Bibliography, p. 168-
174. (Akademiia Nauk SSSR. Sibirskoe Otdelenie. Institut Pochvovedeniia i
Agrokhimii). 1505

RENT'EV, Andrei Terent'evich. Pochvy Amurskoi oblasti i ikh sel'skokhoziaistvennoe ispol'zovanie (Soils of Amur oblast and their agricultural utilization).
Vladivostok, Dal'nevostochnoe Knizhnoe Izdatel'stvo, 1969. 275 p. Bibliography,
p. 262-273. (Ministerstvo Sel'skogo Khoziaistva SSSR. Primorskii Sel'skokhoziaistvennyi Institut). 1506

About 340 references.

1500 - 1506

KREIDA, N. A. Pochvy khvoino-shirokolistvennykh i shirokolistvennykh lesov
Primorskogo kraia (Soils of coniferous-deciduous and deciduous forests of
Primorskii Kray [Maritime Province]). Vladivostok, 1970. 229 p. Bibliography,
p. 220-228 (Dal'nevostochnyi Universitet. Uchënye zapiski, tom 27, chast' 2). 1
    About 200 references.

Kazakhstan and Soviet Middle Asia

POCHVY Kazakhskoi SSR v 16-ti vypuskakh (The soils of the Kazakh SSR in 16 volumes)
Alma-Ata. Izdatel'stvo AN Kazakhskoi SSR, 1960- . 16 v. (Akademiia Nauk
Kazakhskoi SSR. Institut Pochvovedeniia). v. 1 available in English: Soils of
the North-Kazakhstan Region of Kazakh SSR. Yu. V. Fedorin. Jerusalem, Israel
Program for Scientific Translations, 1964. 160 p. IPST 1146.   1
    List of references at end of each volume devoted to an oblast, 50-250 refer-
ences for each volume and oblast.

LOBOVA, Elena Vsevolodnovna. Pochvy pustynnoi zony SSSR (Soils of the desert
zone of the USSR). Moskva, Izdatel'stvo Akademii Nauk SSSR. 1960. 364 p.
Bibliography, p. 351-363. (Akademiia Nauk SSSR. Pochvennyi Institut imeni
V. V. Dokuchaeva). Available in English: Soils of the desert zone of the USSR.
Jerusalem, Israel Program for Scientific Translations, 1967. 416 p. IPST 1911. 1
    About 425 references in alphabetical order on the soil-forming factors and
the soils of Soviet Middle Asia, Kazakhstan, and adjacent areas.

POCHVY doliny reki Chu, ikh priroda i puti ispol'zovaniia dlia sel'skogo khoziaistva
    (Soils of the valley of the Chu river, its natural conditions and the lines of
its utilization for agriculture). A. I. Volkov, M. A. Orlova, and others, eds.
M. A. Orlova, A. N. Osina, E. A. Sokolenko, and others, authors. Alma-Ata,
Izdatel'stvo "Nauka," 1971. 374 p. Bibliography, p. 364-372 (Akakemiia Nauk
Kazakhskoi SSR. Institut Pochvovedeniia).   1

MAMYTOV. Aman Mamytovich. Pochvennye resursy i voprosy zemel'nogo kadastra
Kirgizskoi SSR (Soil resources and questions of the soil cadastre of the Kirgiz
SSR).   Frunze, Izdatel'stvo "Kyrgyzstan," 1971. 110 p. Bibliography, p. 105-
108.   1

MIRZAZHANOV, Kirgizbai Mirzazhanovich. Vetrovaia eroziia oroshaemykh pochv
Uzbekistana i bor'ba s nei (Wind erosion of irrigated poils of Uzbekistan and
the fight against it). Tashkent, Izdatel'stvo "FAN," 1973. 235 p. Bibliography, p. 229-234 (Ministerstvo Sel'skogo Khoziaistva SSSR. Vsesoiuznyi Nauchno-
Issledovatel'skii Institut Khlopkovodstva "SoiuzNIKHI").   1

LAVROV, Arkadii Pavlovich. Pochvennyi ocherk Zaunguzskikh Karakumov (Soil study
of the Trans-Unguz Karakum). A. G. Babaev, ed. Ashkhabad, "Ylym," 1973. 89 p.
Bibliography, p. 83-87. (Akademiia Nauk Turkmenskoi SSR. Institut Pustyn').   1

BIOGEOGRAPHY

Major General Bibliographies

ERATIVNYI zhurnal: biologiia (Reference journal: biology) (Vsesoiuznyi Institut
Jauchnoi i Tekhnicheskoi Informatsii). Moskva. 1954- . monthly. Annual subject
and author indexes.   1514

Each number issued in two parts: I. A-G, T, U; II. D-K, M. N. P. The
major sections and their order in the full edition are: Part I. A. General questions of biology; T. Cytology, general genetics, and genetics of man; U. General
ecology, biocenology, and hydrobiology; B. Virology and microbiology; V. Botany;
*. Physiology of plants. Part II. D. General zoology and invertebrates; E. Ento-
ology; I. Zoology of vertebrates; K. Zooparasitology; M. Morphology of man and
animals; Physical anthropology; N. Physiology of man and animals (general physiology; exchanges of matter and energy; internal organs; blood); P. Physiology of
man and animals (nervous system; sensory organs; internal secretions, reproduc-
ion).

Section U in part 1 is divided into four subsections: general ecology; biocenology and ecosystems; hydrobiology; and conservation of natural biological
resources.

A massive and detailed coverage of the literature in all languages with
abstracts and bibliographical notations.

LOGICHESKIE nauki (Biological sciences), <u>Bibliografiia sovetskoi bibliografii</u>
Bibliography of Soviet bibliography). 1939, 1946- . Annual. Moskva, Izdatel'-
tvo "Kniga," 1941, 1948- . (Vsesoiuznaia Knizhnaia Palata).   1515

Annual list of Soviet bibliographies in the biological sciences. The 1970
nnual volume (1972), p. 104-126 contains 662 entries organized by the following
ections: general questions; paleontology; general biology; microbiology, hydro-
iology; botany; zoology; anthropology; biology of man.

IN, Vladimir Lazarevich. <u>Spravochnoe posobie po bibliografii dlia biologov</u>
Reference aid on bibliography for biologists). Moskva-Leningrad, Izdatel'stvo
kademii Nauk SSSR, 1960. 407 p. (Akademiia Nauk SSSR. Otdelenie Biologicheskikh
auk. Institut Tsitologii).   1516

About 1,000 references. See especially sections, Botany, general bibliography
f native literature, p. 133-138, and Zoology, general bibliography of native
iterature, p. 179-185.

GEOTSENOLOGIIA. Tekushchii ukazatel' literatury (Biogeocenology. Running guide
o the literature). no. 1- (1973-). Novosibirsk (Akademiia Nauk SSSR, Sibir-
koe Otdelenie. Gosudarstvennaia Publichnaia Nauchno-Tekhnicheskaia Biblioteka). 1517

Plants and Vegetation

ICHESKAIA geografiia: annotirovannyi perechen' otechestvennykh bibliografii
zdannykh v 1810-1966 gg. (Physical geography: annotated list of native bibliog-
aphies published 1810-1966). M. N. Morozova and E. A. Stepanova, comps. V. V.
levenskaia, ed. L. G. Kamanin, scientific consultant. Moskva, Izdatel'stvo
Nauka," 1968. 309 p. (Gosudarstvennaia Biblioteka SSSR imeni V. I. Lenina.
nstitut Geografii Akademii Nauk SSSR. Sektor Seti Spetsial'nykh Bibliotek
kademii Nauk SSSR).   1518

Geography of vegetation, p. 172-197, entries 832-988, 157 Soviet bibliog-
aphies on plant geography organized systematically: bibliographies of a general
haracter, indexes to the contents of journals and collections, bibliographies
f the publications of scientific institutions of the USSR; bibliographies of

types of vegetation (tundra; forests; meadows; steppes; deserts; bogs; mountain vegetation); bibliographies on vegetation of individual areas (USSR as a whole; European part of the USSR, North and Northwest, Center, West, Southwest, Volga region, Urals, the Caucasus, the North Caucasus, the Trans-Caucasus; Siberia and the Far East, West Siberia, East Siberia, Far East; Middle Asia and Kazakhstan, Middle Asia, Kazakhstan; foreign countries). Author and geographical indexes.

LEBEDEV, D. V. Vvedenie v botanicheskuiu literaturu SSSR: posobie dlia geobotanikov (Introduction to botanical literature of the USSR: an aid for geobotanists). Leningrad, Izdatel'stvo Akademii Nauk SSSR, 1956. 383 p. (Akademiia Nauk SSSR. Botanicheskii Institut imeni V. L. Komarova).   1⁵

About 3,000 references, including relevant works in geography, p. 266-290. Index of authors, compilers, and editors, and of titles.

_____ . Novye bibliograficheskie posobiia, poleznye dlia botanikov (New bibliographical aids useful for botanists), Botanicheskii zhurnal, v. 43, no. 1 (1958), p. 115-123.   1⁵

PRISTUPA, A. A. Osnovnye syr'evye rasteniia i ikh ispol'zovanie (Basic raw-material plants and their utilization). Leningrad, Izdatel'stvo "Nauka," Leningradskoe Otdelenie, 1973. 412 p. Bibliography, p. 372-394. (Akademiia Nauk SSSR. Nauchnyi Sovet po Probleme "Biologicheskie Osnovy Ratsional'nogo Ispol'zovaniia, Preobrazovaniia i Okhrany Rastitel'nosti Mira"). Index of plant names in Russian and Latin, p. 395-409.   1⁵

BYKOV, Boris Aleksandrovich. Vvedenie v fitotsenologiiu (Introduction to the study of plant communities). Alma-Ata, Izdatel'stvo "Nauka," 1970. 231 p. Bibliography, p. 227-231 (Akademiia Nauk Kazakhskoi SSR. Institut Botaniki). 1⁵

TOPOLOGIIA stepnykh geosistem (Topology of steppe geosystems). V. B. Sochava, V. G. Volkova, G. N. Mart'ianova, and others. Leningrad, Izdatel'stvo "Nauka," Leningradskoe Otdelenie, 1970. 174 p. Bibliography, p. 159-163, and ·at ·end ··· of articles. (Akademiia Nauk SSSR. Sibirskoe Otdelenie. Institut Geografii Sibiri i Dal'nego Vostoka). Resumé in French.   1⁵

148 references.

GEOBOTANICHESKOE izuchenie lugov: sbornik statei (Geobotanical study of meadows: a collection of articles). I. D. Iurkevich and E. A. Kruganovaia, eds. Minsk, Izdatel'stvo Akademii Nauk Belorusskoi SSR, 1962. 147 p. Bibliography, p. 139-146 (Vsesoiuznoe Botanicheskoe Obshchestvo. Belorusskoe Otdelenie. Sbornik botanicheskikh rabot, vypusk 4).   1⁵

About 200 references in alphabetical order in Russian, Belorussian, and foreign languages on the meadows of the USSR.

N'IUBOLD, P. Metody opredeleniia pervichnoi produktivnosti lesov. (Methods of the definition of primary productivity of woods). Novosibirsk, 1971. 68 p. Bibliography, p. 56-68. (Akademiia Nauk SSSR. Sovetskii Natsional'nyi Komitet po MBP SO AN SSSR. Sibirskii Komitet po MBP. Spravochnik MBP no. 2. Materialy po Mezhdunarodnoi Biologicheskoi Programme).   1⁵

BLINOVA, E. I., and VOZZHINSKAIA, V. B. Morskie makrofity i rastitel'nye resursy okeana (Marine macrophytes and plant resources of the ocean. In the book: Osnovy biologicheskoi produktivnosti okeana i ee ispol'zovanie (Basis of biological productivity of the ocean and its utilization). Moskva, 1971. p. 137-171. Bibliography, p. 167-171. Summary in English.   1⁵

SHUL'GIN, I. A. Rastenie i solntse (The plant and the sun). Leningrad, Gidrometeoizdat, 1973. 251 p. Bibliography, p. 219-249. Summary in English.   1⁵

801 references.

NIUKOVICH, Kirill Vladimirovich. Rastitel'nost' gor SSSR: botaniko-geograficheskii
cherk. (Vegetation of mountains of the USSR: a botanical-geographical essay).
. N. Maksumov, ed. Dushanbe, "Donish," 1973. 411 p. Bibliography, p. 386-412
Akademiia Nauk Tadzhikskoi SSR. Sovet po Izucheniiu Proizvoditel'nykh Sil).   1527

Geobotany

KSANDROVA, V. D. Klassifikatsiia rastitel'nosti. Obzor printsipov klassifikatsii
 klassifikatsionnykh sistem v raznykh geobotanicheskikh shkolakh (Survey of the
rincipal classifications and classification systems in the various geobotanical
chools). Leningrad, Izdatel'stvo "Nauka," Leningradskoe Otdelenie, 1969. 275 p.
ibliography, p. 237-274. (Akademiia Nauk SSSR. Nauchnyi Sovet po probleme,
"Biologicheskie Osnovy Ratsional'nogo Ispol'zovaniia, Preobrazovaniia i Okhrany
astitel'nosti Mira.").   1528

    1,567 references.

OV, Boris Aleksandrovich. Geobotanicheskii slovar'. (Geobotanical dictionary).
nd ed. Alma-Ata, "Nauka," 1973. 214 p. (Akademiia Nauk Kazakhskoi SSR.
nstitut Botaniki).   1529

VINA, Fanni Iakovlevna. Geobotanika v Botanicheskom Institute im. V. L. Komarova
AN SSSR, 1922-1964. (Geobotany in the Komarov Botanical Institute of the Academy
of Sciences of the USSR, 1922-1964). Leningrad, Izdatel'stvo "Nauka," Leningradskoe Otdelenie, 1971. 319 p. Bibliography, p. 161-293.   1530

    Review of work on geobotanical research and publications of the section on
geobotany of the Botanical Institute of the Academy of Sciences of the USSR from
its founding in 1922 to 1964. Bibliography includes 4,020 entries alphabetically
by author, and chronologically for each author. The section on problem-thematic
grouping of publications, p. 103-160 lists works by 13 fields, some of which are
subdivided.

ROSHENKO, Pavel Dionis'evich. Geobotanika. Osnovnye poniatii, napravleniia i
metody (Geobotany: basic concepts, directions, and methods). Moskva - Leningrad,
Izdatel'stvo Akademii Nauk SSSR, 1961. 474 p. Bibliography, p. 450-471 (Akademiia
Nauk SSSR. Sibirskoe Otdelenie. Dal'nevostochnyi Filial imeni V. L. Komarova). 1531

    About 750 references in alphabetical order on geobotany and geography of
vegetation, about 570 Soviet and 180 in foreign languages.

KOV, Boris Aleksandrovich. Geobotanika (Geobotany). 2nd ed. Alma-Ata, Izdatel'-
stvo Akademii Nauk Kazakhskoi SSR, 1957. 382 p. Bibliography, p. 352-367.   1532

    More than 500 references on geobotany and geography of vegetation.

RONOV, Anatolii Georgievich. Geobotanika (Geobotany). 2nd ed. Moskva, Izdatel'-
stvo "Vyschaia Shkola," 1973. 384 p. Bibliography, p. 372-376.   1533

    Textbook for biology and geography majors in universities and pedagogical
institutes. Subject index.

KHMAN, Genrietta Isaakovna. Istoriia geobotaniki v Rossii. (History of geobotany
in Russia). Moskva, Izdatel'stvo "Nauka," 1973. 286 p. Bibliography, p. 271-
282. (Moskovskoe Obshchestvo Ispytatelei Prirody. Materialy k poznaniiu fauny i
flory SSSR. Novaia seriia. Otdel botanicheskii, vypusk 17 [25]).   1534

Mapping

AKADEMIIA Nauk SSSR. Botanicheskii Institut imeni V. L. Komarova. Rastitel'nyi pokrov SSSR: poiasnitel'nyi tekst k "Geobotanicheskoi karte SSSR" m. 1:4,000,000 (The vegetative cover of the USSR: explanatory text to the Geobotanical map of the USSR on the scale of 1:4,000,000). E. M. Lavrenko and V. B. Sochava, eds. Moskva, Leningrad, Izdatel'stvo Akakemii Nauk SSSR, 1956. 2 v. 971 p. Bibliography, v. 2, p. 859-912, comp. by V. V. Lipatova. Map sources, v. 2, p. 913-917, comp. by A. N. Lukicheva.   1

About 2,400 references to books and articles, and 105 maps sources. Legend of map with authors of explanatory text for each section, p. 918-926. Index of Latin names of vegetation, p. 927-971. The fullest bibliography of the literature of the geobotany of the USSR up to 1954. The history of basic work on the vegetation of the USSR is treated by V. V. Sochava, v. 1, p. 9-60.

GEOBOTANICHESKOE kartografirovanie (Geobotanical mapping) (Akademiia Nauk SSSR. Botanicheskii Institut imeni V. A. Komarova). Leningrad, Izdatel'stvo "Nauka," Leningradskoe Otdelenie, 1963- . annual. In Russian with supplementary English table of contents and English summary of each volume. Index of articles during 10 years, 1963-1972 in the 1972 volume, in Russian, p. 83-87, and in English, p. 88-92. List of 114 articles with subject index by broad categories.   1

Covers all phases of mapping of vegetation with articles, reviews of major publications and centers of activity, and bibliographies.

LIPATOVA, V. V. Materialy k bibliografii po voprosam kartografirovaniia rastitel'-nosti (Materials on a bibliography on problems of mapping vegetation). In the book: Printsipy i metody geobotanicheskogo kartografirovaniia (Principles and methods of geobotanical mapping). Moskva-Leningrad, Izdatel'stvo Akademii Nauk SSSR, 1962, p. 265-296 (Akademiia Nauk SSSR. Botanicheskii Institut Imeni V. L. Komarova).   1

About 730 references in alphabetical order, about 260 Soviet and about 470 in foreign languages.

LIPATOVA, V. V. Dopolnenie k bibliografii po voprosam kartografirovaniia rastitel'-nosti (Supplement to the bibliography on problems of mapping vegetation), Geobotanicheskoe kartografirovanie, 1963, p. 62-70.   1

188 references, 62 Soviet and 126 in foreign languages, mainly for the period 1960-1961.

LIPATOVA, V. V. Literatura po kartografii rastitel'nosti (Literature on mapping vegetation), Geobotanicheskoe kartografirovanie, 1965, p. 76-84.   1

193 references, 52 Soviet and 141 foreign, mainly for 1962-1964.

_____ . _____ II, Geobotanicheskoe kartografirovanie, 1967, p. 83-91.

203 references, mainly for 1964-1966.

_____ . _____ III, Geobotanicheskoe kartografirovanie, 1971, p. 70-81.

187 references, 117 Soviet and 70 foreign, mainly for 1967-1970.

ISACHENKO, Tat'iana Iosifovna. Slozhenie rastitel'nogo pokrova i kartografirovanie (Composition of vegetation and mapping), Geobotanicheskoe kartografirovanie, 1969, p. 20-33. Bibliography, p. 30-33.   1

About 100 references.

TUPIKOVA, Nataliia Vladimirovna. Zoologicheskoe kartografirovanie (Zoological mapping). Moskva, Izdatel'stvo Moskovskogo Universiteta, 1969. 250 p. Bibliography, p. 238-249.   1

About 300 references.

L'TSOV-BEBUTOV, Aleksandr Mikhailovich. Zoogeograficheskoe kartografirovanie i
_andshaftovedenie (Zoogeographical mapping and landscape studies). In the book:
_andshaftnyi sbornik, ed. by V. G. Konovalenko and N. A. Solntsev. Moskva,
_zdatel'stvo Moskovskogo Universiteta, 1970, p. 49-94. Bibliography, p. 88-94.   1542
    80 references.

Historical Geography of Vegetation

_'F, Evgenii Vladimirovich. Istoricheskaia geografiia rastenii: istoriia flor
_emnogo shara (Historical geography of vegetation: history of the florae of the
_arth). Moskva-Leningrad, Izdatel'stvo Akademii Nauk SSSR. 1944. 546 p.   1543
    Bibliographies at the ends of chapters devoted to the history of development
_f the floral regions of the earth. See especially the chapter, European part of
_he USSR and Siberia, with about 370 references.

L'F, Evgenii Vladimirovich. Istoricheskaia geografiia rastenii (Historical
_eography of vegetation). Moskva-Leningrad, Izdatel'stvo Akademii Nauk SSSR,
1936. 323 p. (Geografiia rastenii, tom 1).   1544
    Bibliographies at ends of chapters: the areal distribution of vegetation
and factors in its formation, 19 references; changes in the geographical distribu-
_ion of vegetation in past geological periods, 38 references; the glacial period
and its influence on vegetation, 152 references; history of development of the
_lora of the USSR, 237 references; and the influence of man on the geographical
_istribution of vegetation, 21 references.

_HNACH, N. A. Etapy razvitiia rastitel'nosti Belorussii v antropogene. K 3-i
Mezhdunarodnoi Palinologicheskoi Konferentsii. Novosibirsk, SSSR, 1971. (Stages
in the development of the vegetation of Belorussia in the Quaternary. To the
3rd International Palynological Conference, Novosibirsk, USSR, 1971). Minsk,
Izdatel'stvo "Nauka i Tekhnika," 1971. 209 p. Bibliography, p. 193-209
(Upravlenie Geologii pri Sovete Ministrov BSSR. Belorusskii Nauchno-Issledovatel'-
_kii Geologo-Razvedochnyi Institut).   1545

_CHUK, Margarita Pavlovna, VOLKOVA, V. S., BUKREEVA, G. V., and others. Istoriia
_azvitiia rastitel'nosti vnelednikovoi zony Zapadno-Sibirskoi nizmennosti v pozdne-
    pliotsenovoe i chetvertichnoe vremia (History of the development of vegetation
_n the nonglaciated zone of the West-Siberian lowland in the post-Pliocene and
_he Quaternary periods). V. N. Saks, ed. Moskva, Izdatel'stvo "Nauka," 1970.
_64 p. Bibliography, p. 348-460. (Akademiia Nauk SSSR. Sibirskoe Otdelenie.
_nstitut Geologii i Geofiziki. Trudy, vypusk 92).   1546

Palynology

_SHTADT, Mark Il'ich. Palinologiia v SSSR, 1952-1957 (Palynology in the USSR,
_952-1957). Moskva, Izdatel'stvo Akademii Nauk SSSR, 1960. 272 p. Bibliography,
_. 47-271. (Akademiia Nauk SSSR. Institut Geografii).   1547
    1,632 entries arranged chronologically. Index by regions of the USSR. Sub-
ject index. Author index.

_SHTADT, Mark Il'ich. Sporovo-pyl'tsevoi metod v SSSR. Istoriia i bibliografiia
_Spore and pollen analysis in the USSR. History and bibliography). Moskva,
_zdatel'stvo Akademii Nauk SSSR, 1952. 223 p. Bibliography, p. 49-221 (Akademiia
_auk SSSR. Institut Geografii).   1548
    Chronological bibliography with 926 entries. Index of the literature by
_egions. Index of data for foreign countries. Subject index. Author index.
_ndex of spots for which there are pollen analyses.

PALINOLOGICHESKIE issledovaniia v Belorussii i drugikh raionakh SSSR. K 3-i
Mezhdunarodnoi palinologicheskoi konferentsii, Novosibirsk, 1971. (Palynological
research in Belorussia and other regions of the USSR. To the 3rd International
Palynological Conference, Novosibirsk, 1971). V. K. Golubtsov, ed. Minsk,
Izdatel'stvo "Nauka i Tekhnika," 1971. 220 p. Bibliography, p. 206-213.
(Upravlenie Geologii pri Sovete Ministrov BSSR. Belorusskii Nauchno-Issledovatel'-
skii Geologo-Razvedochnyi Institut "BelNIGRI." Mezhdunarodnaia Palinologicheskaia
Konferentsiia SSSR, Novosibirsk 1971). 1

Bogs

KATS, Nikolai Iakovlevich. Bolota zemnogo shara (Bogs of the world). Moskva,
Izdatel'stvo "Nauka," 1971. 295 p. Bibliography, p. 257-291. (Akademiia Nauk
SSSR. Moskovskoe Obshchestvo Ispytatelei Prirody). 1

KATS, Nikolai Iakovlevich. O raionirovanii bolot v sviazi s ikh tipizatsiei (On
the regionalization of bogs in relation to their typology), Botanicheskii zhurnal,
v. 52, no. 4 (1967), p. 477-480. 1

About 120 references.

KATS, Nikolai Iakovlevich. Tipy bolot SSSR i Zapadnoi Evropy i ikh geograficheskoe
rasprostranenie (Types of bogs of the USSR and Western Europe and their geographical distribution). Moskva, Geografgiz, 1948. Bibliography, p. 296-317. 1

More than 500 references in alphabetical order, about 260 Soviet and about
240 in foreign languages.

BOGDANOVSKAIA-GIENEF, Ivonna Donatovna. Zakonomernosti formirovaniia sfagnovykh
bolot verkhovogo tipa (Regularities in the formation of sphagnum bogs of the upland
type). Leningrad, Izdatel'stvo "Nauka," 1969. 186 p. Bibliography, p. 182-184.
(Akademiia Nauk SSSR. Vsesoiznoe Botanicheskoe Obshchestvo). 1

About 100 references.

ROMANOVA, Efrosin'ia Andreevna. Geobotanicheskie osnovy gidrologicheskogo izucheniia verkhovykh bolot: s ispol'zovaniem aerofotos"emki (Geobotanical foundations
of the study of upland bogs: with the utilization of air photos). Leningrad,
Gidrometeoizdat, 1961. 244 p. Bibliography, p. 172-177 (Gosudarstvennyi Gidrologicheskii Institut). 1

About 190 references in alphabetical order.

SMOLIAK, L. P. Bolotnye lesa i ikh melioratsiia (Boggy woods and their reclamation).
I. D. Iurkevich, ed. Minsk, "Nauka i Tekhnika," 1969. 209 p. Bibliography,
p. 199-207. (Akademiia Nauk Belorusskoi SSR. Institut Eksperimental'noi Botaniki).
190 references. 1

P'IAVCHENKO, N. I. Lesnoe bolotovedenie: osnovnye voprosy (Forest bog study:
basic questions). Moskva, Izdatel'stvo Akademii Nauk SSSR. 1963. 192 p.
Bibliography, p. 177-185 (Akademiia Nauk SSSR. Sibirskoe Otdelenie. Institut
Lesa i Drevesiny). 1

About 220 references in alphabetical order.

Peat

BIBLIOGRAFICHESKII ukazatel' literatury po torfu (Bibliographical guide to the
literature on peat). Moskva-Leningrad, Gosenergoizdat "Nedra," v. 1-4, Moskva,
"Energiia," v. 5. Kalinin, Tsentr Nauchno-Tekhnicheskoi Informatsii i Propagandy.
v. 6-7. 1960- . Tom 1, 1933-1942, 1960. 135 p. Tom 2, 1943-1955, 1960. 230 p.
Tom 3, 1956-1960, 1962. 182 p. Tom 4, 1961-1963, 1965. 207 p. Tom 5, 1964-
1966. 1969. 232 p. Tom 6. 1967-1969. 1970. 217 p. Tom 7, 1969-1970. 1971.
196 p. (Vsesoiuznyi Nauchno-Issledovatel'skii Institut Torfianoi Promyshlennosti.
VNIITP. Kalininskii Filial). 15

1549 - 1557

Systematic guide. The section peat deposits and their prospecting contains
aterial on the study of peat bogs, the conditions of their formation, vegetation,
ydrology, and geographical distribution. v. 1. 2,530 entries. v. 2. 4,038
ntries. v 3. 2,981 entries. v. 4. 3,176 entries, v. 5. 2,973 entries.

AVCHENKO, N. I. Torfianiki Russkoi lesostepi (Peat bogs of the Russian wooded
teppe). Moskva, Izdatel'stvo Akademii Nauk SSSR, 1958. 191 p. Bibliography,
. 184-190 (Akademiia Nauk SSSR. Institut Lesa).   1558

About 210 references in alphabetical order. Available in English as Peat
ogs of the Russian forest-steppe. Translated from the Russian by A. Gourevich.
erusalem, Israel. Israel Program for Technical Translations. Washington, D.C.
ffice of Technical Services, U.S. Department of Commerce, 1964. 156 p. Bib-
iography, p. 145-156.

Animals

ICHESKAIA geografiia: annotirovannyi perechen' otechestvennykh bibliografii
zdannykh v 1810-1966 gg. (Physical geography: annotated list of native bibliog-
aphies published 1810-1966). M. N. Morozova and E. A. Stepanova, comps. V. V.
levenskaia, ed. L. G. Kamanin, scientific sonsultant. Moskva, Izdatel'stvo
Nauka," 1968. 309 p. (Gosudarstvennaia Biblioteka SSSR imeni V. I. Lenina.
nstitut Geografii Akademii Nauk SSSR. Sektor Seti Spetsial'nykh Bibliotek
kademii Nauk SSSR).   1559

Animal geography, p. 198-217, entries 989-1,102, 114 Soviet bibliographies
n animal geography, organized systematically: bibliographies of a general char-
cter; bibliographies of the animal kingdom of individual areas: USSR as a whole:
ammals, rodents; birds; reptiles and amphibians; fish). European part of the
SSR, North and Northwest, Center, West, Southwest, Moldavia, Ukraine, Volga
egion, Urals and pre-Urals, Caucasus; Asiatic part of the USSR, Siberia, Western
iberia, Eastern Siberia, Far East; Middle Asia and Kazakhstan, Middle Asia,
irgizia, Turkmenia, Uzbekistan, Kazakhstan; Soviet Arctic. Foreign countries.
uthor and geographical index.

RZINA, M. N. comp. Literatura po resursam zhivotnogo mira SSSR za 1965-1968 gg:
geografiia zapasov, ispol'zovanie, vosproizvodstvo (Literature on the resources of animals in the USSR 1965-1968: geography of stocks, utilization, and
reproduction). Itogi nauki: geografiia SSSR, no. 7. Moskva, VINITI, 1969,
p. 164-231.   1560

1,071 entries, organized systematically: (1) animal resources of the USSR
as a whole and general questions, such as the problem of biological productivity
(97) titles); (2) resources of game animals and birds, their reproduction and
utilization (446); (3) fur farming (23); (4) resources of food fish and some
water invertebrates, their reproduction and utilization (469), and supplement
(36). Citation number is given for abstracts of each entry in Referativnyi
zhurnal: geografiia. Index of authors. Regional index.

UMOV, Nikolai Pavlovich. Ekologiia zhivotnykh (Ecology of animals). 2nd ed.
Moskva, Izdatel'stvo "Vysshaia Shkola," 1963. 618 p. Bibliography, p. 578-614. 1561

About 1,000 references in alphabetical order.

LABUKHOV, Nikolai Ivanovich. Metodika eksperimental'nykh issledovanii po
ekologii nazemnykh pozvonochnykh (Methods of experimental research on the
ecology of terrestrial vertebrates). Moskva, Izdatel'stvo "Sovetskaia Nauka,"
1951, 177 p. Bibliography, p. 166-176.   1562

About 250 references arranged systematically by chapter topics: influence
of the temperature of the environment, of radiant energy, of moisture and
precipitation, of atmospheric pressure, and of composition of the air on
animals.

NOVIKOV, Georgii Aleksandrovich. Ekologiia zverei i ptits lesostepnykh dubrav
(Ecology of wild animals and birds in the oak woods of the wooded steppe).
Leningrad, Izdatel'stvo Leningradskogo Universiteta, 1959. 352 p. Bibliog-
raphy, p. 331-351 (Leningradskii Gosudarstvennyi Universitet).   15

About 650 references in alphabetical order on ecological research of the
wooded steppe zone of the European part of the USSR as the environment of
mammals and birds.

FORMOZOV, Aleksandr Nikolaevich. Kratkii obzor rabot po ekologii ptits i
mlekopitaiushchikh za dvadtsatiletie, 1917-1937 gg (Short review of work on
the ecology of birds and mammals during the 20 years, 1917-1937), Zoologicheskii
zhurnal, v. 16, no. 5 (1937), p. 916-949. Bibliography, p. 942-949.   15

About 280 references in alphabetical order.

MAKSIMOV, Anatolii Aleksandrovich. Sel'skokhoziaistvennoe obrazovanie landshafta
i ekologiia vrednykh gryzunov (Agricultural formation of the landscape and the
ecology of harmful rodents). Moskva-Leningrad, Izdatel'stvo "Nauka," 1964.
252 p. Bibliography, p. 218-238 (Akademiia Nauk SSSR. Sibirskoe Otdelenie.
Biologicheskii Institut).   1⁵

About 600 references in alphabetical order on the geographical distribution, ecology, and population dynamics of harmful rodents in agricultural zones
of the USSR, and also on their role in transmission of diseases.

KIRIKOV, Sergei Vasil'evich. Promyslovye zhivotnye, prirodnaia sreda i chelovek
(Commercial wild animals, the natural environment, and man). Moskva, Izdatel'-
stvo "Nauka," 1966. 348 p. Bibliography, p. 320-344. (Akademiia Nauk SSSR.
Institut Geografii).   1⁵

About 1,000 references on commercial fish, game birds, and fur-bearing and
game mammals, the physical environment in which they live, and the role of man.
Arranged alphabetically by name of author. The textual discussion is in two
sections; the first, on changes in the natural environment and distribution of
game and commercial animals from the 14th to the 20th centuries, is by regions
of the Soviet Union; the second, on historical changes in the areal distribution and numbers of commercial animals, is organized by fish, birds, and commercial wild animals.

ABELENTSEV, Viktor Ivanovich, BROMLEI, G. F., KAZARINOV, A. P., and others. Sobol',
kunitsy, kharza. Razmeshchenie, zapasov, ekologiia, ispol'zovanie i okhrana
(Sable and marten: distribution, stock, ecology, utilization, and conservation).
A. A. Nasimovich, ed. Moskva, Izdatel'stvo "Nauka," 1973. 239 p. Bibliography,
p. 229-237. (Akademiia Nauk SSSR. Institut Geografii. Promyslovye zhivotnye
SSSR i sreda ikh obitaniia).   1⁵

GRAKOV, N. N., ed. Okhotnich'e khoziaistvo SSSR (Hunting economy of the USSR).
V. A. Ammosov and others. Moskva, Izdatel'stvo "Lesaia Prom-st'," 1973. 407 p.
Bibliography, p. 386-406.   1⁵

POTAPOVA, G. A. comp. Literatura po rybokhoziaistvennym issledovaniiam v severo-
vostochnoi chasti Tikhogo okeana (Literature on fishing research in the northeast Pacific Ocean). Vsesoiuznyi Nauchno-issledovatel'skii Institut Morskogo
Rybnogo Khoziaistva i Okeanografii, Trudy, v. 58 (1965), p. 311-345.   1⁵

1,065 entries, by fields, including general works (expeditions, general
physical geography, etc.), oceanography, hydrobiology, sea mammals, ichthyology,
and technology.

MIDT, Peter Iulievich. Ryby Okhotskogo moria (Fishes of the Okhotsk Sea).
Moskva-Leningrad, Izdatel'stvo Akademii Nauk SSSR, 1950. 370 p. Bibliography,
p. 337-351 (Akademiia Nauk SSSR. Tikhookeanskii Komitet. Trudy, tom 6).   1570

    About 400 references in alphabetical order, about 180 Soviet and about 220
in foreign languages.

TAPOVA, G. A. comp. Biologiia osetrovykh, ikh razvedenie i promysel. Ukazatel'
literatury (Biology of sturgeon, breeding and fisheries. Guide to the literature). Vsesoiuznyi Nauchno-issledovatel'skii Institut Morskogo Rybnogo
Khoziaistva i Okeanografii. Trudy, v. 52 (1964), p. 348-402.   1571

    1,598 references in Russian 1777-1963.

LEIKOVSKII, Semën Abramovich. Istoriko-bibliograficheskii obzor otechestvennykh
issledovanii morskogo planktona za stoletie (1860-e -- 1960-e gody) (Historical-
bibliographical survey of native research on sea plankton in the century from
the 1860's to the 1960's). Moskva, Izdatel'stvo "Nauka," 1970. 195 p. (Akademiia
Nauk SSSR. Institut Okeanologii imeni P. P. Shirshova).   1572

    Bibliographical handbook. Subject and author indexes.

OROV, Petr Petrovich. Problemy izucheniia nazemnykh ekosistem i ikh zhivotnykh
(Problems of the study of subterranean ecosystems and their animal components).
Frunze, "Ilim," 1971. 95 p. Bibliography, p. 86-94. (Akademiia Nauk Kirgizskoi
SSR. Tian'-Shan'skaia Vysokogornaia Fiziko-geograficheskaia Stantsiia).   1573

AROBOGATOV, Ia. I. Fauna molliuskov i zoogeograficheskoe raionirovanie
kontinental'nykh vodoëmov zemnogo shara (The fauna of mollusks and the zoogeographical regionalization of continental reservoirs of the earth). Leningrad, Izdatel'stvo "Nauka," Leningradskoe Otdelenie, 1970. 372 p. Bibliography,
p. 289-326. Index of Latin names, p. 327-371. (Akademiia Nauk SSSR. Zoo-
logicheskii Institut).   1574

Special Topics

LCHANOV, Aleksandr Alekseevich. Vliianie lesa na okruzhaiushchuiu sredu
(Influence of the forest on the surrounding environment). Moskva, Izdatel'stvo
"Nauka," 1973. 359 p. Bibliography, p. 352-357 (Akademiia Nauk SSSR. Labor-
atoriia Lesovedeniia).   1575

L'KOV, Fedor Nikolaevich. Vozdeistvie rel'efa na rastitel'nost' i zhivotnyi
mir: biogeomorfologicheskii ocherk (The influence of relief on vegetation and
animal life: a biogeomorphological study). Moskva, Geografgiz, 1953. 164 p.
Bibliography, p. 155-163.   1576

    210 references.

Regional: U. S. S. R. as a Whole

ADEMIIA Nauk SSSR. Institut Geografii. Resursy biosfery na territorii SSSR.
Nauchnye osnovy ikh ratsional'nogo ispol'zovaniia i okhrany (Resources of the
biosphere on the territory of the USSR. Scientific foundations of their rational
utilization and conservation). I. P. Gerasimov, ed. Moskva, Izdatel'stvo "Nauka,"
1971. 295 p. Bibliography, p. 285-293. (Akademiia Nauk SSSR. Institut Geografii).   1577

    186 references in alphabetical order.

RNAEV, S. F. Lesorastitel'noe raionirovanie SSSR (Forest-vegetation regionalization of the USSR). Moskva, Izdatel'stvo "Nauka," 1973. 203 p. Bibliography,
p. 191-202 (Akademiia Nauk SSSR. Laboratoriia Lesovedeniia).   1578

RESURSY biosfery na territorii SSSR. Nauchnye osnovy rational'nogo ispol'zovaniia i okrany [The resources of the biosphere on the territory of the USSR: the scientific foundations of rational utilization and protection]. Moskva, 1968. Bibliography, p. 454-464.   15

89 entries.

PAVLOV, Nikolai Vasil'evich. Botanicheskaia geografiia SSSR (Botanical geography of the USSR). Alma-Ata, Izdatel'stvo Akademii Nauk Kazakhskoi SSR, 1948. 704 p. Bibliography, p. 642-666 (Akademiia Nauk Kazakhskoi SSR).   15

About 1,000 references in alphabetical order.

RASTITEL'NOST' SSSR. (Vegetation of the USSR). Moskva-Leningrad, Izdatel'stvo Akademii Nauk SSSR, 1940. v. 1. 664 p. v. 2. 576 p. (Akademiia Nauk SSSR. Botanicheskii Institut).   15

Among the more significant bibliographies are those for the following articles: E. M. Lavrenko, History of the flora and vegetation of the USSR on the basis of the data on the current distribution of vegetation (about 120 references); A. P. Shennikov, Meadow vegetation of the USSR (about 230 references); E. M. Lavrenko, Steppes of the USSR (about 440 references); and A. M. Prozorovskii, Deserts and semideserts of the USSR (about 220 references.

AKADEMIIA Nauk SSSR. Botanicheskii Institut. Flora SSSR. Bibliograficheskaia spravka, I. A. Linchevskii. In the book: Novosti sistematiki vysshikh rastenii. Moskva-Leningrad, Izdatel'stvo "Nauka," 1966. p. 316-330. (Botanicheskii Institut imeni Komarova).   15

BYKOV, Boris Aleksandrovich. Dominanty rastitel'nogo pokrova Sovetskogo Soiuza (Dominants in the vegetation of the USSR). Alma-Ata, Izdatel'stvo Akademii Nauk Kazakhskoi SSR, 1960-1965. 3 v. Bibliographies, v. 1, 1960, p. 295-309, v. 2. 1962, p. 402-423, v. 3, 1965, p. 417-442.   15

About 1,550 references: about 400 references on the vegetation of the USSR in general, in v. 1; about 500 references on bogs, meadows and steppes of the USSR, in v. 2; and about 650 references on forests of the USSR. in v. 3.

KUZIAKIN, Aleksandr Petrovich. Zoogeografiia SSSR (Zoogeography of the USSR), Moskovskii Oblastnoi Pedagogicheskii Institut imeni N. K. Krupskoi, Uchenye zapiski, v. 109 (1962). Biogeografiia, no. 1, p. 3-182. Bibliography, p. 178-182.   15

More than 100 references in alphabetical order.

KIRIKOV, Sergei Vasil'evich. Izmeneniia zhivotnogo mira v prirodnykh zonakh SSSR: XIII-XIX vv. (Changes in the animals in the natural zones of the USSR, 13th-19th centuries). Moskva, Izdatel'stvo Akademii Nauk SSSR, 1959-1960. 2 v. Bibliography, v. 1, p. 161-174, v. 2, p. 142-156. (Akademiia Nauk SSSR. Institut Geografii).   15

About 770 references in alphabetical order in volume 1 for the steppe and forest steppe and in volume 2 for the forest zone and forest steppe.

NASIMOVICH, Andrei Aleksandrovich, and ARSEN'EV, V. A. Obzor issledovanii po zoologii pozvonochnykh v zapovednikakh SSSR za 30 let (Survey of research on the zoology of vertebrates in reservoirs of the USSR during 30 years). Nauchno-metodicheskie zapiski (Glavnoe Upravlenie po Zapovednikam Soveta Ministrov RSFSR), no. 10 (1948), p. 53-111; no. 13 (1949), p. 277-285.   15

About 790 references.

SIMOVICH, Andrei Aleksandrovich. Rol' rezhima snezhnogo pokrova v zhizni kopytnykh zhivotnykh na territorii SSSR (The role of the regime of snow cover in the life of hoofed mammals on the territory of the USSR), Moskva, Izdatel'-stvo Akademii Nauk SSSR, 1955. 404 p. Bibliography, p. 384-402 (Akademiia Nauk SSSR. Institut Geografii).   1587

About 400 references in alphabetical order on the distribution of hoofed mammals in the USSR and the influence of snow cover on their seasonal migration.

MINA, Rimma Petrovna. Zakonomernosti vertikal'nogo rasprostraneniia mlekopitaiushchikh, na primere Severnogo Tian'-Shania (Regularities in the vertical distribution of mammals, a case study of the Northern Tyan'-Shan'). Moskva, Izdatel'stvo "Nauka," 1964. 158 p. Bibliography, p. 151-157.   1588

About 200 references in alphabetical order on the ecological and geographical characteristics of mammals in the mountains of Soviet Middle Asia, particularly in the Tyan'-Shan'.

European U. S. S. R. (See also 2239)

KHOMIROV, Boris Anatol'evich. Sovremennoe sostoianie rastitel'nogo pokrova Krainego Severa SSSR i ocherednye problemy ego izucheniia (Current state of vegetation of the Far North of the USSR and immediate problems of its study), Botanicheskii zhurnal, v. 40, no. 4 (1955), Bibliography, p. 524-527.   1589

About 160 references in alphabetical order.

_____ . Bezlesie tundry, ego prichiny i puti preodoleniia (Treelessness of the tundra, its causes and paths of overcoming it), Moskva-Leningrad, Izdatel'stvo Akademii Nauk SSSR, 1962, 89 p. Bibliography, p. 80-88. (Botanicheskii Institut imeni Komarova).   1590

About 220 references.

OCHVY i rastitel'nost' vostochnoevropeiskoi lesotundry (Soils and vegetation of the East European wooded tundra). Leningrad, Izdatel'stvo "Nauka," Leningradskoe Otdelenie, 1972. 336 p. Bibliography, p. 328-334. (Akademiia Nauk SSSR. Nauchnyi Sovet po Probleme "Biologicheskie Osnovy Ratsional'nogo Ispol'zovaniia, Preobrazovaniia i Okhrany Rastitel'nogo Mira"). Rastitel'nost' Krainego Severa SSSR i ee Osvoenie, B. A. Tikhomirov, ed., no. 2. Opyt Statsionarnogo izucheniia pochv.-rastit. kompleksov lesotundry, part 2.   1591

OLOGIIA i biologiia rastennii vostochno-evropeiskoi lesotundry (Ecology and biology of vegetation of the East-European wooded tundra). Chast' 1. B. N. Norin, A. V. Druzin, I. V. Ignatenko, and others. Leningrad, Izdatel'stvo "Nauka," Leningradskoe Otdelenie. 1970. 356 p. Bibliography, p. 341-352. (Akademiia Nauk SSSR. Nauchnyi Sovet po Probleme "Biologicheskie Osnovy Ratsional'nogo Ispol' zovaniia, Preobrazovaniia i Okhrany Rastitel'nogo Mira"). Rastitel'nost' Krainego Severa SSSR i ee Osvoenie. Opyt statsionarnogo izucheniia pochv.-rastit. kompleksov lesotundry.   1592

OROB'EV, D. V. Tipy lesov Evropeiskoi chasti SSSR (Types of forests of the European part of the USSR), Kiev, Izdatel'stvo Akademii Nauk Ukrainskoi SSR, 1953. 452 p. Bibliography, p. 441-450 (Akademiia Nauk Ukrainskoi SSR. Institut Lesovodstva).   1593

About 330 references in alphabetical order.

UPPE, N. A. Literatura o lesakh Moskovskoi oblasti (Literature on the forests of Moscow oblast). In the book: Lesnoe khoziaistvo Moskovskoi oblasti (Forestry in Moscow oblast). Moskva, Izdatel'stvo Ministerstva Sel'skogo Khoziaistva RSFSR, 1961, p. 116-127.   1594

About 250 references in alphabetical order.

POVARNITSYN, Volodymyr Oleksiiovych. Lisy Ukraïnskoho Polissia (Forests of the
Ukrainian Poles'ye). Kiev, Vid-vo AN Ukraïnskoï RSR, 1959. 208 p. Bibliography,
p. 199-207 (Akademiia Nauk Ukraïnskoï RSR. Instytut Botaniky). In Ukrainian.
Russian title: Povarnitsyn, V. A. Lesa Ukrainskogo Poles'ia. Kiev. (Akademiia
Nauk Ukrainskoi SSR. Institut Botaniki).   15⁰

   About 220 references in alphabetical order in Russian, Ukrainian, and foreign
languages.

ABATUROV, Anatolii Mikhailovich. Poles'ia Russkoi ravniny v sviazi s problemoi
ikh osvoeniia. (Woodlands of the Russian plain in relation to the problems of
their utilization). Moskva, Izdatel'stvo "Mysl'," 1968. 246 p. Bibliography,
p. 235-244.   15⁰

   About 200 references.

KOMAROV, N. F. Etapy i faktory evoliutsii rastitel'nogo pokrova chernozemnykh
stepei (Stages and factors in the evolution of the vegetation of the black-
earth steppe). Moskva, Geografgiz, 1951. 328 p. Bibliography, p. 310-326
(Geograficheskoe Obshchestvo SSSR, Zapiski, n.s. tom 13).   15⁰

   More than 400 references in alphabetical order, about 340 Soviet and about
60 in foreign languages, for the period 1771-1946.

NOSOVA, L. M. Floro-geograficheskii analiz severnoi stepi Evropeiskoi chasti SSSR
(Floristic-geographic analysis of the northern steppe of the European part of
the USSR). Moskva, Izdatel'stvo "Nauka," 1973. 187 p. Bibliography, p. 178-
186 (Akademiia Nauk SSSR. Vsesoiuznoe Botanicheskoe Obshchestvo).   15⁰

RASTITEL'NYI pokrov Belorussii (Vegetation of Belorussia). V. S. Gel'tman, N. F.
Lovchii, L. P. Smoliak, and others. Minsk, "Nauka i Tekhnika," 1969. 175 p.
Bibliography, p. 162-173. (Akademiia Nauk Belorusskoi SSR. Institut
Eksperimental'noi Botaniki. Otdel Geobotaniki).   15⁰

GADZHIEV, Vagid Dzhalal ogly. Vysokogornaia rastitel'nost' Bol'shogo Kavkaza
(v predelakh Azerbaidzhana) i ee khoziaistvennoe znachenie (The high-mountain
vegetation of the Great Caucasus within the Azerbaydzhan SSR and its economic
significance). Baku, "Elm," 1970. 282 p. Bibliography, p. 264-274. (Akademiia
Nauk Azerbaidzhanskoi SSR. Institut Botaniki imeni V. L. Komarova).   16⁰

Siberia and the Soviet Far East

SHUMILOVA, L. V. Botanicheskaia geografiia Sibiri (Botanical geography of Siberia).
L. P. Sergievskaia, ed. Tomsk, Izdatel'stvo Tomskogo Universiteta, 1962. 439 p.
Bibliography, p. 383-425 (Tomskii Gosudarstvennyi Universitet).   16⁰

   About 1,000 references in alphabetical order.

RASTITEL'NYE resursy Sibiri i Dal'nego Vostoka: biulleten' otechestvennoi literatury
(Vegetation resources of Siberia and the Soviet Far East: bulletin of native
literature). Novosibirsk, 1963- quarterly. (Akademiia Nauk SSSR. Sibirskoe
Otdelenie. Gosudarstvennaia Publichnaia Nauchno-Tekhnicheskaia Biblioteka.
Nauchno-Bibliograficheskii Otdel).   16⁰

   Systematic, partly annotated, current bibliography, including sections on
vegetation, geobotany, botanical mapping, the ecology of plant communities, and
vegetation resources. Indexes for names and places from 1964.

SHIMANIUK, A. P. Sosnovye lesa Sibiri i Dal'nego Vostoka: lesovodstvennaia
kharakteristika (Pine forests of Siberia and the Soviet Far East: forestry
characteristics). Moskva, Izdatel'stvo Akademii Nauk SSSR. 1962. 187 p.
Bibliography, p. 173-185 (Akademiia Nauk SSSR. Sibirskoe Otdelenie. Institut
Lesa i Drevesiny).   16⁰

   About 350 references in alphabetical order.

1595 - 1603

OB'EVA, T. A., comp. Rastitel'nost' i rastitel'nye resursy Zapadnoi Sibiri. ibliografiia 1909-1962 gg. (Vegetation and vegetational resources of West iberia. Bibliography, 1909-1962). Moskva, "Nauka," 1964. 154 p. (Akademiia auk SSSR. Sibirskoe Otdelenie. Gosudarstvennaia Publichnaia Nauchno-ekhnicheskaia Biblioteka. Bibliograficheskii Otdel). With the participation f V. P. Sokolovaia and A. A. Konograi.   1604

Systematic bibliography, partly annotated. Index of names and places.

LOV, Georgii Vasil'evich. Lesa Zapadnoi Sibiri: istoriia izucheniia, tipy esov, raionirovanie, puti ispol'zovaniia i uluchsheniia (Forests of Western iberia: history of study, types of forests, regionalization, and lines of tilization and improvement). Moskva, Izdatel'stvo Akademii Nauk SSSR, 1961. 55 p. Bibliography, p. 236-254 (Akademiia Nauk SSSR. Sibirskoe Otdelenie. iologicheskii Institut).   1605

About 700 references in alphabetical order.

ROVSKII, Vul'f Abramovich, ed., and others. Mlekopitaiushchie Iakutii (Mammals f Yakutia). Moskva, Izdatel'stvo "Nauka," 1971. 660 p. Bibliography, p. 643-58 (Akademiia Nauk SSSR. Sibirskoe Otdelenie. Iakutskie Filial. Institut iologii).   1606

About 525 references.

ILLQV, Fëdor Nikolaevich. Ryby Iakutii (Fishes of Yakutia). Moskva, Izdatel'stvo Nauka," 1972. 360 p. Bibliography, p. 348-359. (Akademiia Nauk SSSR. Sibirskoe tdelenie. Iakutskii Filial. Institut Biologii).   1607

About 400 references.

OB'ËV, Konstantin Aleksandrovich. Ptitsy Iakutii (Birds of Yakutia). Moskva, zdatel'stvo Akademii Nauk SSSR, 1963. 336 p. Bibliography, p. 329-335 (Akademiia auk SSSR. Sibirskoe Otdelenie. Iakutskii Filial).   1608

About 250 references.

Kazakhstan

LOV, Nikolai Vasil'evich. Literaturnye istochniki po flore i rastitel'nosti azakhstana (Textual sources for the flora and vegetation of Kazakhstan). oskva-Leningrad, Izdatel'stvo Akademii Nauk SSSR, 1940. 182 p. (Akademiia auk SSSR. Kazakhskii Filial. Trudy, vypusk 19).   1609

2,377 entries in alphabetical order, partly annotated, up to 1936. Subject nd systematic indexes.

PAKHIN, Viktor Mikhailovich. Prirodnoe raionirovanie Kazakhstana: dlia tselei el'skogo khoziaistva (Natural regionalization of Kazakhstan for purposes of griculture). Alma-Ata, Izdatel'stvo "Nauka," 1970. 263 p. Bibliography, p. 55-261 (Akademiia Nauk Kazakhskoi SSR. Sektor Fizicheskoi Geografii).   1610

145 references.

:RATURA i materialy o lesakh Kazakhstana (Literature and materials on the forests f Kazakhstan). In the book: Gudochkin, M. V. and Chaban, P. S. Lesa Kazakhstana 'orests of Kazakhstan). Alma-Ata, 1958. Annex 12, p. 302-321.   1611

About 450 references systematically arranged by forest zones.

IOV, Vsevolod Viacheslavovich. Stepi zapadnogo Kazakhstana v sviazi s dinamikoi :h pokrova (The steppes of western Kazakhstan in relation to the dynamics of ieir cover). Moskva-Leningrad, Izdatel'stvo Akademii Nauk SSSR, 1958. 288 p. .bliography, p. 255-280 (Geograficheskoe Obshchestvo SSSR, Zapiski, n.s. tom 17).   1612

About 750 references in alphabetical order.

KARAMYSHEVA, Zoia Vladimirovna, and RACHKOVSKAIA, Ekaterina Ivanovna. Botanicheskaia
geografiia stepnoi chasti Tsentral'nogo Kazakhstana (Botanical geography of the
steppe section of Central Kazakhstan). Leningrad, Izdatel'stvo "Nauka," Leningradskoe Otdelenie, 1973. 278 p. Bibliography, p. 266-277. (Akademiia Nauk
SSSR. Botanicheskii Institut imeni V. L. Komarova). 16

AFANAS'EV, Aleksandr Vasil'evich. Zoogeografiia Kazakhstana; na osnove rasprostraneniia mlekopitaiushchikh (Zoogeography of Kazakhstan on the basis of the distribution of mammals). Alma-Ata, Izdatel'stvo Akademii Nauk Kazakhskoi SSR, 1960.
259 p. Bibliography, p. 227-244 (Akademiia Nauk Kazakhskoi SSR. Institut
Zoologii). 16

About 475 references in alphabetical order, about 415 Soviet and about 60
in foreign languages.

Soviet Middle Asia

KOROVIN, Evgenii Petrovich. Rastitel'nost' Srednei Azii i Iuzhnogo Kazakhstana
(Vegetation of Soviet Middle Asia and southern Kazakhstan). 2nd ed. v. 2.
Tashkent, Izdatel'stvo Akademii Nauk Uzbekskoi SSR, 1962. 547 p. Bibliography,
p. 436-481 (Akademiia Nauk Uzbekskoi SSR. Institut Botaniki). Summary in
English. 16

About 1,400 references in alphabetical order.

MATERIALY k bibliografii po flore i rastitel'nosti Tsentral'noi Azii (Materials
for a bibliography on the flora and vegetation of Central Asia outside the USSR).
D. V. Lebedev, ed. In the book: Rasteniia Tsentral'noi Azii. Po materialam
Botanicheskogo Instituta imeni V. I. Komarova (Plants of Central Asia, based
on the material in the Komarov Botanical Institute). v. 1. V. I. Grubov.
Introduction. Ferns. Bibliography. Moskva-Leningrad. Izdatel'stvo Akademii
Nauk SSSR, 1963. 167 p. Bibliography, p. 99-166 (Akademiia Nauk SSSR.
Botanicheskii Institut imeni V. L. Komarova). 16

About 850 references in alphabetical order, about 255 Soviet and about 625
in foreign languages.

DEVIATKINA, A. V. Rastitel'nyi i zhivotnyi mir Uzbekistana. Bibliograficheskii
ukazatel', 1917-1952. (Vegetation and animal life of Uzbekistan: bibliographical
guide, 1917-1952), Tashkent, "Fan," 1966. 468 p. (Akademiia Nauk Uzbekskoi SSR.
Fundamental'naia Biblioteka. Priroda i Prirodnye Resursy Uzbekistana, vypusk 3). 16

Flora and vegetation, p. 7-266 and animal life, p. 267-433. Systematic,
partly annotated bibliography.

RASTITEL'NYI pokrov Uzbekistana i puti ego ratsional'nogo ispol'zovaniia (The
vegetation of Uzbekistan and paths of its rational utilization). K. Z. Zakirov,
and others, eds. Tashkent, "Fan," 1971- v. 1- (Akademiia Nauk Uzbekskoi SSR.
Institut Botaniki). 16

ZAKIROV, P. K. Botanicheskaia geografiia nizkogorii Kyzylkuma i khrebta Nuratau
(Botanical geography of the lower mountains of the Kyzylkum and the Nuratau
range). Tashkent, Izdatel'stvo "Fan," 1971. 203 p. Bibliography, p. 187-202
(Akademiia Nauk Uzbekskoi SSR. Institut Botaniki). 16

ZAKHIDOV, T. Z. Biotsenozy pustynia Kyzylkum: opyt ekologo-faunistichego analiza
i sinteza (Biocenosis of the Kyzylkum desert: an attempt at an ecological-
faunistic analysis and synthesis). Tashkent, Izdatel'stvo "Fan," 1971. 303 p.
Bibliography, p. 294-301. (Akademiia Nauk Uzbekskoi SSR. Natsional'nyi Komitet
po Mezhdunarnoi Biologicheskoi Programme. Tashkentskii Gosudarstvennyi Universitet imeni V. I. Lenina. Institut Zoologii i Parazitologii AN Uzbekskoi SSR).
Summary in English. 16

DIN, Leonid Efimovich. Rastitel'nost' pustyn'Zapadnoi Turkmenii (Vegetation of the deserts of Western Turkmenia). Moskva-Leningrad, Izdatel'stvo Akademii Nauk SSSR, 1963. 309 p. Bibliography, p. 273-294 (Akademiia Nauk SSSR. Botanicheskii Institut imeni V. L. Komarova). 1621

About 670 references in alphabetical order.

ANIUKOVICH, Kirill Vladimirovich. Rastitel'nost' vysokogorii SSSR (Vegetation of the high mountains of the USSR). v. 1. [Dushambe], Izdatel'stvo Akademii Nauk Tadzhikskoi SSR, 1960. 169 p. Bibliography, p. 140-167 (Akademiia Nauk Tadzhikskoi SSR. Sovet po Izucheniiu Proizvoditel'nykh Sil Respubliki, Trudy, tom 1). 1622

About 800 references in alphabetical order, about 740 Soviet and about 60 in foreign languages.

Regions Outside the Soviet Union

ADEMIIA Nauk SSSR. Institut Geografii. Sovremennoe sostoianie prirodnoi sredy (biosfery) na territorii Evropy i puti ee sokhraneniia i uluchsheniia (Present status of the natural environment, or the biosphere, in Europe and lines of its conservation and improvement). I. P. Gerasimov and A. A. Nasimovich, eds. Vil'nius, 1972. 263 p. Bibliography, p. 257-262. 1623

About 150 references in all the principal languages of Europe, arranged alphabetically by name of author, first for languages in Cyrillic alphabets and then for languages in Latin alphabets.

VRENKO, Evgenii Mikhailovich. Osnovnye cherty botanicheskoi geografii pustyn' Evrazii i Severnoi Afriki (Basic features of the botanical geography of the deserts of Eurasia and North Africa). Moskva-Leningrad, Izdatel'stvo Akademii Nauk SSSR, 1962. 169 p. Bibliography, p. 156-168 (Akademiia Nauk SSSR. Botanicheskii Institut imeni V. L. Komarova, Komarovskie chteniia, 15). 1624

About 300 references in alphabetical order, about 225 Soviet and about 75 in foreign languages.

ROVSKAIA, M. V. Ukazatel' literatury po biologicheskim issledovaniiam Sovetskoi antarkticheskoi ekspeditsii (Index to literature on biological research of Soviet Antarctic expeditions), Issledovaniia fauny morei. (Akademiia Nauk SSSR. Zoo-Logicheskii Institut) v. 6, no. 14 (1968), p. 255-270. 1625

401 references.

MEDICAL GEOGRAPHY

FERATIVNYI zhurnal. Otdel'nyi vypusk 36: meditsinskaia geografiia (Reference
journal: Separate issue no. 36: medical geography) (Vsesoiuznyi Institut Nauchnoi
i Tekhnicheskoi Informatsii). Moskva. 1962- . monthly . Annual author, subject,
and geographical indexes.   1626
    Extensive coverage of the literature in all languages with abstracts.

OGI nauki i tekhniki: seriia meditsinskaia geografiia (Results of science and
technology: medical geography). (Vsesoiuznyi Institut Nauchnoi i Tekhnicheskoi
Informatsii). Moskva, VINITI, 1- (1967-).   1627
    1. Meditsinskaia geografiia, 1964 (Medical geography, 1964). A. D. Lebedev,
M. A. Savina, eds. 1966. 284 p. 696 references.

    2. _____ . 1966. A. D. Lebedev, ed. 1968. 321 p. 707 references.

    3. _____ . 1967. A. D. Lebedev, ed. 1968. 204 p. 291 references.

    4. _____ . A. D. Lebedev, ed. 1971. 244 p.

    5. _____ . A. D. Lebedev, ed. 1972. 164 p.

    6. _____ . 1973

OKHOROV, B. B. Nekotorye voprosy teorii meditsinskoi geografii (Some questions
of theory of medical geography). Institut Geografii Sibiri i Dal'nego Vostoka,
Doklady, no. 34 (1972), p. 63-73.   1628
    87 references.

LVICH, K. I. Bibliografiia meditsinskoi geografii Dal'nego Vostoka 1917-1968 gg.
Bibliography of medical geography of the Far East 1917-1968), Geograficheskoe
bshchestvo SSSR. Primorskii Filial. _Zapiski_, v. 27 (1968), p. 178-206.   1629

**EBOVICH, Igor' Aleksandrovich. Mediko-geograficheskaia otsenka prirodnykh
ompleksov. Na primere iuzhnykh raionov Srednei Sibiri** (Medical-geographical
valuation of natural complexes. The example of the southern regions of Middle
iberia). Leningrad, Izdatel'stvo "Nauka," Leningradskoe Otdelenie, 1972. 124 p.
ibliography, p. 114-123 (Akademiia Nauk SSSR. Sibirskoe Otdelenie. Institut
eografii Sibiri i Dal'nego Vostoka). Summary in English.   1630

**DROR, Iosif Solomonovich, DEMINA, D. M., and RATNER, E. M. Fiziologicheskie
rintsipy sanitarno-klimaticheskogo raionirovaniia territorii SSSR** (Physiological
rinciples of the medical-climatic regionalization of the USSR). Moskva,
Meditsina," 1974. 176 p. Bibliography, p. 170-175.   1631

**IMOV, Nikolai Nikolaevich. Ocherki po meditsinskoi geografii morei Sovetskogo
oiuza** (Essays on the medical geography of the seas of the Soviet Union). Lenin-
rad, Izdatel'stvo "Nauka," Leningradskoe otdelenie, 1973. 104 p. Bibliography,
. 96-103 (Akademiia Nauk SSSR. Geograficheskoe Obshchestvo SSSR).   1632

CONSERVATION AND UTILIZATION OF NATURAL RESOURCES

General

'OGI nauki i tekhniki: seriia geografiia: okhrana prirody i vosproizvodstvo prirodnykh
resursov (Results of science and technology: geography series: Conservation of
nature and reproduction of natural resources). (Vsesoiuznyi Institut Nauchnoi i
Tekhnicheskoi Informatsii). Moskva, VINITI, 1- (1968-). Irregular.   1633

   1. N. K. Deparma. Sokhranenie prirody za rubezhom. Obzor zarubezhnoi
profilirovannoi periodiki. 1966-1967 gg. (Preservation of nature abroad. Survey of key foreign periodicals). 1968. 159 p. List of 400 periodicals, 224
references.

   2. Chelovek i biosfera (Man and the biosphere).

PARMA, Nataliia Konstantinovna. Sokhranenie prirody za rubezhom. Obzor zarubezhnoi profilirovannoi periodiki 1966-1967 gg. The conservation of nature abroad;
a survey of the principal foreign periodicals. Moskva, VINITI, 1968. 160 p.
(Vsesoiuznyi Institut Nauchnoi i Tekhnicheskoi Informatsii. Itogi Nauki. Seriia
Geografiia.Okhrany prirody i vosproizdvodstvo prirodnykh resursov. 1966-1967 gg.,
vypusk 1).   1634

   Annotated list of about 400 current non-Soviet periodicals from 37 countries
in the field of conservation, arranged by continent and by countries, including
220 general works on conservation, 22 specialized dictionaries and reference
books, 23 periodicals on water and air pollution, and 16 periodicals on game.
List of sources. Alphabetical index of titles. Annotations. General textual
discussion. Titles in original languages and alphabets. Supplementary English
table of contents.

ARTS, A. K. Za Leninskoe otnoshenie k prirode: bibliograficheskii ukazatel'
(On the Leninist relationship to nature: bibliographical guide). Perm', TsBTI,
1960. 68 p. (Perm. sovnarkhoz. TsNTB).   1635

   About 700 entries for books and articles on the protection of forests and
planted green spaces, air, reservoirs, fishing, hunting, and game.

RAZHSKOVSKII, Iurii Nikolaevich. Ocherki prirodopol'zovaniia (Essays on the
utilization of nature). Moskva, Izdatel'stvo "Mysl'," 1969. 268 p. Bibliography,
p. 253-265.   1636

   About 300 references.

NTS, Aleksei Aleksandrovich. Ekonomicheskaia otsenka estestvennykh resursov:
nauchno-metodicheskie problemy ucheta geograficheskikh razlichii v effektivnosti
ispol'zovaniia (Economic appraisal of natural resources: scientific-methodological
problems of taking into account differences in the effectiveness of utilization).
Moskva, Izdatel'stvo "Mysl'," Geografgiz, 1972. 303 p. Bibliography, p. 278-
302. (Akademiia Nauk SSSR. Institut Geografii).   1637

   About 600 references arranged alphabetically by name of author.

RASIMOV, Innokenti Petrovich, ABRAMOV, Lev Solomonovich, and others, eds. Chelovek
obshchestvo i okruzhaiushchaia sreda: geograficheskie aspekty ispol'zovaniia
estestvennykh resursov i sokhraneniia okruzhaiushchei sredy (Man, society, and
the environment: geographical aspects of the utilization of natural resources and
conservation of the environment). Moskva, Izdatel'stvo "Mysl'," Geografgiz, 1973.
436 p. Bibliographies at ends of sections, p. 84-85, 152-154, 298-301, and 413-
417. (Akademiia Nauk SSSR. Institut Geografii).   1638

RASIMOV, Innokenti Petrovich, ed. Priroda i obshchestvo (Nature and society).
Moskva, Izdatel'stvo "Nauka," 1968. 346 p. Includes bibliographies. (Akademiia
Nauk SSSR. Nauchnyi Sovet po Filosofskim Voprosam Estestvoznaniia. Institut
Filosofii. Institut Geografii).   1639

LOPATINA, Elena Brunovna, and NAZAREVSKII, Oleg Rostislavovich. Otsenka prirodnykh
uslovii zhizni naseleniia (Appraisal of natural conditions in the life of people).
Moskva, Izdatel'stvo "Nauka," 1972. 148 p. Bibliography, p. 133-147 (Akademiia
Nauk SSSR. Institut Geografii).   16

About 220 references arranged in alphabetical order by name of author.

SOL'KINA, A. F., and GIMMEL'FARB, B. A. Osvoenie pustyn', polupustyn' i zasushlivykh
raionov: inostrannaia literatura, 1917-1950 gg. (Development of the deserts, semideserts, and arid regions: foreign literature, 1917-1950). Leningrad, 1951. 127 p
(Biblioteka Akademii Nauk SSSR. Bibliograficheskii Otdel. Seriia "V pomoshch'
velikim stroikam kommunizma, no. 15).   16

VORONTSOV, Aleksei Ivanovich, and KHARITONOVA, Nadezhda Zakharovna. Okhrana prirody
(Protection of nature). Moskva, "Vysshaia Shkola," 1971. 359 p. Bibliography,
p. 347-356.   16

VOEIKOV, Aleksandr Ivanovich. Vozdeistvie cheloveka na prirodu (Influence of man
on nature). Moskva, Izdatel'stvo Akademii Nauk SSSR. 1963. 251 p. Bibliography,
p. 247-251.   16

    90 references to work of Voeikov, published 1874-1921, compiled by V. V.
Pokshishevskii.

U. S. S. R.

PARKHOMENKO, Iia Ivanovna, ASLANOVA, G. A., and GAL'TSEVA, T. V. Geograficheskaia
literatura po prirodnym usloviiam i estestvennym resursam SSSR,1968-1971: bibliografiia (Geographical literature on natural conditions and natural resources of
the USSR, 1968-1971). Itogi nauki i tekhniki. Geografiia SSSR, tom 9, Geograficheskoe izuchenie prirodnykh resursov i voprosy ikh ratsional'nogo ispol'-
zovaniia (Results of science and technology. Geography of the USSR, v. 9,
Geographical study of natural resources and questions of their rational utilization). I. I. Parkhomenko, ed. Moskva, VINITI, 1973, p. 168-365. Supplementary
table of contents and abstracts in English.   16

    2,730 entries organized in five main sections: (1) general and theoretical
problems (82 entries); (2) territorial combinations of natural resources (206
entries); (3) the geographical study of individual types of natural resources
(land, mineral raw materials, water, vegetation, and recreational) (2067 entries);
(4) natural conditions of production and human life (118 entries); (5) appraisal
of natural conditions and natural resources (189 entries), with additional sections
on Soviet papers translated abroad (20 entries); and addenda (48 entries). Index
of authors. Associated articles on the Soviet study of land resources, mineral
resources, water resources, forest resources, recreation resources, on the develop
ment of geographical studies of resources in the USSR, and on economic appraisal
of natural resources.

PARKHOMENKO, Iia Ivanovna, and TERNOVSKAIA, G. N. Sovetskaia literatura po voprosam
ekonomicheskoi otsenki prirodnykh uslovii i resursov, 1961-1967 gg. (Soviet
literature on questions of the economic appraisal of natural conditions and resources, 1961-1967), in the book: Mints, Aleksei Aleksandrovich, Ekonomicheskaia
otsenka prirodnykh resursov i uslovii proizvodstva. (Economic appraisal of natura
resources and conditions of production), Itogi nauki: geografiia SSSR, no. 6
Moskva, VINITI, 1968, p. 101-140.

    Systematic guide to the literature, both books and articles in periodicals,
492 entries (reviews noted). Citation number is given for abstracts of each entry
in Referativnyi zhurnal: geografiia. Author index.

RYKIN, Kirill Viacheslavovich, and LEBEDEV, P. N. Zemel'nye resursy SSSR i ikh
hoziaistvennoe ispol'zovanie (The land resources of the USSR and their economic
tilization), in Itogi nauki: geografiia SSSR, no. 1, Zemel'nye resursy i lesnye
esursy SSSR. Moskva, VINITI, 1965, p. 7-87. Bibliography, p. 54-87.   1645

    615 references organized by general literature on land resources of the USSR
95 entries) and by literature for individual natural areas: cold predominantly
oist regions (tundra-taiga), 30 entries: temperate moist continental mainly
Lains regions (taiga-forests), 124 entries; temperate wet-monsoon regions of the
ar East, 10 entries; temperate semiarid and arid plains regions (Dnepr-Irtysh),
48 entries; temperate semiarid regions of Southern Siberia, 49 entries; temperate
nd hot semiarid and arid regions of the Caspian and Middle Asia, 130 entries;
nd temperate and hot regions of irregular moisture in the Caucasus, 29 entries.
ithin each section arranged alphabetically by author. Citation number is given
or abstract of each entry in Referativnyi zhurnal: geografiia.

ERATIVNYI zhurnal: geografiia (Reference journal: geography) (Vsesoiuznyi
nstitut Nauchnoi i Tekhnicheskoi Informatsii). Moskva, 1954- . monthly. E.
eografiia SSSR: geografiia prirodnykh resursov i prirodnye usloviia proizvodstva
Geography of the USSR: geography of natural resources and natural conditions of
roduction).   1646

ASIMOV, Innokenti Petrovich, ed. Resursy biosfery na territorii SSSR: nauchnye
snovy ikh ratsional'nogo ispol'zovaniia i okhrany (Resources of the biosphere in
ne USSR: scientific foundations of their rational utilization and conservation).
oskva, Izdatel'stvo Akademii Nauk SSSR, 1971. 295. Bibliography, p. 285-293.
Akademiia Nauk SSSR. Institut Geografii).   1647

SIMOV, Innokenti Petrovich, ARMAND, David L'vovich, and EFRON, K. M., eds.
rirodnye resursy Sovetskogo Soiuza, ikh ispol'zovanie i vosproizvodstvo (Natural
esources of the Soviet Union: their utilization and regeneration). Moskva,
tdatel'stvo Akademii Nauk SSSR, 1963. 243 p. Bibliographies at ends of articles.
Akademiia Nauk SSSR. Institut Geografii). Available in English as Natural re-
ources of the Soviet Union: their use and renewal. Translated by Jacek I.
omanowski. English edition edited by W. A. Douglas Jackson. San Francisco,
alifornia, W. H. Freeman, 1971. 349 p. [2595].   1648

ANSKAIA, G. N., ed. Pravovye voprosy okhrany prirody v SSSR (Legal problems
f conservation in the USSR). Moskva, Gosiurizdat, 1963. 331 p. Bibliographical
otnotes. (Vsesoiuznyi Institut Iuridicheskikh Nauk).   1649

OVOI, Viktor L'vovich. Lesnye resursy SSSR i ikh ispol'zovanie (Forest resources
f the USSR and their utilization), in Itogi nauki: geografiia SSSR, no. 1,
emel'nye resursy i lesnye resursy SSSR. Moskva, VINITI, 1965, p. 90-150. Bib-
iography, p. 145-150.   1650

    144 entries, arranged alphabetically by author. Citation number is given
or abstract of each entry in Referativnyi zhurnal: geografiia.

ANZEN, Bodo Germanovich, and KRIVOSHCHEKOV, G. M. Novaia literatura po voprosam
khrany prirody Sibiri (za 1951-1960 gg.) (New literature on questions of the
onservation of nature in Siberia, 1951-1960). In the book: Okhrana prirody
ibiri i Dal'nego Vostoka, vypusk 1 (Conservation of nature in Siberia and the
ar East, no. 1). Novosibirsk, Akademiia Nauk SSSR, Sibirskoe Otdelenie, 1962,
. 258-287.   1651

    About 600 references, organized systematically.

MELIORATSIIA bolot i mineral'nykh zabolochennykh zemel'. Bibliograficheskii ukazatel' (Reclamation of bogs and mineralized boggy lands. Bibliographical guide). V. A. Zvereva, comp. Moskva, Izdatel'stvo Ministerstva Sel'skogo Khoziaistva RSFSR, 1959. v. 1. 131 p. v. 2, 156 p. (Respublikanskii Institut po Proektirovaniiu Vodokhoziaistva i Melioratsii Stroitel'stva. Gosudarstvennaia Nauchnaia Biblioteka Ministerstva Vysshego Obrazovaniia SSSR).

About 850 references in volume 1 and about 1,000 references in volume 2, mainly for the period 1945-1956.

PETROV, Mikhail Platonovich. Agrolesomelioratsiia peskov v pustyniakh i polupustyniakh Soiuza SSR: bibliografiia literatury na russkom iazyke 1768-1950 gg. (Agricultural and forest reclamation of sands in the deserts and semideserts of the USSR: bibliography of the literature in Russian, 1768-1950). D. L. Margolina, bibliographical editor. Ashkhabad, Izdatel'stvo Akademii Nauk Turkmenskoi SSR, 1952. 208 p. (Akademiia Nauk Turkmenskoi SSR. Biblioteka Akademiia Nauk SSSR. Leningradskii Gosudarstvennyi Pedagogicheskii Institut imeni M. N. Pokrovskogo).

852 annotated entries in alphabetical order. Introductory essay. Geographical index. Subject index. List of abbreviations with full names of journals and collected volumes.

PETROV, Mikhail Platonovich. Obzor literatury po fitomelioratsii peskov pustyn' i polupustyn' Soiuza SSSR za 1951 g., (Survey of the literature on the plant reclamation of sands of deserts and semideserts of the USSR in 1951). Akademiia Nauk Turkmenskoi SSR, Izvestiia, 1952, no. 2, p. 80-86. 62 entries.

_____ . _____ . 1952. _____ , 1953, no. 3, p. 81-94. 143 entries.

_____ . _____ . 1953. _____ , 1954, no. 2, p. 84-89. 68 entries.

_____ . _____ . 1954. _____ , 1955, no. 2, p. 77-83. 77 entries.

_____ . _____ . 1955. _____ , 1956, no. 2, p. 83-91. 77 entries.

_____ . _____ . 1956. _____ , 1957, no. 3, p. 138-143. 60 entries.

_____ . _____ . 1957. _____ , 1958, no. 2, p. 95-100.

_____ . _____ . 1958. _____ , 1959, no. 2, p. 91-95. 44 entries.

_____ . _____ . 1959. _____ , Seriia biologicheskikh nauk, 1960, no. 2, p. 74-81. 73 entries.

_____ . _____ . 1960. _____ . _____ , 1961, no. 4, p. 76-84. 89 entries.

_____ . _____ . 1961. _____ . _____ , 1962, no. 3, p. 87-94. 74 entries.

_____ . _____ . 1962. _____ . _____ , 1963, no. 5, p. 83-92. 111 entries.

_____ . _____ . 1963. _____ . _____ , 1964, no. 4, p. 83-93. 132 entries.

_____ . _____ . 1964. _____ . _____ , 1965, no. 5, p. 87-97. 132 entries.

_____ . _____ . 1965. _____ . _____ , 1966, no. 6, p. 79-87.

_____ . _____ . [1966-1967]. Problemy osvoeniia pustyn' (Akademiia Nauk Turkmenskoi SSR. Nauchnyi Sovet po Probleme Pustyn'), 1968, no. 2, p. 77-85; no. 4, p. 83-93. 357 entries.

_____ . _____ . 1968. _____ . 1969, no. 4, p. 83-91. 123 entries.

_____ . _____ . 1969. _____ . 1970, no. 4, p. 85-93.

_____ . _____ . 1970. _____ . 1971, no. 4, p. 88.

_____ . _____ . 1971. _____ . 1972, no. 5, p. 82-92.

_____ . _____ . 1972. _____ . 1973, no. 5, p. 81-90.

'BA s zasukhoi. Ukazatel' literatury po polezashchitnym lesonasazhdeniiam,
·avopol'noi sisteme zemledelia, erozii pochvy, stroitel'stvu prudov i vodoemov
stepnykh i lesostepnykh raionakh SSSR (The struggle with drought. Guide to
ıe literature on field protection, planting of woods, grassland agriculture,
·il erosion, the construction of ponds and reservoirs in the steppe and wooded
·gions of the USSR). Leningrad, Gosudarstvennaia Biblioteka imeni Saltykova-
ıchedrina, 1949. 200 p.   1655

ıND, David L'vovich, ed. Regional'nye sistemy protivoerozionnykh meropriiatii
·egional systems of measures against erosion). S. I. Sil'vestrov, E. A.
ɪronova, N. M. Stupina, and others. Moskva, Izdatel'stvo "Mysl'," 1972. 544 p.
.bliography, p. 531-542 (Akademiia Nauk SSSR. Institut Geografii).   1656

ːV, G. A., EIIUBOV, A. Dzh., AKHUNDOV, N. G.   Razvitie nauki ob okhrane prirody
Azerbaydzhanskoi SSR. Reziume. (Development of the science of conservation of
ıture in the Azerbaidzhan SSR. Summary), Akademiia Nauk Azerbaidzhanskoi SSR.
ıstitut Geografii. Trudy, v. 18 (1971), p. 406-439. Bibliography, p. 415-439.
                                                                            1657

                              Nature Reserves

OVSKII, Vadim Borisovich. Zapovedniki Sovetskogo Soiuza (Reserves of the Soviet
ion). A. G. Bannikov, ed. Moskva, Izdatel'stvo "Kolos," 1969. 552 p.   1658

    Annotated catalogue of nature reserves in the USSR, p. 539-549.

VEDNIKI i natsional'nye parki mira: kratkii spravochnik (Reserves and national
·ks of the world: a short handbook), by L. S. Belousova, V. A. Borisov, and A.
Vinokurov. L. K. Shaposhnikov, ed. Moskva, Izdatel'stvo "Nauka," 1969. 239 p.
                                                                            1659

                                                          1655 - 1659

B. ECONOMIC, POPULATION, AND SETTLEMENT GEOGRAPHY

ECONOMIC GEOGRAPHY

General Economic Geography

ERATIVNYI zhurnal: geografiia (Reference journal: geography) (Vsesoiuznyi
nstitut Nauchnoi i Tekhnicheskoi Informatsii). Moskva 1954- . monthly.   1660

References in economic geography occur in the regional sections: E. Geog-
aphy of the USSR (especially a separate section on geography of the economy and
ts branches); Zh. Geography of foreign Europe; I. Geography of foreign Asia and
frica; K. Geography of America, Australia, Oceania, and Antarctica. The regional
ections contain about 12,000 entries annually, the great majority in economic
eography.

3HKIN, Iulian Glebovich. Ekonomicheskaia geografiia: istoriia, teoriia, metody,
raktika (Economic geography: history, theory, methods, practice). Moskva,
zdatel'stvo "Mysl'," 1973. 559 p. Bibliography, p. 519-558.   1661

916 references on the field of economic geography, 801 in Cyrillic, 115 in
atin letters, each arranged alphabetically by name of author. Covers works of
pecial interest on the history of the field or its theory, methods, or practice,
specially as developed in the Soviet Union but with attention also to inter-
ational trends.

'S, Aleksei Aleksandrovich. Ekonomicheskaia geografiia: obzor osnovnykh
endentsii razvitiia v 1966-1970 gg. (Economic geography: a review of the
ain trends of its development in the years 1966-1970), in Itogi Nauki i Tekhniki,
eriia Teoreticheskie voprosy fizicheskoi i ekonomicheskoi geografii. v. 1.
oskva, VINITI, 1972, p. 89-138. Bibliography, p. 126-137. English summary,
. 137-138.   1662

250 references to Soviet work in economic geography 1966-1970, arranged
lphabetically by author.

3HISHEVSKII, Vadim Viacheslavovich, and STEPANOVA, Evgeniia Akimovna. Otrazhenie
ovetskikh ekonomiko-geograficheskikh rabot v zarubezhnykh izdaniiakh (Reflec-
_ons of Soviet work in economic geography in foreign publications). In the book:
konomicheskaia geografiia v SSSR: istoriia i sovremennoe razvitie (Economic
eography in the USSR: history and current development). N. N. Baranskii and
hers, eds. Moskva, Izdatel'stvo "Prosveshchenie," 1965, p. 621-653.   1663

About 420 references on work by foreign authors on Soviet economic geography
d original and translated works by Soviet authors published abroad, from the
d-1950's to 1964, classified by themes.

ROVA, B. T. Razvitie ekonomgeograficheskoi nauki v Azerbaidzhane. Reziume.
evelopment of economic geography in Azerbaydzhan. Summary). Akademiia Nauk
erbaidzhanskoi SSR. Institut Geografii. Trudy, v. 18 (1971), p. 248-279.
bliography, p. 258-279.   1664

I nauki: ekonomgeograficheskaia izuchennost' raionov kapitalisticheskogo mira
esults of science: economic-geographic coverage of the regions of the capitalist
rld). (Vsesoiuznyi Institut Nauchnoi i Tekhnicheskoi Informatsii). Moskva,
NITI, v. 1-3, 1964-1966.   1665

1. Iu. V. Medvedkov. Sostav istochnikov. Razlichiia v intensivnosti
sledovanii (Composition of sources. Diversity in intensity of research).
64. 82 p. List of the most important geographical journals, p. 31-38. 34
ferences.

2. Iu. V. Medvedkov. Prilozheniia matematiki v ekonomicheskoi geografii (Application of mathematics in economic geography). 1965. 162 p. 123 references

3. Iu. V. Medvedkov. Analiz konfiguratsii rasseleniia (Analysis of settlement patterns). 1966. 116 p. 78 references.

FEIGIN, Iakov Grigor'evich. Lenin i sotsialisticheskoe razmeshchenie proizvoditel'nykh sil (Lenin and socialist distribution of productive forces). Moskva, Izdatel'stvo "Nauka," 1969. 223 p. Bibliography, p. 212-221. (Akademiia Nauk SSSR. Institut Ekonomiki).

189 entries, about a fifth from Lenin, the others from well-known Soviet specialists.

AL'BRUT, Moisei Isaakovich. Marksistsko-leninskie teoreticheskie osnovy ekonomicheskoi geografii (Marxist-Leninist theoretical foundations of ekonomic geography). E. S. Solomentsev, ed. Chelyabinsk, Iuzhno-Ural'skoe Knizhnoe Izdatel'stvo, 1971-1973. 2 v. 1. 1971. 201 p. Bibliography, p. 197-200. v. 2. 1973. 156 p. Bibliography, p. 150-155. (Geograficheskoe Obshchestvo SSSR. Cheliabinskii otdel).

Economics and the Economy of the USSR as a Whole

POLITICHESKAIA ekonomiia. Istoriia ekonomicheskikh uchenii. Mezhdunarodnye otnosheniia. Mirovaia ekonomika. Kommunisticheskoe stroitel'stvo v SSSR (Political economy. History of economics studies. International relations. World economy. Communist construction in the USSR), Bibliografiia sovetskoi bibliografii (Bibliography of Soviet bibliography). 1939, 1946- . Annual. Moskva, Izdatel'stvo "Kniga," 1941, 1948- . (Vsesoiuznaia Knizhnaia Palata).

Annual list of Soviet bibliographies in economics and related fields. The 1970 annual volume (1972), p. 41-47, contains 124 main entries. In addition there are detailed sections on branches of the economy: engineering and industry; agriculture; transport; communications; trade; and municipal economy and daily needs, with some 2,627 entries in 1970.

TSAGOLOV, Nikolai Aleksandrovich. ed. Bibliografiia po voprosam politicheskoi ekonomiki 1917-1966 (Bibliography on problems of political economy, 1917-1966). Moskva, Izdatel'stvo Moskovskogo Universiteta, 1969. 552 p.

9,096 entries in Russian on economics organized systematically. Name index, p. 506-543 includes about 5,000 names, in alphabetical order.

GOSUDARSTVENNAIA Biblioteka SSSR imeni V. I. Lenina. Otdel Spravochno-Bibliograficheskoi i Informatsionnoi Raboty. Ekonomika SSSR: annotirovannyi perechen' otechestvennykh bibliografii opublikovannykh v 1917-1964 gg. (Economics of the USSR: an annotated list of native bibliographies published 1917-1964). V. E. Sivolgin, comp. A. D. Rklitskaia, R. V. Shistakovaia, eds. Moskva, Izdatel'stvo "Kniga," 1965. 159 p.

545 entries for bibliographies on the economy of the USSR including separately published works, running bibliographies, bibliographies published in serials or in collected works and bibliographies that are parts of books or articles. Organized systematically. Indexes.

AKADEMIIA Nauk SSSR. Fundamental'naia Biblioteka Obshchestvennykh Nauk imeni V. P. Volgina. Narodnoe khoziaistvo SSSR v 1917-1920 gg.: bibliograficheskii ukazatel' knizhnoi i zhurnal'noi literatury na russkom iazyke, 1917-1963 gg. (Economy of the USSR 1917-1920: bibliography of the book and periodical literature in the Russian language, 1917-1963). E. V. Bazhanova, comp. Moskva, Izdatel'stvo "Nauka," 1967. 619 p.

9,311 entries including books, articles, brochures, government publications, statistical materials, dissertations, partly annotated, arranged by historical sub-periods and by problems. List of sources utilized. Author and title index. Geographic index.

1666 - 1671

DNOE khoziaistvo SSSR v gody velikoi otechestvennoi voiny, iiun' 1941-mai 1945
.: bibliograficheskii ukazatel' knizhnoi i zhurnal'noi literatury na russkom
zyke, 1941-1968 gg: (Economy of the USSR in the years of World War II, June
41-May 1945: bibliography of book and journal literature in the Russian lan-
age, 1941-1968). E. B. Margolina, comp. E. V. Bazhanova, ed. Moskva, Izdatel'-
vo "Nauka," 1971. 460 p. (Akademiia Nauk SSSR. Institut Nauchnoi Informatsii
Fundamental'naia Biblioteka po Obshchestvennym Naukam).   1672

5409 entries arranged systematically. List of 29 sources utilized in the
mpilation of the bibliography, p. 390-391. Index of authors and titles of works
thout authors.

Economic Regionalization and Production Complexes

HISHEVSKII, Vadim Viacheslavovich. Ekonomicheskoe raionirovanie SSSR: obzor
vetskikh issledovanii problem ekonomicheskogo raionirovaniia za 1962-1964 gg.
conomic regionalization of the USSR: survey of Soviet research on the problem
economic regionalization in 1962-1964), in Itogi nauki: geografiia SSSR, no.
Ekonomicheskoe raionirovanie SSSR. Moskva, VINITI, 1965, p. 7-83. Bibliog-
phy, p. 46-82. English abstract, p. 82-83.   1673

529 references organized by history of economic regionalization in the USSR
l); general and theoretical questions of economic regionalization (74); problems
major regions (45); questions of small regions, arranged by general theoretical
rks or by regions (125); economic regionalization by branches of the economy,
:h regional breakdown for agriculture (149); relations of economic regionaliza-
n and regional planning; region-forming role of cities (39); other topics (16)
nslations of Soviet works into foreign languages (24); and supplementary
ries (46). Citation number is given for abstract of each entry in Referativnyi
rnal: geografiia. Author index.

IISHEVSKII, Vadim Viacheslavovich. Bibliografiia sovetskoi literatury po
nomicheskomu raionirovaniiu: obzor opublikovannykh bibliografii (Bibliog-
hy of Soviet literature on economic regionalization: survey of published
liographies). In Itogi nauki: geografiia SSSR, no 2, Ekonomicheskoe
onirovanie SSSR, Moskva, VINITI, 1965, p. 131-145. English abstract.
liography, p. 142-144.   1674
Lists 29 bibliographies of Soviet work on economic regionalization.

L'CHICH, Oleg Alekseevich. Sovremennaia literatura po ekonomicheskomu
ionirovaniiu, 1944-1958 gg. (Current literature on economic regionalization,
44-1958), Voprosy geografii, v. 47 (1959), p. 183-196.   1675

158 entries in Russian and other languages of the USSR for the period 1944-
58, and a supplement for 1958-1959 with 69 entries, for books, articles, and
mphlets.

PIEV, Petr Martynovich. Ekonomicheskoe raionirovanie SSSR (Economic region-
ization of the USSR). Moskva, Gosplanizdat, 1959. 263 p. Bibliography, p.
6-261 by N. K. Glabina. (Nauchno-Issledovatel'skii Ekonomicheskii Institut
splana SSSR).   1676

About 320 entries, arranged by major topics, and within divisions chrono-
gically, of works published in the USSR during the Soviet period up to 1958.

PIEV, Petr Martynovich. Ekonomicheskoe raionirovanie SSSR, kniga 2 (Economic
gionalization, of the USSR, book 2). Moskva, Ekonomizdat, 1963. 248 p. 1963,
236-246. Bibliography, p. 236-246 by P. M. Alampiev and N. K. Glabina.   1677

109 entries, organized systematically, and within sections chronologically,
rtly annotated, in the period 1958-1962.

KOLOTIEVSKII, Anton Mikhailovich. Voprosy teorii i metodiki ekonomicheskogo
raionirovaniia, v sviazi s obshchei teoriei ekonomicheskoi geografii (Problems
of theory and methods of economic regionalization in relation to general theory
in economic geography). Riga, Izdatel'stvo "Zinatie," 1967. 251 p. Bibliography, p. 222-247.(LatviiskiiGosudarstvennyi Universitet). In Russian with
supplementary English table of contents, p. 250-251.   1

     About 600 references arranged alphabetically by name of author, first those
in Cyrillic, then those in Latin letters.

KOLTUN, Mariia Isaakovna. Ekonomicheskoe raionirovanie Sovetskogo Soiuza i
dorevoliutsionnoi Rossii: istoriia i teoriia voprosa; bibliograficheskii ukazatel'
(Economic regionalization of the USSR and prerevolutionary Russia, history and
theory of the problem: bibliographical guide). V. V. Klevenskaia, ed. Moskva,
1959. 45 p. (Gosudarstvennaia Biblioteka SSSR imeni Lenina. Otdel Spravochno-
bibliograficheskoi Informatsionnoi Raboty).   1

     About 400 entries both pre- and post-revolutionary (from 1727), systematical-
ly arranged and within categories chronologically. Annotated. Author index.

PARKHOMENKO, Iia Ivanovna, and TERNOVSKAIA, G. N. Sovetskaia literatura po
proizvodstvenno-territorial'nym kompleksam, 1965-1969 gg (Soviet literature on
production-territorial complexes, 1965-1969), Itogi nauki: geografiia SSSR,
no. 8. Moskva, VINITI, 1970, p. 65-132. English abstract, p. 125.   1

     959 entries, organized systematically: the development of the concept of
production-territorial complexes in Soviet science (21), general and theoretical
aspects (120), complexes in the USSR (698) arranged regionally, economic planning and complexes (67), translations abroad of Soviet works (29), supplement
(24). Citation number is given for abstracts of each entry in Referativnyi
zhurnal: geografiia. Author index.

KISTANOV, Viktor Vasil'evich. Kompleksnoe razvitie i spetsializatsiia ekonomicheskikh raionov SSSR (Integrated development and specialization of economic regions
of the USSR). Moskva, Izdatel'stva "Nauka," 1968. 283 p. Bibliography, p. 275-
282. (Akademiia Nauk SSSR. Gosplan SSSR. Sovet po Izucheniiu proizvoditel'nykh
Sil).

     163 references.

GORBATOV, A. L. Agrarno-promyshlennye kompleksy i ob"edineniia: bibliograficheskii
ukazatel' otechestvennoi literatury za 1960-1973 gg.i inostrannoi za 1962-1973 gg
(Agrarian-industrial complexes and combines: bibliography of native literature
1960-1973 and of foreign literature 1962-1973). 2nd ed. Moskva, 1973. 190 p.
(Vsesoiuznaia Akademiia Sel'skokhoziaistvennykh Nauk imeni V. I. Lenina. Tsentral'
naia Nauchnaia Sel'skokhoziaistvennaia Biblioteka).

KALASHNIKOVA, Tat'iana Mikhailovna, ed. Territorial'nye sistemy proizvoditel'nykh
sil (The territorial system of productive forces). Moskva, Izdatel'stvo "Mysl',"
Geografgiz, 1971. 437 p. Bibliography, p. 427-436.

     About 200 references on the areal organization and regionalization of econom
resources and production.

TEORETICHESKIE osnovy funktsional'noi struktury promyshlennogo kompleksa ekonomiche
skogo raiona (Theoretical foundations of the functional structure of the industrial complex of an economic region). M. M. Palamarchuk, ed. Kiev, "Nauk.
Dumka," 1972. 240 p. Bibliography, p. 234-239. (Akademiia Nauk Ukrainskoi SSR,
Sektor Geografii).

Agriculture

ACHEVSKAIA, Elena Nikolaevna, comp. Bibliografiia po raionirovaniiu, razmesh-
heniiu i spetsializatsii sel'skogo khozizistva, 1960-1966 gg. (Bibliography
n the regionalization, distribution, and specialization of agriculture, 1960-
966). Moskva, Izdatel'stvo "Nauka," 1970. 161 p. (Akademiia Nauk SSSR.
osplan SSSR. Sovet po Izucheniiu Proizvoditel'nykh Sil).   1685

   About 1,600 entries in Russian, arranged regionally by major regions in the
SFSR and by union republics. Index of authors, editors, and titles of books
ithout designated authors.

ACHEVSKAIA, Elena Nikolaevna. comp. Bibliografiia po raionirovaniiu i razmesh-
heniiu sel'skogo khoziaistva SSSR, 1818-1960 (Bibliography on the regionaliza-
ion and distribution of agriculture of the USSR, 1818-1960). Moskva, Izdatel'-
tvo Akademii Nauk SSSR, 1961. 199 p. (Akademiia Nauk SSSR. Sovet po Izucheniiu
roizvoditel'nykh Sil. Sektor Seti Spetsial'nykh Bibliotek).   1686

   More than 2,200 entries arranged regionally. Index of authors, editors,
nd titles of books without designated authors.

SHKIN, Iulian Glebovich. Geograficheskie ocherki prirody i sel'skokhoziaistvennoi
eiatel'nosti naseleniia v razlichnykh raionakh Sovetskogo Soiuza (Geographical
tudies of natural conditions and agricultural activity of the population in
arious regions of the Soviet Union). Moskva, OGIZ Geografgiz, 1947. 423 p.
Moskovskii Gosudarstvennyi Pedagogicheskii Institut imeni V. I. Lenina. Uchënye
apiski, tom XXXVIII. Geograficheskii Fakul'tet. Kafedra Ekonomicheskoi Geo-
rafii).   1687

   566 references at ends of chapters, p. 65-66, 108-110, 152-155, 191-192,
66-272, 312-315, 344-345, 371-372, and 421-422, on agriculture under diverse
hysical conditions of the Soviet Union: on permafrost, on meadow strips, in the
orthern forests, on sandy marshes in Poles'ye, in lacustrine basins and river
valleys, on eroded black soils, in irrigated oases in lowlands, dry farming on
ountain slopes, and irrigated high-mountain valleys.

'SKOKHOZIAISTVENNAIA literatura SSSR (Agricultural literature of the USSR),
Vsesoiuznaia Akademiia Sel'skokhoziaistvennykh Nauk im. V. I. Lenina. Tsentral'-
aia Nauchnia Sel'skokhoziaistvennaia Biblioteka). Moskva, 1948- . 1948-1959
0 nos. a year, 1959- monthly.   1688

   Systematic guide to the current literature, both books and articles. Author
index since 1954.

'SKOE khoziaistvo: rekomendatel'nyi ukazatel' literatury. Kniga 1. Ekonomika
i organizatsiia sel'skogo khoziaistva v SSSR. (Agriculture: guide to recommended
literature. Book 1. Economics and organization of agriculture in the USSR).
L. M. Vadikovskaia and G. K. Donskaia, comps. E. S. Oslikovskaia, ed. Moskva,
Gosudarstvennaia Biblioteka SSSR imeni V. I. Lenina, 1957. 104 p.   1689

   About 280 references, annotated.

ZVITIE i razmeshchenie sel'skogo khoziaistva v zarubezhnykh stranakh (Development and distribution of agriculture in foreign countries). N. A. Utenkov, and
others, eds. Moskva, 1971. 188 p. Bibliography, p. 165-188. (Sovet po Izucheniiu
Proizvoditel'nykh Sil pri Gosplane SSSR. Sektor Nauchnoi Informatsii. Razmeshchenie Proizvoditel'nykh Sil. Sbornik referativnoi informatsii, vypusk 18). 1690

VITAIA, Feofan Farneivich. Klimaticheskie zony vinograda v SSSR (Climatic zones
of vineyards in the USSR) 2nd ed. Moskva, Pishchepromizdat, 1948. 192. Bibliography, p. 184-192. (Vsesoiuznaia Akademiia Sel'skokhoziaistvennykh Nauk
imeni V. I. Lenina. Vsesoiuznyi Institut Rastenievodstva).   1691

   306 numbered references.

LARIN, Ivan Vasil'evich, ed. Senokosy i pastbishcha (Hay-cutting and pastures).
T. A. Rabotnov, I. V. Larin, P. I. Romashov, and others. Leningrad, Izdatel'stvo
"Kolos," Leningradskoe Otdelenie, 1969, 703 p. Bibliography, p. 679-699.   ]
  About 500 references.

Forestry

GOROVOI, Viktor L'vovich, and PRIVALOVSKAIA, Genrietta Alekseevna. Geografiia
lesnoi promyshlennosti SSSR (Geography of the wood industry of the USSR). Moskva,
Izdatel'stvo "Nauka," 1966. 151 p. Bibliography, p. 146-150 (Akademiia Nauk
SSSR. Institut Geografii).   ]

  62 references on forest resources, logging, and wood-processing industries
of the USSR, in alphabetical order by name of author.

Fishing

ROMANOV, N. S. Ukazatel' literatury po rybnomu khoziaistvu iuzhnykh basseinov SSSR
za 1918-1953 gg. Syr'evaia baza i vosproizvodstvo ryb (Guide to the literature
on fishing in the southern basins of the USSR, 1918-1953: source of raw materials
and reproduction of fishes). Moskva, Izdatel'stvo Akademii Nauk SSSR, 1955.
296 p. (Akademiia Nauk SSSR. Otdelenie Biologicheskikh Nauk Ikhtiologicheskaia
Komissiia).   1

  3,906 entries organized regionally: general section, basin of the Caspian
Sea, basin of the Azov and Black seas, basin of the Aral Sea and of lakes Balkhash
and Issyk-Kul'. Subject, geographical, and author indexes.

ROMANOV, N. S. Literatura po rybnomu khoziaistvu Sibiri (Literature on fishing
in Siberia). In the book: Kozhin, N. I. Promyslovye ryby Sibiri i perspektivy
ikh ispol'zovaniia (Commerical fishes of Siberia and prospects for their utilization). Moskva, Pishchepromizdat, 1946, p. 68-79 (Vsesoiuznyi Nauchno-Issledovatel'skii Institut Morskogo Rybnogo Khoziaistva i Okeanografii).   1

  202 references in chronological order, 1775-1945.

PROBLEMY rybnogo khoziaistva vodoëmov Sibiri: Sostoianie rybnykh zapasov v rechnykh
basseinakh Sibiri i meropriiatiia po ikh ukrepleniiu (Problems of fish production of the reservoirs of Siberia: state of fish stock in river basins of Siberia
and measures for strengthening them). A. N. Petkevich and N. P. Votinov, eds.
Tiumen', 1971. 263 p. Bibliography, p. 254-262. (Ministerstvo Rybnogo Khoziaistva RSFSR. Sibirskii Nauchno-Issledovatel'skii Institut Rybnogo Khoziaistva).   1

ROMANOV, N. S. Ukazatel' literatury po rybnomu khoziaistvu Dal'nego Vostoka za
1923-1956 gg. (Guide to the literature on fishing in the Far East, 1923-1956).
Moskva, Izdatel'stvo Akademii Nauk SSSR, 1959. 291 p. Bibliography, p. 283-
288. (Akademiia Nauk SSSR. Otdelenie Biologicheskikh Nauk. Ikhtiologicheskaia
Komissiia).   1

  3,690 entries organized by topics: fish and fishing, sea mammals and their
capture, raw material base and utilization of marine invertebrates, raw material
base and utilization of marine seaweeds. Alphabetical list of Latin names of
fishes. Author index.

Mining and Manufacturing

DEMIIA Nauk SSSR. Gosplan SSSR. Sovet po Izucheniiu Proizvoditel'nykh Sil. ibliografiia po voprosam razmeshcheniia i raionirovaniia promyshlennosti SSSR, 958-1964 (Bibliography on questions of the distribution and regionalization f manufacturing in the USSR, 1958-1964). T. S. Guchek, comp. Ed. by A. E. robst. Moskva, Izdatel'stvo "Nauka," 1966. 259 p.   1698

DEMIIA Nauk SSSR. Sovet po Izucheniiu Proizvoditel'nykh Sil. Sektor Seti petsial'nykh Bibliotek. Bibliografiia po voprosam razmeshcheniia i raioniro-aniia promyshlennosti SSSR, 1901-1957 (Bibliography on questions of the dis-ribution and regionalization of manufacturing in the USSR, 1901-1957). T. S. uchek, comp. Ed. by A. E. Probst. Moskva, Izdatel'stvo Akademii Nauk SSSR, 960. 356 p.   1699

Extensive bibliography with about 6,000 entries arranged by sections: (1) heoretical questions on the geographical distribution of socialist manufactur-ng, (2) the geographical distribution of manufacturing in the USSR, (3) the istribution of individual branches of industry in the USSR, and (4) the ques-ions of economic regionalization and the growth of manufacturing in individual egions of the USSR, and numerous subdivisions. Index of authors, editors, and itles of books without designated authors.

NOMIKA, razmeshchenie i organizatsiia promyshlennogo proizvodstva Sibiri i al'nego Vostoka: bibliografiia 1917-1965. (Economics, distribution and organiza-ion of industrial production of Siberia and the Soviet Far East). M. I. Kirsanova, . P. Latushkina, N. N. Rechkina, and N. P. Galitskaia. Novosibirsk, 1968-1969. v. (Akademiia Nauk SSSR. Sibirskoe Otdelenie. Gosudarstvennaia Publichnaia auchno-Tekhnicheskaia Biblioteka. Otdel Nauchnoi Bibliografii).   1700

Chast' 1. Literatura izdannaia v 1917-1945 gg. 1968. 452 p.

2841 entries in classed arrangement by historical-chronological period, y economic problems, and by branches of industry, partly briefly annotated. In-ludes books and articles. Author and place indexes. List of bibliographical ources.

Chast' 2. 1946-1965. 1969. 387 p.

2267 entries, partly annotated. Author and geographical indexes. List f bibliographical sources and of serials surveyed.

USHCHEV, Anatoli Timofeevich. Geografiia Promyshlennosti SSSR. (Geography of anufacturing in the USSR). Moskva, Izdatel'stvo "Mysl'," 1969. 438 p. Bib-iography, p. 422-436.   1701

About 320 references, arranged by chapters.

USHCHEV, Anatolii Timofeevich. Geografiia promyshlennosti SSSR. (Geography f manufacturing in the USSR). 3rd ed. Moskva, Izdatel'stvo Moskovskogo Univer-iteta, 1960, 183 p. Bibliography, p. 170-183.   1702

About 300 references organized by chapters.

PANOV, Petr Nikolaevich. Geografiia promyshlennosti SSSR (Geography of manu-acturing in the USSR). Moskva, Uchpedgiz, 1950. 223 p. Bibliography, p. 217-22.   1703

158 references grouped by major themes and subdivisions of the field.

SHITS, Raisa Solomonovna. Razmeshchenie chernoi metallurgii SSSR (Distribution f the iron and steel industry of the USSR). Moskva, Izdatel'stvo Akademii Nauk SSR, 1958. 374 p. Bibliography, p. 372-375 (Akademiia Nauk SSSR. Institut konomiki).   1704

115 references

LIVSHITS, Raisa Solomonovna. Razmeshchenie promyshlennosti v dorevoliutsionnoi
Rossii (Distribution of industry in prerevolutionary Russia). Moskva, Izdatel'-
stvo Akademii Nauk SSSR, 1955. 294 p. Bibliography, p. 287-294. (Akademiia
Nauk SSSR. Institut Ekonomiki).   1

    205 references.

ROSTOVTSEV, Mikhail Ivanovich, and RUNOVA, T. G. Dobyvaiushchaia promyshlennost'
SSSR: ekonomiko-geograficheskie ocherki (The mining industry of the USSR:
economic-geographic studies). Moskva, Izdatel'stvo "Mysl'," 1972. 184 p. Bibliography, p. 176-183. (Akademiia Nauk SSSR. Institut Geografii).   1

KRUKOVSKII, Iurii Aleksandrovich. Geografiia chernoi metallurgii sotsialisticheskik
stran zarubezhnoi Evropy (Geography of ferrous metallurgy of the socialist countries of Europe outside the USSR). A. T. Khrushchev, ed. Moskva, Izdatel'stvo
Moskovskogo Universiteta, 1971. 232 p. Bibliography, p. 226-231.   1

SAVENKO, Iu. N. and SHTEINGAUZ, E. O. Energeticheskii balans: nekotorye voprosy
teorii i praktiki (The energy balance: some questions of theory and practice).
A. M. Nekrasov, ed. Moskva, Izdatel'stvo "Energiia," 1971. 183 p. Bibliog-
raphy, p. 179-182.   1

    153 references.

Energy and Electrification

MATSUK, R. V., ed. Energetika narodnogo khoziaistva v plane GOELRO (Energetics of
the plans of GOELRO). Moskva, Izdatel'stvo "Ekonomika," 1966. 255 p. Bibliog-
raphy, p. 153-254.   1

    Bibliography divided into two parts: (1) Lenin on electrification, p. 153-
165, and (2) electrification of branches of the national economy (coal industry,
metallurgy, electro-chemicals, engineering, textiles, food industry, railroads,
and agriculture): bibliography, p. 167-254.

TRUDY Gosudarstvennoi Komissii po Elektrifikatsii Rossii - GOELRO (Works of the
State Commission on the Electrification of Russia - GOELRO). V. S. Kulebakin,
and others, eds. Moskva, Sotsekgiz, 1960. 307 p. Bibliography, p. 275-304.
(Akademiia Nauk SSSR. Energeticheskii Institut imeni G. M. Krizhizhanovskogo
Tsentral'nyi Gosudarstvennyi Arkhiv Oktiabr'skoi Revoliutsii i Sotsialisticheskogo
Stroitel'stva SSSR).   1

    1,029 entries.

Transportation

PARKHOMENKO, Iia Ivanovna, and SOKOLOV, I. L. Sovetskaia literatura po geografii
transporta SSSR, 1966-1972 gg. (Soviet literature on the geography of transport
of the USSR, 1966-1972), Itogi nauki i tekhniki, Geografiia SSSR, tom 10, Geografiia
transporta, Moskva, VINITI, 1973, p. 89-169. Supplementary table of contents and
abstracts in English.   1

    670 numbered entries arranged systematically and regionally. Author index.
Two associated review articles: Geografiia transporta i territorial'naia organizatsiia khoziaistva SSSR (Geography of transportation and the areal organization of
the economy of the USSR), by N. N. Kazanskii, p. 7-20 and Geograficheskie aspekty
izucheniia transporta SSSR (Geographical aspects of the study of the transportation
of the USSR), by I. V. Nikol'skii, p. 21-88.

USSR. Ministerstvo Putei Soobshcheniia. Tsentral'naia Nauchno-Tekhnicheskaia
Biblioteka. Zheleznodorozhnyi transport. Ukazatel' bibliograficheskikh istochnikov
1865-1965). R. I. Dovgard, K. I. Krugman, M. V. Lanshina, and T. A. Ulanova, comps
Moskva, Izdatel'stvo "Transport," 1967. 211 p.   1

Annotated classed list of 1,164 Soviet bibliographies, including hidden
bibliographies in books and articles, arranged chronologically within classes.
ncludes related fields in engineering and economics. Section on persons. In-
ex of authors and titles.

ŻEZNODOROZHNAIA literatura SSSR (Railroad literature of the USSR) Moskva, Izdatel'-
tvo "Transport," 1941- . annual. (Ministerstvo Putei Soobshcheniia SSSR.
sentral'naia Nauchno-Tekhnicheskaia Biblioteka). 1941-1945 in cumulated volume
1950). Alphabetical indexes of names, titles, and names of railroads.   1713

The 1968 volume (1970. 476 p.) has 4,682 entries and a section on economic-
eographic characteristics of railroads, p. 19-22.

ERATIVNYI zhurnal: zheleznodorozhnyi transport (Reference journal: railroad trans-
ort). Moskva, VINITI, 1962- . monthly. Annual author and subject indexes.   1714

OL'SKII, Igor' Vladimirovich. Geografiia transporta SSSR (Geography of trans-
ortation in the USSR). Moskva, Gosudarstvennoe Izdatel'stvo Geograficheskoi
iteratury, 1960. 406 p. Bibliography, p. 390-400.   1715

317 entries, organized in three sections: general informations of the geog-
aphy of transport in the USSR, regional bases of forms of transport in the USSR,
urvey of the geography of transport in the USSR by regions.

OEMIIA Nauk SSSR. Institut Ekonomiki. Razvitie transporta SSSR 1917-1962:
storiko-ekonomicheskii ocherk (Development of transport of the USSR 1917-1962:
n historical-economic essary). Moskva, Izdatel'stvo Akademii Nauk SSSR, 1963.
03 p. Bibliography, p. 396-402.   1716

About 200 references organized by three sections: (1) monographs collected
olumes, and articles; (2) departmental and statistical publications; and (3)
rchival materials.

GO-DONSKOI sudokhodnyi kanal imeni V. I. Lenina: kratkii ukazatel' literatury
The Volga-Don navigable canal named for Lenin: short guide to the literature).
ostov-na-Donu, Gosudarstvennaia Nauchnaia Biblioteka imeni K. Marksa, 1952.
12 p.   1717

AKOV, Serafim Sergeevich, and VASILEVSKII, Leonid Isakovich, eds. Transportnaia
istema mira (Transport system of the world). Moskva, Izdatel'stvo "Transport,"
971. 216 p. Bibliography, p. 209-215 (Institut Kompleksnykh Transportnykh
roblem pri Gosplane SSSR). 87 tables.   1718

233 references.

ILEVSKII, Leonid Isakovich, and SHLIKHTER, Sergei Borisovich, eds. Tendentsii
perspektivy razvitiia transporta i perevozok v stranakh Zapadnoi Evropy
Tendencies and perspecitives of the development of transport and traffic in
he countries of Western Europe). Moskva, Izdatel'stvo "Transport," 1973, 223 p.
ibliography, p. 219-222 (Institut Kompleksnykh Transportnykh Problem pri Gosplane
SSR).   1719

125 references in alphabetical order, separately for Russian (13), and for
oreign languages (112).

Recreation, Tourism, and Health Resorts

HINIAEV, Petr Nikitovich, and FAL'KOVICH, Nikolai Semenovich. Geografiia
ezhdunarodnogo turizma (The geography of international tourism). Moskva,
Mysl'," 1972. 263 p. Bibliography, p. 254-261.   1720

182 references.

AKCHURINA, R. M. Rekreatsionnaia geografiia, in the book: Kraevedenie i turizm
(Local studies and tourism). A. V. Darinskii, ed. Leningrad, 1972. p. 135-144
(Geograficheskoe Obshchestvo SSSR. Institut Obshchego Obrazovaniia Vzroslykh
Akademii Pedagogicheskikh Nauk SSSR).

List of recommended references.

NAIMARK, I. I. ed. Mesta otdykha i turizma. Referativnyi sbornik (Places of
recreation and tourism). Kiev, 1970. 228 p. Bibliography, p. 163-226.
(Gosudarstvennyi Komitet po Grazhdanskomu Stroitel'stvu i Arkhitekture pri
Gosstroe SSSR. Kievniigradostroitel'stva. Zarubezhnyi opyt).

KURASHOV, Sergei Vladimirovich, GOLDFAIL, L. G., and POSPELOVA, G. N., eds. Kurorty
SSSR (Resorts of the USSR). Moskva, Gosudarstvennoe Izdatel'stvo Meditsinskoi
Literatury, 1962. 798 p. Bibliography, p. 773-778. (1st ed. 1951. 502 p.).

Service

KOVALËV, Sergei Aleksandrovich, PARKHOMENKO, Iia Ivanovna, and POKSHISHEVSKII, Vadim
Viacheslavovich. Geograficheskaia literatura po obsluzhivaniiu naseleniia v SSSR,
1968-1972 (Geographic literature on service industries in the USSR, 1968-1972),
Itogi nauki i tekhniki: Geografiia SSSR, tom 11. Geografiia sfery obsluzhivaniia.
Moskva, VINITI, 1974, p. 95-149. (Vsesoiuznyi Institut Nauchnoi i Tekhnicheskoi
Informatsii).

680 references arranged by major topics: general methodological works in
geography of service activities; settlement and service activities; territorial
organization of service in urban areas and agglomerations; territorial organization of services in rural areas; regional studies of service complexes; studies
of specific branches of service activities; mapping of the service sphere; selected
foreign works on geography of services. Supplement lists 44 author's abstracts of
dissertations. Author index.

GEOGRAPHY OF POPULATION

General

ENTEI, Dmitrii Ignat'evich, and BURNASHEV, E. Iu. eds. Bibliografiia po problemam narodonaseleniia. Sovetskaia i perevodnaia literatura, 1965-1968 gg. (Bibliography on problems of demography. Soviet and translated literature, 1965-1968). Moskva, Izdatel'stvo Moskovskogo Universiteta, 1971. 383 p.   1724

More than 2,000 references, organized by 20 sections, including ones on the number, distribution, and composition of the population; problems of resettlement and settlements; and problems of migration of the population.

KHOMENKO, Iia Ivanovna, and TERNOVSKAIA, G. N. Sovetskaia literatura po geografii naseleniia 1961-1965 gg. (Soviet literature on population geography, 1961-1965), in Itogi nauki: seriia geografiia: Geografiia SSSR, vypusk 3, Geografiia naseleniia SSSR. V. V. Pokshishevskii. Moskva, VINITI, 1966, p. 34-168.   1725

1,940 entries on the Soviet literature on population geography for the period 1961-1965, organized by the following fields: bibliography, (23 entries), general and theoretical questions on population geography (104), geography of population of the USSR (589), urban geography and urban population, and problems of city planning with geographical significance (720), geography of rural settlement (134), works in the field of regional planning important for population geography (46), geography of population and ethnic geography of the world as a whole and of foreign countries (241), translations and original Soviet works on geography of population published abroad (36), and additions (47). Location of abstracts in referativny zhurnal: geografiia for each entry is noted. Author index. Index of cities of the USSR. Textual discussion by V. V. Pokshishevskii, p. 7-33.

REV, Boris Sergeevich. Sovetskaia literatura po geografii naseleniia i smezhnym distsiplinam, 1955-1961 gg. (Soviet literature on geography of population and related disciplines, 1955-1961), Materialy I Mezhduvedomstvennogo soveshchaniia po geografii naseleniia (ianvar'-fevral' 1962 g.) (Material for the 1st Inter-Agency conference on geography of population, January-February, 1962), vypusk 6, Obzor Issledovanii po geografii naseleniia v SSSR no. 6, Review of Research on geography of population in the USSR). Moskva-Leningrad, Geograficheskoe obshchestvo Soiuza SSR, 1962. p. 28-144.   1726

More than 1,200 entries arranged systematically: general and theoretical problems; regional studies; distribution of urban population; city planning; other studies relating to cities; rural population distribution and village planning and population distribution; studies of population in economic-geographical complexes; and geography of ethnic groups.

ALEV, Sergei Alekandrovich. Geografiia naseleniia na stranitsakh nauchnykh sbornikov vysshei shkoly, 1948-1957 gg. (Geography of population on the pages of scientific series of institutions of higher learning, 1948-1957). Nauchnye doklady vysshei shkoly. Geologo-geograficheskie nauki, 1959, no. 1, p. 229-237.   1727

In the period 1948-1947 120 volumes on geography were published in the non-periodical scientific series of institutions of higher learning, of which 44 were devoted to geography of population and cities. These are reviewed in detail.

SHISHEVSKII, Vadim Viacheslavovoch. Predmet, sostoianie i zadachi geografii naselenii (The subject, state, and tasks of population geography). In the book: Materialy I mezhduvedomstvennogo soveshchaniia po geografii naseleniia (Material on the First Inter-Agency conference on the geography of population), vypusk 1. Doklady i reziume dokladov na plenume. (Addresses and abstracts of addresses at the plenary session). Moskva-Leningrad, Geograficheskoe obshchestvo Soiuza SSR, 1961. Bibliography, p. 23-25.   1728

91 entries.

VALENTEI, Dmitrii Ignat'evich, ed. <u>Marksistsko-leninskaia teoriia narodonaseleniia</u>.
(Marxist-Leninist theory of demography). 2nd ed., Moskva, Izdatel'stvo "Mysl',"
1974. (1st ed., 1971).   1

SIFMAN, R. L. Konkretnye demograficheskie issledovaniia v SSSR, 1917-1967 (Concrete
demographic research in the USSR, 1917-1967), p. 35-51 in <u>Sovetskaia statistika za
polveka, 1917-1967</u> (Soviet statistics in the half-century, 1917-1967), T. V.
Riabushkin, ed. Moskva, Izdatel'stvo "Nauka," 1970, 315 p. Bibliographical references. (Akademiia Nauk SSSR. Tsentral'nyi Ekonomiko-Matematicheskii Institut.
<u>Uchéhye zapiski po statistike</u>, tom 17).   1

<u>BIBLIOGRAFIIA po problemam narodonaseleniia</u> (Bibliography on problems of demography).
Moskva, Statistika, 1974, in press.   1

     Covers the Soviet literature on demography and related fields over the 11-
year period 1961-1971.

Maps

MOKROVA, N. N., and NEKHOROSHIKH, A. M., comps. <u>Karty naseleniia. Annotirovannyi
ukazatel'</u> (Maps of population: annotated guide). Moskva, 1971-1972. 3 nos.
(Gosudarstvennaia Biblioteka SSSR imeni V. I. Lenina. Otdel Kartografii).   1

     no. 1. Karty razmeshcheniia naseleniia i naselennosti. Mir. SSSR. Evropa
(Maps of the distribution of population and density of population. World. USSR.
Europe). 1971. 105 p.

     no. 2. Karty razmeshcheniia naseleniia i naselennosti. Aziia. Afrika.
Amerika. Avstraliia i Okeaniia. (Maps of the distribution of population and
density of population. Asia. Africa. America. Australia and Oceania). 1970.
116 p.

     no. 3. Karty demograficheskie, sotsial'no-ekonomicheskie, etnograficheskie
i rasseleniia (Demographic, social-economic, ethnographic, and settlement maps).
1972. 203 p.

EVTEEV, Oleg Aleksandrovich. Karty naseleniia v sovetskikh kartograficheskikh
izdaniiakh i geograficheskoi literature za 20 let (1940-1960) (Maps of population in Soviet cartographic publications and geographic literature in the 20
years 1940-1960), in <u>Geografiia naseleniia SSSR</u>. Moskva, Gosudarstvennoe Izdatel'-
stvo Geograficheskoi Literatury, 1962, p. 209-228. (<u>Voprosy geografii</u>, v. 56).   1

     190 entries on Soviet maps of population, organized by major fields: demo-
graphic maps, maps of social-economic characteristics, ethnographic and toponymic
maps, maps of types of settlements.

Regional

<u>REFERATIVNYI zhurnal: geografiia</u> (Reference journal: geography) (Vsesoiuznyi
Institut Nauchnoi i Tekhnicheskoi Informatsii). Moskva, 1954- . monthly. E.
Geografiia SSSR: Geografiia naseleniia i naselénnykh punktov (E. Geography USSR.
Geography of population and settlements).   1

     Monthly abstracts of publications on geography of population and settlements
in the Soviet Union.

VOROB'ĖV, Vladimir Vasil'evich, ed. <u>Geografiia naseleniia Sibiri i Dal'nego Vostoka:
bibliograficheskii ukazatel'</u> (Geography of population of Siberia and the Far
East: a bibliographical guide). Irkutsk, 1968. 223 p. (Akademiia Nauk SSSR.
Sibirskoe Otdelenie. Institut Geografii Sibiri i Dal'nego Vostoka. Nauchnaia
Biblioteka Vostochno-Sibirskogo Filiala).   1

     2,212 entries. Author index. List of bibliographical sources utilized.

ULOV, Avdei Il'ich, and GRIGOR'IANTS, M. G. Narodonaselenie SSSR: statisticheskoe
zuchenie chislennosti, sostava i razmeshcheniia (Demography of the USSR: statis-
ical study of its number, structure, and distribution). Moskva, Izdatel'stvo
Statistika," 1969. 171 p. Bibliography, p. 167-170.   1737

   About 110 references.

DEMIIA Nauk SSSR. Institut Etnografii im. N. N. Miklukho-Maklaia. Naselenie
emnogo shara: spravochnik po stranam (Population of the world: a handbook by
ountries). S. I. Bruk, ed. Moskva, Izdatel'stvo "Nauka," 1965. 374 p.   1738

   Information on population country-by-country, arranged by continents, in-
luding data on ethnic groups, and other characteristics.

RETICHESKIE voprosy geografii, vypusk 1: Geografiia naseleniia za rubezhom
Theoretical problems in geography, no. 1: Population geography abroad), in
togi nauki, seriia geografiia. Moskva, VINITI, 1966. 67 p.   1739

   Includes bibliographies of population geography in Poland by Iu. L.
ivovarov, p. 21-23 (45 entries), in Hungary, by S. A. Kovalëv, p. 34-35 (26
ntries), in France by V. V. Pokshishevskii and K. S. Leikina, p. 50-52 (61
ntries), and in the United States by V. M. Gokhman, p. 63-67 (92 entries).

OVAROV, Iurii L'vovich. Naselenie sotsialisticheskikh stran zarubezhnoi Evropy:
trukturno-geograficheskie sdvigi (Population of socialist countries of Europe,
xcluding the USSR: structural-geographic changes). I. M. Maergoiz, ed. Moskva,
zdatel'stvo "Nauka," 1970. 175 p. Bibliography, p. 163-174. (Akademiia Nauk
SSR. Institut Geografii).   1740

   262 references on population geography of the German Democratic Republic,
oland, Czechoslovakia, Hungary, Romania, Bulgaria, and Yugoslavia since World
ar II, in Russian and in Bulgarian, Czech and Slovak, German, Hungarian, Polish,
omanian, Serbo-Croatian, French, and English.

GRAFIIA trudovykh resursov kapitalisticheskikh i razvivaiushchikhsia stran
Geography of labor force in capitalist and developing countries). Iu. A.
olosova, ed. Iu. A. Kolosova, V. M. Kharitonov, I. F. Antonova, and others,
uthors. Moskva, Izdatel'stvo "Mysl'," 1971. 469 p. Bibliography, p. 454-468. 1741

EVATYI, Ia. N. Problemy narodonaseleniia i sotsial'no-ekonomicheskoe razvitie
tran Azii, Afriki i Latinskoi Ameriki (Problems of demography and social-
conomic development in the countries of Asia, Africa, and Latin America).
oskva, Izdatel'stvo "Nauka," 1970. 365 p. Bibliography, p. 347-363. (Akademiia
auk SSSR. Institut Mirovoi Ekonomiki i Mezhdunarodnykh Otnoshenii).   1742

INSKAIA Amerika: "demograficheskii vzryv"? (Latin America: demographic explosion?).
. V. Vol'skii, and others, eds. Moskva, 1971. 219 p. Bibliography, p. 201-218.
Akademiia Nauk SSSR. Institut Latinskoi Ameriki).   1743

HKOV, Pavel Ivanovich. Naselenie Okeanii: etnograficheskii obzor (The population
f Oceania: an ethnographic survey). Moskva, Izdatel'stvo "Nauka," 1967. 257 p.
ibliography, p. 248-257. (Akademiia Nauk SSSR. Institut Etnografii imeni N. N.
iklukho-Maklaia).   1744

   288 references, 19 Soviet and 269 in foreign languages.

SHISHEVSKII, Vadim Viacheslavovich. Geografiia naseleniia SSSR: ekonomiko-
eograficheskie ocherki (Geography of population of the USSR: economic-geographic
tudies). Moskva, Izdatel'stvo "Prosveshchenie," 1971. 174 p. Bibliographical
ootnotes.   1745

SHISHEVSKII, Vadim Viacheslavovich. Geografiia naseleniia zarubezhnykh stran:
konomiko-geograficheskie ocherki (Geography of population of foreign countries:
conomic-geographic studies). Moskva, Izdatel'stvo "Prosveshchenie," 1971. 174 p.
ibliographical footnotes.   1746

URBAN GEOGRAPHY

NTSEBOVSKAIA, I. V., and PIVOVAROV, Iurii L'vovich. Sovetskaia literatura po
urbanizatsii 1966-1970 gg. (Soviet literature on urbanization 1966-1970), in
Problemy sovremennoi urbanizatsii (Problems of contemporary urbanization), Iu.
L. Pivovarov, ed. Moskva, Statistika, 1972, p. 222-229. (Akademiia Nauk SSSR.
Natsional'nyi Komitet Sovetskikh Geografov). In Russian with supplementary table
of contents and abstracts in English.   1747

    139 entries of Soviet studies on urbanization in the period 1966-1971,
arranged alphabetically by name of author. Entries by year: up to 1966, 11;
1966, 7; 1967, 6; 1968, 13; 1969, 25; 1970, 50; 1971, 27. Of the entries more
than 50 fall primarily within geography, 40 sociology, 21 city planning, 8
economics and demography, and 9 ethnic problems of urbanization.

RKHOMENKO, Iia Ivanovna, and TERNOVSKAIA, G. N. Geografiia gorodov i gorodskogo
naseleniia. Problemy gradostroitel'stva, imeiushchie geograficheskoe znachenie
(Geography of cities and urban population. Problems of city planning having
geographical significance), in their Sovetskaia literatura po geografii naseleniia,
1961-1965 (Soviet literature on population geography, 1961-1965), Itogi nauki:
geografiia SSSR, no. 3. Moskva, VINITI, 1966, p. 83-120.   1748

    720 entries, arranged by USSR as a whole and then by regions. Citation
number is given for abstracts of each entry in Referativnyi zhurnal: geografiia.
Index of authors. Index of cities of the USSR.

OREV, Boris Sergeevich. Sovetskaia literatura po geografii naseleniia i smezhnym
distsiplinam 1955-1961 gg. (Soviet literature on geography of population and
related disciplines 1955-1961), Materialy I Mezhduvedomstvennogo soveshchaniia
po geografii naseleniia (ianvar'-fevral' 1962 g.) (Material for the 1st Inter-
agency Conference on Geography of Population, January-February, 1962), no. 6,
p. 43-58.   1749

    213 entries on geography of cities in the USSR, 1955-1961.

NSTANTINOV, Oleg Arkad'evich. Geograficheskoe izuchenie gorodskikh poselenii v
SSSR (Geographical study of cities in the USSR), in the book: Geografiia
naseleniia v SSSR: osnovnye problemy (Geography of population in the USSR: basic
problems). Moskva-Leningrad, Izdatel'stvo "Nauka," 1964, p. 32-68. Bibliography,
p. 52-65. (Akademiia Nauk SSSR. Geograficheskoe Obshchestvo Soiuza SSR).   1750

    365 entries covering the field of urban geography in the USSR, with textual
discussion of principal contributions. Author index.

RLAMOV, Viktor Sergeevich, KUDRIAVTSEV, Orest Konstantinovich, and SHATSILO, E. S.
oveishaia literatura po probleme gorodov-sputnikov (The most recent literature
on satellite cities), the book: Goroda-sputniki: sbornik statei (Satellite cities:
a collection of articles). Moskva, Geografgiz, 1961, p. 179-193.   1751

    119 entries of Soviet work on satellite cities.

SELENIE v gorodakh: kolichestvennye zakonomernosti. Sbornik statei (Settlement
in cities, quantitative regularities. A collection of articles). V. G. Davidovich
and O. K. Kudriavtsev, eds. Moskva, Izdatel'stvo "Mysl'," Geografgiz, 1968. 215
p. Bibliography of Russian literature, by S. V. Shaposhnikov, p. 210-214. Bibliography of foreign literature, by O. K. Kudriavtsev, p. 214-215. (Geograficheskoe Obshchestvo SSSR. Moskovskii Filial. Komissiia Geografii Naseleniia).   1752

    79 references in Russian and 42 in Western languages on regularities in urban
phenomena and quantitative methods in urban geography.

LAPPO, Georgii Mikhailovich. Geografiia gorodov s osnovami gradostroitel'stva
(Urban geography with the fundamentals of planning). Moskva, Izdatel'stvo Moskovskogo Universiteta, 1969. 184 p. Bibliographies at ends of chapters plus bibliographical footnotes.   17

    72 basic references and 501 bibliographical footnotes arranged by chapters:
problems of contemporary urban planning in relationship to the processes of urbanization; cities and the geographical division of labor; the economic-geographic
situation of cities; functions fulfilled by cities; study of the population of
cities; planned organization of cities; problems of urban settlements and of large
urban agglomerations; conclusions.

KHOREV, Boris Sergeevich. Problemy gorodov: ekonomiko-geograficheskoe issledovanie
gorodskogo rasseleniia v SSSR (Problems of cities: economic-geographic investigation of urban settlement in the USSR). Moskva, Izdatel'stvo "Mysl'," Geografgiz,
1971. 413 p. Bibliographical footnotes.   17

KHOREV, Boris Sergeevich. Gorodskie poseleniia SSSR: problemy rosta i ikh izuchenie.
Ocherki geografii rasseleniia (Urban settlements of the USSR: problems of their
growth and their study. Studies in the geography of settlement). Moskva, Izdatel'-
stvo "Mysl'," Geografgiz, 1968. 254 p. Bibliographical footnotes.   17

DAVIDOVICH, Vladimir Georgievich. Rasselenie v promyshlennykh uzlakh: inzhenerno-
ekonomicheskie osnovy (Settlement in industrial nodes: engineering-economic foundations). Moskva, Gosstroiizdat, 1960. 324 p. Bibliography, p. 314-323. Available
in English: Town planning in industrial districts. Jerusalem, Israel Program for
Scientific Translations, 1968. 320 p. IPST 1904.   17

    266 references.

DAVIDOVICH, Vladimir Georgievich. Planirovka gorodov i raionov: inzhenerno-
ekonomicheskie osnovy (City and regional planning: engineering and economic bases).
2nd ed. Moskva, Stroiizdat, 1964. 326 p. Bibliography, p. 321-324. (1st ed.
Planirovka gorodov, 1947. 316 p.).   17

    104 references.

TSENTRAL'NYI Institut Nauchno-Tekhnicheskoi Informatsii po Grazhdanskomu Stroitel'stvu
i Arkhitekture. Otdel Spravochno-Informatsionnogo Obsluzhivaniia. Chelovek i
sreda, ekologiia goroda: bibliograficheskii ukazatel' otechestvennykh i inostrannykh informatsionnykh materialov (Man and environment, ecology of the city: bibliographical guide to native and foreign information material). Moskva, 1972.
137 p. 1967-1971. G. P. Fadeeva, comp.   17

STRONGINA, M. L. Sotsial'no-ekonomicheskie problemy razvitiia bol'shikh gorodov v
SSSR (Social-economic problems of the growth of large cities in the USSR).
Moskva, Izdatel'stvo "Nauka," 1970. 88 p. Bibliography, p. 80-87. (Akademiia
Nauk SSSR. Institut Ekonomiki).   17

    About 200 references.

VOROB'EV, Vladimir Vasil'evich, ed. Gorodskoe naselenie i goroda (Urban population and cities), in Geografiia naseleniia Sibiri i Dal'nego Vostoka: bibliograficheskii ukazatel' (Geography of population of Siberia and the Far East: a
bibliographical guide). Irkutsk, 1968, p. 23-60. (Akademiia Nauk SSSR. Sibirskoe
Otdelenie. Institut Geografii Sibiri i Dal'nego Vostoka. Nauchnaia Biblioteka
Vostochno-Sibirskogo Filiala).   17

    410 entries on urban geography of Siberia and the Soviet Far East.

OB'ĔV, Vladimir Vasil'evich. Goroda iuzhnoi chasti
Vostochnoi Sibiri: istoriko-geograficheskie ocherki (The cities of the southern
part of Eastern Siberia: historical-geographical essays). Irkutsk, Irkutskoe
nizhnoe Izdatel'stvo 1959. 147 p. Bibliography, p. 136-146 (Akademiia Nauk
SSR. Sibirskoe Otdelenie. Trudy Vostochno-Sibirskogo filiala, vypusk 28.
eriia ekonomiko-geograficheskaia).   1760

   394 references including collected works, archival material, documents,
articles, books, cartographic material, and exploration.

GEOGRAPHY OF RURAL SETTLEMENT

VALEV, Sergei Aleksandrovich. Sel'skoe rasselenie--geograficheskoe issledovanie
Rural settlement--geographical investigation). Iu. G. Saushkin, ed. Moskva,
zdatel'stvo Moskovskogo Universiteta, 1963. 371 p. Bibliography, p. 352-369.   1761

   About 310 entries in Russian and 100 in other languages on the geography
or rural settlement both in general and with specific reference to the Soviet
Union.

KHOMENKO, Iia Ivanovna, and TERNOVSKAIA, G. N. Geografiia sel'skogo rasseleniia
Geography of rural settlements), in their Sovetskaia literatura po geografii
naseleniia, 1961-1965 (Soviet literature on population geography, 1961-1965),
togi nauki: geografiia SSSR, no. 3. Moskva, VINITI, 1966, p. 121-128.   1762

   134 entries, arranged for the USSR as a whole, and then by regions. Citation
number is given for abstracts of each entry in Referativnyi zhurnal: geografiia.
ndex of authors.

ETHNOGRAPHY AND ANTHROPOLOGY

Ethnography

EKSEEV, Valerii Pavlovich. Geografiia chelovecheskikh ras (Geography of human races). Moskva, Izdatel'stvo "Mysl'," 1974. 351 p. Bibliography, p. 311-347.   1763

OVA, Zoia Dmitrievna. Etnografiia. Bibliografiia russkikh bibliografii po etnografii narodov SSSR (1851-1969) (Ethnography. Bibliography of Russian bibliography of peoples of the USSR, 1851-1969). Moskva, Izdatel'stvo "Kniga," 1970. 143 p. (Gosudarstvenna Publichnaia Biblioteka imeni M. E. Saltykova-Shchedrina).   1764

734 annotated entries, arranged by historical-ethnographic fields. Indexes of ethnic names, subjects, geographical names, and periodicals and collected works whose contents are included. List of sources utilized.

POMOGATEL'NYE istoricheskie distsipliny. Arkheologiia. Etnografiia (Auxiliary historical disciplines. Archeology. Ethnography), Bibliografiia sovetskoi bibliografii (Bibliography of Soviet bibliography) 1946- . Annual. Moskva, Izdatel'stvo "Kniga," 1948 - (Vsesoiuznaia Knizhnaia Palata).   1765

Annual list of bibliographies in ethnography and archeology in a section in the chapter on history. In the 1970 annual volume this section, p. 32-34 contained 51 entries.

NAKOVA, O. V., KAMENETSKAIA, R. V. Bibliografiia trudov Instituta Etnografii im. N. N. Miklukho-Maklaia. 1900-1962 (Bibliography of works of the Institute of Ethnography named after N. N. Miklukho-Maklai, 1900-1962). Leningrad, Izdatel'-stvo "Nauka," 1967. 281 p. (Akademiia Nauk SSSR. Biblioteka Akademii Nauk SSSR. Institut Etnografii imeni N. N. Miklukho-Maklaia).   1766

5,195 entries, including all publications of the Institute of Ethnography and the Museum of Anthropology and Ethnography, on the basis of which the Institute was established in 1933. Major sections: general works, 1,088 entries; general and theoretical works on ethnography, 320 entries; ethnography of peoples of the USSR, 2,318 entries; ethnography of peoples of foreign countries, 1,140 entries; anthropology, 262 entries; translations of Soviet works into foreign languages. Indexes: reviews, ethnic names, authors, etc. Has a section Personalia and brief notes on scholars, 129 entries.

KAZATEL' statei i materialov opublikovannykh v 1961-1965 gg. (List of articles and materials on ethnography published in 1961-1965), Moskva, "Nauka," 1966. 41 p. (Sovetskaia etnografiia no. 6).   1767

ELENIN, Dmitrii Konstantinovich. Bibliograficheskii ukazatel' etnograficheskoi literatury o vneshnem byte narodov Rossii 1700-1910 (Bibliographical guide to ethnographic literature on the external life of the peoples of Russia, 1700-1910). St. Petersburg, 1913. 733 p. (Imperatorskoe Geograficheskoe Obshchestvo. Otdelenie Etnografii. Zapiski, v. 40, no. 1).   1768

Covers housing, dress, music, art, and economy.

Anthropology

LADKOVA, Tat'iana Dmitrievna. Spisok vazhneishikh antropologicheskikh rabot, opublikovannykh na russkom iazyke s 1958 po 1967 g. (List of the most important anthropological works published in the Russian language 1958-1967), Voprosy antropologii, no. 30 (1968), p. 167-194. About 800 entries.   1769

_____. Spisok literatury po antropologii, opublikovannoi na russkom iazyke v period s 1937 po 1957 g. (List of literature on anthropology, published in the Russian language in the period 1937-1957), Voprosy antropologii, no. 14 (1963), p. 125-134. About 300 entries. Supplements earlier list by Zalkind and Uryson. 17

_____. Spisok antropologicheskoi literatury na russkom iazyke za 1917-1936 gg. (List of anthropological literature in the Russian language 1917-1936), Voprosy antropologii, no. 2 (1960), p. 147-166. 540 entries.   17

_____. Spisok statei, obzorov, referatov, nekrologov i khroniki, napechatannykh v "Russkom antropologicheskom zhurnale" (1918, 1922-1930 gg.) i "Antropologicheskom zhurnale" (1932-1937 gg.) (List of articles, reviews, papers, obituaries, and notes published in the Russkii antropologicheskii zhurnal, 1918, 1922-1930, and in the Antropologicheskii zhurnal, 1932-1937), Sovetskaia antropologiia, v. 2, no. 1 (1958), p. 115-130.   1

ZALKIND, I. G., and URYSON, M. I. Spisok vazheishei antropologicheskoi literatury, opublikovannoi v SSSR, 1937-1957 (List of the most important anthropological literature published in the USSR, 1937-1957), Sovetskaia antropologiia, v. 2, no. 3 (1958), p. 131-151.   1

      About 520 entries.

ANTROPOLOGIIA (Anthropology), Bibliografiia sovetskoi bibliografii (Bibliography of Soviet bibliography). 1946- . Annual. Moskva, Izdatel'stvo "Kniga," 1948- . (Vsesoiuznaia Knizhnaia Palata).   17

      Annual list of Soviet bibliographies on anthropology. The 1970 annual volume (1972), p. 121, contains 11 entries. A sub-section of the section on biological sciences in the chapter on natural sciences and mathematics.

POTAPOV, L. P. Perechen' statei, pomeshchënnykh v sbornikakh Muzeia antropologii i etnografii s 1900 po 1963 g. tt. I-XXI (List of articles in the volumes of the Musuem of Anthropology and Ethnography from 1900 to 1963 in v. 1-21), Sovetskaia etnografiia, 1964, no. 4, p. 210-217.   1

      About 280 entries.

ALMAZOVA, N. Ia. Bibliografiia antropologicheskikh rabot, opublikovannykh v inostrannykh periodicheskikh izdaniiakh za 1968-1970 gg (Bibliography of anthropological works published in foreign periodicals, 1968-1970) Voprosy antropologii, no. 39 (1971), p. 160-173. About 400 entries.   17

_____. _____. 1964-1969, Voprosy antropologii, no. 35 (1970), p. 182-194, no. 33 (1969), p. 175-191. About 1,000 entries.

_____. _____. 1963-1967, Voprosy antropologii, no. 27 (1967), p. 186-194. About 300 entries.

_____. _____. 1963-1964, Voprosy antropologii, no. 23 (1966), p. 176-185. 361 entries.

_____. _____. 1960-1964, Voprosy antropologii, no. 19 (1965), p. 159-169. About 340 entries.

_____. _____. 1961-1962, Voprosy antropologii, no. 16 (1964), p. 150-153. 120 entries.

_____. _____. 1960-1962, Voprosy antropologii, no. 13 (1963), p. 159-164. About 170 entries.

_____. _____. 1960, Voprosy antropologii, no. 9 (1962), p. 159-163. 157 entries.

_____. _____. 1958-1959, Voprosy antropologii, no. 6 (1961), p. 150-157; no. 8 (1961), p. 172-174. About 300 entries.

1770 - 1776

AZOVA, N. Ia. Spisok rabot, opublikovannykh v izdaniiakh: "Sovetskaia antro-
ologiia" (1957-1959 gg.) i "Voprosy antropologii" (1960-1967 gg.) (List of
orks published in Sovetskaia antropologiia, 1957-1959, and in Voprosy antro-
ologii,(1960-1967), Voprosy antropologii, no. 29 (1968), p. 186-201.   1777
   About 625 entries.

PLANNING

ORAD, Daniil Il'ich. Konstruktivnaia geografiia raiona: osnovy raionnoi
lanirovki (Constructive geography of the region: the bases of regional plan-
ing). Moskva, Izdatel'stvo "Mysl'," Geografgiz, 1965. 407 p. Bibliography,
. 382-405.   1778
   About 450 references on geography and regional planning arranged alpha-
etically by name of author.

TSIK, Evgenii Naumovich. Raionnaia planirovka: geograficheskie aspekti (Regional
lanning: geographical aspects). Moskva, Izdatel'stvo "Mysl'," 1973. 271 p.
ibliography, p. 264-270.   1779
   151 references.

DOSTROITEL'STVO SSSR 1917-1967 (Urban planning and construction in the USSR
917-1967). V. A. Shkvarikov, N. Ia. Kolli, V. A. Lavrov, and others. Moskva,
troizdat, 1967. 397 p. (Gosudarstvennyi Komitet po Grazhdanskomu Stroitel'stvu
Arkhitekture pri Gosstroe SSSR. Tsentral'nyi Nauchno-Issledovatel'skii i
roektnyi Institut po Gradostroitel'stvu).   1780

TOPONYMY

(See also place-name dictionaries)

CHKEVICH, Vadim Andreevich. Toponimika: kratkii geograficheskii ocherk (Toponymy:
 short geographical work). Minsk, Izdatel'stvo "Vysshaia shkola," 1965. 320 p.
ibliography, p. 295-318.   1781
   About 670 references.

ONOV, Vladimir Andreevich, ed. Toponimicheskii minimum: Ukazatel' literatury.
Toponymic minimum: a guide to the literature). Moskva. Geograficheskoe
bshchestvo SSSR. Moskovskii Filial. Toponimicheskaia Komissiia, 1964. 24 p.

   More than 300 references arranged in eleven sections: (1) popular literature;
2) basic literature (theoretical works on toponymy, including foreign work);
3) supplementary literature (conferences, collected works, works in languages
ther than Russian); (4) microtoponymics; (5) toponymic dictionaries and works
n the etymology of place names of individual regions; (6) programs and instruc-
ions for collecting toponyms; (7) history of toponymic science; (8) list of
oponymic journals published abroad; (9) international toponymic congresses and
onferences; (10) transcription works; and (11) bibliography of toponymy.   1782

ONOV, Vladimir Andreevich. Vvedenie v toponimiku (Introduction to toponymics).
oskva, Izdatel'stvo "Nauka," 1965. 177 p. Bibliography, p. 168-174.   1783
   About 150 references.

ONOV, Vladimir Andreevich. Toponimika za piatiletie 1950-1954 gg.: ukazatel'
iteratury (Toponymy in the five-year period 1950-1954: guide to the literature),
eograficheskoe Obshchestvo SSSR. Izvestiia, v. 90, no. 4 (1958), p. 393-401. 1784
   129 annotated entries of books and articles in Russian.

EN'KAIA, Viktoriia Davydovna. Toponimy v sostave leksicheskoi sistemy iazyka
Toponymy in the structure of a lexical system of language). Moskva, Izdatel'stvo
oskovskogo Universiteta, 1969. 168 p. Bibliography, p. 161-167. Part of text
n English.   1785
   About 200 references.

                                                            1777 - 1785

POSPELOV, Evgenii Mikhailovich. Toponimika i kartografiia (Toponymy and cartography)
Moskva, Izdatel'stvo "Mysl'," Geografgiz, 1971. 255 p. Bibliography, p. 243-254.1

GORBACHEVICH, Kirill Sergeevich. Russkie geograficheskie nazvaniia (Russian geographical names). Moskva-Leningrad, Izdatel'stvo "Nauka," 1965. 64 p. Bibliography, p. 61-64.   1
    68 references.

ZHUCHKEVICH, Vadim Andreevich. Toponimika Belorussii (Place names of Belorussia).
Minsk, "Nauka i Tekhnika," 1968. 184 p. Bibliography, p. 178-183.   1
    About 450 place names of Belorussia. Bibliography includes about 150 refer-
ences.

ZHUCHKEVICH, Vadim Andreevich. Proiskhozhdenie geograficheskikh nazvanii(toponimika)
Belorussii (Origin of the geographical names toponymics of Belorussia). Minsk,
Izdatel'stvo Belgosuniversiteta imeni V. I. Lenina, 1961. 78 p. Bibliography,
p. 77-78.   1
    40 references in Russian, Belorussian, and Polish.

IUZBASHEV, Ramzi Mavsun ogly. Dostizheniia toponimiki v Azerbaidzhane za 50 let.
Reziume (Achievements in toponymics in Azerbaidzhan over 50 years. Summary).
Akademiia Nauk Azerbaidzhanskoi SSR. Institut Geografii, Trudy, v. 18 (1971),
p. 302-306. Bibliography, p. 303-306.   1

IUZBASHEV, Ramzi Mavsun ogly. Azerbaidzhanskie geograficheskie terminy: issledovaniia
(Azerbaidzhan geographical terms: investigations). Baku, Izdatel'stvo Akademii
Nauk Azerbaidzhanskoi SSR, 1966. 158 p. Bibliography, p. 152-156. (Akademiia
Nauk Azerbaidzhanskoi SSR. Institut Geografii). (In Azerbaijani).   1
    About 130 references.

TOPONIMIKA Vostoka: issledovaniia i materialy (Toponymics of the East: research and
materials). E. M. Murzaev, and others, eds. Moskva, Izdatel'stvo "Nauka," 1969.
239 p. (Akademiia Nauk SSSR. Institut Vostokovedeniia. Toponimicheskaia Komissiia
Moskovskogo Filiala Geograficheskogo Obshchestva).   1
    Bibliographical footnotes.

MEL'KHEEV, Matvei Nikolaevich. Geograficheskie nazvaniia Vostochnoi Sibiri (Geographical names of Eastern Siberia). Irkutsk, Vostochno-Sibirskoe Knizhnoe
Izdatel'stvo, 1969. 121 p. Bibliography, p. 118-120.   1

LITERATURA po toponimike zarubezhnogo Vostoka (Literature of toponymy in foreign
countries of the East),.in the book: Toponimika Vostoka (Toponymy of the East).
Moskva, Izdatel'stvo Vostochnoi Literatury, 1962. p. 202-209.   1
    128 references on transcriptions of geographical names and on toponymy of
the countries of Asia, Africa, Australasia and Oceaniia, arranged regionally.

TOPONIMIKA Vostoka. Trudy soveshchaniia po toponimike Vostoka, Moskva, 10-13
Aprelia 1961 g. (Toponymy of the East: proceedings of a conference on toponymics
of the East, held in Moscow, April 10-13, 1961). E. M. Murzaev, V. A. Nikonov,
and V. V. Tsybul'skii, eds. Moskva, Izdatel'stvo Vostochnoi Literatury, 1962.
211 p. Bibliography, p. 202-209.   1

NIKONOV, Vladimir Andreevich, and SIMONOVA, A. G., eds. Inostrannaia literatura po
toponimike. Bibliograficheskii obzor. (Foreign literature on toponymics:
bibliographical review). Moskva, "Kniga," 1965. 40 p. (Vsesoiuznaia Biblioteka
Inostrannoi Literatury).   1
    120 entries. Review articles on toponymic literature in England, U.S.A.,
France, Yugoslavia, and Bulgaria

C. HISTORY OF GEOGRAPHY, HISTORY OF EXPLORATION, AND HISTORICAL GEOGRAPHY

HISTORY OF GEOGRAPHY

In The U.S.S.R.

General

DEMIIA Nauk SSSR. Otdelenie Nauk o Zemle. Institut Istorii Estestvoznaniia i
ekhniki. Razvitie nauk o zemle v SSSR (Development of sciences of the earth
n the USSR). Moskva, Izdatel'stvo "Nauka," 1967. 715 p. Bibliography, p. 696-
11. (Sovetskaia Nauka i Tekhnika za 50 let).   1797

    About 390 references organized by large fields corresponding to sections in
he book: the earth as a whole, the core and upper mantle, the land, the ocean,
he atmosphere, and the space near the earth. The section on the land includes
hapters on natural resources and the development of physical and economic geog-
aphy, geographical discovery, geomorphology, climatology, glaciology, permafrost
tudies, waters of the land, soils, vegetation and geobotany, zoogeography, and
urther tasks of geography, p. 291-452, with 87 major synthesizing books listed
n the bibliography, p. 700-703, covering the development of the fields in the
0-year period 1917-1967.

ORIIA estestvoznaniia. Literatura opublikovannaia... (History of science:
iterature published...). Moskva-Leningrad, 1949-1963. (Akademiia Nauk SSSR.
nstitut Istorii Estestvoznaniia. Fundamental'naia Biblioteka Obshchestvennykh
auk).   1798

    v. 1. 1917-1947. O. A. Starosel'skaia-Nikitina and others, eds. Moskva-
eningrad, 1949. 519 p. Chapter IV. Istoriia geologo-geograficheskikh nauk
History of geological and geographical sciences), p. 220-315. About 10,000
eferences in all; 2,311 entries for geology and geography.

    v. 2. 1948-1950. O. A. Starosel'skaia-Nikitina and others, eds. Moskva,
955. 345 p. Chapter IV. Istoriia geologo-geograficheskikh nauk (History of
eological and geographical sciences), p. 163-219. 6,130 references in all;
,047 entries for geology and geography.

    v. 3. 1951-1956. L. V. Kaminer and others, eds. Moskva, 1963. 430 p.
hapter IV. Istoriia geologo-geograficheskikh nauk (History of geological and
eographical sciences), p. 179-269. 7,041 references in all; 1,809 entries for
eology and geography.

TS, Aleksei Aleksandrovich. Istoriia geograficheskoi mysli: obzor sovetskoi
iteratury za 1961-1970 gg. (The history of geographical ideas: a review of
he Soviet literature for 1961-1970), in Itogi nauki i tekhniki,seriia
eoreticheskie voprosy fizicheskoi i ekonomicheskoi geografii, v. 1. Moskva,
INITI, 1972, p. 26-43. Bibliography, p. 39-59. English summary, p. 42-43.   1799

    59 references in Russian, arranged alphabetically by author.

OV, Mikhail Ivanovich. Sovetskie istoriko-geograficheskie issledovaniia
nekotorye itogi i perspektivy) (Soviet historical-geographical research; some
ccomplishments and prospects), Geograficheskoe Obshchestvo SSSR, Izvestiia,
. 99, no. 5 (1967), bibliography, p. 401-402.   1800

    95 entries on Soviet historical geography.

TIKHOMIROV, Georgii Sergeevich. Bibliograficheskii ocherk istorii geografii v
Rossii XVIII v. (Bibliographical essay on the history of geography in Russia in
the 18th century). Moskva, Izdatel'stvo "Nauka," 1968. 135 p. Bibliography,
p. 89-129. (Biblioteka Akademii Nauk SSSR).   18

   Information on publications in Russia of foreign and Russian sources on the
history of geography and the characteristics of Russian historical-geographical
literature. List of native and translated historical-geographical literature on
research.

NEVSKII, Vladimir Vasil'evich. Metodicheskoe posobie po kursu istorii geografii
(Methodological aids for a course on the history of geography). Leningrad,
Izdatel'stvo Leningradskogo Universiteta, 1966. 28 p. Bibliographies, p. 14-28.   1

   74 general references on the history of geography and 225 specific works
for assignments on leading geographers, including lists of references on each of
the following Russian contributors: M. V. Lomonosov, P. I. Rychkov, S. P.
Krasheninnikov, A. F. Middendorf, P. P. Semenov-Tian-Shanskii, N. M. Przheval'skii,
A. I. Voeikov, A. A. Tillo, D. N. Anuchin, V. V. Dokuchaev, G. I. Tanfil'ev,
G. N. Vysotskii, and S. S. Neustruev. All references in Russian.

BODNARSKII, Mitrofan Stepanovich. Ocherki po istorii russkogo zemlevedeniia. v. 1.
(Essays on the history of Russian geography. 1). Moskva, Izdatel'stvo Akademii
Nauk SSSR, 1947. 291 p. Bibliography, p. 240-253 (Akademiia Nauk SSSR. Nauchno-
populiarnaia seriia).   1

   About 305 numbered references, 1627-1946, systematically organized, on
history of Russian geography: (1) bibliography, 24 entries; (2) dictionaries,
15; (3) periodicals, 12; (4) cartography, 28; (5) history of Russian geography,
146; (6) travelers, 50; (7) works in foreign languages, 30. 195 bibliographical
notes. Indexes of about 650 personal names and 1,300 of places. Chronological
table of events.

BERG, Lev Semenovich. Ocherk istorii russkoi geograficheskoi nauki (vplot' do
1923 goda). (Essays on the history of Russian geographical science: up to 1923).
Leningrad, Izdatel'stvo Akademii Nauk SSSR, 1929. 154 p. (Akademiia Nauk SSSR.
Trudy Komissii po Istorii Znanii, tom 4).   1

   Short systematic review of the literature of Russian geographical science:
methodology of geography, institutiins and publications, cartography, research
on the land, research on water, historical geography. Rich bibliographical
details throughout the text and at end of sections.

LEBEDEV, Dmitrii Mikhailovich. Ocherki po istorii geografii v Rossii XV i XVI
vekov. (Essays on the history of geography in Russia, 15th and 16th centuries).
Moskva, Izdatel'stvo Akademii Nauk SSSR, 1956. 240 p. (Akademiia Nauk SSSR.
Institut Geografii).   1

_____ . Geografiia v Rossii XVII veka (dopetrovskoi epokhi): ocherki po istorii
geograficheskikh znanii (Geography in Russia in the 17th century in the pre-
Pretrine period: essays on the history of geographical knowledge). Moskva-
Leningrad, Izdatel'stvo Akademii Nauk SSSR, 1949. 235 p. (Akademiia Nauk SSSR.
Institut Geografii. Seriia "Itogi i problemy sovremennoi nauki").   1

_____ . Geografiia v Rossii petrovskogo vremeni (Geography in Russia in the
period of Peter the Great). Moskva-Leningrad, Izdatel'stvo Akademii Nauk SSSR,
1950. 383 p. (Akademiia Nauk SSSR. Institut Geografii. Seriia "Itogi i
problemy sovremennoi nauki").   1

_____ . Ocherki po istorii geografii v Rossii XVIII v.1725-1800 gg. (Essays on
the history of geography in Russia in the 18th century, 1725-1800). Moskva,
Izdatel'stvo Akademii Nauk SSSR, 1957. 273 p. (Akademiia Nauk SSSR. Institut
Geografii).   1

   Bibliographical footnotes in each of the above four volumes. Indexes of
personal names and geographical names.

1801 - 1808

HOMIROV, G. S. Russkaia literatura po istorii geografii. Vypusk pervyi, A,
, v (Russian literature on the history of geography, no. 1, a, b, v). Moskva,
zdanie MGU, 1948. 121 p. (Moskovskii Gosudarstvennyi Universitet).   1809

   Publications on history of geography in Russian, listed alphabetically by
uthor, for the first three letters of the alphabet only. Over 300 publications
p to 1947, with bibliographical details and annotations.

.PIANSKII, Petr Nikolaevich. Opyt bibliografii knig po geografii, izdannykh v
:aterinskuiu epokhu (A work on the bibliography of books on geography published
iring the period of Catherine II, 1762-1796), Russkoe Geograficheskoe Obshchestvo.
'enburgskii Otdel. Izvestiia, no. 18 (1903), p. 53-144, also in his book, Materialy
istorii russkoi literatury i nauki v XVIII v. Vypusk 1. Opyt bibliograficheskogo
:azatelia knig po geografii, izdannykh v Rossii v tsarstvovanie imperatritsy
:ateriny II (Material on the history of Russian literature and science in the
5th century, no. 1. Work on a bibliographical guide to books on geography, pub-
_shed in Russia during the reign of the Empress Catherine II). Orenburg, 1903,
? p.   1810

   Systematic guide to books and maps. Sections: Geography of Russia; maps;
eneral geography; physical and mathematical geography; travels; and ethnography.
11 publications described, for 220 of which location of copies is noted.

DEMIIA Nauk SSSR. Otdelenie Geologo-Geograficheskikh Nauk SSSR. Pervye russkie
auchnye issledovaniia Ustiurta (First Russian scientific investigation of the
st-Yurt). A. L. Ianshin and L. A. Gol'denberg, eds. Moskva, Izdatel'stvo
kademii Nauk SSSR, 1963. 326 p. Bibliography, p. 301-305.   1811

   About 150 references.

Physical Geography

ZNETSOV, Pavel Savel'evich. Evoliutsiia poniatiia predmeta fizicheskoi geografii
(Evolution of the concepts of the objects of physical geography). Saratov,
zdatel'stvo Saratovskogo Universiteta, 1961. 67 p. Bibliography, p. 60-66.   1812

   About 130 references.

:GOR'EV, Andrei Aleksandrovich. Razvitie teoreticheskikh problem sovetskoi
'izicheskoi geografii, 1917-1934 gg. (The development of theoretical problems
if Soviet physical geography, 1917-1934). Moskva, "Nauka," 1965. 246 p. Bib-
_iography, p. 226-244. (Akademiia Nauk SSSR. Institut Geografii).   1813

   About 325 references on development of physical geography in the Soviet
Union, 1917-1934, arranged alphabetically by name of author.

.DKIN, Naum Grigor'evich. Ocherki po istorii fiziko-geograficheskikh issledovanii
:erritorii SSSR, 1917-1927 gg. (Studies in the history of physical geographical
'esearch of the territory of the USSR, 1917-1927). Moskva, Izdatel'stvo Akademii
Jauk SSSR, 1961. 247 p. Index of personal names, p. 241-246. (Akademiia Nauk
SSSR. Institut Geografii).   1814

   789 bibliographical footnotes. About 700 persons listed in alphabetical
_ndex of personal names. Emphasis on work of geographical institutions. Covers
oth complex physical geographic research and studies on specialized systematic
iranches of physical geography, and both the Soviet Union as a whole and regions.

:GOR'EV, Andrei Aleksandrovich. Razvitie fiziko-geograficheskoi mysli v Rossii:
:IX--nachalo XX v.: kratkii ocherk. (The development of concepts in physical
eography in Russia: 19th to the beginning of the 20th century; short essay).
loskva, Izdatel'stvo Akademii Nauk SSSR, 1961. 91 p. Bibliography, p. 85-90
(Akademiia Nauk SSSR. Institut Geografii. Institut Istorii Estestvoznaniia i
'ekhniki).   1815

131 references (37 in foreign languages), in alphabetical order, on history of development of physical geography in prerevolutionary Russia.

VILENSKII, Dmitrii Germogenovich. Russkaia pochvenno-kartograficheskaia shkola i eë viliianie na razvitie mirovoi kartografii pochv (The Russian soil-cartographic school and its influence on the development of the world cartography of soils). Moskva, Leningrad, Izdatel'stvo Akademii Nauk SSSR, 1945. 143 p. Bibliography, p. 131-141. Available in English as: Russian school of soil cartography and its influence on the soil cartography of the world. Jerusalem, Israel Program for Scientific Translations, 1967. 112 p. IPST no. 1908.   1

187 numbered references, 117 in Cyrillic and 70 in Latin letters.

AKADEMIIA Nauk SSSR. Institut Istorii Estestvoznaniia i Tekhniki. K istorii otkrytiia i izucheniia mestorozhdenii poleznykh iskopaemykh (On the history of discovery and study of mineral deposits). Moskva, Izdatel'stvo, Akademii Nauk SSSR, 1963. 107 p. Bibliography, p. 105-106.   1

KUZIN, Aleksandr Avramievich. Istoriia otkrytii rudnykh mestorozhdenii v Rossii do serediny XIX v. (History of the discovery of ore deposits of Russia up to the middle of the 19th century). Moskva, Izdatel'stvo Akademii Nauk SSSR, 1961. 360 p. Bibliography, p. 318-322. (Institut Istorii Estestvoznaniia i Tekhniki). 1

116 references.

BUBLEINIKOV, Feofan Dmitrievich. Istoriia otkrytiia iskopaemykh bogatstv nashei strany (History of the discovery of the mineral wealth of our country). Moskva, Geografgiz, 1948. 343 p. Bibliography, p. 324-340.   1

BUBLEINIKOV, Feofan Dmitrievich. Geologicheskie poiski v Rossii (Geological exploration in Russia). Moskva, Gosgeoltekhizdat, 1956. 251 p. Bibliography, p. 233-240.   1

162 references.

Economic Geography

KOGAN, S. M. Osnovnaia obshchaia literatura po istorii otechestvennoi ekonomicheskoi geografii i smezhnym voprosam (Basic general literature on the history of native economic geography and related questions), in Ekonomicheskaia geografiia v SSSR: istoriia i sovremennoe razvitie (Economic geography in the USSR: history and present development), ed. by N. N. Baranskii, N. P. Nikitin, V. V. Pokshishevskii, and Iu. G. Saushkin. Moskva, Prosveshchenie, 1965. p. 616-620. See fuller   1 annotation under Biographies of leading geographers.

About 100 references.

SEMEVSKII, Boris Nikolaevich. Formirovanie nauchnoi shkoly sovetskoi ekonomicheskoi geografii kapitalisticheskikh stran. (Formation of the scientific school of Soviet economic geography of capitalist countries), Leningradskii Universitet. Vestnik. Seriia geologii i geografii, 1967, no. 2. Bibliography, p. 18-19.

49 references.

SEMEVSKII, Boris Nikolaevich. Formirovanie nauchnoi shkoly sovetskoi ekonomicheskoi geografii kapitalisticheskikh stran. II (The formation of the scientific school of Soviet economic geography of capitalist countries). Leningradskii Universitet. Vestnik. Seriia Geologiia i geografiia, 1967, no. 3. Bibliography, p. 56-57.

66 entries on the works of Soviet geographers on capitalist countries.

Textbooks

..NSKII, Nikolai Nikolaevich. Istoricheskii obzor uchebnikov geografii, 1876-
'34 (Historical survey of textbooks in geography, 1876-1934). Moskva, Geo-
·afgiz, 1954. 502 p. Index of names, p. 497-501.   1824

   Introduction to geographical textbooks in two periods, 1876-1917 and 1917-
'34 with detailed listing and individual discussion of 233 textbooks, p. 45-354
.d 372-496, averaging nearly two pages for analysis of each listed book.

.N, Leonid Pavlovich. Istoricheskii obzor uchebnikov obshchei i russkoi geografii,
;dannykh so vremeni Petra Velikogo do 1876 goda (Historical review of textbooks
° general and Russian geography published from the time of Peter the Great to
·76, 1710-1876). St. Petersburg, Tip. brat. Lanteleevykh, 1876. 674 p.   1825

   Annotated list of 232 textbooks of geography published in Russia, 1710-1876.
.phabetical list of personal names.

                   Individual Institutions (See also entries 42-73)

)EMIIA Nauk SSSR. Geograficheskoe Obshchestvo SSSR. S. V. Kalesnik, ed.
·ograficheskoe obshchestvo za 125 let (The Geographical society through 125
·ars). Leningrad, Izdatel'stvo "Nauka," 1970. 396 p. Index of personal names,
, 372-384. Index of geographical names, p. 385-395.   1826

   Historical account of the activities of the Geographical Society by regions
·d by systematic fields. About 1,500 persons in index of personal names.
·merous bibliographical footnotes.

[RITS, Igor' Borisovich, and PINKHENSON, Dmitrii Moiseevich. Geograficheskoe
)shchestvo Soiuza SSR 1917-1967 (The Geographical Society of the USSR 1917-
·67). S. V. Kalesnik, ed. Moskva, Izdatel'stvo "Mysl'," Geografgiz, 1968.
·2 p. Bibliography, p. 255-262.   1827

   About 150 references on the history and activity of the Society. List of
·rial and periodical publications of the Society, 1917-1966. Subject index.
·dex of names of about 600 persons associated with the Society in this period.

G, Lev Semenovich. Vsesoiuznoe Geograficheskoe Obshchestvo za sto let, 1845-1945
The All-Union Geographical Society during a hundred years 1845-1945). Moskva-
·eningrad, Izdatel'stvo Akademii Nauk SSSR, 1946. 263 p. Bibliography, p. 249-
·52. (Akademiia Nauk SSSR. Nauchno-populiarnaia Seriia).   1828

   Biographic footnotes.

:HEVA, Vera Fedorovna. Materialy dlia istorii ekspeditsii Akademii Nauk v
/III i XIX vekakh: khronologicheskie obzory i opisaniia arkhivnykh materialov
Materials for the history of expeditions of the Academy of Sciences in the 18th
·d 19th centuries: chronological surveys and descriptions of archival materials).
. S. Berg, B. D. Grekov, and others. eds. Moskva-Leningrad. Izdatel'stvo
·ademii Nauk SSSR, 1940. 310 p. (Akademiia Nauk SSSR. Trudy Arkhiva, vypusk
).   1829

   Historical review of materials preserved in the archives of the Academy of
·ciences of the USSR connected with 142 expeditions, 1707-1903, organized by the
·ademy, on Academy instructions, or whose materials are deposited in its archives.
·ibliography on each expedition. Indexes of geographical names and nationalities;
·ames; scientific disciplines and objectives of study. About 3,000 references.

:HEVA, Vera Fedorovna. Geograficheskii departament Akademii nauk XVIII veka
Geographical Department of the Academy of Sciences in the 18th century). A. I.
·ndreev, ed. Moskva-Leningrad, Izdatel'stvo Akademii Nauk SSSR, 1946. 446 p.
·Akademiia Nauk SSSR. Trudy Arkhiva, vypusk 6).   1830

                                                                        1824 - 1830

Historical study of the activities of the Geographical Department of the Academy of Sciences, 1726-1799 in the geographical study of Russia and in the preparation of geographical maps. Bibliographical materials in the appendix, with 1,114 entries:

Spisok kartograficheskikh izdanii Akademii nauk XVIII v., 1728-1806 gg. (List of cartographic publications of the Academy of Sciences in the 18th century 1728-1806), p. 236-266, 324 entries.

Opisanie rukopisnykh kart XVIII v., khraniashchikhsia v Otdele rukopisnoi knigi Biblioteki Akademii nauk SSSR (Description of manuscript maps of the 18th century, preserved in the Section on manuscript books of the Library of the Academy of Sciences of the USSR), comp. by V. V. Aleksandrov, p. 267-412, 790 entries.

GEOGRAFIIA v Moskovskom Universitete za 200 let: spravochnyi material (Geography in Moscow University during 200 years: reference material). K. K. Markov, Iu. G. Saushkin, and A. I. Solov'ëva, eds. Moskva, 1955. 112 p.

For interpretative essays on the development of geography at Moscow State University by major fields see: Geografiia v Moskovskom Universitete za 200 let, 1755-1955. K. K. Markov and Iu. G. Saushkin, eds. Moskva, Izdatel'stvo Moskovkogo Universiteta, 1955. 287 p.

LENINGRADSKII ordena Lenina gosudarstvennyi universitet imeni A. A. Zhdanova za gody sovetskoi vlasti: kratkii bibliograficheskii ukazatel' (Leningrad State University named for A. A. Zhdanov and awarded the Order of Lenin during the years of Soviet rule: short bibliography). Leningrad, Izdatel'stvo Leningradskogo Universiteta, 1967. 63 p. (Leningradskii Gosudarstvennyi Universitet imeni A. A. Zhdanova).

The Geography Faculty, p. 34-36, entries 307-333, contains 27 references on the history, programs, and faculty in geography in Leningrad University, 1917-1967.

ALEKSANDROV, Igor' Nikolaevich. Nauchno-teoreticheskie i metodiko-pedagogicheskie problemy geografii v Kazanskom universitete v sviazi s razvitiem geograficheskikh idei v Rossii--XIX-nachalo XX veka (Scientific-theoretical and methodological-pedagogic problems of geography in Kazan' university in relation to the development of geographical ideas in Russia, 19th to beginning of the 20th century). Kazan', Izdatel'stvo Kazanskogo Universiteta, 1964. 250 p. Bibliography, p. 235-250.

About 330 references.

BYKOV, Vasilii Dmitrievich, and others, eds. Sovetskaia geografiia v period stroitel'stva kommunizma (Soviet geography in the period of the development of communism). Moskva, Geografgiz, 1963. 487 p. Bibliographies at ends of individual articles.

Foreign: Arabic Geography

KRACHKOVSKII, Ignatii Iulianovich. Arabskaia geograficheskaia literatura. (Arabic geographical literature). Moskva-Leningrad. Izdatel'stvo Akademii Nauk SSSR, 1957. Bibliography, p. 761-814 (Izbrannye sochineniia, tom 4).

More than 1,000 references utilized by the author in the study of the history of geographical sciences and geographical discoveries by the Arabs. Alphabetical order. In Russian and foreign languages. Index of names of Eastern authors (in Russian transcription) and works on them.

PHILOSOPHY AND THEORY OF GEOGRAPHY

GI nauki i tekhniki, seriia geografiia: teoreticheskie voprosy geografii (Results
f science and technology, geography series: theoretical questions of geography).
Vsesoiuznyi Institut Nauchnoi i Tekhnicheskoi Informatsii). Moskva, VINITI, 1
1966). Irregular.   1836

   1. Geografiia naseleniia za rubezhom (Geography of population abroad).
966. 67 p. 224 references.

GI nauki i tekhniki: seriia teoreticheskie voprosy fizicheskoi i ekonomicheskoi
eografii (Results of science and technology: series theoretical problems of
hysical and economic geography). (Vsesoiuznyi Institut Nauchnoi i Tekhnicheskoi
nformatsii). Moskva, VINITI. 1- (1972-).   1837

   1. Obzor osnovnykh tendentsii razvitiia nauki i literatury (Survey of basic
endencies in the development of science and its literature). 1972. 151 p. 707
eferences.

ESNIK, Stanislav Vikent'evich. Predmet geograficheskikh nauk, ikh sistema i
classifikatsii (The object of the geographical sciences, their system and clas-
ification), in Itogi nauk i tekhniki, seriia Teoreticheskie voprosy fizicheskoi
i ekonomicheskoi geografii, v. 1, Moskva, VINITI, 1972, p. 11-25. Bibliography,
. 21-23. English summary, p. 23-25.   1838

   60 references in Russian, arranged alphabetically by author.

CHENKO, Anatolii Grigor'evich. Razvitie geograficheskikh idei (The development
f geographical concepts). Moskva, Izdatel'stvo "Mysl'," Geografgiz, 1971. 416
. Bibliography, p. 380-399.   1839

   About 650 references arranged alphabetically by name of author, those in
ussian first (about half the references) and those in the Latin alphabet later.
ndex of names.

CHIN, Vsevolod Alekseevich. Teoreticheskie problemy geografii (Theoretical
roblems in geography). Moskva, Geografgiz, 1960. 265 p. Bibliography, p. 245-
63 (Moskovskii Gosudarstvennyi Universitet. Geograficheskii Fakul'tet).   1840

   About 600 references on the theory and philosophy of geography from Classical
imes to the present with emphasis on developments in the Soviet Union. A suc-
essor volume by the same author, Teoreticheskie osnovy geografii (Theoretical
oundations of geography), Moskva, Izdatel'stvo "Mysl'," 1972, 430 p., does not
ontain a bibliography; thus for bibliographical purposes the earlier edition
ere listed is the important work.

HISTORY OF GEOGRAPHICAL EXPLORATION AND TRAVEL

General and World-Wide

ZDETSKII, Nikolai Andreevich. Sovetskie geograficheskie issledovaniia i
krytiia. (Soviet geographical research and exploration). Moskva, Izdatel'stvo
fysl'," 1967. 390 p. 1967.   1841

Textual and cartographic sources to each chapter, p. 350-388, with 684
eferences. Chapters 1-10 on exploration and discovery of the USSR and its polar
ossessions, chapters 11-16 on exploration and discovery in other parts of the
orld.

DOVICH, Iosif Petrovich. Ocherki po istorii geograficheskikh otkrytii (Essays
n the history of geographical exploration). Moskva, Izdatel'stvo "Prosveshchenie,"
?67. 714 p. Bibliography, p. 647-653. (Older ed., 1957. 752 p. Bibliography,
684-689, 280 references).   1842
About 400 references.

KIN, Naum Grigor'evich. Geograficheskie otkrytiia i nauchnoe poznanie zemli
geographical discovery and scientific knowledge of the earth). V. S.
eobrazhenskii, ed. Moskva, Izdatel'stvo "Mysl'," Geografgiz, 1972. 132 p.
bliography, p. 125-131. (Akademiia Nauk SSSR. Institut Geografii).   1843

About 110 references to the Russian literature on territorial discovery and
e discovery of physical-geographical regularities, the latter being the dis-
nctive viewpoint and contribution of the text of this book.

IAVLENSKII, Georgii Pavlovich. Russkie geografy i puteshestvenniki:
komendatel'nyi ukazatel' literatury. Vstupitel'naia stat'ia i nauchnaia
nsul'tatsiia D. M. Lebedeva (Russian geographers and travelers: guide to
commended literature. Introductory article and scientific consultation by
M. Lebedev). Moskva, 1955. 122 p. (Gosudarstvennaia Biblioteka SSSR imeni
nina).   1844

210 annotated references, including general works on the history of Russian
ographical discoveries, literature on individual explorations, from the 15th
ntury to 1954 (chronologically), lists of works of scholars and travelers and
terature on them. Geographical index. Index of authors and names.

, Lev Semënovich. Istoriia russkikh geograficheskikh otkrytii (History of
ssian geographical exploration). 2nd ed. Moskva, Izdatel'stvo Akademii Nauk
SR, 1962. 296 p.   1845
Bibliographies at the end of each chapter.

, Lev Semënovich. Ocherki po istorii russkikh geograficheskikh otkrytii.
ssays on the history of Russian geographical discoveries). 2nd ed. Moskva,
ningrad. Izdatel'stvo Akademii Nauk SSSR, 1949. 465 p. (1st ed., 1946.
8 p.). (Akademiia Nauk SSSR. Seriia "Itogi i Problemy Sovrmennoi Nauki").
ailable in German: Geschichte der russischen geographischen Entdeckungen.
sammelte Aufsätze. Leipzig, VEB Bibliographisches Institut, 1954. 283 p.
617].   1846

Bibliographies at ends of chapters and in notes. About 250 references.

INGLEZI, Raisa Markovna, and FRADKINA, Zinaida L'vovna. Glazami sovetskikh liudei: Putevye zapiski o zarubezhnykh stranakh. Rekomendatel'nyi ukazatel' literatury. (Through the eyes of Soviet people: travelers' writings on foreign countries: recommended guide to the literature). Moskva, Gosudarstvennaia Biblioteka SSSR imeni V. I. Lenina, 1956. 54 p.

61 references, annotated, of writings by Soviet travelers abroad, especially in other socialist countries. Alphabetical list of recommended books and articles

GREKOV, Vadim Ivanovich. Ocherki iz istorii russkikh geograficheskikh issledovanii v 1725-1765 gg. (Essays from the history of Russian geographical exploration 1725-1765). Moskva, Izdatel'stvo Akademii Nauk SSSR, 1960. 425 p. Bibliography, p. 389-407. (Akademiia Nauk SSSR. Institut Geografii).

About 620 references. About 800 names of persons listed in index of personal names, p. 409-416.

IATSUNSKII, Viktor Kornel'evich. Istoricheskaia geografiia: istoriia eë vozniknoveniia i razvitiia v XIV-XVIII vekakh (Historical geography: the history of its origins and development in the 14th-18th centuries). Moskva, Izdatel'stvo Akademii Nauk SSSR, 1955. 331 p. Index of names of authors and travelers, p. 319-329. (Akademiia Nauk SSSR. Institut Istorii).

About 950 bibliographical footnotes. Name index has about 1,000 entries.

Regional on the Territory of the U. S. S. R.

Europe

ESAKOV, Vasilii Alekseevich, and SOLOV'ËV, Aleksandr Ivanovich. Russkie geograficheskie issledovaniia Evropeiskoi Rossii i Urala v XIX-nachale XX v. (Russian geographical research on European Russia and the Urals in the 19th and early 20th centuries). Moskva, Izdatel'stvo "Nauka," 1964. 179 p. (Akademiia Nauk SSSR. Institut Istorii Estestvoznaniia i Tekhniki).

438 bibliographical footnotes.

The Arctic

BELOV, Mikhail Ivanovich. Put' cherez Ledovityi okean: ocherki iz istorii otkrytiia i osvoeniia Severnogo morskogo puti (Path through the Arctic Ocean: essays from the history of discovery and utilization of the Northern Sea Route). Moskva, Izdatel'stvo "Morskoi Transport," 1963. 237 p. Bibliography, p. 231-235.

106 references.

BELOV, Mikhail Ivanovich. Osnovnaia literatura po istorii geograficheskikh otkrytii na severe Rossii i russkogo arkticheskogo moreplavaniia: period dokapitalistichesk (Basic literature on the history of geographical discovery in northern Russia and Russian Arctic navigation: pre-capitalist period). Istoriia otkrytiia i osvoeniia Severnogo Morskogo Puti (History of discovery and utilization of the Northern Sea Route). v. 1. Arkticheskoe moreplavanie s drevneishikh vremën do serediny XIX veka (Arctic navigation from ancient times to the middle of the 19th century). Moskva, 1956. Bibliography, p. 534-548. (Arkticheskii Nauchno-Issledovatel'skii Institut). With N. I. Bashmurina.

439 publications in Russian 1745-1955.

PINKHENSON, Dmitrii Moiseevich. Problemy Severnogo Morskogo Puti v epokhu kapitalizma (Problems of the Northern Sea Route in the period of capitalism. v. 2 of Istoriia otkrytiia i osvoeniia Severnogo Morskogo Puti (History of the discovery and utilization of the Northern Sea Route). M. I. Belov. Leningrad, Morskoi Transport, 1962. Bibliography, p. 13-17, 700-722.

About 800 references for 1847-1961.

1847 - 1853

V, Mikhail Ivanovich. <u>Sovetskoe arkticheskoe moreplavanie 1917-1932 gg.</u>
oviet Arctic navigation, 1917-1932). v. 3 of his <u>Istoriia otkrytiia i osvoeniia</u>
<u>vernogo Morskogo Puti</u> (History of the exploration and utilization of the Nor-
ern Sea Route). Leningrad, Morskoi Transport, 1959. Bibliography, p. 465-483
rkticheskii i Antarkticheskii Nauchno-Issledovatel'skii Institut).   1854

About 460 references to the basic literature of Soviet Arctic navigation,
blished 1919-1957, and about 140 references to manuscript material for the
riod 1918-1932 in the archives of the archives of the Arctic and Antarctic
ientific-Research Institute in Leningrad.

V, Mikhail Ivanovich. <u>Nauchnoe i khoziaistvennoe osvoenie Sovetskogo severa</u>
<u>33-1945 gg.</u> (Scientific and economic development of the Soviet North, 1933-
45). v. 4. of his <u>Istoriia otkrytiia i osvoeniia Severnogo Morskogo Puti</u>
istory of the exploration and utilization of the Northern Sea Route). Lenin-
ad, Gidrometeorologicheskoe Izdatel'stvo, 1969. 616 p. Bibliography, p. 554-
4.   1855
About 1.200 references.

MURINA, N. I. Osnovnaia literatura po istorii sotsialisticheskogo osvoeniia
vera i nachal'nogo perioda ekspluatatsii Severnogo morskogo puti, 1933-1945
asic literature on the history of socialist development of the North and the
itial period of the exploitation of the Northern Sea Route, 1933-1945). In
e book: <u>Istoriia otkrytiia i osvoeniia Severnogo morskogo puti</u> (History of
scovery and development of the Northern Sea Route), v. 4. <u>Nauchnoe i</u>
oziaistvennoe osvoenie Sovetskogo Severa, 1933-1945 (Scientific and economic
velopment of the Soviet North, 1933-1945). M. I. Belov. Leningrad, Gidro-
teoizdat, 1969. 616 p. Bibliography, p. 563-584. (Arkticheskii i Antarkti-
eskii Nauchno-Issledovatel'skii Institut).   1856

About 850 references. Name and place index includes bibliography.

IONOV, Aleksandr Fedorovich, and ROMANOVICH, Z. S. Sokrashchennyi spisok
vetskoi literatury po issledovaniiu Arkticheskogo basseina vysokoshirotnymi
speditsiiami i dreifuiushchimi stantsiiami, 1937-1962 gg. (Abridged list of
viet literature on investigation of the Arctic basin by high-latitude expedi-
ons and drifting stations, 1937-1962). <u>Problemy Arktiki i Antarktiki,</u> 1962,
. 11, p. 115-128.   1857

About 340 references in alphabetical order.

GIN, Konstantin Sergeevich. <u>Po studěnym moriam</u> (Through the bitter cold seas).
skva, Geografgiz, 1956. 424 p. Bibliography, p. 308-317, 415-416.   1858

247 references on the history of Russian ice navigation and 22 references
glossaries.

IONOV, Aleksandr Fedorovich. <u>Severnyi polius</u> (The North Pole). 3rd ed.
skva, "Morskoi Transport," 1960. 525 p. Bibliography, p. 503-506.   1859

83 references on history of exploration.

EV, Dm. D. Perechen' literatury (russkoi i inostrannoi) o Severnom morskom
ti iz Evropy v reki Ob' i Enisei (po 1912 g.). (List of Russian and foreign
terature on the Northern Sea Route from Europe to the Ob' and Yenisey rivers up to
12). In the book: Rudnev, D. D. and Kulik, N. A., <u>Materialy k izucheniiu Severnogo</u>
rskogo puti iz Evropy v Ob' i Enisei (Materials for the study of the Northern
a Route from Europe to the Ob' and Yenisey). Petrograd, 1915, p. 61-119.
mmary in English.   1860

767 references (404 in foreign languages) mainly on geographical exploration
the Arctic in the 19th and early 20th centuries, in alphabetical order.

TSKII, Vasilii Mikhailovich. <u>Issledovateli Arktiki-vykhodtsy iz Estonii</u>
xplorers of the Arctic--Estonians). Tallin. Eesti Razmat, 1970. 350 p.
bliography, p. 339-349. (Seriia Opisanii Puteshestvii).   1861

About 210 references.

                                                                      1854 - 1861

Soviet Asia as a Whole

AZAT'IAN, Armen Arshavirovich, BELOV, Mikhail Ivanovich, GVOZDETSKII, Nikolai
Andreevich, and others. Istoriia otkrytiia i issledovaniia sovetskoi Azii
(History of discovery and exploration of Soviet Asia). Moskva, Izdatel'stvo
"Mysl'," 1969. 535 p. Bibliography, p. 495-507 (Otkrytie Zemli).   1

360 references arranged by chapters.

Siberia

NAUMOV, Gurii Vasil'evich. Russkie geograficheskie issledovaniia Sibiri v XIX--
nachale XX v. (Russian geographical exploration of Siberia in the 19th and early
20th centuries). Moskva, Izdatel'stvo "Nauka," 1965. 148 p. (Akademiia Nauk
SSSR. Institut Istorii Estestvoznaniia i Tekhniki).   1

229 references in a historical review of expedition activity of Russian
scholars and travelers in pre-Revolutionary times.

SKALON, Vasilii Nikolaevich. Russkie zemleprokhodtsy--issledovateli Sibiri XVII
veka (Russian land travelers--explorers of Siberia in the 17th century). M. A.
Sergeev, ed. Moskva, 1952. 198 p. Bibliography, p. 190-198 (Moskovskoe
Obshchestvo Ispytatelei Prirody. Istoricheskaia seriia, no. 44).   1

280 references.

NORDEGA, Igor' Glebovich. Istoriia geograficheskogo izucheniia Tuvy vo vtoroi
polovine XIX i pervoi polovine XX v. (History of geographical study of Tuva in
the 2nd half of the 19th century and the 1st half of the 20th), Akademii Nauk SSSR,
Institut Istorii estestvoznaniia i tekhniki, Trudy, v. 27 (1959). Bibliography,
p. 103-111.   1

192 references in Russian and other languages.

The Caucasus

GVOZDETSKII, Nikolai Andreevich. Issledovaniia Kavkaza (Exploration of the
Caucasus), in the book, Russkie geograficheskie issledovaniia Kavkaza i Srednei
Azii v XIX--nachale XX v. (Russian geographical exploration of the Caucasus
and Middle Asia in the 19th and early 20th centuries). Moskva, Izdatel'stvo
"Nauka," 1964, p. 3-60. (Akademiia Nauk SSSR. Institut Istorii Estestvoznaniia
i Tekhniki).

198 references in historical review of expeditions of Russian scholars and
travelers in the pre-revolutionary period.

POLIEVKTOV, Mikhail Aleksandrovich. Evropeiskie puteshestvenniki XIII-XVIII vv.
po Kavkazu (European travelers of the 13th-18th centuries in the Caucasus),
Tiflis, 1935. 231 p. (Akademiia Nauk SSSR. Nauchno-issledovatel'skii Institut
Kavkazovedeniia imeni Marra).

Soviet Middle Asia

AZAT'IAN, Armen Arshavirovich, DONTSOVA, Zoia Nikiforovna, and FEDCHINA, Vera
Nikolaevna. Istoriia issledovaniia Srednei Azii (History of exploration of
Middle Asia), in the book: Russkie geograficheskie issledovaniia Kavkaza i
Srednei Azii v XIX--nachale XX v. (Russian geographical exploration of the
Caucasus and Middle Asia in the 19th and early 20th centuries). Moskva, Izdatel'-
stvo "Nauka," 1964. p. 61-157 (Akademiia Nauk SSSR. Institut Istorii
Estestvoznaniia i Tekhniki).

133 references in historical review of expedition activity of Russian scholar
and travelers in the prerevolutionary period.

1862 - 1868

,OVA, O. V. Obzor russkikh puteshestvii i ekspeditsii v Sredniuiu Aziiu.
terialy k istorii izucheniia Srednei Azii. Chast' 1-3 (Survey of Russian
plorers and expeditions in Middle Asia. Material for the history of the study
Middle Asia, v. 1-3). Tashkent, 1955-1962 (Sredneaziatskii Gosudarstvennyi
iversitet. Materialy k bibliografii, vypuski 5, 7, 9).   1869
  v. 1 ... 1715-1856 gg. 1955. 83 p. (Materialy..vypusk 5)
  v. 2 ... 1856-1869 gg. 1956. 102 p. (Materialy..vypusk 7)
  v. 3 ... 1869-1880 gg. 1962. 181 p. (Materialy..vypusk 9)
  Short descriptions of 77 Russian expeditions and explorations in Middle Asia,
15-1880. Bibliography of published works of participants in each expedition,
works on it, and works related to expeditions and travels. Indexes of names
d places.

HINA, Vera Nikolaevna. Kak sozdavalas' karta Srednei Azii (How the map of
viet Middle Asia was created). Moskva, Izdatel'stvo "Nauka," 1967. 132 p.
bliography, p. 121-128 (Akademiia Nauk SSSR. Institut Istorii Estestvoznaniia
Tekhniki).   1870
  About 280 references.

'IAN, Armen Arshavirovich. Vydaiushchiesia issledovateli prirody Srednei Azii:
oraia polovina XIX v. Chast' I. (Leading investigators of nature in Middle
ia, 2nd half of the 19th century. Part I). Tashkent, "Sredniaia i vysshaia
kola," 1960. 172 p.   1871
  Bibliography of the works of P. P. Semenov-Tyan-Shanskiy, A. P. Fedchenko,
F. Oshanin, and works on them.

'IAN, Armen Arshavirovich. Vydaiushchiesia issledovateli prirody Srednei Azii:
oraia polovina XIX v. Chast' 2. (Outstanding explorers of nature in Middle
ia: second half of the 19th century, Part 2). N. A. Severtsov and A. N. Krasnov.
shkent, "Uchitel'," 1966. 223 p. Bibliography of publications by explorers
on them, p. 202-222.   1872
  About 500 references.

Regional Outside the U.S.S.R.

Navigation on Seas and Oceans

IE moreplavateli (Russian navigators). V. S. Lupach, ed. Moskva, Voenizdat,
3. 671 p.   1873
  Biographical information on about 500 Russian navigators, p. 473-577, and
ssian names found on maps, p. 578-620.

7, Nikolai Nikolaevich. Otechestvennye moreplavateli-issledovateli morei i
anov (Native navigators and explorers of the seas and oceans). Moskva,
grafgiz, 1954. 437 p.   1874
  Biographies of more than 100 persons, arranged by historical periods.

Land Areas of Europe and Asia

OVICH, Iosif Petrovich, and MAGIDOVICH, Vadim Iosifovich. Istoriia otkrytiia
ssledovaniia Evropy (History of the discovery and study of Europe). Moskva,
atel'stvo "Mysl'," 1970. 453 p. Bibliography, p. 402-410. (Otkrytie Zemli). 1875
  About 280 references by sections. Index of place names and persons.

SHCHUKINA, Nina Mikhailovna. Kak sozdavalas' karta Tsentral'noi Azii: raboty russkikh issledovatelei XIX i nachala XX v. (How the map of Central Asia was formed: work of Russian investigators in the 19th and early 20th centuries). Moskva, Geografgiz, 1955. 240 p. Bibliography, p. 226-235.

About 230 references, in alphabetical order.

America, Africa, Australia, and Oceania

MAGIDOVICH, Iosif Petrovich. Istoriia otkrytiia i issledovaniia Tsentral'noi i Iuzhnoi Ameriki (History of discovery and research on Central and South America). Moskva, "Mysl'," 1965. 455 p. Bibliography, p. 418-426. (Otkrytie zemli).

About 240 references, arranged by chapters.

GORNUNG, Mikhail Borisovich, LIPETS, Iulii Grigor'evich, and OLEINIKOV, I. N. Istoriia otkrytiia i issledovaniia Afriki (History of discovery and exploration of Africa). Moskva, Izdatel'stvo "Mysl'," 1973. 454 p. Bibliography, p. 439-451.

ZABRODSKAIA, Mariia Pavlovna. Spisok osnovnykh trudov russkikh puteshestvennikov po Afrike (List of the basic works of Russian travelers in Africa) in her book: Russkie puteshestvenniki po Afrike (Russian travelers in Africa). Moskva, Geografgiz, 1955. 88 p. Bibliography, p. 84-87.

83 references for 1840-1953.

SVET, Iakov Mikhailovich. Istoriia otkrytiia i issledovaniia Avstralii i Okeanii (History of the discovery and exploration of Australia and Oceania). Moskva, Izdatel'stvo "Mysl'," 1966. 400 p. Bibliography, p. 361-373 (Otkrytie Zemli).

293 references, organized by chapters.

Antarctica

TRESHNIKOV, Aleksei Fedorovich. Istoriia otkrytiia i issledovaniia Antarktidy. (History of exploration and research on Antarctica). Moskva, Geografgiz, 1963. 431 p. Bibliography, p. 418-425 (Otkrytie zemli).

206 references (61 in foreign languages), organized by general works and by chapters.

NUDEL'MAN, Aizek Vol'fovich. Sovetskie ekspeditsii v Antarktiku (1955-1963 gg.). (Soviet expeditions in Antarctica, 1955-1963). Moskva, Izdatel'stvo Akademii Nauk SSSR, 1959-1965. (Mezhdunarodnyi Geofizicheskii God. 1957-1959. Akademiia Nauk SSSR. Mezhduvedomstvennyi Geofizicheskii Komitet). Table of contents and summary also in English.

... 1955-1959 gg. 1959. Bibliography, p. 110

... 1958-1960 gg. 1960. Bibliography, p. 99

... 1959-1961 gg. 1962. Bibliography, p. 131

... 1961-1963 gg. 1965. Bibliography, p. 242-244.

About 130 references, on Soviet research in Antarctica under the International Geophysical Year, 1957-1959.

ALEINER, Aron Zalmanovich. Osnovnye etapy geograficheskogo issledovaniia Antarktik (Basic stages in the geographical exploration of Antarctica). In the book: Antarktika. Materialy po istorii issledovaniia i po fizicheskoi geografii (Antarctica. Materials on the history of exploration and on physical geography). Moskva, Geografgiz, 1958. p. 54-94 (Geograficheskoe Obshchestvo SSSR).

134 references (116 in foreign languages) in historical survey of expeditions in Antarctica 1819-1952.

1876 - 1883

HISTORICAL GEOGRAPHY

Historical geography usually is not recognized as a separate category in
)viet bibliographies but valuable materials may be located through a variety
? bibliographies, as for example in retrospective bibliographies of bibliog-
iphies, entries 12-15; retrospective bibliographies in general, entries 23-40;
italogue of the Library of the Geographical Society, entry 41; bibliographies
f historical maps and atlases, entries 127-134; encyclopedias, entries 844-875;
ider gazetteers of the Russian Empire, entries 898-904; paleogeography, entries
300-1302; historical geography of vegetation, entries 1543-1546; history of
?ography, entries 1797-1835; history of geographical exploration and travel,
ntries 1841-1872; history, entries 1892-1894; major general bibliographies of
he Soviet Union as a whole, entries 1951-1962; and historical geography of
rmenia, entries 2237-2238.

3IZHEV, Vladimir Zinov'evich, KOVALCHENKO, Ivan Dmitrievich, and MURAV'ËV,
iatolii Vasil'evich. Istoricheskaia geografiia SSSR. Moskva, "Vysshaia Shkola,"
)73. 320 p.   1884

)RICHESKAIA geografiia Rossii (Historical geography of Russia), B. A. Rybakov,
. P. Nikitin, Ia. E. Bodarskii, eds. Voprosy geografii (Geograficheskoe Obshche-
tvo SSSR. Moskovskii Filial. Nauchnye sborniki), sbornik 83, Moskva, Izdatel'-
tvo "Mysl'," Geografgiz, 1970. 216 p. Bibliographies at ends of articles.
upplementary table of contents and abstracts in English.   1885

)RICHESKAIA geografiia (Historical geography). V. K. Iatsunskii, ed. Voprosy
eografii (Geograficheskoe Obshchestvo SSSR. Moskovskii Filial. Nauchnye sbor-
iki), sbornik 50. Moskva, Geografgiz, 1960. 268 p. Bibliographies at ends of
rticles or in form of footnotes or remarks. Supplementary tables of contents in
nglish and Esperanto.   1886

ORICHESKAIA geografiia (Historical geography), N. N. Baranskii and V. K. Iatsun-
kii, eds. Voprosy geografii (Geograficheskoe Obshchestvo SSSR. Moskovskii
ilial. Nauchnye sborniki), sbornik 31. Moskva, Geografgiz, 1953. 286 p.   1887

ORICHESKAIA geografiia SSSR (Historical geography of the USSR). V. K. Iatsunskii,
d. Voprosy geografii (Geograficheskoe Obshchestvo SSSR. Moskovskii Filial.
auchnye sborniki), sbornik 20. Moskva, Geografgiz, 1950. 356 p. Bibliographical
ootnotes and bibliographical notes.   1888

SUNSKII, Viktor Kornel'evich. Istoricheskaia geografiia kak nauchnaia distsiplina
Historical geography as a scholarly discipline), Voprosy geografii, v. 20 (1950),
. 13-41. Bibliographical footnotes.   1889

IAROV, L. F. Pereselenie i zemleustroistvo v Sibiri v gody stolypinskoi agrarnoi
eformy (Migration and land development in Siberia in the years of the Stolypin
grarian reforms). Leningrad, Izdatel'stvo Leningradskogo Universiteta, 1962.
58 p. Bibliography, p. 562-586 (Leningradskii Gosudarstvennyi Pedagogicheskii
nstitut imeni Gertsena).   1890

   More than 750 references, including archival materials.

SUNSKII, Viktor Kornel'evich. Istoricheskaia geografiia: istoriia eë voznikno-
eniia i razvitiia v XIV-XVIII vekakh (Historical geography: history of its begin-
ing and development in the 14th-18th centuries). S. D. Skazkin, ed. Moskva,
955. 328 p.   1891

AV'ËV, Anatolii Vasil'evich, and SAMARKIN, Viacheslav Viktorovich. Istoricheskaia
eografiia epokhi feodalizma: Zapadnaia Evropa i Rossiia v V-XVII vv. (Historical
eography of the epoch of feudalism: Western Europe and Russia in the 5th-17th
enturies). Moskva, "Prosveshchenie," 1973. 144 p.   1891a

HISTORY

ISTORIIA (History), Bibliografiia sovetskoi bibliografii (Bibliography of Soviet bibliography). 1939, 1946- . Annual. Moskva, Izdatel'stvo "Kniga," 1941, 1948- (Vsesoiuznaia Knizhnaia Palata).

Annual list of Soviet bibliographies in history. The 1970 annual volume (1972), p. 31-41, contains 204 entries under four categories: general questions; auxiliary historical disciplines, archeology, ethnography; general history, history of foreign countries; history of the USSR.

SPRAVOCHNIK po istorii dorevoliutsionnoi Rossii: bibliografiia (Handbook on the history of Prerevolutionary Russia: bibliography). G. A. Glavatskikh and others. P. A. Zaionchkovskii, ed. Moskva, Izdatel'stvo "Kniga," 1971. 515 p. (Gosudarstvennaia Biblioteka SSSR imeni V. I. Lenina...).

RUSSKAIA istoricheskaia bibliografiia, 1855-1864 (Russian historical bibliography, 1855-1864). P. P. Lambin. Sanktpeterburg, Imperatorskaia Akademiia Nauk, 1861-1884. 10 v.

In Part I, History of Russia, a section covers historical-statistical works, geography and topography of oblasts, cities, and remarkable places, and ethnography with varying titles, as follows:

        v. 1.   1855 (1861), p. 80- 98, entries 783- 968.
        v. 2.   1856 (1861), p. 23- 52, entries 281- 682.
        v. 3.   1857 (1865), p. 18- 60, entries 233- 930.
        v. 4.   1858 (1867), p. 16- 46, entries 211- 774.
        v. 5.   1859 (1868), p. 26- 77, entries 349-1409.
        v. 6.   1860 (1869), p. 27- 83, entries 369-1437.
        v. 7.   1861 (1870), p. 25- 70, entries 357-1273.
        v. 8.   1862 (1872), p. 30- 77, entries 411-1343.
        v. 9.   1863 (1877), p. 36- 86, entries 707-2033.
        v. 10. 1864 (1884), p. 53-127, entries 915-2663.

Indexes of personal names and of geographical and other names. Lists of periodicals.

1892 - 1894

D. METHODS IN GEOGRAPHY: CARTOGRAPHIC, STATISTICAL, AND REMOTE SENSING

GENERAL METHODS

, Vasilii Dmitrievich, and others, eds. Metody geograficheskikh issledovanii
»thods of geographical research). Moskva, Geografgiz, 1960. 392 p. Bibliog-
»hies at ends of individual articles. (Moskovskii Gosudarstvennyi Universitet
»ni Lomonosova).   1895

CARTOGRAPHY AND GEODESY

Major General Bibliographies

)EZIIA i kartografiia (Geodesy and cartography). Moskva. Glavnoe Upravlenie
∍odezii i Kartografii SSSR. 1956- . monthly.   1896

    Section "Bibliografiia" (bibliography) covers books, articles, and maps,
ɪth entries arranged alphabetically within each of the sections. In a typical
∍sue the bibliography consists of about three or four pages, for example, p.
ɣ-80.

)EZIIA. Topografiia. Kartografiia (Geodesy, topography, and cartography),
ɪbliografiia sovetskoi bibliografii (Bibliography of Soviet bibliographies).
ɪsesoiuznaia Knizhnaia Palata). 1939, 1946- . Annual. Moskva, Izdatel'stvo
ᵹniga," 1941, 1948- .   1897

    Annual list of Soviet bibliographies on geodesy, topography, and cartography,
ɪblished as separate works or as parts of books or articles. The 1970 annual
ɔlume (1972), p. 56-57, contains 15 entries. A sub-section of the section on
ɦysical-mathematical sciences in the chapter on natural sciences and mathematics.

ɟI nauki i tekhniki: seriia kartografiia. 1- 1962- . Moskva, VINITI, 1964- .
ʀregular. v. 1-4, 1962-1969 as Itogi nauki, seriia geografiia: kartografiia.
. 1 lacks volumber number. v. 5 does not carry a year. v. 1-2 issued by In-
 titut Nauchnoi Informatsii, v. 3-4 by Vsesoiuznyi Institut Nauchnoi i Tekhnicheskoi
ɪformatsii. K. A. Salishchev, ed. [v. 1], 1962 (1964), 906 references; v. 2,
ꝰ63-1964 (1966), 1,014 references; v. 3, 1965-1967 (1968), 912 references; v. 4,
ꝰ67-1969 (1970); v. 5, (1972).   1898

    Review articles and bibliographies in the principal fields of cartography,
few of which are here individually listed.

    v. 6, K. A. Salishchev and Z. G. Riabtseva, eds. Moskva, VINITI, 1974. 239 p.
∍ the following articles, each with bibliographies:

K. A. Salishchev, Complex cartography, p. 6-17 (60 references);
A. M. Berliant, Cartographic methods of research, p. 18-44 (159 references);
A. I. Martynenko, Automation in cartography, p. 45-80 (150 references);
G. A. Meshcheriakov, Mathematical cartography, p. 81-92 (79 references);
O. A. Evteev, Iu. G. Kel'ner, and M. I. Nikishov, Thematic cartography,
      p. 93-122 (166 references);
K. A. Bogdanov and A. V. Pavlova, Marine cartography, p. 123-132 (41 references);
E. M. Pospelov, Cartography of cities, p. 133-144 (54 references);
A. B. Rogov and V. N. Filin, Techniques and technology in the making ofmaps,
      p. 145-177 (232 references);
L. A. Bogomolov, The utilization of air photos and satellite photos in geo-
      graphical research, p. 178-203 (86 references);
O. A. Evteev, Cartographic education, p. 204-217 (101 references); and
K. A. Salishchev, History of cartography, p. 218-238 (112 references).

KARPOV, N. S. Bibliograficheskii ukazatel' literatury po kartografii, izdannoi
  v SSSR v 1938-1958 (Bibliographical guide to the literature of cartography,
  published in the USSR, 1938-1958). V. N. Lysiuk, ed. Moskva, Geodezizdat, 1962.
  208 p. (Glavnoe Upravlenie Geodezii i Kartografii Ministerstva Geologii i Okhrany
  Nedr SSSR. Tsentral'nyi Nauchno-Issledovatel'skii Institut Geodezii, Aeros"ëmki
  i Kartografii, Trudy, vypusk 156).   1
      2,691 entries arranged systematically, partly annotated. Author index.

BIBLIOGRAFICHESKII ukazatel' geodezicheskoi i kartograficheskoi literatury za 20
  let, 1917-1937 gg. (Bibliographical guide to geodetic and cartographic literature for the 20 years 1917-1937). Vypusk 1. Knigi i broshiury (v. 1, books
  and pamphlets). A. S. Chebotarev, ed. Moskva, 1938. 126 p. (Biblioteka
  Moskovskogo Instituta Inzhenerov Geografii, Aerofotos"ëmki i Kartografii
  MIIGAiK).   1

SHIBANOV, Fëdor Anisimovich. Ukazatel' kartograficheskoi literatury, vyshedshei
  v Rossii s 1800 po 1917 god (Guide to cartographic literature published in
  Russia from 1800 to 1917). Leningrad, Izdatel'stvo Leningradskogo Universiteta,
  1961. 223 p.   1
      2,645 entries organized systematically. Alphabetical author index. Accord-
  ing to the Canadian cartographer, v. 10, no. 1 (June, 1973), p. 25, the author
  has an unpublished bibliography of Soviet cartographic literature, 1917-1962,
  with 12,500 titles.

MIROMANOVA, V. M. Bibliograficheskii ukazatel' opublikovannykh knig i statei
  nauchnykh rabotnikov i aspirantov NIIGAiK za 1932-1967 gg. (Bibliographical
  guide to the published books and articles of the scientific workers and graduate
  students of the Novosibirsk Institute of Engineers of Geodesy, Aerial Surveying,
  and Cartography, 1932-1967). Novosibirsk, 1968. 59 p. (Novosibirskii Institut
  Inzhenerov Geodezii, Aerofotos"emki i Kartografii).   1
      607 entries arranged systematically. Includes abstracts of dissertations.
  Author index.

RIABTSEVA, Z. G., VILENSKAIA, G. B., and MYSHEVA, I. A. Periodicheskaia literatura
  po kartografii za rubezhom (Periodical literature on cartography abroad), in
  Itogi nauki, seriia geografiia: kartografiia 1963-1964, v. 2. Moskva, VINITI,
  1966, p. 113-136.   1
      108 geographical or other periodicals outside the USSR with substantial
  material on cartography.

Special Topics

50 LET sovetskoi geodezii i kartografii (50 years of Soviet geodesy and cartography).
  Moskva, Izdatel'stvo "Nedra," 1967. 447 p. (Glavnoe Upravlenie Geodezii i Karto-
  rafii pri Sovete Ministrov SSSR. Voenno-Topograficheskaia Sluzhba Sovetskoi Armii)
                                                                               1

FEL' , Sergei Efimovich. Kartografiia Rossii XVIII veka (Cartography of Russia in
  the 18th century). Moskva, Izdatel'stvo Geodezicheskoi Literatury, 1960. 226 p. 1

CHURKIN, Vladimir Gerasimovich. Atlasnaia kartografiia (Atlas cartography).
  Leningrad, Izdatel'stvo "Nauka," 1974. 140 p. Bibliography, p. 133-139.   1

VRILOVA, Serafima Abramovna. Istoriia kartografii (History of cartography), in Itogi nauki [Seriia geografiia]: kartografiia 1962 [v. 1] Moskva, VINITI, 1964, p. 83-111.   1907

404 entries on history of cartography, published 1958-1963, organized by a general section, then regionally by countries, followed by history of special cartography, of atlases, of globes, individual elements on maps, and descriptions of collections and catalogues of ancient maps.

___ . ___ . ___ 1963-1964, v. 2. 1966, p. 137-162.

434 references.

LISHCHEV, Konstantin Alekseevich. Tematicheskaia kartografiia (Thematic cartography), Itogi nauki, seriia geografiia: kartografiia 1965-1967, v. 3. Moskva, VINITI, 1968, p. 127-146.   1908

171 references on thematic maps, 72 in Cyrillic, mainly Russian, and 99 in Western languages in Latin letters.

___ . ___ . ___ 1967-1969, v. 4, 1970, p. 81-87.

33 references for the period 1967-1969.

TEEV, Oleg Aleksandrovich, KEL'NER, Iurii Georgievich, NIKISHOV, Maksim Ivanovich. Tematicheskaia kartografiia, in Itogi nauki i tekhniki: seriia kartografiia, v. 5. Moskva, VINITI, 1972, p. 67-98.   1909

153 references.

MATICHESKAIA kartografiia (Thematic cartography). M. M. Mekler, and others, eds. Moskva, 1973. 55 p. (Geograficheskoe Obshchestvo SSSR. Moskovskii Filial).   1910

List of doctoral and candidate dissertations in cartography, defended during the period 1961-1970, compiled by S. M. Gul', p. 43-54. 70 entries.

NZBURG, Georgii Aleksandrovich. Matematicheskaia kartografiia (Mathematical cartography), in Itogi nauki [Seriia geografiia]: kartografiia 1962 [v. 1] Moskva, VINITI, 1964, p. 5-13.   1911

29 references.

___ . ___ . ___ 1963-1964, v. 2. 1966, p. 6-18

58 references for 1963-1964.

___ . ___ . ___ 1965-1967, v. 3. 1968, p. 7-22

93 references for 1965-1967.

VLOV, Aleksandr Andreevich. Matematicheskaia kartografiia (Mathematical cartography), in Itogi nauki. Seriia geografiia: kartografiia 1967-1969. v. 4. Moskva, VINITI, 1970, p. 32-42.   1912

58 references for 1967-1969.

___ . ___ . in Itogi nauki i tekhniki: seriia kartografiia, v. 5. Moskva, 1972. p. 53-66.   1913

80 references.

LISHCHEV, Konstantin Alekseevich. Ob avtomatizatsii v kartografii (On automation in cartography), in Itogi nauki, seriia geografiia: kartografiia 1963-1964, v. 2, Moskva, VINITI, 1966, p. 19-29.   1914

31 references.

VASMUT, A. S. Avtomatizatsiia v kartografii (Automation in cartography), in
Itogi nauki, seriia geografiia: kartografiia 1965-1967, v. 3. Moskva, VINITI,
1968, p. 34-56.   19.
     86 references.

_____ . _____ . _____ 1967-1969, v. 4, 1970, p. 43-64.   19
     90 references.

MARTYNENKO, A. I. Avtomatizatsiia v kartografii (Automation in cartography), in
Itogi nauki i tekhniki: seriia kartografiia, v. 5. Moskva, VINITI, 1972, p. 24-
52.   19
     150 references.

MEKLER, Morits Maksovich. Redaktirovanie kompleksnykh atlasov zarubezhnykh stran
(Editing complex atlases of foreign countries). Moskva, Izdatel'stvo "Nedra,"
1968. 132 p. Bibliography, p. 127-131.   19
     162 entries, including Soviet and foreign geographical atlases and maps.

KEL'NER, Iuri Georgievich, ASMUS, Ingrid Gerbertovna, and RIKHTER, Tamara Iakovlevna.
Redaktirovanie i sostavlenie kart rel'efa, gidrologicheskikh i zhivotnogo mira
dlia atlasov mira i materikov (Editing and compiling maps of relief, hydrology,
and animal life for atlases of the world and of the continents). N. V. Leont'ev,
ed. Moskva, Izdatel'stvo "Nedra," 1974. 121 p. Bibliography, p. 112-120.
(Glavnoe Upravlenie Geodezii i Kartografii pri Sovete Ministrov SSSR. Tsentral'nyi
Nauchno-Issledovatel'skii Institut Geodezii, Aeros"ëmki i Kartografii, Trudy,
vypusk 182).   19

INZHENERNO-GEOLOGICHESKOE kartirovanie: rekomandatel'nyi ukazatel' literatury
(Engineering-geological mapping: guide to recommended literature). N. I. Zhaldak,
comp. Kiev, 1969. 54 p. (Ukrainskii Gosudarstvennyi Institut Inzhenerno-
Tekhnicheskikh Izyskanii BNTI).   19
     About 210 references, systematically arranged, partly annotated, both Soviet
and foreign.

NIKOL'SKAIA, E. A. K voprosu o metodakh sostavleniia gidrogeologicheskikh kart
(On the question of methods of compilation of hydrogeological maps). Itogi nauki.
Seriia geologiia. Gidrogeologiia. Inzhenernaia geologiia. Moskva, VINITI,
1970. p. 5-47.   19
     267 references.

GIDROGEOLOGICHESKOE kartirovanie: rekomendatel'nyi ukazatel' literatury (Hydrogeological mapping: guide to recommended literature). N. I. Zhaldak, comp. Kiev,
1969. 64 p. (Ukrainskii Gosudarsvennyi Institut Inzhenerno-Tekhnicheskikh
Izyskanii. BNTI).   19
     About 260 references, arranged systematically, partly annotated, both Soviet
and foreign.

BERLIANT, A. M. Kartograficheskii metod issledovaniia (The cartographic method
of research), in Itogi nauki, seriia geografiia: kartografiia 1967-1969, v. 4.
Moskva, VINITI, 1970, p. 65-80.   19
     107 references on the utilization of cartographic methods in research,
mainly in Russian, especially for geomorphology.

BERLIANT, A. M. Kartograficheskii metod issledovaniia prirodnykh iavlenii: teksty
lektsii (The cartographic method of investigation of natural phenomena: text of
lectures). Moskva, 1971. 121 p. Bibliography, p. 116-121. (Moskovskii Gosudarstvennyi Universitet imeni M. V. Lomonosova. Geograficheskii Fakul'tet.
Kafedra Geodezii i Kartografii).   1

ANIKASHVILI, Aleksandr Fedorovich. Kartografiia: voprosy obshchei teorii
Cartography: problems of general theory). Tbilisi, "Metsniereba," 1968. 298 p.
ibliography, p. 284-292. (Institut Geografii imeni Vakhushti). In Georgian.
able of Contents also in Russian and English.   1924
    About 260 references.

EV, Shirbek Dzhafar ogly. Razvitie kartografii v Azerbaidzhane za 50 let.
eziume (Development of cartography in Azerbaidzhan during 50 years. Summary).
kademiia Nauk Azerbaidzhanskoi SSR. Institut Geografii, Trudy, v 18 (1971),
. 291-297. Bibliography, p. 294-297.   1925

Geodesy

ERATIVNYI zhurnal: geodeziia i aeros"emka (Reference journal: geodesy and
erial surveying). 1963-   monthly. Moskva, Vsesoiuznyi Institut Nauchnoi i
ekhnicheskoi Informatsii. Until 1970 the title was: Referativnyi zhurnal:
eodeziia. Annual author and subject indexes. For earlier years see
eferativnyi zhurnal: astronomiia. Geodeziia (Reference journal: astronomy
nd geodesy) (Institut Nauchnoi Informatsii), 1953-1962. monthly. 1953 sub-
itle was simply: astronomiia.   1926

    Systematically arranged, including headings for general, practical astronomy,
urveying, aerial surveying and photogrammetry, aerophototopography, geodetic
urveys, theory of the figure of the earth and geodetic gravimetry, theory and
ethods of mathematical treatment of results of geodetic measurements and survey.
onthly author index. Annual author and subject indexes.

IKOV, Evgenii Fedorovich, and SOLOV'EV, L. L. Bibliograficheskii ukazatel'
eodezicheskoi literatury za 5 let 1962-1966. (Bibliographical guide to geo-
etic literature for the 5 years 1962-1966). L. S. Khrenov, ed. Moskva,
zdatel'stvo "Nedra," 1972. 471 p. (Glavnoe Upravlenie Geodezii i Kartografii
ri Sovete Ministrov SSSR. Tsentral'nyi Nauchno-Issledovatel'skii Institut
eodezii, Aeros"emki i Kartografii).   1927

    Basic guide to geodetic literature 1962-1966 with additions also to the
arlier bibliographies for the periods 1917-1956 and 1957-1961.

IKOV, Evgenii Fedorovich. Bibliograficheskii ukazatel'geodezicheskoi literatury
a 40 let (1917-1956 gg.) (Bibliographical guide to geodetic literature for
orty years, 1917-1956), L. S. Khrenov, ed. Moskva, Geodezizdat, 1961. 535 p. 1928

    8,868 entries, systematically arranged, on geodesy and related disciplines
ublished in the USSR in Russian and other languages of the peoples of the USSR.
uthor and title indexes.

DEZIIA (Geodesy). Itogi nauki. 1965. Moskva, VINITI, 1967. p. 49-68,
18-134.   1929

    658 references at ends of articles on the utilization of artifical earth
atellites in geodesy and on analytical photogrammetry.

NTRAL'NYI Nauchno-Issledovatel'skii Institut Geodezii, Aeros"emki i Kartografii.
ibliograficheskii ukazatel' trudov za 1929-1955 g. (Bibliographical guide to
ublications 1929-1955). E. G. Pchelko, comp. Moskva, Geodezidat, 1956. 60 p.
Glavnoe Upravlenie Geodezii i Kartografii MVD SSSR).   1930

    168 entries in chronological order, including contents of collected works.
ystematic index to articles with 613 entries.

_ . _____ . za 1956-1960. E. G. Pchelko, comp. Moskva, Geodezizdat, 1961.
 p. (Glavnoe Upravlenie Geodezii i Kartografii Ministerstva Geologii i Okhrany
dr SSSR.   1931

89 books and collected works with analysis of contents and 111 articles, systematically organized.

_____ . _____ . za 1961-1966 g. E. G. Pchelko, comp. Moskva, Geodezizdat, 1967.
47 p.   1

110 books and collected works and 192 articles.

STATISTICAL METHODS IN GEOGRAPHY

MATHEMATICAL methods and modern geography. V. S. Preobrazhenskii (editor-in-chief), V. V. Annenkov, N. I. Blazhko, V. M. Gokhman, E. N. Ivanovskaia (secretary), Iu. G. Saushkin, and A. M. Trofimov, eds. Kazan', Publishing House of the Kazan University, 1972. 235 p. Bibliography, p. 221-235 (Kazan State University. National Committee of Soviet Geographers).   1

229 references in alphabetical order by name of author. Titles of works in Russian are given in English translation and in Russian transliterated into Latin letters. Based on papers presented at the 2nd All-Union meeting on the application of mathematical methods in geography at the regional meeting of the Commission on Quantitative Methods of the International Geographical Union, held in Kazan' at the Kazan' State University, December, 1971, and published for the 22nd International Geographical Congress to make available in English Russian research in the application of mathematical methods in geography.

BOCHAROV, Mikhail Kuz'mich. Metody matematicheskoi statistiki v geografii (Methods of mathematical statistics in geography). Moskva, Izdatel'stvo "Mysl'," Geografgiz, 1971. 371 p. Bibliography, p. 361-365.   1

76 references arranged in alphabetical order by name of author, 72 in Russian and 4 in English.

BORSUK, O. A., and SPASSKAIA, I. I. Matematicheskie metody v geomorfologii (Mathematical methods in geomorphology), Itogi nauki i tekhniki: Teoreticheskie i obshchie voprosy geografii. Tom 1. Sovremennye metody geograficheskikh issledovanni. Moskva, VINITI, 1974. p. 65-153. Bibliography, p. 139-151. Summary in English. (Vsesoiuznyi Institut Nauchnoi i Tekhnicheskoi Informatsii).   19

243 references: 146 in Russian, 97 in foreign languages.

1932 - 1934a

REMOTE SENSING METHODS AND AERIAL PHOTOGRAPHS

EDOVANIE prirodnoi sredy kosmicheskimi sredstvami: geologiia i geomorfologiia.
klady sovetskikh uchënykh na Soveshchanii sovetsko-amerikanskoi rabochei gruppy
 issledovaniiu prirodnoi sredy s pomoshch'iu kosmicheskogo sredstva. Moskva,
'-17 fevralia 1973 g. (Research on the natural environment by means of artificial
rth satellites: geology and geomorphology. Reports of Soviet scientis at the
nference of the Soviet-American working group on research on the natural environ-
nt by artifical earth satellites, Moscow, February 12-17, 1973). Moskva, 1973.
'6 p. References at ends of articles. (Akademiia Nauk SSSR. Komissiia po
sledovaniiu Prirodnykh Resursov s Pomoshch'iu Kosmicheskogo Sredstva).   1935

)GRADOV, Boris Veniaminovich, and GRIGOR'EV, Aleksei Alekseevich. Kosmicheskie
tody izucheniia prirodnoi sredy (Remote sensing methods for the study of the
tural environment), Itogi nauki i tekhniki: Teoreticheskie i obshchie voprosy
ografii, tom 1: Sovremennye metody geograficheskikh issledovanii. Moskva,
NITI, 1974, p. 5-64. Bibliography, p. 50-62. English summary. (Vsesoiuznyi
stitut Nauchnoi i Tekhnicheskoi Informatsii).   1936

   292 references: 67 in Russian, 225 in Western languages.

PODINOV, Georgii Valentinovich. Primenenie aerofotos"ëmki v geograficheskikh
ssledovaniiakh (Utilization of air photographs in geographical research), Itogi
auki [seriia geografiia]: kartografiia 1962 [v. 1]. Moskva, VINITI, 1964, p.
8-48.   1936a

   70 references.

____. ____. ____ 1963-1964, v. 2, 1966, p. 96-112.

   138 references for 1963-1964.

OMOLOV, L. A. Primenenie aerofotos"ëmki v geograficheskikh issledovaniiakh
Utilization of air photographs in geographical research), in Itogi nauki, seriia
eografiia: kartografiia 1965-1967, v. 3, 1968, p. 102-126.   1937

   134 references 1965-1967.

____. Primenenie aeros"emki i kosmicheskoi s"emki v geograficheskikh issledov-
niiakh (Utilization of air photographs and satellite photographs in geographical
esearch), Itogi nauki, seriia geografiia: kartografiia 1967-1969, v. 4. 1970,
. 148-163.

   77 references on geographical utilization of air and satellite photography.

____. ____. Itogi nauki i tekhniki; seriia kartografiia, v. 5. Moskva, VINITI,
972, p. 163-187.

   83 references.

RNOV, Leonid Evgen'evich. Teoreticheskie osnovy i metody geograficheskogo
leshifrirovaniia aerosnimkov (Theoretical foundations and methods of the geographical interpretation of air photographs). Leningrad, zdatel'stvo Leningradskogo Universiteta, 1967. 214 p. Bibliography, p. 206-212.   1938

   156 references.

OGRADOV, Boris Veniaminovich, and KONDRAT'EV, K. Ia. Kosmicheskie metody
emlevedeniia, puti ikh razrabotki i primeneniia dlia izucheniia prirodnykh
esursov Zemli (Cosmic, i.e. remote sensing, methods in geography, paths of their
elaboration and application to the study of natural resources of the earth),
kademiia Nauk SSSR, Izvestiia, seriia geograficheskaia, 1970, no. 2, p. 101-103. 1939

   75 references.

1935 - 1939

VINOGRADOV, Boris Veniaminovich. Osnovnye napravleniia razrabotki i primeneniia
aerometodov pri geograficheskikh issledovaniiakh v SSSR (Basic directions of
elaboration and utilization of aerial methods in geographic research in the USSR),
Geograficheskoe Obshchestvo SSSR. Izvestiia. v. 99, no. 1 (1967), p. 93-96.   1

   123 entries on geographical applications of aerial methods for the period
1962-1964.

KUDRITSKII, Dmitrii Mikhailovich, and SAMOILOVICH, Georgii Georgievich, eds.
Aerometody izucheniia prirodnykh resursov (Aerial methods for the study of natural
resources). Moskva, Geografgiz. 1962. 328 p. Bibliography, p. 317-328.   1
   About 250 references.

LABORATORIIA AEROMETODOV, Leningrad. Bibliograficheskii ukazatel' opublikovannykh
rabot sotrudnikov Laboratorii aerometodov 1947-1968 (Bibliography of the
publications of collaborators of the Laboratory of Aerial Methods 1947-1968).
P. Ia. Raizer. N. N. Chizhova, ed. Leningrad, 1969. 166 p. (Ministerstvo
Geologii SSSR. Laboratoriia Aerometodov).   1

   1,091 entries, systematically arranged. Author index.

RAIZER, Petr Iakovlevich, comp. Aerometody i ikh primenenie: bibliograficheskii
ukazatel' 1836-1955 (Aerial methods and their utilization: a bibliographical guide,
1836-1955). D. M. Kudritskii and K. L. Mafranovskii, eds. Leningrad, Izdatel'-
stvo Akademii Nauk SSSR, Leningradskoe Otdelenie, 1959. 724 p. (Akademiia Nauk
SSSR. Laboratoriia Aerometodov). (Earlier edition: Aerofotos"ëmka i aerofoto-
grammetriia: bibliograficheskii ukazatel' otechestvennoi literatury, 1864-1949 gg.
Leningrad, IKVVIA, 1951, 337 p.).

   A bibliography with more than 10,000 entries, classified into 10 basic sections with 49 divisions and 22 subdivisions.

RAIZER, Petr Iakovlevich. Opyt sistematizatsii nauchno-tekhnicheskoi literatury po
aerofotos"ëmke i aerometodam (An attempt at a systematization of the scientific
technical literature on air photography and aerial methods), Geodeziia i kartografiia, 1962, no. 3, p. 72-76.

AL'TER, Sani Peisakhovich. Landshaftnyi metod deshifrirovaniia aerofotosnimkov:
obshchie polozheniia i printsipy (The landscape method of interpreting air photographs: general state and principles). Moskva-Leningrad. Izdatel'stvo "Nauka,"
1966. 88 p. Bibliography, p. 77-87. (Institut Geografii Sibiri i Dal'nego
Vostoka).

   About 300 references.

AL'TER, Sani Peisakhovich. Landshaftnyi metod geomorfologicheskogo deshifrirovaniia
aerofotosnimkov na primere Nizhnego Priirtysh'ia (The landscape method of geomorphological interpretation of air photographs based on the example of the lower
Irtysh area), Sibirskii geograficheskii sbornik, v. 7 (1971), p. 143-198.
Bibliography, p. 195-198.

PART IV. REGIONAL GEOGRAPHY: THE SOVIET UNION

THE COUNTRY AS A WHOLE

Bibliographies of Bibliographies

LIOGRAFIIA sovetskoi bibliografii (Bibliography of Soviet bibliography)
Vsesoiuznaia Knizhnaia Palata). 1939, 1946- annual. Moskva, Izdatel'stvo
Kniga," 1941, 1948- .   1947
    The annual inventory of bibliographies published in the Soviet Union in all
ields. Geography, a sub-section under the section geological and geographical
ciences in the chapter on natural sciences and mathematics, includes 1,894 bib-
iographies in Russian and other languages of the USSR for the 26 years 1939,
946-1970, many, but not all, concerned with the Soviet Union itself. The sub-
ection on hydrology, meteorology, and climatology in the section of geological
nd geographical sciences and the subsection on geodesy, topography, and cartog-
aphy in the section on physical-mathematical sciences of the chapter on natural
ciences and mathematics, also contain many useful bibliographies on the Soviet
nion. Other chapters, sections, or subsections of particular geographic interest
n study of the Soviet Union are history, political economy, international rela-
ions, communist construction, biological sciences, anthropology, engineering
nd industry, agriculture, and transport.

ICHESKAIA geografiia: annotirovannyi perechen' otechestvennykh bibliografii,
zdannykh v 1810-1966 gg. (Physical geography: annotated list of native bib-
iographies published 1810-1966). comp. by M. N. Morozova, E. A. Stepanova, comps.,
J. V. Klevenskaia, ed., L. G. Kamanin, scientific consultant. Moskva, Izdatel'-
stvo "Kniga," 1968. 309 p. (Gosudarstvennaia Biblioteka SSSR imeni V. I. Lenina.
Institut Geografii Akademii Nauk SSSR.   Sektor Seti Spetsial'nykh Bibliotek
Akademii Nauk SSSR).   1948

    Lists 1,336 bibliographies on physical geography produced in the USSR 1810-
1966, the majority dealing particularly with the physical geography of the USSR.
Covers general bibliographies of the USSR, systematic fields of physical geography, and regional physical geography. The section on regional physical geography of the USSR, p. 218-266, entries 1,103-1,304, includes 202 bibliographies
arranged: USSR as a whole; European part and Caucasus; European part; North-West,
North-East, Center, West, South-West, Volga region, Urals, Caucasus, North
Caucasus; Asiatic part of the USSR, Siberia, the Far East; Middle Asia and
Kazakhstan, Middle Asia, Kazakhstan. Some of these regions, such as Siberia,
the Caucasus, and Middle Asia are further subdivided. Indexes of names and places.

VRILOVA, Serafima Abramovna; PARKHOMENKO, Iia Ivanovna; and KUZNETSOVA, N. K.
Bibliografii po geografii, vyshedshie v SSSR 1946-1961 gg. (Bibliographies of
geography, published in the USSR 1946-1961), Geograficheskii sbornik (Institut
Nauchnoi Informatsii Akademii Nauk SSSR). [1]. Moskva, VINITI, 1963, p. 32-45. 1949

    266 separate bibliographies of geography published in the USSR 1946-1961,
mostly on the Soviet Union itself. Organized systematically and regionally (geography of the USSR as a whole, regional bibliographies of the RSFSR and its individual regions, republics of the Soviet East (as a whole), union republics--
Armenia, Belorussia, Georgia, Kazakhstan, Kirgizia, Latvia, Lithuania, Turkmenistan,
Uzbekistan, and Ukraine--and regional geography of foreign countries. A key source
for Soviet bibliographies of geography for the period covered.

BAGROV, Lev Semënovich [Leo]. Spisok bibliograficheskikh ukazatelei po obshchei
geografii i etnografii Rossii (List of bibliographical guides for general geography and ethnography of Russia). St. Petersburg, 1913. 16 p. (Also in: Kratkii
obzor deiatel'nosti Pedagogicheskago muzeia voenno-uchebnykh zavedenii za 1910-
1911, St. Petersburg, 1913, p. 16-29).   1

About 120 bibliographies of Russia arranged by general bibliographies
(alphabetically by author) and local bibliographies (alphabetically by name of
area covered).

Major General Bibliographies

REFERATIVNYI zhurnal: geografiia (Reference journal: geography) (Vsesoiuznyi
Institut Nauchnoi i Tekhnicheskoi Informatsii). Moskva. 1954- . 12 numbers a
year. Annual author index (separate volume) and annual subject and geographic
index (another separate volume).   1ᵃ

The most comprehensive listing of the substantive works in geography on the
Soviet Union, its regions, its topical fields, and its major problems.

Section E. SSSR (USSR) lists more than 2,000 items a year, mainly in economic
geography, including geography of population and settlement. In general works in
physical geography are listed under the individual fields of physical geography
such as geomorphology, hydrology, glaciology, meteorology and climatology, biogeography, geography of soils, and oceanography.

ITOGI nauki i tekhniki: geografiia SSSR (Results of science and technology: geography of the USSR). (Vsesoiuznyi Institut Nauchnoi i Tekhnicheskoi Informatsii).
Moskva, VINITI, 1- (1965-). Irregular. No. 5 never published.   1

1. Zemel'nye i lesnye resursy SSSR. (Land and forest resources of the
USSR). I. I. Parkhomenko, ed. 1965. 150 p. 759 references.

2. Ekonomicheskoe raionirovanie SSSR, 1962-1964 (Economic regionalization
of the USSR). I. I. Parkhomenko, ed. 1965. 145 p. 529 references.

3. V. V. Pokshishevskii. Geografiia naseleniia v SSSR, 1961-1965 gg.
(Geography of population of the USSR). I. I. Parkhomenko, ed. 1966. 168 p.
1,940 references.

4. N. I. Mikhailov. Fiziko-geograficheskoe raionirovanie 1963-1965
(Physical — geographical regionalization 1963-1965). I. I. Parkhomenko, ed.
1967. 146 p. 1,136 references.

6. A. A. Mints. Ekonomicheskaia otsenka prirodnykh resursov i uslovii
proizvodstva, 1961-1967 (Economic appraisal of natural resources and conditions
of production, 1961-1967). I. I. Parkhomenko, ed. 1968. 140 p. 492 references.

7. Resursy zhivotnogo mira SSSR: geografiia zapasov, ispol'zovanie,
vosproizvodstvo. (Animal resources of the USSR: geography of reserves, utilization, reproduction). A. A. Nasimovich, ed. 1969. 231 p. 1,071 references.

8. Proizvodstvenno-territorial'nye kompleksy (Production-territorial complexes). I. I. Parkhomenko. ed. 1970. 134 p. 959 references.

9. Geograficheskoe izuchenie prirodnykh resursov i voprosy ikh ratsional'nogo
ispol'zovaniia (Geographical study of natural resources and problems of their
rational utilization). I. I. Parkhomenko, ed. 1973. 365 p. Bibliography, 1968-
1971, p. 174-341. 2,730 references. Author index. Table of contents and abstracts also in English.

10. Geografiia transporta (Geography of transportation). I. I. Parkhomenko,
ed. 1973. 170 p. Bibliography, p. 96-161. 670 references. Author index.
Table of contents and abstracts also in English.

11. Geografiia sfery obsluzhivaniia (Geography of service activities). S.
A. Kovalëv, I. I. Parkhomenko, and V. V. Pokshishevskii, eds. 1974. 680 references plus 44 dissertation abstracts.

1950 - 1952

IAVLENSKII, Georgii Pavlovich, comp. Geografiia SSSR: Annotirovannyi ukazatel'
teratury v pomoshch' uchiteliu (Geography of the USSR: annotated guide to the
terature as an aid to teachers). Moskva, Uchpedgiz, 1957. 168 p.   1953

   About 400 selected books on all aspects of the geography of the USSR, arranged
~ systematic fields and by regions with sections also on leading Russian geog-
▪phers and explorers, on methods of teaching, and on handbooks and atlases.
▪jor reviews of cited works are listed. The works cited are from the period
▪45-1957 with a preponderance from the later years.

▪GRAFICHESKAIA literatura SSSR: bibliograficheskii ezhegodnik. 1954 (Geographi-
al literature of the USSR: bibliographical annual. 1954). Iu. A. Barinov, A.
. Kruglakovskii, and L. Ia. Shraiber, comps. Moskva, 1961. 302 p. (Akademiia
▪auk SSSR. Biblioteka Instituta Geografii). Rotaprint.   1954

   2,427 entries of books and articles in Russian and other languages of the
▪SSR (if a Russian abstract is included). Index of authors, editors, and per-
ons. List of abbreviations and of full form of names of periodicals analyzed
▪n 1954 and 1955. Organized systematically by fields, and regionally for the
▪SSR, and for foreign countries.

▪GRAFICHESKAIA literatura SSSR (Bibliograficheskii ezhegodnik). 1955. (Geo-
▪raphical literature on the USSR: bibliographical annual. 1955). Iu. A. Barinov,
▪. N. Kruglakovskii, and L. Ia. Shraiber, comps. Moskva, 1962. 2 v. 332 p.
Akademiia Nauk SSSR. Sektor seti spetsialnykh bibliotek. Institut geografii.
▪iblioteka).   1955

   2,903 entries on the geography of the USSR in 1955 including books, abstracts
▪f dissertations, articles in serials and collected works in the Russian language
▪r in other languages of the USSR if a Russian abstract is provided. Organized
▪y systematic then regional fields; under regional geography works in physical
▪nd economic geography are separate. Index of names.

▪VENSKAIA, Valentina V. Spisok bibliograficheskikh istochnikov po geografii,
▪assmotrennykh na seminare (List of bibliographical sources for geography examin-
▪d in a seminar). In the book: Biblioteki SSSR: opyt raboty (Libraries of the
▪SSR: an experiment in work), no. 4. Moskva, 1955, p. 29-39 (Gosudarstvennaia
▪iblioteka SSSR imeni Lenina. Nauchno-meteodologicheskii Kabinet Biblioteko-
▪edeniia). Section 3a. Kraevedcheskaia literatura (regional literature), p. 33-
▪5, entries 56-88.   1956

   List of 33 general regional bibliographies for parts of the Russian Empire
▪nd the Soviet Union up to 1955.

▪GEEV, Ivan V. Nasha strana: knigi po geografii SSSR (Our country: books on the
▪eography of the USSR). G. N. Cherdantsev, ed. Moskva, 1946. 78 p. Bibliog-
▪aphy, p. 76-78 (Gosudarstvennaia Biblioteka SSSR imeni V. I. Lenina. Besedy
▪ Knigakh).   1957

   Discussion of 32 books published 1936-1945.

▪FIL'EV, Gavriil Ivanovich. Geografiia Rossii, Ukrainy i primykaiushchikh k nim
▪ zapada territorii v predelakh Rossii 1914 goda. Chast' 1-2. (Geography of
▪ussia and the Ukraine and territory adjoining on the west within the boundaries
▪f Russia in 1914. Parts 1-2). Odessa, 1916-1923. Part 1 has title: Geografiia
▪osii.   1958

   Part 1 (1916. 212 p.) devoted to an introduction, history of research,
▪nstitutions and publications,and cartography, includes bibliographical notes,
▪. 193-212, and geographical institutions and publications, p. 177-183.

   Part 2, no. 1 (1922) devoted to relief of European Russia and the Caucasus,
includes a bibliography, p. 321-330, with about 200 references.

Part 2, no. 2 (1923, 334 p.), devoted to the relief of Asiatic Russia, has about 1,200 references at ends of chapters.

Of interest primarily for the prerevolutionary history of the geographical investigation of the relief of the USSR.

MEZHOV, Vladimir Izmailovich. Russkaia istoricheskaia bibliografiia. Ukazatel' knig i statei po russkoi i vseobshchei istorii i vspomogatel'nym naukam za 1800-1854 gg. vkliuchitel'no. (Russian historical bibliography. Guide to books and articles on Russian and general history and auxiliary sciences 1800-1854 inclusive. Bibliographie des livres et articles russes d'histoire et sciences auxiliares de 1800-1854 inc. Tom 3. St. Petersburg, I. M. Sibiriakov, 1893. Geografiia, p. 1-194.   1

4,926 annotated entries on geography (nos. 21,492-26,417) cover geography, hydrography, orography, cartography, travels. The whole series of three volumes contains about 35,000 entries on history and related fields in the broadest sense.

STUCKENBERG, Johann Ch. Versuch eines Quellen-Anzeigers alter und neuer Zeit für das Studium der Geographie, Topographie, Ethnographie und Statistik des Russischen Reiches. St. Petersburgh, Druckerey der Militair-Lehr-Anstalten, 1849-1852. 2 v in 4.   1

Books, maps and atlases in Russian and foreign languages from the 14th century to 1848 and articles from the 18th century to 1851 on the geography and ethnography of Russia. v. 1 has 846 entries for maps and atlases. v. 2 has 241 entries for collections, periodicals, and newspapers. Contents of collected works analyzed. More than 4900 items altogether.

A basic source on geography of Russia up to the mid-19th century. Still of value.

CHERTKOV, Aleksandr Dmitrievich. Vseobshchaia biblioteka Rossii, ili Katalog knig dlia izucheniia nashego otechestva vo vsekh otnosheniiakh i podrobnostiakh (General library of Russia, or Catalogue of books for the study of our fatherland in all its relations and details). Moskva, Universitetskaia Tipografiia, 1838. 631 p. Otdelenie 2. Geografiia i statistika (Part 2. Geography and statistics), p. 127-212, addendum, p. 560-578. Second addendum, Moskva, Tipografiia A. Semena, 572 p., part 2, 1845 [1844], p. 175-238.   1

Sections: geographical and statistical description of all Russia; descriptions of guberniyas cities, monasteries, and villages; descriptions of rivers, lakes, and the like; natural history of Russia; geographical and statistical dictionaries; travel in Russia and adjacent seas. 4,701 works listed in main volume, 2,200 in second supplement.

A second edition appeared under the title, Katalog kniz Chertkovskoi biblioteki Otdelenie vtoroe. Zemleopisanie (Catalogue of books of the Chertkov library. Part 2. Geography). Moskva, 1864, with additions up to 1864 by P. I. Bartenev after the death of A. D. Chertkov. Issued in individual numbers. no. 1, description of Russia as a whole; no. 3, including sections on meteorology, botany, and zoology; no. 5, travels, with annotated descriptions of many works.

ZIABLOVSKII, E. F. Zemleopisanie Rossiiskoi Imperii dlia vsekh sostoianii (Geography of the Russian Empire for all circumstances). Tom 1. S.-Peterburg, 1810. Survey of the literature, p. xxxvii-lxiv.   1

Apparently the first review and bibliography of the geographic literature on Russia. Now of historical interest only.

MAJOR REGIONAL SETS COVERING THE ENTIRE SOVIET UNION

The Soviet Union: Geographical Description in 22 Volumes

VETSKII Soiuz: Geograficheskoe opisanie v 22-kh tomakh (The Soviet Union: geographical description in 22 volumes). S. V. Kalesnik, and others, eds. Moskva,
Izdatel'stvo "Mysl'," 1966-1972. 22 v.   1963

Up-to-date series covering all regions of the Soviet Union with balanced
treatment of physical, population, and economic geography. Written by leading
geographers of the Soviet Union, specialists on specific regions, for a wide
audience (90,000-100,000 copies printed), to celebrate the fiftieth anniversary
of the Revolution. Extensively illustrated with photographs (some in color) and
maps (some in color). Statistical tables in an appendix at the back of each
volume. Index of place names and lists of Russian and scientific names of plants
and animals. Lists of local geographical terms in each region.

Brief bibliographies list the outstanding books on each region and provide
in total 1,039 references. The volumes are listed below but information common
to all, such as series title, place of publication, and name of publisher, is
not repeated.

___ . Obshchii obzor (General overview). S. V. Kalesnik and V. F. Pavlenko,
eds. 1972. 813 p. Bibliography, p. 797-799.   1964

84 references arranged alphabetically by name of author. A well-selected
up-to-date list of the major works on the Soviet Union as a whole with emphasis
on systematic fields, such as subdivisions of physical geography and economic
geography, and on natural resources and their utilization.

This volume provides a good systematic geography of the Soviet Union as a
whole covering physical geography, historical geography, population, and economic
geography in turn.

___ . Rossiiskaia Federatsiia. Obshchii obzor. Evropeiskii Sever (Russian
Federation. General survey. European North). S. V. Kalesnik, ed. Obshchii
obzor. A. V. Darinskii, ed., Evropeiskii Sever. 1971. 565 p. Bibliography,
p. 553.   1965

31 references on the European North.

___ . ___ . Tsentral'naia Rossiia: Tsentral'nyi raion, Volgo-Viatskii raion,
Chernozemnyi Tsentr (Central Russia: Central region, Volga-Vyatka region, and
Black-Earth Center). G. M. Lappo, ed., Tsentral'nyi raion. B. S. Khorev, ed.,
Volgo-Viatskii raion. F. N. Mil'kov, ed., Chernozemnyi Tsentr. 1970. 907 p.
Bibliography, p. 891-892.   1966

84 references on Central Russia.

___ . ___ . Evropeiskii Iugo-Vostok: Povolzh'e. Severnyi Kavkaz (The
European Southeast: Volga region. North Caucasus). K. V. Dolgopolov, ed.,
Povolzh'e. S. A. Vodovozov, ed., Severnyi Kavkaz. 1968. 795 p. Bibliog-
raphy, p. 363-364, 772-773.   1967

51 references on the Volga region, p. 363-364, and 53 references on the North
Caucasus, p. 772-773.

___ . ___ . Ural (The Urals). I. V. Komar, ed. 1969. 404 p. Bibliography,
p. 394.   1968

38 references on the Urals.

_____ . _____ . <u>Zapadnaia Sibir'</u> (Western Siberia). M. I. Pomus, ed. 1971.
429 p. Bibliography, p. 417.

   37 references on Western Siberia.

_____ . _____ . <u>Vostochnaia Sibir'</u> (Eastern Siberia). V. V. Pokshishevskii
and V. V. Vorob'ev, eds. 1969. 493 p. Bibliography, p. 482.

   42 references on Eastern Siberia.

_____ . _____ . <u>Dal'nii Vostok</u> (Soviet Far East). A. B. Margolin, ed. 1971.
397 p. Bibliography, p. 391.

   40 references on the Soviet Far East.

_____ . <u>Ukraina. Obshchii obzor</u> (The Ukraine: general survey). A. M. Marinich,
ed. 1969. 309 p. Bibliography, p. 302-303.

   44 references on the Ukraine as a whole in Russian and Ukrainian.

_____ . <u>Ukraina. Raiony</u> (The Ukraine. Regions). I. A. Kugukalo and A. M.
Marinich, eds. 1969. 357 p. Bibliography, p. 344.

   22 references on regions of the Ukraine in Russian and Ukrainian.

_____ . <u>Estoniia</u> (Estonia). M. I. Rostovtsev, ed. 1967. 254 p. Bibliography,
p. 248-249.

   68 references on Estonia in Russian and Estonian.

_____ . <u>Latviia</u> (Latvia). M. I. Rostovtsev, ed. 1968. 238 p. Bibliography,
p. 231-232.

   55 references on Latvia in Russian and Latvian, titles in Latvian
being listed in Russian with notation that the work is in Latvian.

_____ . <u>Litva</u> (Lithuania). K. K. Bieliukas and M. I. Rostovtsev, eds. 1967.
286 p. Bibliography, p. 279.

   26 references on Lithuania in Russian and Lithuanian, titles for the latter
listed in Russian translation with the note that they are in Lithuanian.

_____ . <u>Belorussiia</u> (Belorussia). K. I. Lukashev, ed. 1967. 309 p. Bibliography, p. 308-309.

   49 references on Belorussia in Russian and Belorussian.

_____ . <u>Moldaviia</u> (Moldavia). Ia. S. Grosul and M. M. Radul, eds. 1970. 253 p.
Bibliography, p. 249.

   37 references on Moldavia.

_____ . <u>Gruziia</u> (Georgia). F. F. Davitaia, ed. 1967. 318 p. Bibliography,
p. 308.

   39 references on Georgia in Russian or Georgian, titles for the latter
listed in Russian translation.

_____ . <u>Armeniia</u> (Armenia). A. B. Bagdasarian, ed. 1966. 342 p. Bibliography, p. 335.

   34 references on Armenia in Russian and Armenian, titles for the latter
given in Russian translation.

_____ . <u>Azerbaidzhan</u>. G. A. Aliev, ed. 1971. 317 p. Bibliography, p. 314.

   39 references on Azerbaidzhan.

1969 - 1981

__ . Kazakhstan. N. N. Pal'gov and M. Sh. Iarmukhamedov, eds. 1970. 405 p.
ibliography, p. 396.   1982

    30 references on Kazakhstan.

__ . Kirgiziia (Kirgizia). K. O. Otorbaev and S. N. Riazantsev, eds. 1970.
86 p. Bibliography, p. 278.   1983

    40 references on Kirgizia.

__ . Uzbekistan. L. N. Babushkin, ed. 1967. 318 p. Bibliography, p. 310.   1984

    33 references on Uzbekistan.

__ . Turkmenistan. V. N. Kunin, ed. 1969. 277 p. Bibliography, p. 271.   1985

    34 references on Turkmenistan in Russian or Turkmen.

__ . Tadzhikistan. D. A. Chumichev, ed. 1968. 238 p. Bibliography, p. 238.   1986

    29 references on Tadzhikistan.

Natural Conditions and Natural Resources of the USSR in 15 volumes

DEMIIA Nauk SSSR. Institut Geografii. Prirodnye usloviia i estestvennye resursy
SSR (Natural conditions and natural resources of the USSR), I. P. Gerasimov,
nd others, eds. Moskva, Izdatel'stvo "Nauka," 1963-1972. 15 v.   1987

    About 7,800 references on the physical geography of the Soviet Union are
iven in the 15 volumes of this series. A massive and detailed survey, with
xtensive text, bibliographies, maps, and illustrations of the major physical
egions of the Soviet Union. Each volume, written by a large number of special-
sts, is divided into two parts: the general characteristics of the region with
ystematic chapters on relief and geological structure, climate, waters, soils,
egetation, and animal life; regional characteristics with descriptions of physical-
eographic subregions, and a discussion of natural resources and problems of
tilization of natural resources. Lists of Russian and Latin names of plants.
ndex of place names. The separate volumes are listed below. Sponsor, series
itle, place of publication, and publisher, common to all volumes, are not re-
eated.

ETSKAIA Arktika (Moria i ostrova Severnogo Ledovitogo okeana) (The Soviet
rctic with the seas and islands of the Arctic Ocean). Ia. Ia. Gakkel' and L.
. Govorukha, eds. 1970. 526 p. Bibliography, p. 494-512 (Arkticheskii i
ntarkticheskii Nauchno-Issledovatel'skii Institut).   1988

    610 references by authors in alphabetical order, names in Cyrillic letters
irst, then those in Latin letters.

ER Evropeiskoi chasti SSSR (The North of the European part of the USSR). G. D.
ikhter, ed. 1966. 452 p. Bibliography, p. 423-440. (Akademiia Nauk SSSR.
nstitut Geografii).   1989

    About 550 references, in alphabetical order.

DNIAIA polosa Evropeiskoi chasti SSSR. (The central belt of the European USSR).
. V. Zonn and A. A. Mints, eds. 1967. 440 p. Bibliography, p. 409-427.
Akademiia Nauk SSSR. Institut Geografii).   1990

    About 600 references, in alphabetical order. The central belt includes the
elorussian SSR and the three Baltic republics as well as the central part of the
uropean RSFSR.

IUGO-VOSTOK Evropeiskoi chasti SSSR (South-eastern European part of the USSR).
A. G. Doskach and K. V. Dolgopolov, eds. 1971. 457 p. Bibliography, p. 430-448.
(Akademiia Nauk SSSR. Institut Geografii).   1
    569 references, in alphabetical order.

URAL i Priural'e (The Urals and the Pre-Urals). I. V. Komar and A. G. Chikishev,
    eds. 1968. 459 p. Bibliography, p. 432-447. (Akademiia Nauk SSSR. Institut
    Geografii).   1
    497 references, in alphabetical order.

ZAPADNAIA Sibir' (Western Siberia). G. D. Rikhter, ed. 1963. 486 p. Bibliog-
    raphy, p. 461-475. (Akademiia Nauk SSSR. Institut Geografii).   1
    About 430 references, in alphabetical order.

SREDNIAIA Sibir' (Middle Siberia). L. G. Kamanin and B. N. Likhanov, eds. 1964.
    478 p. Bibliography, p. 451-467. (Akademiia Nauk SSSR. Institut Geografii). 1
    About 500 references in alphabetical order.

PREDBAIKAL'E i Zabaikal'e (Pre-Baykal and Trans-Baykal regions). V. S. Preobrazhen-
    skii, M. I. Pomus, and V. B. Sochava, eds. 1965. 490 p. Bibliography, p. 457-
    478. (Akademiia Nauk SSSR. Institut Geografii. Institut Geografii Sibiri i
    Dal'nego Vostoka).   1
    About 660 references, arranged alphabetically on Eastern Siberia, both east
    and west of Lake Baykal.

IAKUTIIA (Yakutiya). S. S. Korzhuev, ed. 1965. 465 p. Bibliography, p. 441-455
    (Akademiia Nauk SSSR. Institut Geografii. Institut Geografii Sibiri i Dal'nego
    Vostoka).   1
    About 500 references arranged in alphabetical order.

IUZHNAIA Chast' Dal'nego Vostoka (Southern part of the Soviet Far East). V. V.
    Nikol'skaia and A. S. Khomentovskii, eds. 1969. Bibliography, p. 396-408
    (Akademiia Nauk SSSR. Institut Geografii).   1
    384 references, in alphabetical order.

SEVER Dal'nego Vostoka (The Northern part of the Soviet Far East). N. A. Shilo,
    ed. 1970. 486 p. Bibliography, p. 454-474 (Akademiia Nauk SSSR. Severo-
    Vostochnyi Kompleksnyi Institut).   1
    About 675 references in alphabetical order.

UKRAINA i Moldaviia (The Ukraine and Moldavia). A. M. Marinich and M. M. Palamar-
    chuk, eds. 1972. 438 p. Bibliography, p. 422-429 (Kievskii Gosudarstvennyi
    Universitet. Akademiia Nauk Ukrainskoi SSR. Akademiia Nauk Moldavskoi SSR). 1
    273 references in alphabetical order.

KAVKAZ (The Caucasus). N. V. Dumitrashko, ed. 1966. 480 p. Bibliography, p.
    446-470. (Akademiia Nauk SSSR. Institut Geografii. Institut Geografii AN
    Azerbaidzhanskoi SSR. Institut Geografii AN Gruzinskoi SSR. Institut Geo-
    logicheskikh Nauk AN Armianskoi SSR. Sektor Geografii).   2
    About 750 references in alphabetical order.

KAZAKHSTAN (Kazakhstan). B. A. Fedorovich and O. R. Nazarevskii, eds. 1969.
    480 p. Bibliography, p. 457-469 p. (Akademiia Nauk SSSR. Institut Geografii). 2
    About 400 references in alphabetical order.

EDNIAIA Aziia (Middle Asia). E. M. Murzaev, ed. 1968. 482 p. Bibliography,
p. 455-469. (Akademiia Nauk SSSR. Institut Geografii).   2002
    429 references in alphabetical order.

The Studies in Nature Series of 12 Volumes

CHERK prirody" (Studies of nature) series. The geographical publishing house
Glavnaia Redaktsiia Geograficheskoi Literatury, or Geografgiz, for short),
Later part of "Mysl'," published in 1961-1971 a series of small monographs on
12 major natural regions of the Soviet Union. Most are by a single author.
Each is only about a quarter the size of the larger series on natural conditions
and natural resources of the USSR, but they form a handy and less detailed introduction to the physical geographical regions of the USSR and provide a selected
bibliography with about 1,500 references on the physical geography of the regions
of the Soviet Union. A map of the area covered by each regional volume is on the
reverse of the title page of each volume. Lists of Russian and scientific names
of plants and animals. Index of geographical names. These volumes are listed
below.   2003

KHTER, Gavriil Dmitrievich, and CHIKISHEV, Anatolii Grigor'evich. Sever Evropei-
koi chasti SSSR: ocherk prirody (The northern part of the European USSR: a
ketch of its natural conditions). Moskva, Izdatel'stvo "Mysl'," 1966. 237 p.
Bibliography, p. 218-223.   2004
    About 100 references in alphabetical order by name of author.

'KOV, Fedor Nikolaevich. Sredniaia polosa Evropeiskoi chasti SSSR: ocherk
prirody (The Central zone of the European part of the USSR: a sketch of its
atural conditions). Moskva, Geografgiz, 1961. 216 p. Bibliography, p.191-201. 2005
    About 230 references in alphabetical order by name of author.

EL'NIKOV, Vasilii Leont'evich. Iuzhnaia polosa Evropeiskoi chasti SSSR: ocherk
prirody (The southern belt of the European USSR: a sketch of its natural con-
itions). Moskva, Geografgiz, 1963. 222 p. Bibliography, p. 206.   2006
    24 references in alphabetical order by name of author.

NEV, Andrei Mikhailovich. Ural i Novaia Zemlia: ocherk prirody (The Urals and
ovaia Zemlya: a sketch of natural conditions). Moskva, Izdatel'stvo "Mysl',"
965. 215 p. Bibliography, p. 195-198.   2007
    About 60 references in alphabetical order by name of author.

AMOVICH, David Iosifovich, KRYLOV, Georgii Vasil'evich, NIKOLAEV, Vladimir
leksandrovich, and TERNOVSKII, Dmitrii Vladimirovich. Zapadno-Sibirskaia
izmennost': ocherk prirody (The West Siberian plain: a sketch of its natural
onditions). Moskva, Geografgiz, 1963. 262 p. Bibliography, p. 235-238.   2008
    About 60 references arranged alphabetically by name of author.

MUZIN, Iurii Pavlovich. Sredniaia Sibir': ocherk prirody (Middle Siberia: a
ketch of its natural conditions). Moskva, Izdatel'stvo "Mysl'," 1964. 310 p.
ibliography, p. 287-293.   2009
    135 references by sections, general works, Northern Middle Siberia, Central
iberia, and Central Yakutiya, and within each section alphabetically by name of
uthor.

HAILOV, Nikolai Ivanovich. Gory Iuzhnoi Sibiri: ocherk prirody (The mountains
f Southern Siberia: a sketch of their natural conditions) Moskva, Geografgiz,
961. 238 p. Bibliography, p. 219-223.   2010
    About 115 references, arranged alphabetically by name of author.

NIKOL'SKAIA, Vera Vasil'evna. Dal'nii Vostok: ocherk prirody iuzhnoi poloviny
Dal'nego Vostoka (The Far East: an essay on natural conditions in the southern
half of the Soviet Far East). Moskva, Geografgiz, 1962. 215 p. Bibliography,
p. 201-207.   2

About 120 references, arranged alphabetically by name of author.

PARMUZIN, Iurii Pavlovich. Severo-Vostok i Kamchatka; ocherk prirody (The Northeast and Kamchatka: a sketch of their natural conditions). Moskva, Izdatel'stvo
"Mysl'," 1967. 368 p. Bibliography, p. 339-348.   2

About 220 references on natural conditions of the Northeast of the Asiatic
part of the Soviet Union, arranged alphabetically by author under three headings:
general works, Verkhoyano-Kolyma country and the Far Northeast, and Kamchatka.

GVOZDETSKII, Nikolai Andreevich. Kavkaz: ocherk prirody (The Caucasus: a sketc'.
of natural conditions). Moskva, Geografgiz, 1963. 262 p. Bibliography, p. 234-
243.   2

207 references, arranged alphabetically by name of author.

GVOZDETSKII, Nikolai Andreevich, and NIKOLAEV, Vladimir Aleksandrovich. Kazakhstan:
ocherk prirody (Kazakhstan: a study of its natural conditions). Moskva, Izdatel'-
stvo "Mysl'," Geografgiz, 1971. 295 p. Bibliography, p. 282-287.   2

138 references arranged alphabetically by name of author.

MURZAEV, Eduard Makarovich. Sredniaia Aziia: ocherki prirody (Soviet Middle
Asia: sketches of its natural conditions). Moskva, Geografgiz, 1961. 248 p.
Bibliography, p. 235-239.   2

About 100 references arranged alphabetically by name of author.

The Blue Series of Economic Geographies in 25 Volumes

In the period 1954-1966 the Institute of Geography of the Academy of Sciences
of the U.S.S.R. prepared a series of 25 monographs on the economic geography of
the larger regions of the USSR; 18 volumes appeared in the four-year period 1955-
1958; the other 7 volumes were spread over the following eight years. These
were published as the "blue series" by the geographical publishing house (Gosudarstvennoe Izdatel'stvo Geograficheskoi Literatury, or Geografgiz), or its successor,
"Mysl'." Ten volumes cover the larger regions of the Russian Soviet Federated
Socialist Republic and 15 volumes the other fourteen union republics (2 volumes
for the Ukraine and 1 for each other republic). The text is written by specialists
on each area. There are maps, tables, illustrations, and bibliographies. The
25 volumes contain about 5,900 references on the economic geography of the regions
of the Soviet Union.

AKADEMIIA Nauk SSSR. Institut Geografii, and Leningradskii Gosudarstvennyi Universitet. Geografo-ekonomicheskii Nauchno-issledovatel'skii Institut. Severo-
zapad RSFSR: ekonomiko-geograficheskaia kharakteristika (The Northwest of the
RSFSR: its economic-geographic characteristics). V. V. Pokshishevskii, G. S.
Nevel'shtein, and E. B. Lopatina, eds. Moskva, Izdatel'stvo "Mysl'," 1964.
652 p. Bibliography, p. 639-650.   2

About 300 references arranged alphabetically by name of author.

AKADEMIIA Nauk SSSR. Institut Geografii. Tsentral'nyi raion: ekonomiko-
geograficheskaia kharakteristika (The Central region: economic-geographic
characteristics). V. V. Pokshishevskii, S. N. Riazantsev, and N. I. Lialikov,
eds. Moskva, Geografgiz, 1962. 800 p. Bibliography, p. 770-797.   2

About 450 references arranged alphabetically by name of author in two
sections: general overview, and regional.

'VEEV, Gennadii Petrovich, PRIVALOVSKAIA, Genrieta Aleksandrovna, and KHOREV,
oris Sergeevich. <u>Volgo-Viatskii raion: ekonomiko-geograficheskaia kharakteristika</u>
The Volga-Vyatka region: economic-geographic characteristics). N. I. Lialikov,
'. V. Pokshishevskii, and S. N. Riazantsev, eds. Moskva, Geografgiz, 1961. 535 p.
ibliography, 524-534. (Akademiia Nauk SSSR. Institut Geografii).   2018

About 250 references, arranged regionally, Volga-Vyatka region as a whole,
or'kiy oblast, Kirov oblast, Kostroma oblast, Mary ASSR, Mordov ASSR, Chuvash
SSR, and within each area alphabetically by name of author.

GOPOLOV, Konstantin Vasil'evich. <u>Tsentral'no-chernozemnyi raion: ekonomiko-
eograficheskaia kharakteristika</u> (The central black-earth region: economic-
eographic characteristics). Moskva, Geografgiz, 1961. 415 p. Bibliography,
. 396-413. (Akademiia Nauk SSSR. Institut Geografii).   2019

About 450 references arranged alphabetically by name of author.

DEMIIA Nauk SSSR. Institut Geografii. <u>Povolzh'e: ekonomiko-geograficheskaia
harakteristika</u> (The Volga region: economic-geographic characteristics). K. V.
olgopolov, V. V. Pokshishevskii, and S. N. Riazantsev, eds. Moskva, Geografgiz,
957. 464 p. Bibliography, p. 456-461.   2020

About 130 references arranged alphabetically by name of author.

DEMIIA Nauk SSSR. Institut Geografii. <u>Severnyi Kavkaz</u> (The North Caucasus).
. P. Maslov, A. I. Gozulov and S. N. Riazantsev, eds. Moskva, Geografgiz, 1957.
07 p. Bibliography, p. 500-505.   2021

About 140 references arranged alphabetically by name of author.

PANOV, Petr Nikolaevich. <u>Ural</u> (The Urals). Moskva. Geografgiz, 1957. 164 p.   2022

This volume, lacking a bibliography and much smaller than others in the same
eries, was judged to be inadequate in various respects and the Institute of Geog-
aphy of the Academu of Sciences arranged for preparation of two volumes by I. V.
omar, which though different from the Blue series in format are part of its over-
ll planning (entries 2128-2129).

US, Moisei Isaakovich. <u>Zapadnaia Sibir': ekonomiko-geograficheskaia kharakteristika</u>
Western Siberia: economic-geographic characteristics). Moskva, Geografgiz, 1956.
43 p. Bibliography, p. 624-641. (Akademiia Nauk SSSR. Institut Geografii).   2023

About 425 references arranged alphabetically by name of author.

DEMIIA Nauk SSSR. Institut Geografii. Institut Geografii Sibiri i Dal'nego
ostoka. <u>Vostochnaia Sibir': ekonomiko-geograficheskaia kharakteristika</u> (Eastern
iberia: economic-geographic characteristics). V. A. Krotov, M. I. Pomus, G. D.
ikhter, and V. B. Sochava, eds. Moskva, Geografgiz, 1963. 888 p. Bibliography,
. 866-886.   2024

About 500 references arranged alphabetically by name of author.

DEMIIA Nauk SSR. Institut Geografii. <u>Dal'nii Vostok: ekonomiko-geograficheskaia
harakteristika</u> (The Soviet Far East: economic—geographic characteristics).
. V. D'iakonov, V. V. Pokshishevskii, and A. S. Khomentovskii, eds. Moskva,
zdatel'stvo "Mysl'" Geografgiz, 1966. 494 p. Bibliography, p. 470-492.   2025

About 550 references arranged alphabetically by name of author and 8 periodi-
als.

DEMIIA Nauk Ukrainskoi SSR. Institut Ekonomiki. <u>Ukrainskaia SSR</u> (Ukrainian
SR). A. A. Nesterenko, A. S. Koroed, and G. L. Gradov, eds. Moskva, Geografgiz,
957-1958. 2 v. 557 p., 314 p. Bibliography, v. 1, p. 550-556.   2026

About 160 references in Russian and Ukrainian arranged alphabetically by name
f author.

ROSTOVTSEV, Mikhail Ivanovich, and TARMISTO, Vello Iuliusovich. Estonskaia SSR:
ekonomiko-geograficheskaia kharakteristika (The Estonian SSR: economic-geographic
characteristics). A. T. Veimer and S. N. Riazantsev, eds. 2nd ed., Moskva, Geografgiz, 1957. 367 p. Bibliography, p. 356-366. (Akademiia Nauk Estonskoi SSR.
Institut Ekonomiki. Akademiia Nauk SSSR. Institut Geografii). (1st ed. Tallin,
Estonskoe Gosudarstvennoe Izdatel'stvo, 1955. 311 p.).   202

    About 220 references arranged alphabetically by name of author separately for
about 120 references in Russian and for about 100 in Estonian.

VEIS, E. E., PURIN, Valentin Rudol'fovich. Latviiskaia SSR: ekonomiko-geograficheskaia
kharakteristika (Latvian SSR: economic-geographic characteristics). Moskva, Geografgiz, 1957. 440 p. Bibliography, p. 433-438.   20

    About 150 references arranged alphabetically by name of author, separately
for about 90 references in Russian and about 60 in Latvian.

AKADEMIIA Nauk Litovskoi SSR. Institut Geografii Akademii Nauk SSSR. Litovskaia
SSR (Lithuanian SSR). K. K. Bieliukas, J. Bulavas, and I. V. Komar, eds. Moskva,
Geografgiz, 1955. 391 p. Bibliography, p. 378-390.   20

    228 numbered references arranged in alphabetical order, separately for 135
references in Russian and for 93 titles in Lithuanian or other languages using
the Latin alphabet.

AKADEMIIA Nauk BSSR. Institut Ekonomiki. Belorusskaia SSR (Belorussian SSR).
G. T. Kovalevskii and F. S. Martinkevich, eds. Moskva, Geografgiz, 1957. 488 p.
Bibliography, p. 480-487.   20

    About 200 references arranged in alphabetical order by name of author.

ODUD, Afanasii Lukich. Moldavskaia SSR (Moldavian SSR). Moskva, Geografgiz, 1955.
224 p. Bibliography, p. 220-223. (Moldavskii Filial Akademii Nauk SSSR. Institut
Geografii Akademii Nauk SSSR).   2

    About 75 references arranged alphabetically by name of author.

AKADEMIIA Nauk Gruzinskoi SSR. Institut Geografii imeni Vakhushti, and Akademiia
Nauk SSSR. Institut Geografii. Gruzinskaia SSR: ekonomiko-geograficheskaia
kharakteristika (Georgian SSR: economic-geographic characteristics). A. N.
Dzhavakhishvili, and S. N. Riazantsev, eds. 2nd ed. Moskva, Geografgiz, 1958.
400 p. Bibliography, p. 396-399. (1st ed., Moskva, Izdatel'stvo Akademii Nauk,
1956. 350 p.).   2

    About 100 references in alphabetical order by name of author, all in Cyrillic
(i.e. Georgian works are given in Russian translation, with notation that the work
is in Georgian).

AKADEMIIA Nauk Armianskoi SSR. Institut Ekonomiki, and Akademiia Nauk SSSR. Institut Geografii. Armianskaia SSR (Armenian SSR). O. A. Marukhian, E. M. Murzaev,
and S. N. Riazantsev, eds. Moskva, Geografgiz, 1955. 284 p. Bibliography, p.
274-283.   2

    About 200 references in alphabetical order by name of author in Cyrillic
alphabet (i.e. Armenian titles are given in Russian translation with note that
the work is in Armenian).

AKADEMIIA Nauk Azerbaidzhanskoi SSR. Institut Geografii, and Akademiia Nauk SSSR.
Institut Geografii. Azerbaidzhanskaia SSR: ekonomiko-geograficheskaia kharakteristika (Azerbaidzhan SSR: economic-geographic characteristics). M. A. Kashkai
and P. M. Alampiev, eds. Moskva, Geografgiz, 1957. 445 p. Bibliography, p. 434-
443.   2

    About 200 references in alphabetical order by name of author.

2027 - 2034

KADEMIIA Nauk SSSR. Institut Geografii, and Akademiia Nauk Kazakhskoi SSR. Sektor
Geografii. Kazakhskaia SSR: ekonomiko-geograficheskaia kharakteristika (Kazakh
SSR: economic-geographic characteristics). N. N. Baranskii and O. R. Nazarevskii,
eds. Moskva, Geografgiz, 1957. 734 p. Bibliography, p. 701-732.   2035
    About 600 references in alphabetical order by name of author.

IAZANTSEV, Sergei Nikolaevich, and PAVLENKO, Viktor Fedorovich. Kirgizskaia SSR:
ekonomiko-geograficheskaia kharakteristika (Kirgiz SSR: economic-geographic
characteristics). Moskva, Geografgiz, 1960. 485 p. Bibliography, p. 479-484.   2036
    About 150 references in alphabetical order by name of author.

REDNEAZIATSKII Gosudarstvennyi Universitet. Geograficheskii Fakul'tet. Uzbekskaia
SSR (Uzbek SSR). N. L. Korzhenevskii, ed. Moskva, Geografgiz, 1956. 471 p.
Bibliography, p. 464-469.   2037
    About 125 references in alphabetical order by name of author.

REIKIN, Zakhar Grigor'evich. Turkmenskaia SSR: ekonomiko-geograficheskaia
kharakteristika (Turkmen SSR: economic-geographic characteristics). V. N. Kunin,
ed. 2nd ed. Moskva, Geografgiz, 1957. 451 p. Bibliography, p. 442-449. (1st
ed., Moskva, Geografgiz, 1954. 316 p.).   2038
    About 150 references in alphabetical order by name of author.

KADEMIIA Nauk Tadzhikskoi SSR, and Akademiia Nauk SSSR. Institut Geografii.
Tadzhikskaia SSR: ekonomiko-geograficheskaia kharakteristika (Tadzhik SSR:
economic-geographic characteristics). I. K. Narzikulov and S. N. Riazantsev,
eds. Moskva, Geografgiz, 1956. 228 p. Bibliography, p. 225-227.   2039
    57 references in alphabetical order by name of author.

Development and Location of Productive Forces in 15 volumes
(In progress)

ZVITIE i razmeshchenie proizvoditel'nykh sil SSSR (Development and location of
productive forext of the USSR). Moskva, Izdatel'stvo "Nauka," 1967- . In progress.
(Akademiia Nauk SSSR; Gosplan SSSR. Sovet po Izucheniiu Proizvoditel'nykh Sil). 2040
    The Council for the Study of Productive Forces (Sovet po Izucheniiu Proiz-
voditel'nykh Sil), a geographically oriented preplanning agency under the State
Planning Committee (Gosplan SSSR) began in 1967 the publication of a series of
monographs, of about 300 pages each, on the official economic planning regions of
the USSR. The first volume was sponsored by SOPS alone, but later volumes were
sponsored by both the Academy of Sciences of the USSR and SOPS. Each volume is
prepared by a group of authors and contains a discussion of the regional economy
by branches of the economy, maps, statistical tables, a bibliography of 100 or so
references, and a place-name index. The volumes are listed below in order of date
of publication.

VERO-ZAPADNYI ekonomicheskii raion (The Northwest Economic Region). G. I. Granik,
ed. 1967. 302 p. Bibliography, p. 295-297. (Gosplan SSSR. Sovet po Izucheniiu
Proizvoditel'nykh Sil).   2041
    66 references in alphabetical order.

ADNO-SIBIRSKII ekonomicheskii raion (West Siberian Economic Region). A. B.
Margolin, ed. 1967. 251 p. Bibliography, p. 245-246.   2042
    32 references in alphabetical order.

RAZVITIE i razmeshchenie proizvoditel'nykh sil Kazakhskoi SSR (Development and location of productive forces of the Kazakh SSR). S. B. Baishev, ed. 1967. 259 p. Bibliography, p. 251-253. (Akademiia Nauk SSSR: Gosplan SSSR, Sovet po Izucheniiu Proizvoditel'nykh Sil; Gosplan Kazakhskoi SSR. Nauchno-Issledovatel'-skii Institut; Akademiia Nauk Kazakhskoi SSR. Institut Ekonomiki).   2(

    72 references in alphabetical order.

PRIBALTIISKII ekonomicheskii raion (Baltic Economic Region). A. B. Margolin, ed. 1970. 311 p. Bibliography, p. 304-306.   2(

    About 90 references on the three Baltic republics of Estonia, Latvia, and Lithuania and Kaliningrad oblast of the RSFSR.

UKRAINSKAIA SSR. Ekonomicheskie raiony (Ukrainian SSR. Economic Regions). V. N. Kal'chenko and F. N. Sukhopara, eds. 1972. 315 p. Bibliography, p. 305-308.   2(

    109 references in Russian and Ukrainian on the three economic planning regions in the Ukrainian SSR: Donets-Dnepr, Southwest, and South economic regions).

SREDNEAZIATSKII ekonomicheskii raion (The Middle Asian Economic Region). V. B. Zhmuida, ed. 1972. 298 p. Bibliography, p. 289-291.   2(

    72 references in Russian on the four republics of Soviet Middle Asia: Kirgizia, Uzbekistan, Turkmenia, and Tadzhikistan.

ZAKAVKAZSKII ekonomicheskii raion (Transcaucasian Economic Region). A. A. Adamesku and E. D. Silaev, eds. 1973. 245 p. Bibliography, p. 238-240.   2(

    About 80 references in Russian on the three Transcaucasian republics of Georgia, Armenia, and Azerbaidzhan.

TSENTRAL'NYI ekonomicheskii raion (Central Economic Region). E. B. Alaev, ed. 1973. 269 p. Bibliography, p. 251-253.   2(

    About 90 references in alphabetical order.

    The following volumes appear to continue the series but with the titles changes to begin with the word Problemy (Problems), without the recorded sponsorship of the Academy of Sciences of the USSR and of Gosplan SSSR, Sovet po Izuchenii Proizvoditel'nykh Sil, and with a change of publisher to "Mysl'," Geografgiz.subdivision.

PROBLEMY razvitiia i razmeshcheniia proizvoditel'nykh sil Povolzh'ia (Problems in the development and location of productive forces of the Volga region). A. A. Adamesku, ed. Moskva, Izdatel'stvo "Mysl'," Geografgiz, 1973. 272 p. Bibliography, p. 270-271.   2(

    56 references in alphabetical order.

DETINA, Samuil Isaakovich, OVCHINNINSKII, Nikolai Vladimirovich, and SHAKHOVA, Ol'ga Tikhonovna. Problemy razvitiia i razmeshcheniia proizvoditel'nykh sil Tsentral'-nochernozemnogo raiona (Problems of the development and location of productive forces of the Central Black-earth region). Moskva, Izdatel'stvo "Mysl'," Geografgiz, 1973. 183 p.   2

GLADYSHEV, Anatolii Nikolaevich, KULIKOV, Aleksandr Vasil'evich, and SHAPALIN, Boris Fedorovich. Problemy razvitiia i razmeshcheniia proizvoditel'nykh sil Dal'nego Vostoka (Problems in the development and location of productive forces of the Soviet Far East). Moskva, Izdatel'stvo "Mysl'," Geografgiz, 1974. 215 p. Bibliography, p. 208-213.   2(

    123 references in alphabetical order.

Economic-Administrative Regions of the USSR in 14 Volumes

OMICHESKIE administrativnye raiony SSSR: ukazatel' novoi literatury po prirode,
sursam i khoziaistvu (Economic-administrative regions of the USSR: guide to
w literature on nature, resources, and economy). V. V. Pokshishevskii, ed.
skva, Izdatel'stvo Akademii Nauk SSSR, 1957-1958. 14 v. (Akademiia Nauk SSSR.
stitut Nauchnoi Informatsii).   2052

 v. 1. Pokshishevskii, Vadim Viacheslavovich. Raiony Severa i Severo-Zapada
egions of the North and Northwest). 1957. 59 p.   2053

 v. 2. Gavrilova, Serafima Abramovna, and Parkhomenko, Iia Ivanovna. Raiony
entra RSFSR. (Central Regions of the RSFSR). 1958. 88 p.   2054

 v. 3. Pokshishevskii, V. V. Raiony Povolzh'ia (Volga Regions). 1958. 84 p. 2055

 v. 4. Parkhomenko, I. I. Raiony Nizhnego Dona i Severnogo Kavkaza (Regions
' the Lower Don and North Caucasus). 1958. 100 p.   2056

 v. 5. Parkhomenko, I. I. Raiony Urala (Regions of the Urals). 1957. 48 p. 2057

 v. 6. Gavrilova, S. A. Raiony Zapadnoi Sibiri (Regions of Western Siberia).
58. 62 p.   2058

 v. 7. Pokshishevskii, V. V. Raiony Vostochnoi Sibiri (Regions of Eastern
beria). 1958. 75 p.   2059

 v. 8. Gavrilova, S. A. Raiony Dal'nego Vostoka (Regions of the Far East).
58. 43 p.   2060

 v. 9. Parkhomenko, I. I. Ukrainskaia SSR. Moldavskaia SSR. (The Ukrainian
R. The Moldavian SSR). 1958. 163 p.   2061

 v. 10. Gavrilova, S. A. Belorusskaia SSR (Belorussian SSR). 1957. 35 p.   2062

 v. 11. Gavrilova, S. A., and Parkhomenko, I. I. Uzbekskaia SSR, Kirgizskaia
R, Tadzhikskaia SSR, Turkmenskaia SSR (Uzbekistan, Kirgizia, Tadzhikistan, and
rkmenistan). 1958. 161 p.   2063

 v. 12. Parkhomenko, I. I. Kazakhskaia SSR (Kazakhstan). 1958. 143 p.   2064

 v. 13. Gavrilova, S. A. Gruzinskaia SSR, Azerbaidzhanskaia SSR, Armianskaia
R (Georgia, Azerbaidzhan, and Armenia). 1957. 122 p.   2065

 v. 14. Gavrilova, S. A. Litovskaia SSR, Latviiskaia SSR, Estonskaia SSR
Latvia, Lithuania, and Estonia). 1957. 52 p.   2066

Detailed regional bibliographies covering together the entire Soviet Union
nd physical features, resources, and economic geography. Most of the liter-
ture listed is for the period 1953-1956 but include some for 1950-1952 or in
957. Citations are from Referativnyi zhurnal: geologiia i geografiia (1954-1955)
r geografiia (1956-1957). Each volume consists of two parts, part 1, references
n the region as a whole, part 2, references on individual oblasts, krays, and
utonomous republics in each region.

Russia: A Full Geographic Description in 11 Published Volumes

ENOV-TIAN'-SHANSKII, Veniamin Petrovich, ed. Rossiia: polnoe geograficheskoe
pisanie nashego otechestva: nastol'naia i dorozhnaia kniga dlia russkikh liudei
Russia: a full geographic description of our native land: a reference book for
he Russian people for the office and for travel). Under the general direction
f P. P. Semenov-Tian-Shanskii and V. I. Lamanskii. St. Peterburg, Izdanie A.
. Devriena, 1899-1913. 11 v. published out of projected 20 v.: v. 1-3, 5-7, 9,
4, 16, 18, 19.   2067

v. 1. Moskovskaia promyshlennaia oblast' i verkhnee Povolzh'e (Moscow industrial region and the upper Volga). 1899. 484 p. Bibliography, p. 445-453.

v. 2. Srednerusskaia chernozemnaia oblast' (Central Russian black-earth region). 1902. 717 p. Bibliography, p. 640-652.

v. 3. Ozërnaia oblast' (Lake region). 1900. 456 p. Bibliography, p. 419-426.

v. 5. Ural i Priural'e (Urals and the Pre-Urals). 1914. 669 p. Bibliography, p. 598-619.

v. 6. Srednee i nizhnee Povolzh'e i Zavolzh'e (Middle and lower Volga and trans-Volga). 1901. 559 p. Bibliography, p. 550-558.

v. 7. Malorossiia (Little Russia, i.e. the Ukraine). 1903. 518 p. Bibliography, p. 464-478.

v. 9. Verkhnee Poddneprov'e i Belorussia (Upper Dnepr and Belorussia). 1905. 620 p. Bibliography, p. 575-582.

v. 14. Novorossiia i Krym (New Russia and the Crimea). 1910. 983 p. Bibliography, p. 911-941.

v. 16. Zapadnaia Sibir' (Western Siberia). 1907. 591 p. Bibliography, p. 543-551.

v. 18. Kirgizskii krai (Kirgiz district). 1903. 478 p. Bibliography, p. 436-452.

v. 19. Turkestanskii krai (Turkestan district). 1913. 861 p. Bibliography, p. 781-803.

The most important and detailed of the geographical treatises of pre-Revolutionary Russia. Extensive bibliographies of from 500 to 1,000 references at end of each volume.

BIBLIOGRAPHIES OF REGIONAL AND LOCAL STUDIES

VEDCHESKAIA bibliografiia (Regional bibliography), Bibliografiia sovetskoi
bliografii (Bibliography of Soviet bibliography). (Vsesoiuznaia Knizhnaia
lata). 1962- . Annual. Moskva, Izdatel'stvo "Kniga," 1964- .   2079

    Annual list of general Soviet regional bibliographies for the parts of the
SR, either those published as separate works or as parts of books or serials.
e annual volume (1970), p. 12-15, contains 66 entries. A sub-section of the
ction on general bibliography (Obshchaia bibliografiia).

UNISTICHESKOE stroitel'stvo i ekonomika soiuznykh respublik, oblastei, raionov
gorodov (Communist development and the economy of union republics, oblasts,
yons, and cities), in the chapter Kommunisticheskoe stroitel'stvo (Communist
nstruction) in Bibliografiia sovetskoi bibliografii. Annual.   2080

    This section contains many bibliographies on the economy of specific regions
the USSR. The 1970 volume on pages 45-47, for instance, lists bibliographies
the economy of Sverdlovsk Oblast (1958-1968 with 446 entries), the Crimea (1917-
67 with 1076 entries), Perm' Oblast, 1938-1966, with 3905 entries), Eastern
beria (from 1917, 1200 entries), Central Chernozem oblasts (1966-1970 with 700
tries), Siberia and the Far East (1946-1965, with 2267 entries), and Belorussia
10 entries).

IOGRAFIIA kraevedcheskoi bibliografii RSFSR: annotirovannyi ukazatel'. (Bib-
ography of regional bibliographies of the RSFSR: annotated guide) (Gosudar-
vennaia Publichnaia Biblioteka imeni M. E. Saltykova-Shchedrina). Leningrad.
63- .   2081

    1. Ozerova, G. A. Obshchie i otraslevye bibliografii kak istochnik
aevedcheskoi bibliografii RSFSR. (General and systematic bibliographies as
urces or regional bibliographies of the RSFSR). 1964. 467 p. 1,030 entries.   2082

    2. Bibliografiia geograficheskikh i ekonomicheskikh raionov, kraev,oblastei
avtonomnykh respublik RSFSR (Bibliography of geographical and economic regions,
ays, oblasts, and autonomous republics). Leningrad, 1963-1966.   2083

    no. 1. Ozerova, G. A. Raiony Krainego Severa RSFSR v tselom, Evropeiskii
ver i Severo-Zapad (Regions of the Far North of the RSFSR as a whole, European
rth and Northwest). 1966. 413 p. 1,396 entries.   2084

    no. 3. Gorbachevskaia, N. Volgo-Viatskii raion (Volga-Vyatka region).
64. 118 p. 348 entries.   2085

    no. 4. Gorbachevskaia, N. Tsentral'no-Chernozemnyi raion (Central
ack-earth region). 1965. 92 p. 219 entries.   2086

    no. 6. Ozerova, G. A. Severnyi Kavkaz (The North Caucasus). 1963. 445 p.
259 entries.   2087

NIZATSIIA i metodika kraevedeniia: annotirovannyi ukazatel' novoi literatury
rganization and methodology of local studies: annotated guide to new literature).
ningrad, Gosudarstvennaia Publichnaia Biblioteka imeni M. E. Saltykova-
chedrina, 1958- . (v. 1-11, 1958-1960, as: Organizatsiia, metodika i bibliog-
fiia kraevedeniia).   2088

    Consists of two parts: books and articles on methods of organizing local
udies; bibliographical guides to individual published works on parts of the
viet Union.

STRANOVEDENIE. Kraevedenie (Regional geography. Local geography), in Gosudarstvennaia Biblioteka SSSR imeni V. I. Lenina. Svodnyi ukazatel' bibliograficheskik]
spiskov i kartotek sostavlennykh bibliotekami Sovetskogo Soiuza v 1960- godu.
Obshchestvennye nauki, khudozhestvennaia literatura, isskustvo (Full guide to
bibliographical lists and card indexes compiled by libraries of the Soviet Union
in the year 1960- . Social sciences, belles lettres, and art. Moskva, Izdatel'stvⁱ
"Nauka," 1961- . Annual.

1960. 1961, p. 75-75, entries 1,330-1,394. 65 regional bibliographies, 45 on the
Soviet Union, 20 on foreign regions.

1961. 1962, p. 60-63, entries 966-1,019. 54 regional bibliographies, 41 on the
Soviet Union, 13 on foreign regions.

1962. 1963, p. 63-66, entries 1,076-1,139. 64 regional bibliographies, 4 general,
52 on the Soviet Union, 8 on foreign regions.

1963. 1964, p. 71-74, entries 1,140-1,175. 36 regional bibliographies, 2 on the
whole world, 25 on the Soviet Union, 9 on foreign regions.

1964. 1965, p. 68-70, entries 1,068-1,095. 28 regional bibliographies, 19 on
the USSR, 9 on foreign areas.

1965. 1967, p. 66-67, entries 1,066-1,086. 21 regional bibliographies, 16 on
the USSR, 5 on foreign areas.

1966. 1967, p. 72-74, entries 1,283-1,307. 25 regional bibliographies, 19 on
the USSR, 6 on foreign areas.

1967. 1968, p. 68-70, entries 1,269-1,303. 35 regional bibliographies, 25 on
the USSR, 10 on foreign areas.

1968. 1969, p. 79-81, entries 1,276-1,295. 20 regional bibliographies, 13 on
the USSR, 7 on foreign areas.

1969. 1970, p. 75-76, entries 1,325-1,347. 23 regional bibliographies, 17 on
the USSR, 6 on foreign areas.

GAVRILOVA, Serafima Abramovna; PARKHOMENKO, Iia Ivanovna; and KUZNETSOVA, N. K.
Bibliografii po geografii, vyshedshie v SSSR (1946-1961 gg.) (Bibliographies of
geography, published in the USSR 1946-1961), Geograficheskii sbornik (Institut
Nauchnoi Informatsii Akademii Nauk SSSR). 1. Moskva, VINITI, 1963, p. 32-45.

The section on regional bibliographies of the USSR, p. 36-44, entries 71-
233, contains a rich and comprehensive listing of 163 geographical and general
bibliographies of regions, oblasts, and even individual cities, of geographic
value, 1946-1961, many of which are running bibliographies published currently
at regular intervals, such as yearly. Bibliographies are listed by union republics; those within the RSFSR alphabetically by name of autonomous republic,
kray, or oblast.

STRANOVEDENIE i kraevedenie (Regional studies), in Akademiia Nauk SSSR. Biblioteka
Akademii Nauk SSSR. Ukazatel' osnovnykh otechestvennykh bibliografii i spravochny
izdanii po estestvennym i fiziko-matematicheskim naukam (List of basic native
bibliographies and reference works in the natural and physical sciences). R. L.
Baldaev, comp. A. I. Mankevich, ed. Leningrad, BAN, 1966, p. 175-189.

105 regional bibliographies, 79 for regions in the Soviet Union, etnries
981-1059, p. 175-185, and 26 for foreign countries or areas, entries 1060-1085,
p. 185-189.

OZEROVA, Galina Aleksandrovna. V pomoshch' izucheniiu rodnogo kraia: Ukazatel'
kraevedcheskoi raboty (Aid in the study of the home region: guide to the literature on the organization, methods, and practice of local-studies work).
Leningrad, 1960. 100 p. (Gosudarstvennaia Publichnaia Biblioteka imeni Saltykova
Shchedrina).

291 entries on the basic types, forms, and branches of local-studies work.
view of sources and supplementary literature with 166 entries. Alphabetical
dex of authors and titles, subject index.

LAEV, Valerii Alekseevich. Spisok osnovnykh bibliograficheskikh posobii dlia
iavleniia kraevedcheskoi literatury (List of basic bibliographical aids for
king known local-studies literature). In the book: Kraevedcheskaia rabota
lastnykh bibliotek (Local-studies work of oblast' libraries). Moskva, 1961,
165-174 (Gosudarstvennaia Biblioteka SSSR imeni Lenina. Moskovskii Gosudar-
vennyi Bibliotechnyi Institut).   2093

126 references.

RATIVNYI zhurnal: geografiia (Reference journal: geography) (Vsesoiuznyi
stitut Nauchnoi i Tekhnicheskoi Informatsii). Moskva, 1954- monthly. E.
ografiia SSSR: raionirovanie strany (E. Geography of the USSR: regionaliza-
on of the country).   2094

*UBLIKI, kraia i oblasti SSSR: ukazatel' literatury (Republics, krays, and
·lasts of the USSR: index of the literature). Ia. S. Artiukhnov, and N. P.
.kitin. Moskva, Vsesoiuznaia Knizhnaia Palata, 1943. 93 p.   2095

About 1,600 entries, mostly for the years 1938-1940.

*NOV, Nikolai Vasil'evich. Sinkhronisticheskie tablitsy russkoi bibliografii,
*00-1928, so spiskom vazhneishikh bibliograficheskikh trudov (Synchronic tables
: Russian bibliography, 1700-1928, with lists of the most important biblio-
·aphical works). Moskva, Vsesoiuznaia Knizhnaia Palata, 1962. 192 p.   2096

1693 entries. The column "Kraevaia bibliografiia" (regional bibliography),
, 13-127 lists many regional bibliographies. Another column "Otraslevaia bib-
.ografiia" (subject bibliographies) is also useful.

REGIONS OF THE SOVIET UNION

R. S. F. S. R.

THE ARCTIC AND THE SUB-ARCTIC

ATSKII, Gerasim Vasil'evich. Fiziko-geograficheskoe raionirovanie Arktiki.
hysical-geographic regionalization of the Arctic). Leningrad, Izdatel'stvo
ningradskogo Universiteta, 1967-1973. 3 parts.   2097

    Chast' 1. Polosa materikovykh tundr (Part 1. The belt of mainland tundras).
67. 136 p. Bibliography, p. 125-135. About 310 entries.

    Chast' 2. Polosa okrainnykh morei s ostrovami (Part 2. Belt of the outlying
as and islands). 1970. 120 p. Bibliography, p. 99-115. About 450 entries.

    Chast' 3. Arkticheskii bassein (Part 3. The Arctic basin). 1973. 71 p.
bliography, p. 60-68. About 180 references (120 in Russian and 60 foreign).

OR'EV, Andrei Aleksandrovich. Subarktika: opyt kharakteristiki osnovnykh tipov
ograficheskoi sredy (The Subarctic: an attempt at a characterization of basic
pes of the geographical environment). 2nd ed. Moskva-Leningrad, Izdatel'stvo
ademii Nauk SSSR, 1956. 222 p. Bibliography, p. 210-220. (Akademiia Nauk
SR. Institut Geografii). (1st ed. 1946. Bibliography, p. 161-167).   2098

    250 references, arranged alphabetically by author. Reprinted in Grigor'ev,
A. Tipy geograficheskoi sredy: izbrannye teoreticheskie raboty (Types of
ographical environment: selected theoretical works, Moskva, Izdatel'stvo "Mysl',"
70, references, p. 457-467.

CHKOV, Vasilii Vasil'evich. Krainii Sever: problemy ratsional'nogo ispol'zovaniia
irodnykh resursov (The Far North: problems of the rational utilization of natal resources). Moskva, Izdatel'stvo "Mysl'," Geografgiz, 1973. 184 p. Bibliog-
phy, p. 178-183.   2099

    About 115 references arranged in alphabetical order by name of author.

OV, Ivan Markelovich. Ledianaia zona: fiziko-geograficheskoe opisanie poliarnogo
ktora SSSR (The icy zone: physical-geographical description of the Polar sector
the USSR). Arkangel'sk, Sevkraigiz, 1933. Bibliography, p. 99-116.   2100

    About 500 references in alphabetical order, 1805-1931, on history of explora-
on and on physical geography.

THE RUSSIAN PLAIN

KOV, Fedor Nikolaevich. Fiziko-geograficheskii raion i ego soderzhanie: na
imere Russkoi ravniny (The physical-geographic region and its content, on
e example of the Russian plain). Moskva, Geografgiz, 1956. 221 p. Bibliog-
phy, p. 210-219.   2101

    About 180 references.

TER, Gavriil Dmitrievich, and MIL'KOV, Fëder Nikolaevich, eds. Lesostep' i step'
sskoi ravniny (The wooded steppe and the steppe of the Russian plain). Moskva,
datel'stvo Akademii Nauk SSSR, 1956. 296 p. Bibliography, p. 274-285
kademiia Nauk SSSR. Institut Geografii).   2102

    402 references arranged alphabetically. Index of geographical names. List
native plants mentioned in the text with Russian and Latin names.

MIL'KOV, Fedor Nikolaevich. Lesostep' Russkoi ravniny: opyt landshaftnoi kharakteristiki (The wooded steppe of the Russian plain: an experiment in landscape characterizations). Moskva, Izdatel'stvo Akademii Nauk SSSR, 1950. 296 p. Bibliography, p. 264-294 (Akademiia Nauk SSSR. Institut Geografii).

About 750 references in Russian and Ukrainian at ends of chapters.

ZONN, Sergei Vladimirovich, and CHIKISHEV, Anatolii (Grigor'evich, eds. Voprosy preobrazovaniia prirody Russkoi ravniny (Problems of the transformation of nature on the Russian plain). Moskva, 1973. 266 p. Bibliographies at ends of articles. (Akademiia Nauk SSSR. Institut Geografii).

THE NORTHWEST

ISACHENKO, Anatolii Grigor'evich, DASHKEVICH, Zoia Vasil'evna, and KARNAUKHOVA, Ekaterina Vasil'evna. Fiziko-geograficheskoe raionirovanie Severo-Zapada SSSR (Physical-geographic regionalization of the Northwest of the USSR). Leningrad, Izdatel'stvo Leningradskogo Universiteta, 1965. 248 p. Bibliography, p. 241-246.

About 130 references.

KOL'SKII poluostrov: bibliograficheskii ukazatel' sovetskoi literatury 1930-1939, vypusk 1. Priroda (Kola peninsula: bibliographical guide to the Soviet literature 1930-1939, no. 1. Natural conditions). E. I. Mikhailova, S. S. Gurevich, K. Ia. Ratner, V. M. Rimskaia-Korsakova, E. P. Faidel', K. I. Shafranovskii, and I. G. Moroziuk, comps. A. V. Sidorenko, ed. Apatity, 1960. 324 p. (Biblioteka AN SSSR. Biblioteka Kol'skogo filiala imeni Kirova).

2,753 entries on natural conditions in the Kola peninsula, systematically arranged, including physical geography, geodesy, geology, geophysics, hydrology, soils, botany, and zoology. Subject and geographical indexes. List of sources.

TIKHOMIROV, I. K. Etapy izucheniia Khibin (Stages in the study of the Khibiny), Trudy Khibinskoi geograficheskoi stantsii, vypusk 1 (1960), bibliography, p. 51-64.

193 references in Russian and other languages.

AKADEMIIA Nauk SSSR. Karel'skii Filial, and Akademiia Nauk SSSR. Institut Geografii. Karel'skaia ASSR. (Karelian ASSR). A. A. Grigor'ev and A. V. Ivanov, eds. Moskva, Geografgiz, 1956. 335 p. Bibliography, p. 327-333.

136 references in alphabetical order by name of author.

PRIRODNYE usloviia i resursy iuga tsentral'noi chasti Vologodskoi oblasti (Natural conditions and resources of the south-central part of Vologda Oblast). Iu. D. Dmitrevskii, ed. Vologda, 1970. 336 p. Bibliography, p. 327-335 (Leningradskii Pedagogicheskii Institut imeni A. I. Gertsena, Uchéhye zapiski, tom 408).

About 180 references.

SHISHKIN, Nikolai Ivanovich. Komi ASSR: ekonomiko-geograficheskaia kharakteristika (Komi ASSR: economic-geographic characteristics). Moskva, Geografgiz, 1959. 224 p. Bibliography, 208-216.

About 175 references arranged in alphabetical order by name of author.

THE CENTRAL REGION

GVOZDETSKII, Nikolai Andreevich, and ZHUCHKOVA, Vera Kapitonovna, eds. Fiziko-geograficheskoe raionirovanie Nechernozemnogo tsentra (Physical-geographic regionalization of the non-blackearth center). Moskva, Izdatel'stvo Moskovskogo Universiteta, 1963. 451 p. 1963. Bibliography, p. 439-451. (Prirodnoe i Ekonomiko-geograficheskoe Raionirovanie SSSR dlia Tselei Sel'skogo Khoziaistva. Seriia Fiziko-geograficheskaia).

About 290 references in alphabetical order.

TS, Aleksei Aleksandrovich. Podmoskov'e: ekonomiko-geograficheskaia kharakteri-
tika (The Moscow region: economic-geographic characteristics). Moskva, Geo-
rafgiz, 1961. 302 p. Bibliography, p. 296-302.   2112

About 170 references on the economic geography of the Moscow region arranged
lphabetically by author.

ERATURA po geografii Moskvy i Podmoskov'ia za 10 let, 1950-1960 (Literature on
he geography of Moscow and the Moscow region during ten years, 1950-1960),
oprosy geografii, v. 51 (1961), p. 182-186.   2113

About 100 references in alphabetical order.

RODA goroda Moskvy i Podmoskov'ia (Natural conditions of Moscow and the Moscow
egion). Moskva-Leningrad, Izdatel'stvo Akademii Nauk SSSR, 1947. 379 p.
Akademiia Nauk SSSR. Institut Geografii. Nauchno-Populiarnaia Seriia. K 800-
etiiu Moskvy).   2114

About 260 references at ends of sections on elements of nature: relief and
eological structure, hydrographic network, climate, vegetation, soils and fauna.

CHAROVA, Z. P. Moskva -- stolitsa SSSR, 1918-1946 gg., rekomendatel'nyi ukazatel'
iteratury (Moscow -- capital of the USSR, 1918-1946, guide to recommended lit-
rature). Moskva, Gosudarstvennaia Publichnaia Biblioteka SSSR imeni Lenina,
947. 135 p. (Chto chitat' k 800-letiiu Moskvy).   2115

EEV, P. I. Peschanye porody Meshchërskoi nizmennosti: v sviazi s ee osusheniem
Sandy deposits of the Meshchërskaia lowland in relation to their drainage).
loskva, Izdatel'stvo Moskovskogo Universiteta, 1969. 273 p. Bibliography, p.
61-272.   2116

About 330 references.

THE VOLGA-VYATKA REGION

GO-VIATSKII raion: ekonomiko-geograficheskii obzor (Volga-Vyatka region: eco-
iomic-geographic survey). A. A. Barysheva, I. K. Orfanov, S. I. Prokhorov, and
. L. Trube, eds. Gorkiy, Volgo-Viatskoe Knizhnoe Izdatel'stvo, 1964. 287 p.
3ibliography, p. 281-286.   2117

About 140 references arranged alphabetically by name of author.

DREV, Boris Sergeevich. Gor'kovskaia oblast': ekonomiko-geograficheskie ocherki.
Priroda, naselenie, khoziaistvo (Gor'kiy oblast: economic-geographic sketches.
Natural conditions, population, economy). Gor'kiy, Volgo-Viatskoe Knizhnoe
Izdatel'stvo, 1967. 368 p. Bibliography, p. 360-367.   2118

About 180 references alphabetically by name of author.

THE BLACK-EARTH REGION

L'KOV, Fedor Nikolaevich, ed. Fiziko-geograficheskoe raionirovanie tsentral'nykh
chernozemnykh oblastei (Physical-geographic regionalization of the central black-
earth oblasts). Voronezh, Izdatel'stvo Voronezhskogo Universiteta. 1961. 263 p.
Bibliography, p. 248-255 (Voronezhskii gosudarstvennyi universitet.
Materialy po prirodnomu i ekonomiko-geograficheskomu raionirovaniiu SSSR dlia
tselei sel'skogo khoziaistva).   2119

About 160 references in alphabetical order on physical geography and physical-
geographical regionalization.

THE VOLGA REGION

STUPISHIN, Aleksandr Vladimirovich, ed. Fiziko-geograficheskoe raionirovanie Srednego Povolzh'ia (Physical-geographic regionalization of the Middle Volga region). Kazan', Izdatel'stvo Kazanskogo Universiteta, 1964. 197 p. Bibliography, p. 189-194. (Prirodnoe i Ekonomiko-geograficheskoe Raionirovanie SSSR dlia Tselei Sel'skogo Khoziaistva. Seriia Fiziko-geograficheskaia).   2

About 180 references.

MIL'KOV, Fedor Nikolaevich. Srednee Povolzh'e: fiziko-geograficheskoe opisanie (The Middle Volga region: physical-geographical description). Moskva, Izdatel'stvo Akademii Nauk, 1953. 262 p. Bibliography, p. 246-253. (Akademiia Nauk SSSR. Institut Geografii).   2

177 references, arranged alphabetically. Indexes of geographical names and of plant names.

KAZAN'. Universitet. Biblioteka. Geograficheskie rukopisi, karty i plany. (Geographical manuscripts, maps, and plans). A. N. Varlamov, comp. Kazan', Izdatel'-stvo Kazanskogo Universiteta, 1969. 76 p. (Opisanie rukopisei, vypusk 14).   2

SAVENKOV, Sergei Ivanovich. Prirodnye usloviia Nizhnego Zavolzh'ia: ekonomiko-geograficheskaia kharakteristika (Natural conditions of the Lower Trans-Volga region: economic geographical characteristics). Saratov, Izdatel'stvo Saratovskogo Universiteta, 1962. 160 p. Bibliography, p. 147-160. (Saratovskii Gosudar-stvennyi Universitet. Materialy po ekonomiko-geograficheskomu raionirovaniiu SSSR dlia tselei sel'skogo khoziaistva).   2

About 170 references.

KUZNETSOV, Pavel Savel'erich, ed. Fiziko-geograficheskie raiony Nizhnego Povolzh'ia (Physical-geographic regions of the Lower Volga region). Saratov, Izdatel'stvo Saratovskogo Universiteta, 1961. 156 p. Bibliography, p. 149-154. (Saratovskii Gosudarstvennyi Universitet. Materialy po fiziko-geograficheskomu Raionirovaniiu SSSR dlia Tselei Sel'skogo Khoziaistva).   2

About 130 references in alphabetical order.

THE NORTH CAUCASUS

MASLOV, Evgenii Petrovich. Proizvoditel'nye sily Severnogo Kavkaza (problemy geo-grafii khoziaistva i osvoeniia resursov). (Productive forces of the North Caucasus: problems of the geography of production and the development of resources). V. V. Pokshishevskii, ed. Moskva, Izdatel'stvo "Nauka," 1966. 264 p. Bibliography, p. 257-262. (Akademiia Nauk. Institut Geografii).   2

About 175 references, arranged alphabetically.

KALININ, S. D. Prirodnye resursy Rostovskoi oblasti: Ukazatel' knig, zhurnalnykh i gazetnykh statei, za 1948-1955 g. vypusk 2. (Natural resources of Rostov oblast: guide to books, journal, and newspaper articles, 1948-1955. no. 2, Rostov na Donu, Knizhnoe Izdatel'stbo, 1957. 159 p. (Rostovskaia Gosudarstven-naia Nauchnaia Biblioteka imeni Karla Marksa).   2

1,353 entries, systematically arranged, and annotated. Sections include: maps and mapping, physical geography, paleogeography, geology, geomorphology, climate and weather, waters of the land, Volgo-Don, underground waters, soils, vegetation, animal life.

GIUL', Kasum Kiazimovich, VLASOVA, S. V., KISIN, I. M., and TERTEROV, A. A. Fizicheskaia geografiia Dagestanskoi ASSR (Physical geography of the Dagestan ASSR). Makhach-Kala, Dagknigoizdat, 1959. 249 p. Bibliography, p. 231-234.   2

About 115 references in alphabetical order.

THE URALS

AR, Igor' Valer'ianovich. Ural: ekonomiko-geograficheskaia kharakteristika
The Urals: economic-geographic characteristics). Moskva, Izdatel'stvo Akademii
auk SSSR, 1959. 367 p. Bibliography, p. 352-366. (Akademiia Nauk SSSR. In-
titut Geografii).   2128

About 575 references on the economic geography of the Urals as a whole
rranged alphabetically by name of author. Text discusses the region as a whole
y systematic fields.

AR, Igor' Valer'ianovich. Geografiia khoziaistva Urala: poraionnaia ekonomiko-
eograficheskaia kharakteristika (Geography of the economy of the Urals: regional
conomic-geographic characteristics). Moskva, Izdatel'stvo "Nauka," 1964. 395 p.
ibliography, p. 386-394. (Akademiia Nauk SSSR. Institut Geografii).   2129

About 280 references on the regional economic-geography of the Urals in
lphabetical order by name of author. Text is organized by regions and subregions.

HIPOVA, Nina Pavlovna. Kratkii ocherk istorii izucheniia prirody Urala v
ovetskoe vremia (Short essay on the history of the study of nature in the Urals
n the Soviet period). In the book: Doklady na sektsiiakh uchënogo soveta, vypusk
(Papers in sections of the scientific council, no. 2). Sverdlovsk, 1960,
. 111-119. (Sverdlovskii oblastnoi kraevedcheskii muzei).   2130

About 200 references, including maps and atlases.

A, Leonid Evgen'evich. Goroda Urala. Chast' pervaia. Feodal'nyi period (Cities
f the Urals. Part 1. Feudal period). Moskva, Geografgiz, 1951. 422 p. Bib-
iography, p. 385-421.   2131

About 700 references on the historical geography of urbanization in the Urals
p to 1861 under the following classes of material: classics of Marxism-Leninism,
. 385, documents, p. 386-388; statistical material, p. 388-390, maps, p. 390,
esearch and literature, p. 391-421. These are summarized in a short survey of
ources and literature, p. 27-44.

MSKAIA oblast': annotirovannyi bibliograficheskii ukazatel' (Perm' oblast:
nnotated bibliographic guide). L. P. Vershinina, L. A. Spiridonova, L. V.
erëkhina, K. S. Shilova, and S. D. Shirokova, comps. Perm', Knizhnoe Izdatel'stvo,
961. 214 p. (Permskaia Gosudarstvennaia Publichnaia Biblioteka imeni M. Gor'kogo.
auchno-Bibliograficheskii Otdel).   2132

Bibliography on physical geography is in the section on natural conditions and
atural resources, p. 69-97, with 220 entries systematically organized under:
eography, regional studies, geology, climate, water resources, soils, living
ature. Author and title index.

IL'NIKOV, I. P., and others, eds. Fiziko-geograficheskoe raionirovanie Bashkirskoi
SSR (Physical-geographic regionalization of the Bashkir ASSR). Ufa, 1964.
ibliography, p. 198-208 (Bashkirskii Gosudarstvennyi Universitet Uchenye zapiski,
om 16. Seriia geograficheskaia, no. 1).   2133

About 250 references in alphabetical order on the Bashkir ASSR and adjacent
reas of the South Urals.

HAEV, Khalil' Ianovich. Prirodnye usloviia i resursy Bashkirskoi ASSR: ekonomiko-
eograficheskaia kharakteristika (Natural conditions and resources of the Bashkir
SSR: economic-geographic characteristics). Ufa, 1959. Bibliography, p. 279-
95. (Akademiia Nauk SSSR. Bashkirskii filial).   2134

About 400 references, grouped by sections of the book: general questions;
eology, relief; mineral deposits; climate, waters; soils and land resources;
egetation; animal life.

KHISTMATOV, Mukhamed'ian Fazylianovich. Ocherki po geografii Bashkirii (Essays on
the geography of Bashkiria). Ufa, 1963. 104 p. Bibliography, p. 98-104.
(Bashkirskii Gosudarstvennyi Universitet. Bashkirskii Institut Usovershenstvovanii.
Uchitelei).   21

    About 140 references.

PROKAEV, Vasilii Ivanovich. Fiziko-geograficheskaia kharakteristika iugo-zapadnoi
chasti Srednego Urala i nekotorye voprosy okhrany prirody etoi territorii (Physical-geographic characteristics of the southwest part of the Middle Urals and some
problems of the conservation of nature in this territory). Sverdlovsk, 1963.
Bibliography, p. 181-185. (Akademiia Nauk SSSR. Ural'skii Filial.Komissiia po
okhrane prirody, Trudy, vypusk 2).   21

    About 160 references in alphabetical order.

MAKUNINA, Aleksandra Aleksandrovna. Regional'naia fizicheskaia geografiia SSSR.
Ural'sko-Novozemel'skaia strana. Teksty lektsii. vypusk 2. Landshafty Urala
(Regional physical geography of the USSR. Ural-Novaya Zemlya country. Text of
lectures. no. 2. Landscapes of the Urals). Moskva, 1969. 129 p. Bibliog-
raphy, p. 118-128. (Moskovskii Gosudarstvennyi Universitet imeni M. V. Lomonosova.
Geograficheskii Fakul'tet).   21

SIBERIA AND THE SOVIET FAR EAST AS A WHOLE
(See also articles in Sibirskaia sovetskaia entsiklopediia, entry 851)

VOROB'EV, Vladimir Vasil'evich, and VERSHINSKAIA, N. I. Publikatsii sibirskikh i
dal'nevostochnykh organizatsii Geograficheskogo Obshchestva SSSR. 1945-1963 gg.
(Publications of the Siberian and Far Eastern organizations of the Geographical
Society of the USSR, 1945-1963). Irkutsk, 1966. 167 p. (Akademiia Nauk SSSR.
Sibirskoe Otdelenie. Institut Geografii Sibiri i Dal'nego Vostoka, and Geo-
graficheskoe Obshchestvo SSSR, Biuro Sibirskikh i Dal'nevostochnykh Organizatsii). 21

    1,495 entries, arranged systematically. Author index.

PUBLIKATSII sibirskikh i dal'nevostochnykh organizatsii geograficheskogo obshchestva
SSSR, tom 2 (1964-1969 gg.) vypusk 1 (Publications of the Siberian and Far
Eastern organizations of the Geographical Society of the USSR, v. 2, 1964-1969,
no. 1). D. A. Timofeev, ed. V. V. Vorob'ev and G. F. Chernova, comps. Irkutsk,
1971. 177 p. (Akademiia Nauk SSSR. Sibirskoe Otdelenie. Institut Geografii
Sibiri i Dal'nego Vostoka. Biuro Sibirskikh i Dal'nevostochnykh Organizatsii
Geograficheskogo Obshchestva SSSR).   21

    2,418 entries, arranged systematically for physical geography: geology and
mineral deposits; geomorphology; karst and caves; permafrost and hydrology;
climatology and meteorology; hydrology.

NARODNOE khoziaistvo Sibiri i Dal'nego Vostoka: informatsionnyi-bibliograficheskii
biulleten' (Economy of Siberia and the Far East: informational-bibliographic
bulletin). Novosibirsk, Sibirskoe Otdelenie Akademii Nauk SSSR. 1970- . 6 nos.
a year. (Gosudarstvennaia Publichnaia Nauchno-Tekhnicheskaia Biblioteka Sibirskogo
Otdeleniia Akademii Nauk SSSR. Otdel Nauchnoi Bibliografii).   2

TOMASHEVSKII, V. V. Materialy k bibliografii Sibiri i Dal'nego Vostoka: XV-pervaia
polovina XIX v. (Materials for a bibliography of Siberia and the Far East: 15th-
first half of the 19th centuries). Vladivostok, Izdatel'stvo Akademii Nauk SSSR,
1957. 213 p. (Akademiia Nauk SSSR. Dal'nevostochnyi Filial).   21

    About 3,500 entries.

2135 - 2141

SIA. Pereselencheskoe Upravlenie. <u>Aziatskaia Rossiia</u>. S.-Peterburg, Izdanie
Pereselencheskogo Upravleniia Glavnogo Upravleniia Zemleustroistva i Zemledeliia.
1914. 3 v. Bibliography, v. 3, p. lxxi-cxli.   2142

    Bibliography by topics, such as agriculture, emigration, natural history,
general geography, and cartography.

HOV, Vladimir Izmailovich. <u>Sibirskaia bibliografiia. Ukazatel' knig i statei
v Sibiri na russkom iazyke i odnykh tol'ko knig na inostrannykh iazykakh za ves'
period knigopechataniia</u> (Siberian bibliography. Guide to books and articles
in Siberia in the Russian language and books only in foreign languages for the
entire period of book printing). v. 2. <u>Istoriia, biografii, geografiia,
puteshestviia, statistika, etnografiia, kartografiia</u>. (History, biographies,
geography, travels, statistics, ethnography, cartography). St. Petersburg, 1891.
170 p.   2143

    Geography is included in the section, Geography, travels, statistics,
ethnography, and cartography of Siberia, entries 11,925-17,316. Systematic
arrangement. Comprehensive bibliography up to 1890.

    Reprinted 1903. Reprinted, Vaduz, Liechenstein, Kraus Reprint, 1963.

EEV, Stepan Nikolaevich. <u>Materialy dlia bibliografii Sibiri</u> (Materials for a
bibliography of Siberia). Tobol'sk. 1892-1895. 4 v.   2144

    Each volume covers one calendar year in alphabetical order. Indexes for
v. 1 and 2.

MOFEEV, Dmitrii Andreevich. Poverkhnosti vyravnivaniia Sibiri i Dal'nego Vostoka:
sostoianie izuchennosti i problemy. (Peneplain surfaces of Siberia and the Far
East: status of study of the problem). In the book: <u>Geografiia i geomorfologiia
Azii</u>. Moskva, Izdatel'stvo "Nauka," 1969, p. 100-105 (Akademiia Nauk SSSR.
Geomorfologicheskaia Komissiia. Institut Geografii).   2145

    138 references.

AFONOV, Nikolai Timofeevich. <u>Osnovnye problemy formirovaniia promyshlennykh
kompleksov v vostochnykh raionakh SSSR. Chast' pervaia. Osobennosti razvitiia
i razmeshcheniia promyshlennosti</u> (Basic problems in the formation of industrial
complexes in the eastern regions of the USSR. Part 1. Characteristics of
development and distribution of manufacturing). Leningrad, Izdatel'stvo Leningradskogo Universiteta, 1970. 168 p. Bibliography, p. 159-167. (Leningradskii
Gosudarstvennyi Universitet).   2146

    About 260 references.

                              WESTERN SIBERIA

ROB'EV, Vladimir Vasil'evich, ed. <u>Geograficheskie osobennosti osvoeniia
taezhnykh raionov Zapadno-Sibirskoi nizmennosti: materialy ekspeditsii</u> (Geographical features of the development of the tayga regions of the West-Siberian
lowland: expedition materials). Irkutsk, Institut Geografii Sibiri i Dal'nego
Vostoka, 1969. 272 p. Bibliography, p. 263-269. (Akademiia Nauk SSSR. Sibirskoe Otdelenie. Institut Geografii Sibiri i Dal'nego Vostoka. <u>Materialy Ob'-
Irtyshskoi Ekspeditsii</u>, vypusk 1).   2147

    About 140 references.

LIMONOVA, V. A., comp. Okhrana prirody Zapadnoi Sibiri. Ukazatel' literatury
(Conservation in Western Siberia: guide to the literature). In the book: <u>Priroda
Tomskoi oblasti i ee okhrana</u> (Nature in Tomsk oblast and its conservation),
vypusk 2. Tomsk, Tomskii Universitet, 1965, p. 123-151.   2148

    519 references, mostly for the period 1950-1963.

                                                             2142 - 2148

MISEVICH, Kornei Nikolaevich, and CHUDNOVA, Valentina Ivanovna. <u>Naselenie raionov sovremennogo promyshlennogo osvoeniia severa Zapadnoi Sibiri</u> (Population of the regions of modern industrial development in the north of West Siberia). V. V. Vorov'ev, ed. Novosibirsk, Izdatel'stvo "Nauka," Sibirskoe Otdelenie, 1973. 209 p. Bibliography, p. 197-207 (Akademiia Nauk SSSR. Sibirskoe Otdelenie. Institut Geografii Sibiri i Dal'nego Vostoka): Supplementary title page and abstract in English.

About 200 references in alphabetical order.

<u>PRIRODA i ekonomika Privasiugan'ia</u> (Nature and the economy of the Vasyugan'ye). Tomsk, Izdatel'stvo Tomskogo Universiteta, 1966. 347 p. Bibliography at ends of chapters and sections and p. 333-345, comp. by Iu. K. Mal'tseva, V. A. Filimonova, and G. M. Sazhneva.

About 320 references, organized by systematic fields, on the Vasyugan swamp area in the southern part of the West Siberian plain.

AKSENOVA, N. N. Bibliografiia literatury o prirode poimy rek Obskogo basseina (Bibliography of the literature on nature of the bottomlands of the rivers of the Ob' basin). B. G. Ioganzen, ed. In the book: <u>Priroda poimy reki Obi i ee khoziaistvennoe osvoenie</u> (Nature in the bottomlands of the Ob' river and its economic utilization). Tomsk, 1963. p. 350-407. (Tomskii gosudarstvennyi universitet imeni Kuibysheva, <u>Trudy</u>, Tom 152).

About 1,500 entries for the Russian literature from the 18th century to 1959 on general questions of the history of research on organic and inorganic nature and the economy of the bottomlands of the Ob' river basin.

SERGEEV, Grigorii Maksimovich. <u>Ostrovnye lesostepi i podtaiga Prieniseiskoi Sibiri</u> (Islands of wooded steppe and sub-tayga of Siberia west of the Yenisey River). Irkutsk, Vostochno-Sibirskoe Knizhnoe Izdatel'stvo, 1971. 264 p. Bibliography, p. 238-248. (Akademiia Nauk SSSR. Sibirskoe Otdelenie. Institut Geografii Sibiri i Dal'nego Vostoka).

ROZEN, Mikhail Fedorovich. <u>Ocherki i bibliografiia issledovanii prirody Altaia</u> (Studies and bibliography of research on natural conditions of the Altay). Barnaul, Altaiskoe Knizhnoe Izdatel'stvo, 1970. 255 p. (Geograficheskoe Obshchestvo SSSR. Altaiskii Otdel. <u>Izvestiia</u>, vypusk 12).

2,258 entries grouped by three periods: 1667-1750, 1851-1917, and the Soviet period. Author index.

ROZEN, Mikhail Fedorovich. <u>Istoriia issledovaniia prirody Gornogo Altaia</u> (History of research on natural conditions of Gornyi Altai). Gorno-Altaisk, Knizhnoe Izdatel'stvo, 1961. 95 p. (Gorno-Altaiskii Nauchno-Issledovatel'skii Institut Istorii, Iazyka i Literatury).

About 620 entries on scientific literature on the study of natural conditions and productive forces of the Altay mountains. Author index.

SPIDCHENKO, Konstantin Ivanovich. <u>Goroda Kuzbassa: ekonomiko-geograficheskii ocherk</u> (Cities of the Kuzbass: a study in economic geography). Moskva, Geografgiz, 1947. 147 p. Bibliography, p. 138-145.

217 numbered references in alphabetical order by name of author.

FILIMONOV, M. R., MASLOV, D. P., and AKSENOVA, N. A. Ukazatel' literatury o prirode Tomskoi oblasti (Guide to the literature on nature in Tomsk oblast), in the book: <u>Prirodnye biologicheskie resursy Tomskoi oblasti i perspektivy ikh ispol'zovaniia</u>, B. G. Ioganzen, ed. Tomsk, Izdatel'stvo Tomskogo Universiteta, 1966, p. 207-263. (Tomskii Gosudarstvennyi Universitet).

About 1,500 entries for books and articles on natural resources of Tomsk oblast in Western Siberia and their utilization.

2149 - 2156

ANZEN, Bobo Germanovich. Priroda Tomskoi oblasti (Natural conditions in Tomsk
olast). 2nd ed., Tomsk, Knizhnoe Izdatel'stvo, 1959. 151 p. Bibliography,
. 132-150.   2157

About 360 references, organized systematically.

ROM, Samuil Gdal'evich, VOSTRIAKOVA, Natal'ia Vasil'evna, and SHIRIKOV, Viacheslav
lkailovich. Izmenenie prirodnykh uslovii v Srednei Obi posle sozdaniia Novosibir-
koi GES (Changes in natural conditions in the Middle Ob' after the construction
f the Novosibirsk hydro-electric station). V. S. Mezentsev, ed. Novosibirsk,
lbliography, p. 135-143. (Akademiia Nauk SSSR. Sibirskoe Otdelenie. Geografiche-
koe Obshchestvo SSSR. Novosibirskii Otdel).   2158

189 references in alphabetical order.

EASTERN SIBERIA

APOVA, N. K. Razvitie proizvoditel'nykh sil Vostochnoi Sibiri: materialy k bib-
lografii (Development of productive forces in Eastern Siberia: materials for a
lbliography). Irkutsk, 1970. 101 p. (Vostochno-Sibirskii Sovet Koordinatsii i
lanirovaniia Nauchno-Issledovatel'skikh Rabot po Gumanitarnym Naukam. Irkutskii
osudarstvennyi Universitet imeni Zhdanova. Vostochno-Sibirskoe Otdelenie Geo-
raficheskogo Obshchestva SSSR).   2159

Classed, partly annotated bibliography of books and articles from 1917 on.
200 entries. Introductory article by V. A. Krotov.

HOVA, Natal'ia Georgievna. Fiziko-geograficheskie issledovaniia Vostochnoi
ibiri v XIX veke (Physical-geographic research on Eastern Siberia in the 19th
entury). Moskva-Leningrad, Izdatel'stvo "Nauka," 1964. 191 p. Bibliography,
. 171-189. (Akademiia Nauk SSSR. Institut Istorii Estestvoznaniia i Tekhniki). 2160

About 400 references, organized by chapters: geographical concepts of
astern Siberia up to the 19th century; geographical investigations of Eastern
iberia in the first half of the 19th century; geographical investigations of
astern Siberia in the 1850's and 1860's; geographical investigations of Eastern
iberia, the 1870's to the 1890's. Reviews of references listed are cited.

HTSINSKII, Iu. B., BUDZ, M. D., and ZARUBIN, N. E. Opolzni, seli, termokarst
r Vostochnoi Sibiri i ikh inzhenerno-geologicheskoe znachenie (Landslides, mud
lows, and thermokarst in Eastern Siberia and their significance in engineering
eology). Moskva, Izdatel'stvo "Nauka," 1969. 136 p. Bibliography, p. 127-
33. (Akademiia Nauk SSSR. Sibirskoe Otdelenie. Institut Zemnoi Kory).   2161

213 references.

YKO, N. N., and MORACHEVSKAIA, Elena Nikolaevna. Bibliografiia Krasnoiarskogo
raia, 1924-1960 (Bibliography of Krasnoyarsk Kray, 1924-1960). v. 1 Prirodno-
konomicheskie usloviia i razvitie narodnogo khoziaistva (Physical and economic
onditions and growth of the economy). Krasnoyarsk, Krasnoiarskoe Knizhnoe
zdatel'stvo, 1963. 569 p. (Krasnoiarskaia Kraevedcheskaia Biblioteka. Biblioteka
oveta po Izucheniiu Proizvoditel'nykh Sil pri Gosplane SSSR).   2162

Literature on physical geography is included in the section, natural condi-
ions and resources, p. 7-247, under the headings: general works; geomorphology,
laciology, permafrost studies; hydrography and hydrology; climate; soils; vege-
ation; and animal life, with about 700 entries. Indexes of authors and of geo-
raphical names.

ILLOV, Mikhail Vasil'evich, and SHCHERBAKOV, Iurii Adrianovich, eds. Krasnoiarskii
rai: prirodnoe i ekonomiko-geograficheskoe raionirovanie (Krasnoyarsk Kray:
hysical and economic-geographic regionalization). Krasnoyarsk, Knizhnoe Izdatel'-
tvo, 1962. 404 p. Bibliography, p. 384-402.   2163

About 400 references.

KIRILLOV, Mikhail Vasil'evich. Geografiia Krasnoiarskogo kraia i istoriia razvitiia ego prirody (Geography of Krasnoyarsk kray and history of the development of its natural conditions). Krasnovarsk, 1970. 210 p. Bibliography, p. 202-208. (Krasnoiarskii Gosudarstvennyi Pedagogicheskii Institut).

TAIMYRO-SEVEROZEMEL'SKAIA oblast'. Fiziko-geograficheskie kharakteristiki (Taymyr and Severnaya Zemlya region: physical-geographic characteristics). L. S. Govorukha I. M. Averina, I. V. Semenov, and others. R. K. Sisko, ed. Leningrad, Gidrometeoizdat, 1970. 374 p. Bibliography, p. 358-373. (Glavnoe Upravlenie Gidrometeorologicheskoi Sluzhby pri Sovete Ministrov SSSR. Arkticheskii i Antarkticheskii Nauchno-Issledovatel'skii Institut).

    346 references.

BIBLIOGRAFIIA Irkutskoi oblasti, vypusk 4. Fizicheskaia geografiia (Bibliography of Irkutsk oblast no. 4. Physical geography). L. E. Zubasheva, comp. B. V. Zonov and L. K. Zhilkinaia, eds. Leningrad, Izdatel'stvo Leningradskogo Universiteta, 1957. 173 p. (Irkutskii Gosudarstvennyi Universitet imeni Zhdanova. Trudy nauchnoi biblioteki, vypusk 13).

    2,358 entries, 1720-1950, of books and articles, arranged chronologically. Indexes to authors, titles, persons mentioned in text, places, subjects.

AKADEMIIA Nauk SSSR. Sibirskoe Otdelenie. Institut Geografii Sibiri i Dal'nego Vostoka. Fizicheskaia geografiia Irkutskoi oblasti (bibliografiia), vypusk 2. (1951-1962) (Physical geography of Irkutsk oblast: bibliography, no. 2, 1951-1962). V. I. Iakhnenko, comp. Irkutsk, Vostochno-Sibirskoe Knizhnoe Izdatel'stvo. 1965. 140 p.

    Chronological list of 1,139 entries of books, articles, chapters in collections, abstracts of dissertations, maps, and atlases on the physical geography of Irkutsk oblast in Eastern Siberia, 1951-1962. Indexes of authors, place names, and subjects.

BIBLIOGRAFIIA Irkutskoi oblasti. Okhota i rybolovstvo (Bibliography of Irkutsk Oblast'. Hunting and fishing). L. E. Zubasheva, comp. Irkutsk, Vostochno-Sibirskoe Knizhnoe Izdatel'stvo, 1966. 180 p. (Irkutsk. Universitet. Nauchnaia biblioteka. Trudy, vypusk 17).

GEOGRAFIIA Irkutskoi oblasti. Uchebnoe posobie (Geography of Irkutsk oblast: textbook), no. 1- Irkutsk, 1971- (Irkutskii Gosudarstvennyi Universitet).

    vypusk 1. Istoriia geograficheskogo izucheniia Irkutskoi oblasti. (History of geographical investigation of Irkutsk oblast). V. M. Boiarkin. 1971. 134 p. Bibliography, p. 128-132.

    vypusk 2. Ocherki po fizicheskoi geografii Irkutskoi oblasti. (Studies in the physical geography of Irkutsk oblast). V. M. Boiarkin. 1972. 292 p. Bibliographies at end of each study.

    vypusk 3. Fiziko-geograficheskoe raionirovanie Irkutskoi oblasti (Physical-geographic regionalization of Irkutsk oblast). V. M. Boiarkin. 1973. 328 p. Bibliography, p. 306-325.

BIBLIOGRAFIIA Iakutskoi ASSR, 1931-1959. Tom. 2. Prirodnye usloviia, resursy i narodnoe khoziaistvo (Bibliography of the Yakut ASSR, 1931-1959. v. 2, Natural conditions, resources, and economy), T. S. Guchek, V. G. Zemlianskaia, M. N. Karavaev, A. I. Sytina, V. M. Sentsov, and V. P. Tiuliaeva, comps. Moskva, Izdatel'stvo Akademii Nauk SSSR. 1962. 255 p. (Akademiia Nauk SSSR. Sektor Seti Spetsial'nykh Bibliotek. Gosekonomsovet SSSR. Biblioteka SOPS. Iakutskaia Respublikanskaia Biblioteka imeni A. S. Pushkina).

    About 4,000 entries, systematically organized. Index of authors, editors and titles.

ERNAIA Iakutiia: fiziko-geograficheskaia kharakteristika (Northern Yakutiya:
hysical-geographic characteristics). I. M. Averina, V. G. Agapitov, N. A.
oronina, and others. Ia. Ia. Gakkel' and E. S. Korotkevich, eds. 2nd ed.,
eningrad, "Morskoi Transport," 1962. 280 p. Bibliography, p. 274-280.
rkticheskii i Antarkticheskii Nauchno-issledovatel'skii Institut, Trudy, tom
36).   2171
    224 entries.

LIOGRAFIIA Buriat-Mongolii za 1890-1936 gg. (Bibliography of Buryat-Mongolia,
890-1936). v. 1. Estestvoznanie (Natural sciences). V. A. Obruchev, V. V.
lekhin, G. Iu. Vereshchagin, S. I. Ognev, and M. I. Pomus, eds. Moskva-
eningrad, Izdatel'stvo Akademii Nauk SSSR, 1939. 625 p. (Buriat-Mongol'skii
osudarstvennyi Nauchno-Issledovatel'skii Institut Iazyka, Literatury i Istorii). 2172

    Material on geography is listed in part 1, Geography, with sections on works
f a general character, history and organization of geographical research; part
, physical geography, with sections on orography and geomorphology, hydrogeology,
ydrology, meteorology and climatology, soil science, persons, and bibliography;
nd in part 4, vegetation; and in part 5, animal life. Indexes.

    For references through 1890 see V. I. Mezhov, Sibirskaia bibliografiia.

OBRAZHENSKII, Vladimir Sergeevich, and others. Tipy mestnosti i prirodnoe
aionirovanie Buriatskoi ASSR (Types of localities and regionalization by
atural conditions of the Buryat ASSR). Moskva, Izdatel'stvo Akademii Nauk
SSSR, 1959. 218 p. Bibliography, p. 212-216. (Akademiia Nauk SSSR. Institut
eografii).   2173
    About 120 references.

HENTSEV, V. N. Ekonomiko-geograficheskii analiz spetsializatsii raiona: na
rimere Vostochnogo Zabaikal'ia (Economic-geographic analysis of the specializa-
ion of a region: the case study of Eastern Trans-Baikalia). Novosibirsk, Izdatel'-
stvo "Nauka," Sibirskoe otdelenie. 1972. Bibliography, p. 92-94. (Akademiia
auk SSSR. Sibirskoe Otdelenie. Institut Geografii Sibiri i Dal'nego Vostoka).   2174

HEEV, Valerii Sergeevich. Verkhnecharskaia kotlovina: opyt topologicheskogo
zucheniia landshafta (The upper Chara basin: an experiment in the topological
nvestigation of the landscape). V. B. Sochava, ed. Novosibirsk, Izdatel'stvo
"Nauka," Sibirskoe Otdelenie, 1973. 143 p. Bibliography, p. 137-142 (Akademiia
auk SSSR. Sibirskoe Otdelenie, Institut Geografii Sibiri i Dal'nego Vostoka).   2175

                            THE SOVIET FAR EAST

PITSA, Andrei Petrovich, ed. Priroda i chelovek (Nature and man). Vladivostok,
1973. 317 p. Bibliographies at ends of articles. (Akademiia Nauk SSSR.
Dal'nevostochnyi Nauchnyi Tsentr. Tikhookeanskii Institut Geografii).   2176

ADEMIIA Nauk SSSR. Institut Geografii. Dal'nii Vostok: fiziko-geograficheskaia
harakteristika (The Far East: physical-geographic characteristics). G. D.
ikhter, ed. Moskva, Izdatel'stvo Akademii Nauk SSSR, 1961. 439 p. Bibliog-
aphy, p. 412-420.   2177
    About 270 references, arranged alphabetically. Index of geographical names.
ist of Russian and Latin names of plants. List of Russian and Latin names of
nimals. Text organized mainly by systematic fields of physical geography.

LIOGRAFIIA Dal'nevostochnogo Kraia 1890-1931. Tom 1. Fizicheskaia geografiia
Bibliography of the Far Eastern district, 1890-1931. v. 1. Physical geography).
. N. Asatkin and V. A. Samoilov, eds. Moskva. Izdatel'stvo Vsesoiuznaia
ssotsiatsiia Sel'skokhoziaistvennoi Bibliografii, 1935. 378 p.   2178

2,106 annotated entries, organized systematically: physical geography, orography and geomorphology, hydrology, meteorology and climatology, soils, personalia. Indexes of authors and editors, collective authors and institutions, personal names, subjects, places.

PRIRODA Amurskoi oblasti (Natural conditions in Amur oblast). N. K. Shul'man, I. A. Andreeva, I. A. Palenko, I. E. Kositsyn, A. P. Til'ba, and L. M. Barancheev. Blagoveshchensk, Amurskoe Knizhnoe Izdatel'stvo, 1959. 311 p.

Bibliographies at end of each chapter. About 240 references on the physical-geographic characteristics of the oblast: structure, surface, climate, waters, soils, vegetation, and animal life.

KHABAROVSKII Krai, Dal'nii Vostok. Ukazatel' trudov Priamurskogo (Khabarovskogo) filiala Geograficheskogo Obshchestva SSSR. 1962-1967 gg. (Khavarovsk Kray, the Far East. Guide to work of the Amur [Khabarovsk] branch of the Geographical Society of the USSR, 1962-1967). A. A. Stepanov, comp. Khabarovsk, Priamurskii (Khabarovskii) Filial Geograficheskogo Obshchestva SSSR, 1968. 54 p.

540 entries of books, articles, manuscript reports, and references of collaborators of the branch on nature, history, population, and the economy of Khabarovsk kray and the Soviet Far East, organized regionally. Systematic and author indexes.

MASIUK, M. S., and SUTYGINA, P. T. Chto chitat' o Khabarovskom krae: ukazatel' literatury (What to read on Khabarovsk Kray: guide to the literature). With the collaboration of V. G. Pletnevaia, A. A. Stepanov, N. A. Rakov, B. I. Iankele-vich, and S. S. Kotel'nikov. Khabarovsk, 1948. 114 p. (Khabarovskaia Kraevedche-skaia Biblioteka).

Literature on geography is in the section on natural conditions and natural resources under the headings: land surface, seas and rivers; climate; soils, mineral deposits; vegetation; animal life, resources of the waters. About 270 references.

LIVEROVSKII, Iurii Alekseevich, and KOLESNIKOV, Boris Pavlovich. Priroda iuzhnoi poloviny sovetskogo Dal'nego Vostoka: fiziko-geograficheskaia kharakteristika (Natural conditions in the southern half of the Soviet Far East: physical-geographical characteristics). Moskva, Geografgiz, 1949. 383 p. Bibliography, p. 355-366 (Akademiia Nauk SSSR. Institut Geografii. Priroda SSSR. Nauchno-populiarnye ocherki).

About 200 references arranged alphabetically.

LIUBIMOVA, Elena L'vovna. Kamchatka: fiziko-geograficheskii ocherk (Kamchatka: a physical-geographic study). Moskva, Geografgiz, 1961. 190 p. Bibliography, p. 178-189. (Akademiia Nauk. Institut Geografii).

About 200 references in alphabetical order.

SERGEEV, Mikhail Alekseevich. Narodnoe khoziaistvo Kamchatskogo-kraia (Economy of Kamchatka kray). Moskva, Leningrad, Izdatel'stvo Akademii Nauk SSSR, 1936. 815 p. Bibliography, p. 799-809.

399 references.

ALEKSANDROV, Sergei Mikhailovich. Ostrov Sakhalin (Sakhalin island). Moskva, Izdatel'stvo "Nauka," 1973. 183 p. Bibliography, p. 168-182. (Akademiia Nauk SSSR. Dal'nevostochnyi Nauchnyi Tsentr. Sakhalin. Kompleksnyi Nauchno-Issledovatel'skii Institut. Istoriia razvitiia rel'efa Sibiri i Dal'nego Vostoka).

KORSUNSKAIA, Galina Vasil'evna. Kuril'skaia ostrovnaia duga: fiziko-geograficheskii ocherk (The Kuril island arc: a physical-geographic study).Moskva, Geografgiz, 1958 224 p. Bibliography, p. 218-223.

About 150 references.

OV'EV, Aleksandr Ivanovich. <u>Kuril'skie ostrova</u>   (The Kuril Islands). 2nd ed.
oskva-Leningrad, Izdatel'stvo Glavsevmorputi, 1947.   307 p.   Bibliography, p.
27-235. (Akademiia Nauk SSSR. Institut Geografii).   2187

About 230 references arranged alphabetically, from 1755 to 1946, of which
3 are in foreign languages. For manuscripts place of preservation is noted.
ainly devoted to the history of exploration and development.

GEEV, Mikhail Alekseevich. <u>Kuril'skie ostrova</u> (The Kuril Islands). Moskva,
eografgiz, 1947.   152 p.   Bibliography, p. 141-152.   2188

231 references, arranged alphabetically, in Russian and other languages.

UKRAINE

ALENKO, Kateryna Olefirivna, and VYSOTS'KA, Polina Iuliïvna. Fizychna heohrafiia
RSR, 1840-1958 rr., bibliohrafichnyi pokazhchyk (Physical geography of the
krainian SSR, 1840-1958, bibliographical guide). P. K. Zamorii, ed. Kyïv,
ydavnytstvo Akademiï Nauk Ukraïns'koï RSR, 1960. 402 p. (Akademiia Nauk
kraïns'koi RSR. Derzhavna Publichna Biblioteka). In Ukrainian. Russian title:
ovalenko, Ekaterina Elifer'evna, and Vysotskaia, Polina Iul'evna. Fizicheskaia
eografiia USSR, 1840-1958 gg., bibliograficheskii ukazatel'. Kiev, Izdatel'stvo
kademii Nauk Ukrainskoi SSR. (Akademiia Nauk Ukrainskoi SSR. Gosudarstvennaia
ublichnaia Biblioteka).   2189

About 5,000 entries in Ukrainian or Russian. Divided into two sections:
kraine as a whole and physical geographical regions and zones, Polissia wooded
teppe, steppe, the Carpathians, and the Crimea. Within each of these regions
he entries are arranged under general physical-geographical characteristics,
eological-geomorphological conditions, climate, soils, vegetation, and animal
ife. Alphabetical author and place indexes.

ENCHUK, Kalenyk Ivanovych, ed. Heohrafichni landshafty Ukraïny (Geography of
he landscape of the Ukraine). Kyïv, Naukova Dumka, 1966. 131 p. Bibliography,
. 121-129, by K. I. Herenchuk and I. M. Hunevs'kyi. In Ukrainian with summary
n Russian. (Akademiia Nauk Ukraïns'koï RSR. Sektor Heohrafiï. Heohrafichne
ovarystvo URSR. Respublikans'kyi Mizhvidomchyi Zbirnyk Fizychna Heohrafiia ta
eomorfolohiia). Russian title: Geograficheskie landshafty Ukrainy. (Akademiia
auk Ukrainskoi SSR. Sektor Geografii. Geograficheskoe Obshchestvo Ukrainskoi
SR. Fizicheskaia Geografiia i Geomorfologiia), K. I. Gerenchuk, ed.   2190

About 190 references in Ukrainian or Russian on natural regions, landscapes,
hysical geography, and applied landscape studies in Ukraine.

OV, V. P., and others, eds. Fiziko-geograficheskoe raionirovanie Ukrainskoi SSR
Physical-geographic regionalization of the Ukrainian SSR). Kiev, Izdatel'stvo
ievskogo Universiteta, 1968. 683 p. Bibliography, p. 670-680. (Materialy po
rirodnomu i ekonomiko-geograficheskomu raionirovaniiu SSSR dlia tselei sel'skogo
hoziaistva. Seriia fiziko-geograficheskaia).   2191

About 300 references.

VNIA, P. V. ed. Ukraïns'ka RSR: ekonomiko -heohrafichna kharakterystyka
Ukrainian SSR: economic-geographic characteristics). Kyïv, Vydavnytstvo
yïvskoho Universytetu, 1961. 210 p. Bibliography, p. 206-209. In Ukrainian.
ussian title: Krivnia, P. V. Ukrainskaia SSR: ekonomiko-geograficheskaia
harakteristika.   2192

About 150 references in Ukrainian and Russian.

YK, Kateryna Osypivna, and TSAPENKO, I. I. Istoriia rozvytku heolohichnykh ta
eohrafichnykh znan' v Ukraïns'kii RSR: bibliohrafichnyi pokazhchyk 1944-1960 rr.
History of the development of geological and geographical knowledge in the
krainian SSR: bibliographical guide, 1944-1960). In the book: Narysy z istoriï
ekhniky i pryrodoznavstva, vypusk 2. (Studies on the history of technology and
atural science, no. 2). Kyïv, Vydavnytstvo Akademiï Nauk Ukraïns'koi RSR, 1962.
. 121-139. In Ukrainian. Russian title: Novik, E.O., and Tsapenko, I. I.
storiia razvitiia geologicheskikh i geograficheskikh znanii v Ukrainskoi SSR:
ibliograficheskii ukazatel' 1944-1966 gg., in the book: Ocherki po istorii
ekhniki i estestvoznaniia, vyp.2.   2193

394 references in Ukrainian and Russian.

CHYZHOV, Makar Panasovych. Ukraïns'kyĭ lisostep (Ukrainian wooded steppe). Kyïv. "Radians'ka Shkola," 1961. 204 p. Bibliography, p. 193-202. In Ukrainian. Russian title: Chizhov, Makar Afanas'evich. Ukrainskaia lesostep: fiziko-geograficheskii ocherk. 2

280 references on the physical geography of the wooded steppe of the Ukraine.

MARYNYCH, Oleksandr Mefodievych, ed. Pryroda Kyïvskoï oblasty. (Natural conditions of Kiev oblast). Kyïv, Vydavnytstvo Kyïvs'koho Universytetu 1972. 235 p. Bibliography, p. 227-234 (Kyïvs'kyi Derzhavnyi Universytet imeni T. H. Shevchenka). In Ukrainian. 2

PRYRODA i pryrodni resursy Pivnichnoho Prychornomor'ia: pokazhchyk literatury za 1965 r. (Natural conditions and natural resources of the area north of the Black Sea: guide to the literature for 1965). Odessa, "Maiak," 1969. 68 p. (Odes'ka Derzhavna Naukova Biblioteka im. Hor'koho, and other libraries). In Ukrainian. Russian title: Priroda i prirodnye resursy Severnogo Prichernomor'ia: ukazatel' literatury za 1965 g. 2

Systematic guide to scientific and popular-scientific literature, both books and articles, partly annotated. About 600 entries in Russian or Ukrainian. Geographic index.

PREOBRAZHENSKII, Vladimir Sergeevich. Ocherki prirody Donetskogo kriazha (Essays on natural conditions of the Donets ridge). Moskva, Izdatel'stvo Akademii Nauk SSSR, 1959. 199 p. Bibliography, p. 191-198. (Akademiia Nauk SSSR. Institut Geografii). 2

About 190 references.

KOSTRITSKII, Mikhail Emelianovich, and ENA, Vasilii Georgievich. Issledovaniia prirody Krymskogo poluostrova v sovetskoe vremia (Investigation of natural conditions in the Crimean peninsula in the Soviet period), Geograficheskoe Obshchestvo SSSR. Krymskii Otdel. Izvestiia, no. 5 (1958), p. 71-82. 2

376 references in chronological order.

PAVLOVA, Nina Nikolaevna. Fizicheskaia geografiia Kryma (Physical geography of the Crimea). Leningrad, Izdatel'stvo Leningradskogo Universiteta, 1964. 106 p. Bibliography, p. 100-105. (Leningradskii Gosudarstvennyi Universitet imeni Zhdanova. Otdel Zaochnogo Obucheniia). 2

About 130 references arranged systematically: paleogeography, geology, and geomorphology; climate; waters; soils, vegetation, animal life; physical geography.

LITERATURA po heohrafii zakhidnykh oblastei URSR (Literature on the geography of the western region of Ukraine). Heohrafichnyi zbirnyk (L'vivs'kyi viddil Heohrafichnoho Tovarystva SRSR). v. 5 (1959), p. 156-177. A. T. Vashchenko, K. I. Herenchuk, P. M. Tsys', and P. V. Klymovych, comps. v. 6 (1961), p. 161-184. A. T. Vashchenko, K. I. Herenchuk, and I. M. Hunevs'kyi. In Ukrainian. Russian title: Literatura po geografii zapadnykh oblastei Ukrainskoi SSR. Geograficheskii sbornik (L'vovskii otdel Geograficheskogo Obshchestva SSSR). 2

About 1,200 publications in Russian and Ukrainian arranged systematically: general geographic literature, geology, geomorphology, meteorology and climatology, hydrology, vegetation, animal life, economic geography and the economy.

HERENCHUK, Kalenyk Ivanovych, KOÏNOV, Mykhaïlo Matviïovych, and TSYS', Petro Mykolaevych. Pryrodno-heohrafichnyi podil L'vivs'koho ta Podil's'koho eko-nomichnykh raioniv (Physical-geographic division of the L'vov and Podolian economic regions). L'viv, Vydavnytstvo L'vivs'koho Universytetu, 1964. 220 p. Bibliography, p. 214-220. (L'vivs'kyi Derzhavnyi Universytet im. Iv. Franka. L'vivs'kyi Viddil Heohrafichnoho Tovarystva URSR). In Ukrainian. Russian title: Gerenchuk, Kalinik Ivanovich; Koinov, Mikhail Matveevich; and Tsys', Petr Nikolaevich. Prirodno-geograficheskoe delenie L'vovskogo i Podol'skogo konomicheskikh raionov. L'vov, Izdatel'stvo L'vovskogo Universiteta. (L'vovskii osudarstvennyi Universitet imeni Franko. Geograficheskoe Obshchestvo Ukrainskoi ЅR. L'vovskii Otdel).   2200

About 160 references on the physical geography of the L'vov and Podolian egions in the Southwest part of the Ukraine.

MENKO, O. S. Pryroda i pryrodni bahatstva Chernihivshchyny. Anot. pokazhchyk iteratury. (Natural conditions and natural resources of the Chernigov region: nnotated guide to the literature). Chernihiv, 1961. 113 p. (Chernihiv. Derzh. bl. Biblioteka. Materialy do Bibliohrafiï Chernihivshchyny. vypusk 1). In krainian. Russian title: Klimenko, O. S. Priroda i prirodnye bogatstva hernigovshchiny: annotirovannyi ukazatel' literatury. Chernigov. (Chernigovskaia osudarstvennaia Biblioteka imeni Korolenko. Materialy k Bibliografii Chernigo- shchiny, vypusk 1).   2201

376 references systematically organized, partly annotated, of the principal ublications on the Chernigov region of the Southwest Ukraine, 1851-1959 in krainian and Russian. Major divisions: general physical-geographical description, eological structure and relief, mineral deposits, climate, internal waters, soils, egetation, and animal life. List of main bibliographical sources. Subject- eographical and author indexes.

ENCHUK, Kalenyk Ivanovych, ed. Pryroda ukraïns'kykh Karpat (Natural conditions n the Ukrainian Carpathians). L'viv. Vydavnytstvo L'vivs'koho Universytetu. 968. 265 p. Bibliography, p. 252-263. (L'vivs'kyi Derzhavnyi Universytet im. v. Franka. L'vivs'kyi Viddil Heohrafichnoho Tovarystva URSR). In Ukrainian. ussian title: Gerenchuk, Kalinik Ivanovich. Priroda ukrainskikh Karpat. L'vov. Izdatel'stvo L'vovskogo Universiteta imeni Franko. Geograficheskoe Obshchestvo Jkrainskoi SSR. L'vovskii Otdel).   2202

335 references in alphabetical order, first in Russian or Ukrainian, then n languages using the Latin alphabet.

KIENKO, V. V. Khoziaistvennaia otsenka estestvennykh resursov i prirodnykh slovii Karpatskikh oblastei USSR (Economic evaluation of natural resources and natural conditions of the Carpathian oblasts of the Ukrainian SSR). In the ook: Uchenye zapiski Ivano-Frankovskogo otdela Ukrainskogo Geograficheskogo Obshchestva, Kiev-Vil'nius, 1962, p. 330-339 (Akademiia Nauk SSSR. Ukrainskoe Geograficheskoe Obshchestvo. Ivano-Frankovskii Otdel. Akademiia Nauk Litovskoi SSR. Institut Geologii i Geografii).   2203

211 references in alphabetical order on the Carpathian oblasts in the South- western Ukraine.

JCHIN, Vsevolod Aleksandrovich. Geografiia Sovetskogo Zakarpat'ia (Geography of Soviet Trans-Carpathia). Moskva. Geografgiz, 1956. 296 p. Bibliography, p. 278-288.   2204

243 references in alphabetical order in Russian, Ukrainian, and foreign languages.

IRODA, naselenie i khoziaistvo iugo-zapadnogo ekonomicheskogo raiona (Natural conditions, population, and the economy of the Southwest Economic Region). A. I. Tokmakov, and others, eds. Chernovtsy, 1973. 155 p. Bibliography, p. 142-153. (Chernovitskii Gosudarstvennyi Universitet. Geograficheskii Fakul'tet).   2205

THE BALTIC STATES

ΚADEMIIA Nauk SSSR. Institut Geografii. Sovetskaia Pribaltika: problemy ekonomiche-
skoi geografii (Soviet Baltic republics: problems of economic geography). A. A.
Mints and M. I. Rostovtsev, eds. Moskva, Izdatel'stvo "Nauka," 1966. 278 p.
Bibliography, p. 273-276.   2206
     76 references in Russian.

ESTONIA

ΚADEMIIA Nauk Estonskoi SSR. Estonskoe Geograficheskoe Obshchestvo. O razvitii
geografii v Estonskoi SSR 1960-1968 (On the development of geography in the
Estonian SSR 1960-1968). E. Varep, L. Merikal'iu, A. Raukas, and V. Tarmisto,
eds. Tallin, 1970. 237 p. (Text in Russian). (Eesti NSV Teaduste Akadeemia.
Eesti Geograafia Selts).   2207
     A review of the development of geography in Estonia 1960-1968 in a volume
dedicated to the 125th anniversary of the Geographical Society of the USSR. 1,586
references in extensive bibliographies. Titles of works in Estonian given both
in Estonian and in Russian translation.

     Varep, E. Istoriia geografii (History of geography), p. 20-34. 21 refer-
ences in Russian and 89 in Estonian.

     Oiavere, E. (Ojavere E.) Kartografiia (Cartography), p. 35-39. 5 references
in Russian and 23 in Estonian.

     Liblik, T., Raukas, A., and Khang, E. (Hang, E.). Geomorfologiia i paleo-
geografiia Chetvertichnogo perioda (Geomorphology and paleogeography of the
Quaternary period), p. 40-72. 106 references in Russian and 140 in Estonian or
other languages.

     Int, L., Raik, A., and Ross, Iu. Obshchaia i prikladnaia klimatologiia
(General and applied climatology), p. 73-86. 76 references in Russian and 40 in
Estonian.

     Vel'ner, Kh. (Velner, H.), Kullus, L.-P., and Mardiste, Kh. (Mardiste, H.).
Gidrologiia sushi i moria. Vodnoe khoziaistvo (Hydrology of the land and the sea.
The water economy), p. 87-100. 71 references in Russian and 60 in Estonian or
other languages.

     Kongo, A., Rooma, I. Geografiia i kartografiia pochv (Geography and cartog-
raphy of soils), p. 101-108. 20 references in Russian and 20 in Estonian.

     Mazing, V. (Masing, V.). Biogeografiia (Biogeography), p. 109-121. 36
references in Russian and 51 in Estonian or other languages.

     Kil'dema, K. (Kildema, K.). Landshaftovedenie (Landscape studies), p. 122-
134. 27 references in Russian and 63 in Estonian or other languages.

     Eilart, Ia (Eilart, J.). Okhrana prirody (Conservation of nature), p. 135-
140. 8 references in Russian and 42 in Estonian or other languages.

     Kaufman, V. (Kaufmann, V.), Laas, K., and Paal'berg, Kh. (Paalberg, H.),
Naselenie, trudovye resursy i poseleniia (Population, labor force, and settlements), p. 141-152. 28 references in Russian and 53 in Estonian or other
languages.

     Tarmisto, V. Razmeshchenie proizvodstva (Distribution of production), p.
153-182. 83 references in Russian and 106 in Estonian or other languages.

     Valma, A., and Pragi, U. Geografiia transporta (Geography of transport),
p. 183-188. 8 references in Russian and 33 in Estonian.

     Nymmik, S. (Nõmmik, S.). Ekonomicheskoe raionirovanie (Economic region-
alization), p. 189-192. 9 references in Russian and 9 in Estonian or other
languages.

Tul'p, L. (Tulp, L.), and Vabar, M. Ekonomicheskie sviazi i geografiia vneshnei torgovli (Economic relations and the geography of external trade), p. 193-198. 22 references in Russian and 14 in Estonian.

Vabar, M. Ekonomicheskaia geografiia zarubezhnykh stran (Economic geography of foreign countries), p. 199-202. 6 references in Russian and 19 in Estonian.

Tiits, Kh. (Tiits, H.). Shkol'naia geografiia (School geography), p. 203-214. 2 references in Russian and 81 in Estonian.

Varep, E. Istoricheskaia geografiia (Historical geography), p. 215-222. 3 references in Russian and 67 in Estonian.

Varep, E. Toponimika (Toponymics), p. 223-227. 1 reference in Russian and 45 in Estonian.

Mikhel'soo, A. Nauchno-populiarnaia literatura po geografii (Scientific-popular literature on geography), p. 228-233. 13 references in Russian and 71 in Estonian.

Varep, E. Bibliografiia (Bibliography), p. 234-236. 3 references in Russian and 12 in Estonian.

ON the development of geography in the Estonian S.S.R. 1940-1960. Tallinn. Academy of Sciences of the Estonian SSR, 1960. 109 p. (Estonian Geographical Society. Publication, no. 2): in English. (Eesti Geograafia Selts. Publikatsioonid, 2; Estonskoe Geograficheskoe Obshchestvo, Publikatsiia, 2). Russian title: O razvitii geografii v Estonskoi SSR 1940-1960. 2

More than 500 publications mainly in Estonian and Russian, organized by sections: scientific research institutions, p. 10-11; history of geography, p. 15-19; geomorphology, p. 24-28; climatology, p. 32-34; hydrology, p. 39-41; geography of soils, p. 44-46; biogeography, p. 52-59; study of landscapes, p. 62-66; phenology, p. 67; economic geography, p. 71-75; division of the territory of Estonia into economic regions, p. 78; study of urban areas, p. 79-80; geography in Soviet Estonian schools, p. 85-89; study of local lore, p. 92-93; nature conservation, p. 96-97; geographical literature, p. 100-106; and Estonian periodical publications quoted in the bibliography, p. 107. By 20 different authors. Titles of works given in English translation.

KOORITS, V. Bibliografiia trudov prepodavatelei, sotrudnikov i aspirantov Geograficheskogo otdeleniia Tartuskogo gosudarstvennogo universiteta (Bibliography of works of teachers, collaborators, and graduate students of the Geographical section of Tartu State University). Tartuskii Gosudarstvennyi Universitet, Uchenye zapiski, no. 237 (1969), Trudy po geografii, no. 6, p. 215-232. Title in Estonian: Koorits, V. Tartu Riikliku Ülikooli geograffiaosakonna õppejõudude, teaduslike töötajate, abiõppejõudude ja aspirantide trükis ilmunud tööde bibliograafia, Tartu Riikliku Ülikooli Toimetised. vihik 237. Geograafia-alaseid tõid. VI. 2

KILDEMA. K. Maastikulise uurimise põhisuundi Eesti (Basic directions of the development of landscape studies in Estonia). Tallinn. 1968. 114 p. Bibliography, p. 31-99. (Eesti NSV Teaduste Akadeemia. Tallinna Botaanikaad). In Estonian with summaries in Russian and German. Russian title: Kil'dema, K. Osnovnye napravleniia razvitiia landshaftovedeniia v Estonii (Akademiia Nauk Estonskoi SSR. Tallinskii Botanicheskii Sad). 2

About 760 entries on the study of the physical geography of the landscape in Estonia.

LATVIA

CERA Stučkas Latvijas Valsts Universitātes mācību spēku publikācijas. 1945-1965.
Bibliogrāfija. II. Ķīmija. Ģeogrāfija. Biologija (Latvian State University
named for Peter Stuckas. Publications of the faculty, 1945-1965. Bibliography,
CI. Chemistry. Geography. Biology). Rīgā, Izdevniecība "Zvaigzne," 1971. 122 p.
Petera Stučkas Latvijas Valsts Universitāte. Zinātniskā Bibliotēka). Russian
title: Publikatsii prepodavatelei Latviiskogo Gosudarstvennogo Universiteta im.
P. Stuchki 1945-1965. Bibliografiia II. Khimiia. Geografiia. Biologiia.
Latviiskii Gosudarstvennyi Universitet im. P. Stuchki. Nauchnaia Biblioteka]).
in Latvian and Russian.   2211

   Geography, p. 55-78, entries 675-984 includes 310 entries organized systematically: general questions; physical geography (general questions; geomorphology,
geology of the Quaternary; landscape studies; climatology and hydrology; physical
geography of foreign countries); economic geography (general questions and
regionalization; economic geography of the USSR; economic geography of Latvia;
economic geography of foreign countries); travel notes. Headings in Latvian and
Russian. Entries mostly in Latvian or Russian. Author indexes first in Latvian,
then in Russian.

VIISKAIA SSR. 1940-1960. Ukazatel' literatury (Latvia, 1940-1960: a guide to
the literature). 3 v. v. 1, comp. by O. Putse, L. Veinberg. Riga, Gosudarstvennaia Biblioteka Latviiskoi SSR, 1961. 578 p. (Gosudarstvennaia Biblioteka
Latviiskoi SSR).   2212

   8,200 entries in Russian and Latvian, arranged systematically then by
language and year.

IBER, Ian Fritsevich, and ALAMPIEV, Petr Martynovich, eds. Latviiskaia SSR:
ocherki ekonomicheskoi geografii. (Latvian SSR: studies in economic geography).
Riga, Izdatel'stvo Akademii Nauk Latviiskoi SSR, 1956. 394 p. Bibliography,
p. 374-380. (Akademiia Nauk Latviiskoi SSR. Institut Ekonomiki).   2213

   172 references in Russian and 56 in Latvian. Index of geographical names.

KOLAEVA-SEREDINSKAIA, G. F. Istoriia issledovaniia prirodnykh uslovii territorii
Latviiskoi SSR. 1710-1917 (History of research on natural conditions in the
territory of the Latvian SSR, 1710-1917). Leningrad, Izdatel'stvo "Nauka,"
Leningradskoe Otdelenie, 1970. 116 p. Bibliography, p. 108-115. (Akademiia
Nauk SSSR. Geograficheskoe Obshchestvo SSSR).   2214

KEVITS, Ia. Ia. Geografiia sel'skogo khoziaistva Latviiskoi SSR (Geography of
agriculture of the Latvian SSR). Riga, "Zinatne," 1973. 267 p. Bibliography,
p. 263-266.   2215

LITHUANIA

SALYKAS, Alfonsas, ed. Lietuvos TSR fizinė geografija. (Physical geography of
the Lithuanian SSR). v. 1, Vilnius, Valstybinė Politinės ir Mokslinės Literatūros
Leidykla, 1958. 504 p. Bibliography, p. 467-478. v. 2. Vilnius, Mintis, 1965.
496 p. Bibliography, p. 440-450. (Lietuvos TSR Mokslų Akademija. Geologijos
ir Geografijos Institutas ir Vilniaus Valstybinis Universitetas, Gamtos Mokslų
Fakultetas). In Lithuanian. Table of contents also in Russian and German.
Russian title: Basalikas, Alfonsas Bronislavovich. Fizicheskaia geografiia
Litovskoi SSR. Vil'nius, Gospolitnauchizdat. (Akademiia Nauk Litovskoi SSR.
Institut Geologii i Geografii. Fakul'tet Estestvoznaniia Vil'niusskogo
Gosudarstvennogo Universiteta imeni Kapsukasa).   2216

   About 270 references in volume 1 on the physical geography of Lithuania as
a whole and 332 references in volume 2 on physical-geographic regions of Lithuania.

BALTRAMAITIS, Silvestr Iosifovich. Sbornik bibliograficheskikh materialov dlia
geografii, etnografii i statistiki Litvy (Collection of bibliographical materials
for geography, ethnography, and statistics of Lithuania). Imperatorskoe Russkoe
Geograficheskoe Obshchestvo, Zapiski, v. 21, no. 1. S.-Peterburg, 1891. 289 p.   2

    4,096 entries.

_____. Sbornik bibliograficheskikh materialov dlia geografii, istorii, istorii
prava, statistiki i etnografii Litvy (Collection of bibliographical materials for
geography, history, history of law, statistics, and ethnography of Lithuania),
2nd ed. ...Zapiski, v. 25, no. 1. S.-Peterburg, 1904. 616 p.   2

    8,541 entries, organized by topics such as geography, hydrography, geology,
orography, travels, atlases, and maps. Author and subject indexes.

BELORUSSIA

ASTASHKIN, Nikolai Dmitrievich. Prirodnye resursy BSSR (Natural resources of the
Belorussian SSR). Minsk, "Nauka i Tekhnika," 1970. 459 p. Bibliography, p. 450-
457.   2

LIARSKII, Petr Alekseevich. Posobie po kraevedeniiu (Aid for local studies).
Minsk, "Vysheishaia shkola," 1966. 239 p. Bibliography, p. 214-218.   2

    About 150 references, organized by chapters, on aids to the study of the
regional geography of Belorussia.

MOLDAVIA

SUKHOPARA, Fëdor Nikolaevich, ed. Ekonomicheskie podraiony MSSR (Economic sub-
regions of the Moldavian SSR). A. A. Gudym, N. P. Korableva, L. F. Nikul, and
others. Kishinev, "Shtiintsa," 1973. 131 p. Bibliography, p. 127-130.
(Akademiia Nauk Moldavskoi SSR. Institut Ekonomiki).   2

THE CAUCASUS AND TRANS-CAUCASUS

GRAFIIA khoziaistva respublik Zakavkaz'ia (Geography of the economy of the
epublics of the Trans-Caucasus). A. A. Mints, ed. Moskva, Izdatel'stvo "Nauka,"
966. 283 p. Bibliography, p. 278-281 (Akademiia Nauk SSSR. Institut Geo-
rafiia. Akademiia Nauk Gruzinskoi SSR. Institut Geografii imeni Vakhushti.
kademiia Nauk Azerbaidzhanskoi SSR. Institut Geografii. Akademiia Nauk
rmianskoi SSR. Sektor Geografii).   2222

    102 references arranged in alphabetical order in the Cyrillic alphabet by
ame of author. References mostly in Russian and those in Armenian, Azerbaidzhani,
nd Georgian are given with titles translated into Russian.

MESKU, Aleko Aleksandrovich, and SILAEV, Evgenii Dmitrievich, and others, eds.
akavkazskii ekonomicheskii raion (Trans-Caucasian economic region). E. D.
ilaev, V. G. Postnikov, G. N. Mekhalev, and others. Moskva, Izdatel'stvo "Nauka,"
973. 245 p. Bibliography, p. 238-240. (Akademiia Nauk SSSR. Gosplan SSSR.
ovet po Izucheniiu Proizvoditel'nykh Sil. Razvitie i Razmeshchenie Proizvoditel'-
ykh Sil SSSR). Index of geographical names, p. 241-244.   2223

ISASHVILI, Vasilii Zakharovich. Prirodnye zony i estestvenno-istoricheskie
blasti Kavkaza (Natural zones and natural-historical regions of the Caucasus).
oskva, Izdatel'stvo "Nauka," 1964. 327 p. Bibliography, p. 314-325.   2224
    About 310 references.

ZDETSKII, Nikolai Andreevich. Fizicheskaia geografiia Kavkaza (Physical geog-
aphy of the Caucasus). Moskva, Izdatel'stvo Moskovskogo Universiteta. vypusk 1.
bshchaia chast'. Bol'shoi Kavkaz (no. 1. General part. Great Caucasus).
954. 208 p. Bibliography, p. 202-205. vypusk 2. Predkavkaz'e. Zakavkaz'e
no. 2. Fore-Caucasus, Trans-Caucasus). 1958. 264 p. Bibliography, p. 257-262.
    About 165 references, organized by chapters. Includes cartographic material. 2225

CHENKO, Mikhail Afanas'evich. K istorii fiziko-geograficheskikh i osobenno
liatsiologicheskikh issledovanii Bol'shogo Kavkaza (On the history of physical-
eographic and especially glaciological research of the Great Caucasus).
har'kov, Khar'kovskii Universitet, Uchenye zapiski, v. 81. Trudy Geograficheskogo
akul'teta, v. 3 (1957). Bibliography, p. 270-284.   2226
    462 references.

GEORGIA

UASHVILI, Levan Iosifovich. Sak'art'velos phizikuri geographis (Physical geog-
aphy of Georgia). Tbilisi, T'SU Gamomtsemloba, 1969-1970. 2 v. v. 1, 1969.
70 p. v. 2, 1970. 347 p. In Georgian. Russian form of title: Fizicheskaia
eografiia Gruzii. Tbilisi, Izdatel'stvo Tbilisskogo Universiteta.   2227
    97 references in v. 1 on systematic physical geography of Georgia. 496
eferences in v. 2 on individual landscape zones of Georgia.

EBA, D. B. Aghmosavlet' Sak'art'velos phizikur-geographiuli daraioneba (Physical-
eographic regionalization of East Georgia), v. 1. Tbilisi, "Metsniereba," 1968.
50 p. Bibliography, p. 330-348 . (Sak'art'velos SSR Metsnierebat'a Akademia.
akhushtis Sakhelobis Geographis Instituti). In Georgian. Russian form of title:
iziko-geograficheskoe raionirovanie Vostochnoi Gruzii (dlia tselei sel'skogo
hoziaistva). (Akademiia Nauk Gruzinskoi SSR. Institut Geografii imeni Vakhushti).
                                                                             2228
    288 references on physical geographic characteristics and evaluation of
limatic conditions for agriculture.

LANDSHAFTNYI sbornik. M. F. Sakhokia. Tbilisi, Izdatel'stvo Tbilisskogo Universiteta, 1972. 169 p. Bibliography, p. 147-156. In Russian or Georgian.   2

QIPHIANI, Sh., TSINTSILOZOVI, Z., OK'RODZHANASHVILI, A., and DZHISHKARIANI, V. Sak'art'velos karstsuli mghvimeebis kadastri (Register of karst caves of Georgia). Tbilisi, "Metsniereba," 1966. 259 p. Bibliography, p. 230-252. In Georgian. Russian form of listing: Kipiani, Sh. Ia., Tintilozov, Z. K. Okrodzhanashvili, A. A., and Dzhishkariani, V. M. Kadastr karstovykh peshcher Gruzii (Institut Geografii imeni Vakhushti).   2

    383 references.

TINTILOZOV, Z. K. Anakopiiskaia propast': opyt kompleksnoi speleologicheskoi kharakteristiki (Anakopia chasm: an experiment in complex speleological characterization). Tbilisi, "Metsniereba," 1968. 72 p. Bibliography, p. 66-69. In Russian with summaries in English and Georgian. (Akademiia Nauk Gruzinskoi SSR. Institut Geografii imeni Vakhushti).   2

    66 references on the largest cave in Geoegia.

NEMANISHVILI, S. Mt'iani mkhareebis mdinareuli tserasebi: das. Sak'art'velos magalit'ze (Fluvia terraces in mountain regions: based on Georgian examples). Tbilisi, "Metsniereba," 1973. 186 p. (Sak'art'velos SSR Metsnierebat'a Akademia. Vakhushtis Sakhelobis Geographiis Instituti). In Georgian.   2

    155 references.

KORDZAKHIA, Mitrofan Otievich. Sak'art'velos hava (Climate of Georgia). Tbilisi, Sak'art'velos SSR Metsnierebat'a Akademiis Gamomtsemloba, 1961. 249 p. Bibliography, p. 243-246. (Vakhushtis Sakhelobis Geographiis Instituti). In Georgian. Russian listing: Klimat Gruzii. Tbilisi, Izdatel'stvo Akademii Nauk Gruzinskoi SSR. (Akademiia Nauk Gruzinskoi SSR. Institut Geografii imeni Vakhushti).   2

    81 references.

TSERETELI, David Vissarionovich. Pleistotsenovye otlozheniia Gruzii (Pleistocene deposits of Georgia). Tbilisi," Metsniereba," 1966. 584 p. Bibliography, p. 535-580. Summary in English. (Akademiia Nauk Gruzinskoi SSR. Institut Geografii imeni Vakhushti).   2

    452 references on basic geomorphological types of relief, genetic types of Pleistocene deposits, stratigraphy, and paleogeography of Georgia.

KAVRISHVILI, Vissarion Ivanovich. Landshaftno-gidrologicheskie zony Gruzinskoi SSR (Landscape-hydrological zones of the Georgian SSR). Tbilisi, Izdatel'stvo Akademii Nauk Gruzinskoi SSR, 1955. 170 p. Bibliography, p. 163-170.   2

    183 references in alphabetical order by name of author. Titles of publications in Georgian are given in Russian translation with notation that the work is in the Georgian language.

VLADIMIROV, Lev Aleksandrovich. Pitanie rek i vnutrigodovoe raspredelenie rechnogo stoka na territorii Gruzii (Flow into rivers and the annual distribution of stream flow in Georgia). Tbilisi, "Metsniereba," 1964. 250 p. Bibliography, p. 245-249. (Akademiia Nauk Gruzinskoi SSR. Institut Geografii imeni Vakhushti).   2

    100 references.

VLADIMIROV, Lev Aleksandrovich. Vodnyi balans Bol'shogo Kavkaza: bez Azerbaidzhanskoi i Dagestanskoi chastei (Water balance of the Great Caucasus excluding the Azerbaidzhan and Dagestan parts). Tbilisi, "Metsniereba," 1970. 142 p. (Akademiia Nauk Gruzinskoi SSR. Institut Geografii imeni Vakhushti).   2

    72 references.

KHARIIA,Vladimir Konstantinovich. Isparenie s vodnoi poverkhnosti vodoëmov
avkaza (Evaporation from the surface of reservoirs of the Caucasus). Tbilisi,
Metsniereba," 1973. 186 p. Bibliography, p. 182-185. (Akademiia Nauk Gruzinskoi
SR. Institut Geografii imeni Vakhushti).   2238
   67 references.

HADZE, Elena Viktorovna. Botaniko-geograficheskii ocherk izvestniakovykh gor
apadnoi Gruzii. F. F. Davitaia, ed. (Botanical-geographical sketch of limestone
ountains of Western Georgia). Tbilisi, "Metsniereba," 1968. 138 p. Bibliography,
. 125-137.(Akademiia Nauk Gruzinskoi SSR. Institut Geografii im. Vakhushti). 2239
   237 references on natural conditions and vegetation of limestone mountains
nd karst of Western Georgia and similar areas in other countries.

LESIANI, Georgii Grigor'evich. Razvitie i razmeshchenie sotsialisticheskogo
roizvodstva v Gruzinskoi SSR (Development and distribution of socialist produc-
ion in the Georgian SSR). Tbilisi, Izdatel'stvo "Metsniereba," 1965. 374 p. 2240
   331 bibliographical footnotes organized by four chapters: (1) factors in the
evelopment and territorial distribution of production, (2) economic development
nd forms of manifestation of general principles of socialist distribution of
roduction in the Georgian SSR, (3) economic regionalization and distribution of
roduction, and (4) some future problems of the development and distribution
f production in the Georgia SSR.

'ART'VELOS SSR ekonomiuri geographiis dzirit'adi sakit'khebi (Basic problems of
he economic geography of the Georgian SSR). Tbilisi, "Ganat'leba," part 1, 1970.
38 p. part 2, 1973. 440 p. In Georgian. (Sak'art'velos SSR Metsnierebat'a
kademia. Vakhushtis Sakhelobis Geographiis Institut).   2241
   Part 1 contains references on climatic resources, history, population,
ndustry, agriculture, transport, and resorts and tourism. Part 2 contains
eferences on the economic regionalization of the Georgian SSR.

BZHLI: Sak'art'velos SSR ekonomiuri geographia (Collected volume: the economic
eography of the Georgian SSR). Tbilisi, "Tsodna," 1961. 388 p. Bibliography.
n Georgian. (Sak'art'velos SSR Metsnierebat'a Akademia. Vakhushtis Sakhelobis
eographiis Instituti).   2242
   168 references.

AOSHVILI [Jaoshvili], Vakhtang Shalvovich. Naselenie Gruzii: ekonomiko-
eograficheskoe issledovanie (Population of Georgia: economic-reographical
nvestigation) Tbilisi, Izdatel'stvo "Metsniereba," 1968. 398 p. Bibliography,
. 381-393. (Akademiia Nauk Gruzinskoi SSR. Institut Geografii im. Vakhushti). 2243
n Russian with supplementary table of contents in English, p. 396-398.
   346 references in Russian, Georgian, and other languages, arranged in alpha-
etical order by name of author, first in Cyrillic, including those in Georgian
ith titles translated into Russian, then those with Latin letters.

'ART'VELOS SSR Metsnierebat'a Akademia. Vakhushtis Sakhelobis Geographiis
nstituti. T'bilisi: ekonomiur-geographiuli dakhasiat'eba (Tbilisi: economic-
eographic characteristics). A. N. Javakhishvili, ed. Tbilisi, Sak'art'velos
SR Metsnierebat'a Akademiis Gamomtsemloba, 1957. 440 p. Bibliography, by
. Sharashidze, p. 418-436. In Georgian with supplementary title and table of
ontents in Russian. Russian title: Akademiia Nauk Gruzinskoi SSR. Institut
eografii imeni Vakhushti. Tbilisi: ekonomiko-geograficheskaia kharakteristika.
. N. Dzhavakhishvili, ed. Tbilisi, Izdatel'stvo Akademii Nauk Gruzinskoi SSR,
957. 440 p.   2244
   180 numbered references in Georgian and 237 numbered references in Russian
n separate lists. Published on the occasion of the 1500th anniversary of the
ounding of the city of Tbilisi.

KAVRISHVILI, Ketevana Vissarionovna. Fiziko-geograficheskaia kharakteristika okrestnostei Tbilisi (Physical-geographical characteristics of the environs of Tbilisi). Tbilisi, "Metsniereba," 1965. 165 p. Bibliography, p. 132-146. (Akademiia Nauk Gruzinskoi SSR. Geograficheskoe Obshchestvo Gruzinskoi SSR). 2

About 270 references.

JAOSHVILI, Vakhtang. K'ut'aisi: ekonomiur-geographiuli dakhasiat'eba (Kutaisi: economic-geographic description). Tbilisi, Sak'art'velos SSR Metsnierebat'a Akademiis Gamomtsemloba, 1962. 352 p. Bibliography, p. 344-348. (Vakhushtis Sakhelobis Geographiis Instituti). In Georgian. Russian form of author and title: Dzhaoshvili, Vakhtang Shalvovich. Kutaisi: ekonomiko-geograficheskaia kharakteristika. Tbilisi, Akademiia Nauk Gruzinskoi SSR, 1962. 352 p. (Akademiia Nauk Gruzinskoi SSR. Institut Geografii imeni Vakhushti). 2

127 references on natural conditions, resources, history, population, industry agriculture, and transport of Kutaisi and its region.

NIZHARADZE, Nadima Izetovicha. Sovetskaia Adzhariia: ekonomiko-geograficheskaia kharakteristika (Soviet Adzharia: economic-geographic characteristics). Batumi, Godudarstvennoe Izdatel'stvo, 1961. 262 p. Bibliography, p. 256-260. 2

117 references in alphabetical order in Russian; works in Georgian are listed with titles translated into Russian but with notation that the publication is in the Georgian language.

KARBELASHVILI, Luarsab Andreevich. Iugo-Osetiia. Ekonomiko-geograficheskii obzor i bibliografiia (South Osetia: an economic-geographic survey and bibliography). Tbilisi, Akademiia Nauk Gruzinskoi SSR, 1962. 460 p. In Russian and Georgian. 2

6,608 references on South Osetia, 4,345 in Russian and 2,363 in Georgian.

KVERENCHKHILADZE, R. Samkhlet' Oset'i (South Osetia). Tskhinvali, 1968. 215 p. In Georgian with Russian summary. Russian title: Iugo-Osetiia: priroda, naselenie, khoziaistvo. 2

241 references on natural resources, population, and the economy.

ARDZHEVANIDZE, I. Voenno-Gruzinskaia doroga: kraevedcheskii ocherk (The Georgian military highway: a regional study). Tbilisi, "Tekhnika da shroma," 1950. 161 p. Bibliography, p. 151-159. 2

About 230 references in Russian and Georgian.

ARMENIA

PETROSIAN, Ov. Armianskaia Sovetskaia Sotisalisticheskaia Respublika: annotirovannaia bibliografiia (The Armenian Soviet Socialist Republic: annotated bibliography). S. S. Mkrtchian, ed. v. 1. Yerevan, Gosudarstvennaia Knizhnaia Palata Armianskoi SSR, 1958. 268 p. In Armenian and Russian. 2

v. 1 covers natural conditions of Armenia with 1,751 entries, arranged systematically: geography; geology; paleontology; the Sevan-Zanga problem, water resources, hydrogeology, and hydrology; meteorology and climatology. In each section the entries are listed in two chronological orders, in Armenian and in Russian. Indexes of authors, editors, and reviewers in Armenian and in Russian.

VALESIAN, Lemvel Akopovich. Proizvodstvenno-territorial'nyi kompleks Armianskoi SSR: ekonomiko-geograficheskoe issledovanie problem formirovaniia i razvitiia (The production-territorial complex of the Armenian SSR: economic geographic study of the problems of formation and growth). Erevan, Izdatel'stvo "Aiastan," 1970. 375 p. Bibliography, p. 358-373. 2

About 320 references arranged alphabetically by name of author. Titles of
rks in Armenian are given in Russian translation with notation that the pub-
cation is in the Armenian language.

BYAN, T'adewos Khach'atowri. Hayastani patmakan ashkhar'hagrowt'yown:
rvagtser (Historical geography of Armenia: studies). 2nd ed. Erevan, Mitq,
68. 509 p. Bibliography, p. 447-452. (Erevani Petakai Hamalsaran). In
menian with summaries in Russian, p. 453-461 and in English, p. 462-465.
ssian title: Akopian, Tadevos Khachaturovich. Istoricheskaia geografiia Armenii:
herki. Erevan, Izdatel'stvo "Mitk," (Erevanskii Gosudarstvennyi Universitet).
t. ed. Owrvagtser hayastani patmakan ashkhar'hagrowt'yown (Russian title: Ocherki
istoricheskoi geografii Armenii). 1960. 480 p. Bibliography, p. 432-439.   2253

141 references, 107 in Armenian, 32 in Russian, and 2 in English.

BYAN, T'adewos Khach'atowri. Hayastani patmakan ashkhar'hagrowt'yown (Historical
ography of Armenia). Erevan, 1951-1953. 3 nos. In Armenian. Russian title:
opian, Tadevos Khachaturovich. Istoricheskaia geografiia Armenii: materialy.
osudarstvennyi Zaochnyi Pedagogicheskii Institut).   2254

    no. 1. 1951. Bibliography, p. 184-187. 56 references.
    no. 2. 1952. Bibliography, p. 215-220. 93 references.
    no. 3. 1953. Bibliography, p. 302-307. 88 references.

    237 references in Armenian, Russian, and foreign languages.

                                AZERBAIDZHAN

KAI, Mir-Ali, and ALIEV, Gady Badalovich, eds. Fizicheskaia geografiia
erbaidzhanskoi SSR (Physical geography of the Azerbaidzhan SSR). A. F. Liaister,
 A. Kashkai, N. V. Malinovskii, and others. Baku, 1945. 279 p. Bibliography,
 263-269 (Akademiia Nauk SSSR. Azerbaidzhanskii Filial. Sektor Ekonomiki i
ografii).
                                                                           2255
    184 references arranged systematically.

EMOV, N. K., and BUDAGOV, B. A. O landshaftnykh issledovaniiakh Azerbaidzhanskoi
SR i ikh razvitii. Reziume (Landscape investigations in the Azerbaidzhan SSR and
heir development. Summary), Akademiia Nauk Azerbaidzhanskoi SSR. Institut Geo-
rafii, Trudy, v. 18 (1971), p. 20-51. Bibliography, p. 32-51.   2256

                                 KAZAKHSTAN

KHSTAN. Obshchaia fiziko-geograficheskaia kharakteristika (Kazakhstan: general
ysical-geographic characteristics). A. A. Grigor'ev, ed. Moskva-Leningrad,
:datel'stvo Akademii Nauk SSSR, 1950. 492 p. Bibliography, p. 462-473.(Akademiia
uk SSSR. Institut Geografii. Akademiia Nauk Kazakhskoi SSR. Seriia "Itogi i
oblemy Sovremennoi Nauki").   2257

    About 260 references at ends of chapters: history of geographical exploration
 Kazakhstan, study of relief and geological structure, climate, rivers and lakes,
getation and soils, animal life.

JPAKHIN, Viktor Mikhailovich. Prirodnoe raionirovanie Kazakhstana: dlia tselei
sel'skogo khoziaistva (Natural regionalization of Kazakhstan: for the purposes
of agriculture). Alma-Ata, Izdatel'stvo "Nauka," 1970. 263 p. Bibliography,
p. 255-261. (Akademiia Nauk Kazakhskoi SSR. Sektor Fizicheskoi Geografii).   2258

    145 references.

                                                        2253 - 2258

KUSTANAISKII ekonomicheskii administrativnyi raion: bibliograficheskii ukazatel'
literatury (The Kustanai economic administrative region: bibliographic guide to
the literature). M. Abdullina, G. Demesheva, M. Urazova, and E. Ivanchikova,
comps. In the book: Bibliotechno-bibliograficheskii biulleten', vypusk. 2.
Kustanaiskii ekonomicheskii administrativnyi raion. Alma-Ata. Izdatel'stvo
Akademii Nauk Kazakhskoi SSR, 1959. p. 13-260. (Akademiia Nauk Kazakhskoi SSR.
Tsentral'naia Nauchnaia Biblioteka).

Systematic bibliography with Russian and Kazakh titles in separate order.
In the section natural conditions, p. 23-87, are headings for physical-geographic
characteristics of the region: general questions, geomorphology, climate,
hydrology, soils, flora and vegetation, and animal life. About 420 references.
Indexes for authors and geographical names. 2,678 entries in the entire volume.

MORACHEVSKAIA, Elena Nikolaevna. Bibliografiia po razvitiiu proizvoditel'nykh
sil Kustanaiskogo ekonomicheskogo administrativnogo raiona i Bol'shogo Turgaia
v tselom, 1801-1959 gg. (Bibliography on the development of productive forces
of Kustanay economic administrative region and the Bol'shoy Turgay as a whole,
1801-1959). Moskva, 1959. 135 p. (Akademiia Nauk SSSR. Sovet po Izucheniiu
Proizvoditel'nykh Sil. Mezhduvedomstvennaia Komissia po Regionu Bol'shogo Turgaia
Sektor Seti Spetsial'nykh Bibliotek).

Systematic guide on the Turgay lowland and adjacent regions. Material on
physical geography is in the section of natural conditions and natural resources
under the headings: general questions (geography), hydrology and hydrogeology,
climate, soils, vegetation, and animal life. About 180 references. Index of
authors, editors, and titles. 1,033 entries in the whole volume.

PRIRODNOE raionirovanie Severnogo Kazakhstana: Kustanaiskaia, Severo-Kazakhstanskaia,
Kokchetavskaia, Akmolinskaia i Pavlodarskaia oblasti (Natural regionalization
of North Kazakhstan: Kustanay, North Kazakhstan, Kokchetav, Akmolinsk, and Pavlodar
oblasts). B. A. Fedorovich, and others, eds. Moskva-Leningrad, Izdatel'stvo
Akademii Nauk SSSR, 1960. 468 p. Bibliography, p. 457-466 (Akademiia Nauk SSSR.
Sovet po Izucheniiu Proizvoditel'nykh Sil. Institut Geografii. Botanicheskii
Institut imeni V. L. Komarova. Pochvennyi Institut imeni V. V. Dokuchaeva.
Akademiia Nauk Kazakhskoi SSR. Institut Pochvovedeniia).

About 300 references in alphabetical order.

KARAGANDINSKII ekonomicheskii administrativnyi raion. Bibliograficheskii ukazatel'
literatury (Karaganda economic administrative region: bibliographical guide to
the literature). E. I. Ivanchikova, M. T. Kolesnikova, E. M. Konobritskaia, M.
M. Kudriashova, Sh. N. Kul'baeva, and S. G. Medvedeva, comps. Alma-Ata, Izdatel'-
stvo Akademii Nauk Kazakhskoi SSR, 1959. 458 p. (Akademiia Nauk Kazakhskoi SSR.
Tsentral'naia Nauchnaia Biblioteka). In Russian and Kazakh.

Systematic guide for Karaganda, Pavlodar, and Akmolinsk oblasts. Entries
for Russian and Kazakh languages separately. In the section on natural conditions
are headings for physical-geographic characteristics of the region: general
material, geomorphology, climate, hydrology, soils, flora and vegetation, and
animal life. Separate indexes in Russian and Kazakh for authors, collected works,
and geographical names. Large section on the economics of the region. 4,488
entries.

LOMONOVICH, Mikhail Ivanovich, ed. Iliiskaia dolina, eë priroda i resursy. (Ili
valley, its natural conditions and resources). Alma-Ata, Izdatel'stvo Akademii
Nauk Kazakhskoi SSR, 1963. 341 p. Bibliographies, p. 219-224, 334-339. (Akademiia
Nauk Kazakhskoi SSR. Institut Geologicheskikh Nauk. Institut Pochvovedeniia).

172 references on the Ili trough and 187 references on the lower reaches of
the river.

SOVIET MIDDLE ASIA

EMIIA Nauk SSSR. Institut Geografii. Sredniaia Aziia: fiziko-geograficheskaia
arakteristika (Middle Asia: physical-geographic characteristics). E. M.
rzaev, ed. Moskva, Izdatel'stvo Akademii Nauk SSSR. 1958. 648 p. Bibliog-
ohy, p. 611-624.   2264
About 500 references, arranged alphabetically. Index of geographical names.
iex of Latin, Russian, and local names of plants.

EMIIA Nauk SSSR. Institut Geografii. Sredniaia Aziia: ekonomiko-geograficheskaia
arakteristika i problemy razvitiia khoziaistva (Soviet Middle Asia: economic-
>graphic characteristics and problems of the growth of the economy). A. A.
its, ed. Moskva, Izdatel'stvo "Mysl'," Geografgiz, 1969. 504 p. Bibliog-
ohy, p. 493-502.   2265
About 175 references in alphabetical order by name of author.

JKIN, Ivan Semenovich. Ocherki fizicheskoi geografii Srednei Azii chast' 1.
shchii obzor (Essays on the physical geography of Middle Asia. v. 1. General
erview). Moskva, 1956. 407 p. Bibliography, p. 390-404 and 406-407 (Moskovskii
sudarstvennyi Universitet imeni Lomonosova. Geograficheskii Fakul'tet. Kafedra
>morfologii).   2266
     206 references.

I, N. A. Turanskaia fiziko-geograficheskaia provintsiia (Turanian physical-
>graphical province). L. N. Babushkin, ed. Tashkent, 1969. 138 p. Bibliog-
ohy, p. 124-137 (Tashkentskii Gosudarstvennyi Universitet imeni V. I. Lenina.
ichnye trudy, vypusk 353).   2267
     About 410 references.

I, N. A. Fiziko-geograficheskoe raionirovanie Turanskoi chasti Srednei Azii
iysical-geographical regionalization of the Turanian part of Soviet Middle Asia).
shkent, "Fan," 1969. 131 p. Bibliography, p. 117-130. (Tashkentskii Gosudar-
vennyi Universitet imeni Lenina).   2268
     282 references.

IKHIN, Viktor Mikhailovich. Fizicheskaia geografiia Tian'-Shania: prirodno-
>graficheskie osobennosti, osnovnye voprosy landshaftnogo kartirovaniia i
mpleksnogo fiziko-geograficheskogo raionirovaniia (The physical geography of
e Tyan'-Shan': natural-geographic features, basic questions of landscape mapping
i complex physical-geographic regionalization). Alma-Ata, Izdatel'stvo Akademii
ik Kazakhskoi SSR, 1964. 373 p. Bibliography, p. 361-372. (Akademiia Nauk
zakhskoi SSR. Otdel Geografii).   2269
     About 270 references in alphabetical order.

)ROV, Sergei Vasil'evich. Pustynia Ustiurt i voprosy ee osvoeniia (The desert
the Ustyurt and its development). Moskva, Izdatel'stvo "Nauka," 1971. 134 p.
oliography, p. 124-133. (Akademiia Nauk SSSR. Moskovskoe Obshchestvo Ispytatelei
irody, Trudy, tom 44).   2270

IANOV, Boris Vasil'evich. Drevnie orositel'nye sistemy Priaral'ia: v sviazi s
itoriei vozniknoveniia i razvitiia oroshaemogo zemledeliia (Ancient irrigation
stems of the Aral area: in relation to the history of origin and growth of
rigated land). Moskva, Izdatel'stvo "Nauka," 1969. 254 p. Bibliography, p.
4-250. (Akademiia Nauk SSSR. Institut Etnografii imeni N. N. Miklukho-Maklaia).
     About 900 references.   2271

ATAEV, A. A., BATYROV, A. B., and FREIKIN, Z. G. Osnovnye puti povysheniia
ekonomicheskoi effektivnosti sel'sko-khoziaistvennogo osvoeniia pustyn' Srednei
Azii (Basic paths of raising the economic effectiveness of agricultural development of the deserts of Middle Asia). N. T. Nechaeva, ed. Ashkhabad, "Ylym,"
1973. 261 p. Bibliography, p. 256-260. (Akademiia Nauk Turkmenskoi SSR.
Institut Pustyn'. Akademiia Nauk SSSR. Institut Geografii).

MEZHOV, Vladimir Izmailovich. Turkestanskii sbornik sochinenii i statei otnosia-
shchikhsia do Srednei Azii voobshche i Turkestanskogo kraia v osobennosti.
Sistematicheskii i azbuchnyi ukazateli sochinenii i statei na russkom i inostrann
iazykakh (Turkestan collection of works and articles relating to Middle Asia as
whole and to Turkestan in particular. Classed and alphabetical indexes of books
and articles in Russian and foreign languages). Sankt-Peterburg, V. Bezobrazov,
1878-1888. 3 v. v. 1-2 have added title page in French: Recueil du Turkestan,
comprenant des livres et des articles sur l'Asie centrale en général et le (sic)
province du Turkestan en particulier.

    4,706 items included in the 416 volumes of Turkestanskii sbornik.

KIRGIZIA

UMURZAKOV, Sadybakas Umurzakovich. Literatura (Literature). In the book: Priroda
Kirgizii: kratkaia fiziko-geograficheskaia kharakteristika (Natural conditions o
Kirgizia: a short physical-geographical characterization). M. N. Bol'shakov, I.
V. Vykhodtsev, E. V. Nikitina, and others. Frunze, Kirgizgosizdat. 1962.
298 p. Bibliography, p. 279-297 (Geograficheskoe Obshchestvo SSSR. Kirgizskii
Filial).

    About 400 references in alphabetical order on the physical geography of
Kirgizia, 1825-1961.

LOSEV, Davyd Semenovich. Prirodnye usloviia i prirodnye resursy Kirgizii, 1946-
1955: Ukazatel' literatury (Natural conditions and natural resources of Kirgizia
guide to the literature). K. O. Otorbaev, ed. Frunze, 1963. 275 p. (Gosudarstvennaia Respublikanskaia Biblioteka Kirgizskoi SSR im. N. G. Chernyshevskogo.
Bibliografiia Kirgizii v 4-kh tomakh. Tom 3, vypusk 3).

    2,142 entries organized systematically by fields: geography, hydrology,
meteorology, climatology, geology, paleontology, flora and vegetation, fauna and
animal life. Biographical notes on scholars and scientists in each field. Index
of names and places. List of publications consulted.

_____ . _____ . 1956-1960. Frunze, "Kyrgyzstan," 1972. 378 p.

NOVICHENKO, E. I. Bibliografiia bibliografii o Kirgizii, 1852-1967. Annotirovanny
ukazatel' literatury (Bibliography of bibliographies on Kirgizia, 1852-1967.
Annotated guide to the literature). Frunze, Izdatel'stvo "Kirgyzstan," 1969.
190 p. (Ministerstvo Kul'tury Kirgizskoi SSR. Gosudarstvennaia Respublikanskaia
Biblioteka Kirgizskoi SSR imeni N. G. Chernyshevskogo).

    1,242 entries in systematic arrangement (chronological within categories) in
Russian and Kirgiz. Index of names.

ZABIROV, Rashit Dzhamalievich, and BLAGOOBRAZOV, Vladimir Alekseevich, eds. Basseir
reki Naryn: fiziko-geograficheskaia kharakteristika (The bassin of the Naryn
river: physical-geographic characteristics). Frunze, Izdatel'stvo Akademii Nauk
Kirgizskoi SSR, 1960. 230 p. Bibliography, p. 223-229 (Akademiia Nauk Kirgizskc
SSR. Otdel Geografii).

    155 references.

BLAGOOBRAZOV, Vladimir Alekseevich. Tian'-Shan'skaia fiziko-geograficheskaia
stantsiia: bibliograficheskii ukazatel' (Tyan'-Shan' Physical-Geographic Station:
a bibliographical guide). Frunze, "Ilim," 1965. 225 p. (Akademiia Nauk
Kirgizskoi SSR).

2272 - 2278

426 entries in alphabetical order of work at the station 1947-1964, p. 25-
4. Author, subject, and geographical indexes.

OZHOEV, Bektur Orozgozhoevich. Priroda vysokogornykh pastbishch vnutrennego
an'-Shania [Ak-Sai i Arpa]. (Natural conditions in the high-mountain pastures
the inner Tyan'-Shan' [Ak-Say and Arpa]). Frunze, "Ilim," 1968. 147 p.
oliography, p. 139-146. (Akademiia Nauk Kirgizskoi SSR. Tian'-Shanskaia
sokogornaia Fiziko-geograficheskaia Stantsiia).   2279

About 180 references.

AROVA, Ol'ga Dmitrievna. Landshafty Alaiskoi doliny i ee raionirovanie dlia
elei sel'skogo khoziaistva (Landscapes of the Alay valley and its regionaliza-
on for the purposes of agriculture). Frunze, "Ilim," 1973. 132 p. Bibliog-
ohy, p. 123-131. (Akademiia Nauk Kirgizskoi SSR. Institut Geologii).   2280

UZBEKISTAN

SHKIN, Leonid Nikolaevich, and KOGAI, N. A. Fiziko-geograficheskoe raionirovanie
bekskoi SSR (Physical-geographic regionalization of the Uzbek SSR). In the
ok: Voprosy geografii raionirovaniia Srednei Azii i Uzbekistana (Problems in
e geography of regionalization of Middle Asia and Uzbekistan). Tashkent, 1964,
238-247 (Tashkentskii gosudarstvennyi universitet imeni Lenina.
uchnye trudy, vypusk 231. Geograficheskie nauki, no. 27).   2281

About 200 references.

ATKINA, A. V. Geografiia Uzbekistana: ukazatel' literatury, kniga I. 1917-1960
. (Geography of Uzbekistan: guide to the literature, v. 1, 1917-1960). L. N.
bushkin, ed. Tashkent, 1972. 435 p. (Akademiia Nauk Uzbekskoi SSR. Funda-
ntal'naia Biblioteka. Priroda i prirodnye resursy Uzbekistana, vypusk 1).   2282

RINOVA, Mariia Mikhailovna, and EREMIANTS, N. K., comps. Nizov'ia Amu-Dar'i:
irodnye usloviia i sel'skoe khoziaistvo. Ukazetel' osnovnoi literatury (The
wer Amu-Dar'ya: natural conditions and agriculture. Guide to the basic liter-
ure). V. B. Gussak, ed. Tashkent, 1961. 235 p. (Ministerstvo Sel'skogo
oziaistva Uzbekskoi SSR. Tsentral'naia Nauchnaia Sel'skokhoziaistvennaia
blioteka).   2283

1,901 entries, organized systematically, partly annotated, published from
e second half of the 19th century to 1961, in Russian, Karakalpak, and Uzbek
nguages.

850 references on physical geography systematically arranged under natural
nditions, p. 4-106, with headings: general, climate, relief and geomorphology,
drology (Amu-Dar'ya, Aral Sea), soils, vegetation, and animal life. Indexes
names and places. List of sources.

OK, Asamitdin Saidovich. Landshafty pravoberezh'ia Srednego Zarafshana (Land-
apes of the right bank of the middle Zeravshan). Tashkent, "Fan," 1972. 132 p.
bliography, p. 126-131 (Akademiia Nauk Uzbekskoi SSR. Otdel Geografii).   2284

OV, M. Prirodnye resursy nizov'ev reki Zarafshan i ikh ispol'zovanie. (Natural
sources of the lower course of the Zeravshan river and their utilization).
shkent, "Fan," 1967. 173 p. Bibliography, p. 153-162. (Samarkandskii Gosudar-
vennyi Universitet imeni Navoi).   2285

About 180 entries.

DOV, E. Peski Vnutrennikh Kyzylkumov (The sands of the inner Kyzyl-Kum),
shkentskii Universitet, Nauchnye trudy, no. 269; geograficheskie nauki, no. 32
964), bibliography, p. 148-154.   2286

About 160 references on the physical geography of the sands of one of the
serts of Soviet Middle Asia.

TURKMENISTAN

BABUSHKIN, Leonid Nikolaevich, and KOGAI, N. A. Fiziko-geograficheskoe raionirovan: Turkmenskoi SSR (Physical-geographic regionalization of the Turkmen SSR). Tashkent, "Fan," 1971. 184 p. Bibliography, p. 179-184. (Tashkentskii Gosudarstvennyi Universitet imeni V. I. Lenina).

AKADEMIIA Nauk Turkmenskoi SSR. Tsentral'naia Nauchnaia Biblioteka. Pustyni Turkmenii i ikh khoziaistvennoe osvoenie: ukazatel' literatury 1950-1965 (The desert: of Turkmenia and their economic development: a guide to the literature 1950-1965). T. B. Berdyev and M. P. Petrov, eds. A. Iazberdyev, L. K. Karadzhaeva, L. I. Klippa and A. Ia. Stepanov, comps. Ashkhabad, Izdatel'stvo "Ylym," 1972. 435 p.

3,963 entries organized systematically under earth sciences including cartography and geodesy, geophysics (including hydrology, meteorology, and climatology), geology (including minerals and hydrogeology), and geography; biology; and technology (including economic development of resources)--each with a finely articulated set of subdivisions. Geography, p. 173-195, entries 1,586-1,811, includes 231 entries in general section and in sections on general physical geography, physical-geographic regionalization, landscape studies, and geomorphology. In each subsection works in Russian are listed first, then those in Turkmen (224 in Russian, 7 in Turkmen). Indexes of authors and of works listed by title. Entries provide full bibliographical information including the number of entries in the bibliographies of articles or monographs listed.

BABAEV, Agadzhan Gel'dyevich. Oazisnye peski Turkmenistana i puti ikh osvoeniia. (Oasis sands of Turkmenistan and paths of mastering them). M. P. Petrov, ed. Ashkhabad, "Ylym," 1973. 353 p. Bibliography, p. 334-343 (Akademiia Nauk Turkmenskoi SSR. Institut Pustyn').

KUNIN, Vladimir Nikolaevich, ed. Ocherki prirody Kara-Kumov (Studies on natural conditions of the Kara-Kum [desert]). Moskva, Izdatel'stvo Akademiia Nauk SSSR, 1955. 407 p. Bibliography, p. 388-397. (Akademiia Nauk SSSR. Institut Geografii).

190 references in alphabetical order.

MAKEEV, Pavel Semenovich. Fiziko-geograficheskii ocherk Nizmennykh Karakumov (Physical-geographic study of the low-lying Karakum). In the book: Prirodnye resursy Karakumov: fiziko-geograficheskie opisanie (Natural resources of the Karakum: physical-geographical description). v. 2. Nizmennye Karakumy (Lowlying Karakum). Moskva-Leningrad, Izdatel'stvo Akademii Nauk SSSR, 1940. Bibliography, p. 86-97 (Akademiia Nauk SSSR. Sovet po Izucheniiu Proizvoditel'nykh Sil. Institut Geografii).

About 400 references in alphabetical order.

KNIAZHETSKAIA, E. A. Literatura o Zapadnom Uzboe, 1714-1950: bibliograficheskii ukazatel' (Literature on the Western Uzboy, 1714-1950: a bibliographical guide). K. I. Shafranovskii, ed. Ashkhabad, Izdatel'stvo Akademii Nauk Turkmenskoi SSR. 1956. 136 p. (Akademiia Nauk Turkmenskoi SSR. Biblioteka Akademiia Nauk SSSR).

551 entries, annotated, chronologically arranged. Name index.

TADZHIKISTAN

OGRAFIIA Tadzhikistana (Bibliography of Tadzhikistan). v. 1. Geografiia i
rologiia (Geography and hydrology). Leningrad, Izdatel'stvo Akademii Nauk
R, 1933. 66 p. (Akademiia Nauk SSSR. Sovet po Izucheniiu Proizvoditel'nykh
. Trudy, Seriia Tadzhikistana. vypusk 1).   2292

Two basic sections: general geography, physical geography. Subject and
graphical indexes.

K, A. G. Literatura po geologii, geomorfologii i gidrogeologii Tadzhikistana,
1-1955 (Literature on the geology, geomorphology, and hydrogeology of Tadzhiki-
n 1951-1955). Vsesoiuznoe Mineralogicheskoe Obshchestvo. Tadzhikskoe Otdelenie.
iski, vypusk 1 (1959), p. 163-177 (Akademiia Nauk Tadzhikskoi SSR, tom 104).   2293
350 references.

ANIANTS, Okmir Egishevich, and SIN'KOVSKAIA, Antonia Semenovna. Bibliografiia
ira. Ukazatel' literatury 1920-1964 gg. (Bibliography of the Pamir: guide to
literature 1920-1964). R. B. Baratov, ed. vypusk 1. Priroda (v. 1. Nature).
hanbe, "Donish," 1968. 266 p. (Akademiia Nauk Tadzhikskoi SSR. Tsentral'naia
chnaia Biblioteka).   2294

1,826 entries on history of research on natural conditions, geology, physical
graphy, soils, vegetation, animal life, anthropology, and local medicine of
Pamir mountains in Soviet Middle Asia organized topically, with some annota-
ns. Includes books, articles, and abstracts of dissertations. Author index.

ANIANTS, Okmir Egishevich. Osnovnye problemy fizicheskoi geografii Pamira
sic problems of the physical geography of the Pamir). v. 1. Dushanbe,
atel'stvo Akademii Nauk Tadzhikskoi SSR, 1965. 240 p. Bibliography, p. 220-
(Akademiia Nauk Tadzhikskoi SSR. Botanicheskii Institut). v. 2. Dushanbe,
nish," 1966. 244 p. Bibliography, p. 219-233 (Akademiia Nauk Tadzhikskoi
. Pamirskaia Biologicheskaia Stantsiia Pamirskoi Bazy). English summary. 2295

About 830 references, 470 in volume 1 and 360 in volume 2.

ANIANTS, Okmir Egishevich. Mezhdu Gindukushem i Tian'-Shanem: Istoriia
zheniia prirody Pamira (Between the Hindu-Kush and the Tyan'-Shan: History
the study of natural conditions in the Pamirs). Dushanbe, Tadzhikgosizdat,
2. 127 p. Bibliography, p. 116-126.   2296

About 220 references.

ENKOVA, Valentina Konstantinovna. Prirodnye osobennosti i rastitel'nyi pokrov
hnogo Pamira (Natural characteristics and vegetation of the southern Pamir).
ingrad, Izdatel'stvo "Nauka," Leningradskoe Otdelenie, 1971. 136 p. Bibliog-
ny, p. 126-133. (Akademiia Nauk SSSR. Geograficheskoe Obshchestvo SSSR). 2297

PART V.   REGIONAL GEOGRAPHY OF AREAS OUTSIDE THE SOVIET UNION

WORLD AS A WHOLE

DEMIIA Nauk SSSR. Gosplan SSSR. Sovet po Izucheniiu Proizvoditel'nykh Sil.
egional'nye issledovaniia za rubezhom (Regional research abroad). Iu. M. Pavlov
nd E. B. Alaev, eds. Moskva, Izdatel'stvo "Nauka," 1973. 303 p. Bibliography,
. 292-299.   2298

About 190 references on regional studies of areas outside the Soviet Union,
O in Russian and about 120 in Western languages. Author and subject indexes.

EV, Enrid Borisovich. Regional'noe planirovanie v razvivaiushchikhsia stranakh
Regional planning in developing countries). Moskva, "Nauka," 1973. 216 p.
ibliography, p. 193-201 (Akademiia Nauk SSSR. Sovet po Izucheniiu Proizvoditel'-
ykh Sil pri Gosplane SSSR).   2299

Indexes of personal names, subjects, and geographical names.

'SKII, Viktor Vatslavovich, POKSHISHEVSKII, Vadim Viacheslavovich and RIABCHIKOV,
leksandr Maksimovich. Vyborochnaia bibliografiia regional'nykh i stranovedche-
kikh rabot, izdannykh tsentral'nymi i oblastnymi izdatel'stvami posle 1945 g.
Selected bibliography of regional work, published by central and regional presses
fter 1945), in the book: Sovremennye problemy geografii (Current problems in
eography). Moskva, Izdatel'stvo "Nauka," 1964. Bibliography, p. 369-372
Akademiia Nauk SSSR. Natsional'nyi komitet sovetskikh geografov). Summary in
nglish.   2300

121 references of selected Soviet works on regional geography, both of Soviet
nion and of foreign countries in the period 1945-1963.

OIAVLENSKII, Georgii Pavlovich. Annotirovannyi ukazatel' literatury po geografii
arubezhnykh stran. Posobie dlia uchitelei (An annotated guide to the literature
f the geography of foreign countries: an aid for teachers). Moskva, Uchpedgiz,
960. 174 p.   2301

About 380 entries, arranged by continent and country. Handbooks, atlases,
eriodicals, and methodological literature are distinguished.

ATOVA, Galina Petrovna. Strany mira. Rekomendatel'nyi ukazatel' literatury
Countries of the world: guide to recommended literature). Moskva, Gosudarstven-
aia Biblioteka SSSR imeni V. I. Lenina, 1957. 132 p.   2302

About 600 references arranged by countries. Author, title, and geographical
ndexes.

NORTH AMERICA

MOVA, Nina Aleksandrovna, VOIAKINA, Svetlana Mikhailovna, and GORBUNOV, A. M.
. Sh. A. Kanada: rekomendatel'nyi ukazatel' literatury (U. S. A. and Canada:
uide to recommended literature). N. N. Solovevaia, ed. Moskva, Izdatel'stvo
Kniga," 1964. 169 p. (Gosudarstvennaia Biblioteka SSSR imeni V. I. Lenina).   2303

About 500 annotated entries on the political, economic, and cultural life
f the United States and Canada.

BKOVA, A. N. Aleutskie ostrova: fiziko-geograficheskii ocherk (The Aleutian
islands: a physical-geographical study). Moskva, Geografgiz, 1948. 288 p.
Bibliography, p. 275-286 (Geograficheskoe Obshchestvo SSSR. Zapiski, n.s. tom 4).
                                                                              2304
333 references.

ANTIPOVA, Angelina Vasil'evna. Kanada: priroda i estestvennye resursy (Canada: nature and natural resources). Moskva, Izdatel'stvo "Mysl'," 319 p. 1965. Bibliography, p. 307-317. (Akademiia Nauk SSSR. Institut Geografii).

About 250 references, organized by chapters, general works, structure of the surface, mineral deposits, climatic conditions, coastal and internal waters, soils and vegetation, animal life, and conclusions.

IGNAT'EV, Grigorii Mikhailovich. Grenlandiia (Greenland). Moskva, Geografgiz, 1956. 248 p. Bibliography, p. 240-246.

   154 references.

ANDREEVA, Vera Mikhailovna, GOKHMAN, Veniamin Maksovich, KOVALEVSKII, Vladimir Pavlovich, and POLOVITSKAIA, Mariia Efimovna. Ekonomicheskie raiony SShA: Sever (Economic regions of the USA: the North). K. M. Popov and M. G. Solov'eva, eds. Moskva, Geografgiz, 1958. 830 p. Bibliography, p. 801-807. (Akademiia Nauk SSSR. Institut Geografii).

   About 150 references in Russian and English.

POLOVITSKAIA, Mariia Efimovna. Ekonomicheskie raiony SShA: Iug (Economic regions of the United States: the South). K. M. Popov and M. G. Solov'eva, eds. Moskva, Geografgiz, 1956. 609 p. Bibliography, p. 478-494 (Akademiia Nauk SSSR. Institut Geografii).

   About 350 references in Russian and English.

POLOVITSKAIA, Mariia Efimovna. Ekonomicheskie raiony SShA: Zapad (Economic regions of the United States: the West). K. M. Popov and M. G. Solov'eva, eds. Moskva, Izdatel'stvo "Mysl'," Geografgiz, 1966. 534 p. Bibliography, p. 509-516. Supplementary English table of contents, p. 533-534 (Akademiia Nauk SSSR. Institut Geografii).

   About 190 references in Russian and English.

LATIN AMERICA

LATINSKAIA Amerika v sovetskoi pechati: annotirovannyi bibliograficheskii ukazatel' knig i statei na russkom iazyke o sovremennom politicheskom polozhenii, ekonomike, kul'ture, geografii i istorii stran Latinskoi Ameriki 1967-1968 gg. (Latin America in Soviet publications: annotated bibliography of books and articles in the Russian language on the contemporary political situation, the economy, culture, geography, and history of the countries of Latin America 1967-1968). Z. G. Vorob'ëva and K. A. Surova. Moskva, 1969. 151 p. (Akademiia Nauk SSSR. Institut Nauchnoi Informatsii i Fundamental'naia Biblioteka po Obshchestvennym Naukam. Nauchnaia Biblioteka Instituta Latinskoi Ameriki).

   1,135 entries, arranged systematically, and within categories chronologically. Index of authors and titles.

____ 1965-1966. Z. G. Vorob'ëva, I. M. Kaptsova, and K. A. Surova. Moskva, 1967. 136 p. (Fundamental'naia Biblioteka Obshchestvennykh Nauk imeni V. P. Volgina. Nauchnaia Biblioteka Instituta Latinskoi Ameriki). (Title varies slightly).

   1,114 entries.

____ 1963-1964. Z. G. Vorob'ëva, S. A. Kolomiitseva, and K. A. Surova. Moskva, 1965. 114 p.

   829 entries.

____ 1946-1962. E. B. Anikina, and others. Moskva, 1964. 132 p. (Akademiia Nauk. Institut Latinskoi Ameriki).

   1,928 entries by country and within countries chronologically.

.VSKII, Boris Nikolaevich. Ekonomicheskaia geografiia Kuby: khoziaistvenno-
rritorial'nye problemy Respubliki Kuba (Economic geography of Cuba: economic-
rritorial problems of the Republic of Cuba). P. M. Alampiev, ed. Leningrad,
datel'stvo "Nauka," Leningradskoe Otdelenie, 1970. 219 p. Bibliography,
211-217 (Akademiia Nauk SSSR. Geograficheskoe Obshchestvo SSSR).   2311

     245 references mostly in Spanish or Russian organized by 13 chapters.

'VA, T. N. Meksika. Vodnoe khoziaistvo i ekonomicheskoe razvitie (Mexico.
ne water management and economic development). Moskva, Izdatelstvo "Nauka,"
273. 210 p. Bibliography, p. 203-209. (Akademiia Nauk SSSR. Institut
tinskoi Ameriki).   2312

SKII, Viktor Vatslavovich, ed. Peru. Nekotorye aspekty ekonomicheskogo
zvitiia (Peru. Some aspects of economic development). Moskva, 1969. 174 p.
bliography, p. 168-173. (Akademiia Nauk SSSR. Institut Latinskoi Ameriki).   2313

NOVA, Raisa Andreevna. Argentina: ekonomiko-geograficheskaia kharakteristika
rgentina: economic-geographic characteristics). Moskva, Izdatel'stvo "Mysl',"
74. 239 p. Bibliography, 230-238.   2314

EUROPE

ANY EVROPY. Rekomendatel'nyi ukazatel' literatury. Chast' 1. Sotsialisticheskie
trany (The countries of Europe: guide to recommended literature, v. 1, socialist
ountries). N. A. Bespalova, S. M. Voiakina, comps. Moskva, "Kniga," 1965. 92 p.
Gosudarstvenna Biblioteka SSSR imeni V. I. Lenina).   2315

     About 340 annotated entries on the economic, political, and cultural life
f the socialist countries of Europe, excluding the USSR. Index.

MOV, Ruben Artemovich. Fizicheskaia geografiia zarubezhnoi Evropy (Physical
eography of Europe, outside the USSR). Moskva, "Mysl'," 1973. 272 p. Bib-
iography, p. 265-271.   2316

AND, David L'vovich. Rumyniia: fiziko-geograficheskoe opisanie (Romania: physical-
ographic description). Moskva, Leningrad, Izdatel'stvo Akademii Nauk SSSR. 1946.
59 p. Bibliography, p. 249-256. (Akademiia Nauk SSSR. Institut Geografii.
uchno-populiarnaia seriia).   2317

     185 entries, 1856-1945.

RIN, Vladimir Vladimirovich, and AVDEICHEV, Lev Alekseevich. Iugoslaviia.
conomiko-geograficheskaia kharakteristika (Yugoslavia: economic-geographic
haracteristics). Moskva, Izdatel'stvo "Mysl'," 1970. 239 p. Bibliography,
. 232-238.   2318

     136 references.

RYNIN, Boris Fedorovich. Fizicheskaia geografiia Zapadnoi Evropy (Physical
ography of Western Europe). Moskva, Uchpedgiz, 1948. 416 p. Bibliography,
. 398-402.   2319

     165 references, arranged by regions: Western Europe as a whole, Southern
urope, Balkan Peninsula, Italy, Iberian Peninsula, Middle Europe, the Alps, the
arpathians, Hercynian France, Hercynian Middle Europe, the Northern Plain of
iddle Europe, the British Isles, Northern Europe, the Scandinavian Pennisula,
inland, and Iceland.

EBRIANNYI, Leonid Ruvimovich. Fizicheskaia geografiia i chetvertichnaia geologiia
anii. (Physical geography and Quaternary geology of Denmark). Moskva, 1967.
ibliography, p. 260-271.   2320

     About 340 references.

GRATSIANSKII, Andrei Nikolaevich. Priroda Sredizemnomor'ia (Natural conditions
in the Mediterranean). Moskva, Izdatel'stvo "Mysl'," 1971. 510 p. Bibliography,
p. 456-470. Index of geographical names, p. 497-509.

ASIA AND AFRICA TOGETHER

NOVAIA sovetskaia i inostrannaia literatura po stranam Azii i Afriki (New Soviet
and foreign literature on the countries of Asia and Africa). Moskva. Fundamental'naia Biblioteka Obshchestvennykh Nauk AN SSSR. 1964- monthly. Supersedes Novaia sovetskaia i inostrannaia literatura po stranam zarubezhnogo Vostoka.

LITERATURA o stranakh Azii i Afriki. Ezhegodnik 1961 (Literature on the countries
of Asia and Africa. Yearbook 1961), E. R. Bochkareva, A. M. Grishina, V. P.
Zhuravlëva, and others, comps. Moskva, "Nauka," 1964. 224 p. (Akademiia Nauk
SSSR. Institut Narodov Azii. Later also: Institut Nauchnoi Informatsii. Fundamental'naia Biblioteka Obshchestvennykh Nauk imeni V. P. Volgina).

_____. 1962. A. M. Grishina, E. F. Bochkareva, V. P. Zhuravlëva, and N. V.
Shiriaeva, comps. Moskva, 1965. 276 p.

_____. 1963. A. M. Grishina, I. V. Aleksandrova, V. P. Zhuravlëva, and N. V.
Shiriaeva, comps. Moskva, 1967. 268 p.

_____. 1964-1965. A. M. Grishina, I. V. Aleksandrova, V. P. Zhuravlëva, and
others, comps. Moskva, 1972. 519 p.

Bibliography of works published in the USSR either originally in Russian
or in Russian translation. Arranged by regions and countries. Author indexes.

AFRICA

MILIAVSKAIA, S. L. and SINITSYNA, N. E. Bibliografiia Afriki: Dorevoliutsionnaia
i sovetskaia literatura na russkom iazyke, original'naia i perevodnaia (Bibliography of Africa. Prerevolutionary and Soviet literature in the Russian
language, both original publications and translations). no. 1. Moskva, Izdatel'-
stvo "Nauka," 1964. 276 p. (Akademiia Nauk SSSR. Institut Afriki. Fundamental'-
naia Biblioteka Obshchestvennykh Nauk).

2,506 entries from the second half of the 19th century to 1961, arranged
regionally: general, North Africa, Northwest Africa, East Africa, Southern Africa,
West and Central Africa. Each has a section for geography, including exploration
and travelers notes. Author index. Appendices include a list of candidate and
doctoral dissertations, 1941-1961, with 101 entries, and a list of geographical
maps, 103 entries.

AKIMOVA, Nina Aleksandrovna, and VOIAKINA, Svetlana Mikhailovna. Strany Afriki:
rekomendatel'nyi ukazatel' literatury (The countries of Africa: a guide to
recommended literature). Moskva, 1961. 109 p. (Gosudarstvennaia Biblioteka
SSSR imeni Lenina).

About 250 annotated entries on the political, economic, and cultural life
of the countries of Africa. Index of authors and names.

EKONOMIKA nezavisimykh stran Afriki (Economy of independent countries of Africa).
V. G. Solodovnikov, and others, eds. V. G. Solodovnikov, I. A. Svanidze, A. A.
Demidova, and others, authors. Moskva, Izdatel'stvo "Nauka," 1972. 364 p.
Bibliography, p. 348-353. (Akademiia Nauk SSSR. Institut Afriki).

212 references. Subject and name index.

NIDZE, I. A. Sel'skoe khoziaistvo tropicheskoi Afriki (Agriculture in tropical
frica). Moskva, Izdatel'stvo "Mysl'," 1972. 352 p. Bibliography, p. 339-349
Akademiia Nauk SSSR. Institut Afriki).   2327

RILOV, Nikolai Ivanovich. Problemy planirovaniia i razvitiia sel'skogo
hoziaistva v stranakh Afriki (Problems of planning and development in the
ountries of Africa). Moskva, Izdatel'stvo "Nauka," 1973. 388 p. Bibliography,
. 369-378. (Akademiia Nauk SSSR. Institut Afriki).   2328

    Indexes of subjects, geographical names, and personal names.

)KAL'SKAIA, Zinaida Iul'evna. Pochvenno-geograficheskii ocherk Afriki: usloviia
ochvoobrazovaniia, pochvy i ikh raspredelenie (A soil-geographical study of
.frica: conditions of soil formation, soils, and their distribution). Moskva-
.eningrad, Izdatel'stvo Akademii Nauk SSSR, 1948. 408 p. Bibliography, p. 382-
98 (Akademiia Nauk SSSR. Pochvennyi Institut imeni Dokuchaeva).   2329

    455 references, regionally organized.

NUNG, Mikhail Borisovich, and UTKIN, Georgii Nikolaevich. Marokko: ocherki po
izicheskoi i ekonomicheskoi geografii (Morocco: essays in physical and economic
eography). Moskva, Izdatel'stvo "Mysl'," 1966. 319 p. Bibliography, p. 305-
16. (Akademiia Nauk SSSR. Institut Geografii).   2330

    About 320 entries, organized by chapters.

NUNG, Mikhail Borisovich. Alzhiriia: fiziko-geograficheskaia kharakteristika
Algeria: physical-geographic characteristics). Moskva, Geografgiz, 1958. 288 p.
ibliography, p. 280-287 (Akademiia Nauk SSSR. Institut Geografii).   2331

    About 180 references, mainly in French and Russian.

TREVSKII, Iurii Dmitrievich. Nil: ocherki khoziaistvennoi ispol'zovaniia (The
ile: essays on economic utilization). Vologda, Knizhnoe Izdatel'stvo, 1958.
52 p. Bibliography, p. 146-151. (Geograficheskoe Obshchestvo SSSR).   2332

    163 references in Russian and other languages.

ZHOV, Nikolai Nikolaevich. Tanzaniia: ekonomiko-geograficheskaia kharakteristika
Tanzania: economic-geographic characteristics). Moskva, Izdatel'stvo "Mysl',"
972. 296 p. Bibliography, p. 290-295.   2333

                                    ASIA

AND, David L'vovich; DOBRYNIN, Boris Fedorovich; EFREMOV, Iurii Konstantinovich;
IMAN, Lev Iakovlevich; MURZAEV, Eduard Makarovich; and SPRYGINA, Liudmila Ivanovna.
arubezhnaia Aziia: fizicheskaia geografiia. (Asia, excluding the USSR: physical
eography). B. F. Dobrynin and E. M. Murzaev, eds. Moskva, Uchpedgiz, 1956. 608 p.
                                                                          2334
    467 references, arranged regionally at the ends of major sections: Southwest
sia, p. 86-87 (17), Asia Minor and the Middle East, p. 185-188 (134), Central
sia, p. 297-300 (82), East China, p. 376-377 (48), Northeast China and Korea,
. 424-425 (61), Japan, p. 483-485 (69), South Asia, p. 557 (11), and for Asia
s a whole, p. 584 (47).

ZAEV, Eduard M., and PULIARKIN, Valerii Alekseevich, eds. Zemel'nye i vodnye
esursy zarubezhnoi Azii (Land and water resources of Asia, except USSR). Itogi
auki i tekhniki:Geografiia zarubezhnykh stran (Results of science and technology.
eries on geography of foreign countries), tom 1. Moskva, VINITI, 1972. 132 p.
Vsesoiuznyi Institut Nauchnoi i Tekhnicheskoi Informatsii).   2335

                                                        2327 - 2335

1,778 entries organized regionally: Central and East Asia; South-East Asia; and the Near and Middle East. Separate sections for land and water resources together and for land resources alone and water resources alone. Works in Cyrillic listed first in alphabetical order of author, then works in Latin alphabet within each section. Works in Chinese and Japanese are listed in Russian translation in the Cyrillic sections. Location of abstracts noted in Referativnyi zhurnal: geografiia by citation number.

NOVAIA sovetskaia i inostrannaia literatura po stranam zarubezhnogo Vostoka (New Soviet and foreign literature on the countries of the East outside the USSR). Moskva, Fundamental'naia Biblioteka Obshchestvennykh Nauk AN SSSR, 1953-1963. Monthly. Supersedes Novaia inostrannaia literatura po vostokovedeniiu and Novaia sovetskaia literatura po vostokovedeniiu. Superseded by Novaia sovetskaia i inostrannaia literatura po stranam Azii i Afriki.   2

NOVAIA sovetskaia literatura po vostokovedeniiu (New Soviet literature on Oriental studies). Moskva, Fundamental'naia Biblioteka Obshchestvennykh Nauk AN SSSR, 1949-1952. Mainly monthly. Superseded by Novaia sovetskaia i inostrannaia literatura po stranam zarubezhnogo Vostoka.   2

VOIAKINA, Svetlana Mikhailovna. Strany Azii. Rekomendatel'nyi ukazatel' literatury (Countries of Asia: guide to the recommended literature). Moskva. 1960. 137 p. (Gosudarstvennaia Biblioteka SSSR imeni V. I. Lenina).   1

About 400 annotated entries on political, economic, and cultural life of the countries of Asia. Index of authors and names.

AKADEMIIA Nauk SSSR. Knigi glavnoi redakstii vostochnoi literatury izdatel'stva "Nauka," 1967-1971: annotirovannyi katalog (Books of the publishing house of science, section on Eastern literature, 1967-1971: annotated catalogue). Moskva, Izdatel'stvo "Nauka," Glavnaia Redaktsiia Vostochnoi Literatury, 1973. 227 p.   2

859 entries arranged first by fields such as history, geography, literature, and languages, then by regions. Geography and travelers, p. 110-121, entries 451-495 includes 45 publications.

MEZHOV, Vladimir Izmailovich. Bibliografiia Azii: ukazatel' knig i statei ob Azii na russkom iazyke i odnikh iazykakh, kasaiushchikhsia otnoshenii Rossii s aziatskim Gosudarstvami (Bibliography of Asia: list of books and articles on Asia in the Russian language and some other languages, touching on the relations of Russia and Asiatic states). S.-Peterburg, 1891-1894. 3 v.   2

Covers the East as a whole, China, Manchuria, Mongolia, Dzungaria, Korea, Tibet, Japan, Indochina, India, Persia, Baluchistan, Turkey, Arabia, Afghanistan, the Middle Asian khanates, and the Russian possessions in Middle Asia. About 15,000 entries.

SISTEMATICHESKII ukazatel' statei, kasaiushchikhsia materika Azii, pomeshchënnykh v izdanniakh Russkogo geograficheskogo obshchestva s 1846 do 1897 goda (Classed list of articles concerning the continent of Asia, located in publications of the Russian Geographical Society from 1846 to 1897). Irkutsk, 1898. 245 p. (Russkoe Geograficheskoe Obshchestvo. Vostochno-Sibirskii otdel).   2

Southwest Asia

SVERCHEVSKAIA, A. K., and CHERMAN, T. P. Bibliografiia Turtsii, 1917-1958 (Bibliography of Turkey, 1917-1958), B. M. Dantsig, ed. Moskva, Izdatel'stvo Vostochnoi Literatury, 1959. 190 p. (Akademiia Nauk SSSR. Institut Vostokovedeniia).   2

Systematic guide. The section on geography and travel works has 72 entries.

CHEVSKAIA, A. K., and CHERMAN, T. P., comps. <u>Bibliografiia Turtsii, 1713-1917</u>
Bibliography of Turkey, 1713-1917). B. M. Dantsig, ed. Moskva, Izdatel'stvo
stochnoi Literatury, 1961. 267 p. (Akademiia Nauk SSSR. Institut Narodov
ii).   2343

The section, Geography and travel works, has 522 entries.

EEV, Sergei Nikolaevich. <u>Turtsiia (Aziatskaia chast'--Anatoliia)</u>. Fiziko-
ograficheskoe opisanie. (The Asiatic part, Anatolia, of Turkey: physical-
ographic description). Moskva-Leningrad, Izdatel'stvo Akademii Nauk SSSR.
46. 215 p. Bibliography, p. 207-212. (Akademiia Nauk SSSR. Institut
ografii. Nauchnaia-populiarnaia seriia).   2344

125 references.

HQEBIA, Nodar. <u>Turk'et'i</u> (Turkey). Tbilisi, T'SU Gamomtsemloba, 1970. 135 p.
Georgian. Russian form of author's name: Nachkhebiia, Nodar Valerianovich.   2345

ELIAN, Zh. S., comp. <u>Bibliografiia po kurdovedeniiu</u> (Bibliography on Kurdish
udies). Moskva, Izdatel'stvo Vostochnoi Literatury, 1963. 184 p. (Akademiia
uk SSSR. Institut Narodov Azii).   2346

2,690 entries on studies of the Kurds, including books, monographs, articles,
tes, pamphlets, dissertation abstracts, in Russian and West European languages,
sed on the collections of state libraries in the USSR, basically from the 18th
ntury to 1960. Includes also works in Georgian, Armenian, and Azerbaidzhani
blished in the Soviet Union. Organized in 12 sections: (1) geography, (2)
onomy, (3) history, ethnography, (4) national politics, (5) social movements,
) present status of Kurds outside the USSR, (7) Kurds in the USSR, (8) language,
) literature, folklore, (10) religion, (11) Kurdish studies in Russia and the
SR, and (12) Bibliography. Within each section entries are arranged chrono-
gically. Alphabetical index of authors, editors, compilers, translaters, com-
ntators, reviewers, and titles of works. At end of the work a list of Kurdish
udies 1961-1963 in Russian.

OV, Mikhail Platonovich. <u>Bibliografiiapo geografii Irana: Ukazatel' literatury
russkom iazyke, 1720-1954</u> (Bibliography on the geography of Iran: guide to
e literature in the Russian language, 1720-1954). V. N. Kunin, ed. Ashkhabad,
datel'stvo Akademii Nauk Turkmenskoi SSR, 1955. 235 p. (Akademiia Nauk Turk-
nskoi SSR. Biblioteka Akademii Nauk SSSR).   2347

960 annotated entries in alphabetical order. Indexes for geographical names,
bjects. List of periodicals, serials, and collected works utilized, with
breviation and full title. Introductory essay on Russian geographical research
Iran.

CHEVSKAIA, A. K. <u>Bibliografiia Irana. Literatura na russkom iazyke, 1917-1965</u>
. (Bibliography of Iran. Literature in the Russian language, 1917-1965).
A. Kuznetsova, ed. Moskva, Izdatel'stvo "Nauka," 1967. 391 p. (Akademiia
uk SSSR. Institut Narodov Azii).   2348

7,959 entries organized systematically. Index.

HQEBIA, Nodar. <u>Irani</u> (Iran). Tbilisi, T'SU Gamomtsemloba, 1972. 202 p. In
orgian. Russian form of author's name: Nachkhebiia, Nodar Valerianovich.   2349

61 references.

TINA, Tat'iana Ivanovna, comp. <u>Bibliografiia Afganistana. Literatura na
sskom iazyke</u> (Bibliography of Afghanistan. Literature in the Russian language).
. V. Gankovskii, ed. Moskva, "Nauka," 1965. 272 p. (Akademiia Nauk SSSR.
stitut Narodov Azii).   2350

Systematic guide. Section on geography includes headings: cartographic material; general works, travels; toponymics; physical geography; economic geography. Within each section works are listed in chronological order. 546 references.

POLIAK, A. A. Fizicheskaia geografiia Afganistana. Uchebnoe posobie (Physical geography of Afghanistan. A teaching aid). M. G. Aslanov, ed. Moskva, 1953. 276 p. Bibliographies, p. 179, 208, 269-274. (Moskovskii Institut Vostokovedeniia

About 150 references.   2

South Asia

India

BIRMAN, D. A., and KOTOVSKII, Grigorii Grigor'evich. Bibliografiia Indii. Dorevoliutsionnaia i sovetskaia literatura na russkom iazyke i iazykakh narodov SSSR, original'naia i perevodnaia (Bibliography of India. Prerevolutionary and Soviet literature in Russian and in languages of the peoples of the USSR, original and translated). G. G. Kotovskii, ed. Moskva, Izdatel'stvo "Nauka;" Glavnaia Redaktsiia Vostochnoi Literatury, 1965. 608 p. (Akademiia Nauk SSSR. Institut Narodov Azii. Fundamental'naia Biblioteka Obshchestvennykh Nauk im. V. P. Volgina. AN UzSSR. Institut Vostokovedeniia im. Abu Raikhana Biruni). With the collaboration of A. Es-Kh. Vafa and N. V. Untilova.   2

Bibliographical guide to 9,073 works through 1961, arranged systematically. Headings for geography and regional studies include: general works on India, handbooks; works on individual regions and cities of India; discovery and geographical study of India, history of geography; writings of travelers; physical geography. Author index. Supersedes earlier edition. 1959. 219 p. 3,858 entries, ed. by G. G. Kotovskii, and others.

RIABCHIKOV, Aleksandr Maksimovich. Priroda Indii (Natural conditions in India). Moskva, Geografgiz, 1950. 291 p. Bibliography, p. 282-287 (Geograficheskoe Obshchestvo SSSR. Zapiski, n.s. tom 12).   2

160 references.

SHIROKOV, G. K. Industrializatsiia Indii. (Industrialization of India). Moskva, Izdatel'stvo "Nauka," 1971. 390 p. Bibliography, p. 380-388. (Akademiia Nauk SSSR. Institut Vostokovedeniia).   2

SAKHAROV, I. V. Izuchenie Indii v geograficheskom obshchestve SSSR, 1917-1968 gg. (The study of India in the Geographical Society of the USSR, 1917-1968), Geograficheskoe Obshchestvo SSSR. Izvestiia, v. 100, no. 5 (1968), p. 414-418.   2

138 references on works by members of the Geographical Society of the USSR.

Other Countries

BIRMAN, D. A., and KAFITINA, M. N. Bibliografiia Pakistana,1947-1967 (Bibliography of Pakistan 1947-1967). Moskva, "Nauka," 1973. 54 p. (Akademiia Nauk SSSR. Institut Vostokovedeniia. Institut Nauchnoi Informatsii po Obshchestvennym Naukam).   2

PULIARKIN, Valerii Alekseevich. Kashmir. Moskva, Geografgiz, 1956. 227 p. Bibliography, p. 223-226 (Akademiia Nauk SSSR. Institut Geografii).   2

103 references.

MAN, D. A., and KAFITINA, M. N. Bibliografiia Nepala. 1917-1967 (Bibliog-
aphy of Nepal, 1917-1967). Moskva, "Nauka," 1973. 22 p. (Akademiia Nauk SSSR.
istitut Vostokovedeniia. Institut Nauchnoi Informatsiia po Obshchestvennym
aukam).   2358

MAN, D. A., and KAFITINA, M. N. Bibliografiia Tseilona, 1917-1967 (Bibliog-
aphy of Ceylon, 1917-1967). Moskva, "Nauka," 1973. 38 p. (Akademiia Nauk
SSR. Institut Vostokovedeniia. Institut Nauchnoi Informatsii po Obshchestvennym
aukam).   2359

Southeast Asia

.IOGRAFIIA Iugo-Vostochnoi Azii. Dorevoliutsionnaia i sovetskaia literatura
a russkom iazyke, original'naia i perevodnaia (Bibliography of Southeast Asia.
'erevolutionary and Soviet literature in the Russian language, original and
ranslated). A. M. Grishina, M. I. Nefedov, D. A. Birman, and others, comps.
oskva, Izdatel'stvo Vostochnoi Literatury, 1960. 256 p. (Akademiia Nauk SSSR.
istitut Narodov Azii. Fundamental'naia Biblioteka Obshchestvennykh Nauk).   2360

   Systematic guide to 3,752 books and articles published through 1958 by
ountries and fields in Russian alphabetical order: Southeast Asia as a whole,
irma, British North Bornea, Vietnam, Indonesia, Cambodia, Laos, Malaya and
ingapore, Sarawak, Thailand, Timor, Philippines. Within the larger countries
ere is a special section on geography, including history of geographical
xploration and travelers. Author index.

AKOVA, Lina Ivanovna. Birma. Prirodnye raiony i landshafty (Burma; natural
egions and landscapes). Moskva. Izdatel'stvo "Mysl'," 1967. 272 p. Bibliog-
aphy, p. 260-267.   2361
   About 210 entries on the physical geography of Burma.

OKONNIKOVA, Anna Andreevna. Birma: fiziko-geograficheskaia kharakteristika
Burma: physical-geographic characteristics). Moskva, Geografgiz, 1959. 158 p.
bliography, p. 153-157.   2362
   About 110 references in English and Russian.

HEGLOVA, Tat'iana Nikolaevna. V'etnam: fiziko-geograficheskaia kharakteristika
Viet-Nam: physical-geographic characteristics). Moskva, Geografgiz, 1957.
33 p. Bibliography, p. 165-173. (Akademiia Nauk SSSR. Institut Geografii).   2363
   About 200 references.

OLAND, Vladimir Markovich. Priroda Severnogo V'etnama (Natural conditions in
orth Viet-Nam). Moskva, Izdatel'stvo Akademii Nauk SSSR, 1961. 175 p. Bib-
lography, p. 172-174. (Akademiia Nauk SSSR. Nauchno-populiarnaia seriia).   2364
   About 50 references in Russian and other languages.

East Asia

China: Core Area

CHKOV, Petr Emel'ianovich. Bibliografiia Kitaia (Bibliography of China).
oskva, Izdatel'stvo Vostochnoi Literatury, 1960. 692 p. (Akademiia Nauk SSSR.
istitut Narodov Azii).   2365

   Bibliographical guide to 19,551 works in Russian on China 1730- - 1957,
rranged systematically. An earlier edition, 1730-1930, published in 1932,
as reprinted in 1948.

Section on general geography, p. 283-329, includes headings for travels, studies, general works on geography, general works on regions of China, and cities.

Section on the natural-geographic environment, p. 330-347, includes headings for relief, hydrography; geology, paleontology; climate; vegetation; animal life; cartography.

More than 2,000 references in these geographical sections.

KITAISKAIA Narodnaia Respublika. 1949-1959. Rekomendatel'nyi ukazatel' literatury (People's Republic of China. 1949-1959. Guide to recommended literature). G. P. Bogatova, S. M. Volkina, M. I. Davydova, and M. E. Zelenina, comps. Moskva, Tipografiia Biblioteki imeni Lenina, 1959. 63 p.   23

180 references, annotated. Index of authors and titles.

KAZAKOV, Sergei Vasil'evich. Uspekhi Kitaiskoi Narodnoi Respubliki. Rekomendatel'-nyi ukazatel' literatury (Progress in the People's Republic of China: guide to recommended literature). M. F. Iurev, ed. 2nd ed., Moskva, Gosudarstvennaia Biblioteka SSSR imeni Lenina, 1955. 68 p. (1st ed., 1952. 79 p.).   23

200 references, annotated, systematically organized covering history, political struggles, economics, culture.

ZAICHIKOV, Vladimir Timofeevich, ed. Fizicheskaia geografiia Kitaia (Physical geography of China). Moskva, Izdatel'stvo "Mysl'," 1964. 739 p. Bibliography, p. 694-722. (Akademiia Nauk SSSR. Institut Geografii).   23

About 550 references, listed alphabetically, covering the period 1861-1962 and including works in Russian, Chinese (listed in Russian translation), and Western European languages.

Available in English as The physical geography of China. New York, F. A. Praeger, 1969. 2 v.

ZAICHIKOV, Vladimir Timofeevich, ed. Vostochnyi Kitai. Primorskie provintsii (Eastern China. Maritime provinces). Moskva, Geografgiz. 1955. 312 p. Bibliography, p. 305-311. (Akademiia Nauk SSSR. Institut Geografii).   23

About 160 references in Russian, Chinese, and other languages.

MURZAEV, Eduard Makarovich. Severo-Vostochnyi Kitai: fiziko-geograficheskoe opisanie (North-East China: a physical-geographic description). Moskva, Izdatel'stvo Akademii Nauk SSSR. 1955. 252 p. Bibliography, p. 228-239. (Akademiia Nauk SSSR. Institut Geografii).   23

About 250 references in Russian and other languages.

Central Asia

IUSOV, Boris Vasil'evich. Tibet: fiziko-geograficheskaia kharakteristika (Tibet: physical-geographic characteristics). Moskva, Geografgiz, 1958. 224 p. Bibliography, p. 219-223.   23

About 125 references in Russian and foreign languages.

ZHURAVLËV, Iu. Novye izdaniia o Tibete i tibettsakh (Recent literature on Tibet and Tibetans), Sovetskaia etnografiia, 1959, no. 3, p. 171-177.   23

Covers literature 1954-1958.

ROV, Mikhail Platonovich. Pustyni Tsentral'noi Azii. Tom 1. Ordos, Alashan',
eishan' (Deserts of Central Asia v. 1. Ordos, Ala Shan, and Bey Shan). Moskva-
eningrad, Izdatel'stvo "Nauka," 1966. 274 p. Bibliography, p. 256-263.
Akademiia Nauk SSSR. Akademiia Nauk Turkmenskoi SSR. Leningradskii Gosudar-
tvennyi Universitet). 2373

About 220 references on the deserts of Inner Mongolia and of northeast
entral Asia, arranged alphabetically by name of author, names in Cyrillic first
ncluding works in Russian and Chinese (with titles translated into Russian), then
n the Latin alphabet.

ROV, Mikhail Platonovich. Pustyni Tsentral'noi Azii. Tom 2. Koridor Khesi,
saidam, Tarimskaia vpadina. (The deserts of Central Asia, v. 2. Khesi cor-
idor [northern foothills of the Nan Shan],Tsaidam, and Tarim Basin). Leningrad,
zdatel'stvo "Nauka," 1967. 288 p. Bibliography, p. 273-279 (Akademiia Nauk
SSR. Akademiia Nauk Turkmenskoi SSR. Leningradskii Gosudarstvennyi Universitet). 2374

About 220 references on the deserts of western China, in alphabetical order
y name of author, first in Cyrillic including both Russian and Chinese (with
itles translated into Russian) and then in Latin letters.

NETSOV, Nikolai Timofeevich. Vody Tsentral'noi Azii (The waters of Central
sia). Moskva, Izdatel'stvo "Nauka," 1968. 272 p. Bibliography, p. 260-271.
Akademiia Nauk SSSR. Institut Geografii). 2375

About 300 references.

IITSYN, Vasilii Mikhailovich. Tsentral'naia Aziia (Central Asia). Moskva,
eografgiz, 1959. 456 p. Bibliography, p. 450-455. 2376

About 140 references on the physical geography of Central Asia outside the
SSR.

RZAEV, Eduard Makarovich. Priroda Sin'tsziana i formirovanie pustyn' Tsentral'noi
Azii (Natural conditions in Sinkiang and the formation of the deserts of Central
Asia). Moskva, Izdatel'stvo "Nauka," 1966. 382 p. Bibliography, p. 358-372.
(Akademiia Nauk SSSR. Institut Geografii). 2377

About 480 references arranged by languages: in Russian, about 375 references,
in Chinese (titles given in Russian translation), 61 references, and in Western
languages, 46 references, alphabetically by author within each group.

LIVANOV, Evgenii Ivanovich. Geomorfologiia Dzhungarii (Geomorphology of Dzungaria).
Moskva, Izdatel'stvo "Nedra," 1965. 156 p. Bibliography, p. 150-155. 2378

About 150 references.

Mongolia

ALDAEV, Rodion Lazarevich, and VASIL'EV, N. N. Bibliografiia Mongol'skoi Narodnoi
Respubliki. Knigi i stat'i na russkom iazyke, 1951-1961 (Bibliography of the
Mongolian People's Republic. Books and articles in the Russian language, 1951-
1961). Moskva, Izdatel'stvo Vostochnoi Literatury, 1963. 120 p. (Akademiia
Nauk SSSR. Biblioteka Akademii Nauk SSSR. Institut Narodov Azii). 2379

Systematic guide. The section on geography, p. 19-24, has headings: general
works; hydrography; soils, vegetation; animal life. 103 references. The general
section has a heading on Russian scientists and explorers of Mongolia and works
on them, p. 14-19. 75 references. Author index.

TIULIAEVA, V.P. Mongol'skaia Narodnaia Respublika; bibliografiia knizhnoi i zhurnal'noi literatury na russkom iazyke, 1935-1950 gg. (The Mongolian People's Republic: bibliography of book and journal literature in the Russian language, 1935-1960). Moskva, Izdatel'stvo Akademii Nauk SSSR, 1953. 88 p. (Akademiia Nauk SSSR. Komissiia Nauk Mongol'skoi Narodnoi Respubliki. Mongol'skaia Komissiia. Trudy, vypusk 42). E. M. Murzaev, ed.

Systematic, partly annotated, bibliography. Material for physical geography in the section on natural conditions and resources, p. 36-56 with headings: travels and expeditions; geography, climate, hydrology; geology and paleontology; soils, vegetation; animal life. 225 references. Author index.

IAKOVLEVA, Ekaterina Nilovna. Bibliografiia Mongol'skoi Narodnoi Respubliki: sistematicheskii ukazatel' knig i zhurnal'nykh statei na russkom iazyke (Bibliography of the Mongolian People's Republic: systematic guide to books and journal articles in the Russian language). F. E. Telezhnikov, ed. Moskva, 1935. 230 p. (Nauchno-Issledovatel'skaia Assotsiatsiia po Izucheniiu Natsional'nykh i Kolonial'nykh Problem, Trudy, vypusk 18).

Systematic guide to the literature from the 18th century through 1934. Material on physical geography in the section on natural-historical conditions, p. 85-102, 191-192 with headings: general works; surface, minerals, irrigation; climate; vegetation; animal life. Also in the general section under the headings: expeditions and travels, p. 17-38, 187. 391 references.

MURZAEV, Eduard Makarovich. Mongol'skaia Narodnaia Respublika: fiziko-geograficheskoe opisanie (The Mongolian People's Republic: physical-geographic description). 2nd ed. Moskva, Geografgiz, 1952. 472 p. Bibliography, p. 448-463 (Akademiia Nauk SSSR. Institut Geografii). (1st ed., 1948. Bibliography, p. 293-302. 220 references).

    350 references

MURZAEV, Eduard Makarovich. Geograficheskie issledovaniia Mongol'skoi narodnoi respubliki (Geographical study of the Mongolian People's Republic). Moskva-Leningrad, Izdatel'stvo Akademii Nauk SSSR, 1948. 211 p. Bibliography, p. 185-203. (Akademiia Nauk SSSR. Institut Geografii i Mongol'skaia Komissiia. Seriia "Itogi i Problemy Sovremennoi Nauki").

    About 380 references in alphabetical order.

SELIVANOV, Evgenii Ivanovich. Neotektonika i geomorfologiia Mongol'skoi Narodnoi Respubliki. (Neotectonics and geomorphology of the Mongolian People's Republic). Moskva, Izdatel'stvo "Nedra," 1972. 293 p. Bibliography, p. 280-291. (Ministerstvo Geologii SSSR. Nauchno-Issledovatel'skaia Laboratoriia Geologii Zarubezhnykh Stran "NILZarubezhgeologiia.").

OBRUCHEV, Vladimir Afanas'evich. Vostochnaia Mongoliia. Geograficheskoe i geologicheskoe opisanie. v. 1-2. Obzor literatury. Orograficheskie i gidrograficheskie ocherki (Eastern Mongolia. Geographical and geological descriptions, v. 1-2. Review of the literature. Orographic and hydrological studies). Moskva-Leningrad, Izdatel'stvo Akademii Nauk SSSR, 1947, p. 9-138. (Akademiia Nauk SSSR. Mongol'skaia Komissiia. Geograficheskoe Obshchestvo SSSR).

    Annotated chronological list with 574 entries, 18th century to 1940. Author index.

LAVRENKO, Evgenii Mikhailovich, and others, eds. Biologicheskie resursy i prirodnye usloviia Mongol'skoi Narodnoi Respubliki (Biological resources and natural conditions of the Mongolian People's Republic). Tom 2. Kalinina, A. V. Osnovnye tipy pastbishch Mongol'skoi Narodnoi Respubliki: ikh struktury i produktivnost' (Basic types of pastures of the Mongolian People's Republic: their structure and productivity). V. M. Poniatovskaia, ed. Leningrad, Izdatel'stvo "Nauka," Leningradskoe otdelenie, 1974. 185 p. Bibliography, p. 173-179. List of Latin names of plants, p. 180-184.

Japan

LIOGRAFIIA Iaponii. Literatura, izdannaia v Sovetskom Soiuze na russkom iazyke
1917 po 1958 g. (Bibliography of Japan. Literature published in the Soviet
nion in the Russian language, 1917-1958). V. A. Vlasov, V. S. Grivnin, I. P.
uznetsova, I. L. Kurant, and M. V. Sutiagina, comps. Moskva, Izdatel'stvo
ostochnoi Literatury, 1960. 328 p. (Akademiia Nauk SSSR. Institut Narodov
zii. Vsesoiuznaia Gosudarstvennaia Biblioteka Inostrannoi Literatury).   2387

    Systematic guide to Russian literature on Japan, 1917-1958, with 6,249
ntries. The section on geography and ethnography, p. 30-39, has headings:
eneral questions, physical geography, travelers notes and impressions. About
65 references. Index of names.

VNIN, Vladimir Sergeevich, LESHCHENKO, N. F., and SUTIAGINA, M. V. Bibliog-
afiia Iaponii. Literatura, izdannaia v Rossii s 1734 po 1917 g. (Bibliog-
aphy of Japan. Literature published in Russia 1734-1917). Moskva, Izdatel'-
tvo "Nauka," 1965. 379 p. (Akademiia Nauk SSSR. Institut Narodov Azii.
sesoiuznaia Gosudarstvennaia Biblioteka Inostrannoi Literatury).   2388

    Systematic, partly annotated, guide to Russian literature on Japan, 1734-
917, with 7,897 entries. Section on geography and ethnography, p. 17-49 has
eadings for general works, physical and economic geography, travelers notes
nd impressions. About 350 references. Index of names.

KEVICH, Anatolii Iosifovich. Ocherki ekonomiki sovremennoi Iaponii (Studies
f the economy of contemporary Japan). Moskva, Izdatel'stvo "Nauka," 1972.
76 p. Bibliography, p. 367-375 (Akademiia Nauk SSSR. Institut Vostokovedeniia).
                                                                                      2389
    215 references.

AND, David L'vovich. Ostrov Khokkaido: fiziko-geograficheskoe opisanie (The
sland of Hokkaido: a physical-geographical description). Moskva-Leningrad,
zdatel'stvo Akademii Nauk SSSR, 1947. 147 p. Bibliography, p. 130-134
Akademiia Nauk SSSR. Institut Geografii. Nauchno-populiarnaia seriia).   2390

    120 references organized systematically, in Russian and other languages,
rom 1883 to 1944.

AUSTRALIA AND OCEANIA

AKHOVSKII, Kim Vladimirovich, ed. Novoe v izuchenii Avstralii i Okeanii (Recent
esearch on Australia and Oceania). Moskva, Izdatel'stvo "Nauka," 1972. 263 p.
ibliography, p. 237-262 by O. S. Larionova. References also at ends of articles.
Akademiia Nauk SSSR. Institut Vostokovedeniia).   2391

L'MAN, Nikolai Karlovich. Priroda Novoi Zelandii (Natural conditions in New
ealand). Moskva, Geografgiz, 1955. 128 p. Bibliography, p. 121-127.   2392
    144 references.

POLAR AREAS

ISKII, Vasilii Vasil'evich. Zhizn' cheloveka v Arktike i Antarktike. (Life of
an in the Arctic and Antarctic). Leningrad, Izdatel'stvo "Meditsina," Lenin-
radskoe Otdelenie, 1973. 199 p. Bibliography, p. 187-198.   2393

KOROTKEVICH, Evgenii Sergeevich. Poliarnye pustyni (Polar deserts). Leningrad. Gidrometeoizdat, 1972. 420 p. Bibliography, p. 388-410 (Arkticheskii i Antarkticheskii Nauchno-Issledovatel'skii Institut).   2

612 references arranged alphabetically, separately for Cyrillic with 375 entries and for Latin letters with 237 entries, on the physical geography of Polar regions.

KATALOG dannykh i publikatsii po Arktike i Antarktike, vypusk 2. ed. L. G. Fevraleva. Moskva, 1969. 166 p. (Mirovoi tsentr dannykh v SSSR. World Data Center in the USSR).   2

KATALOG dannykh i publikatsii po Arktike i Antarkitike (Catalogue of data and publications on the Arctic and Antarctic). Moskva, 1962. 209 p. (SSSR. Mirovoi Tsentr Dannykh B).   2

1,602 publications in Russian and foreign languages, separately for the Arctic and the Antarctic, each arranged systematically, for material of the International Geophysical Year.

The Arctic

(See also the Arctic and Subarctic of the RSFSR in Part V, Regional Geography: the Soviet Union, entries 2097-2100, and Arctic Climatology and Meteorology, entries 1195-1196).

GORBATSKII, Gerasim Vasil'evich. Severnaia poliarnaia oblast': obshchaia fiziko-geograficheskaia kharakteristika(The Northern polar region: general physical-geographic characteristics). Leningrad, Izdatel'stvo Leningradskogo Universiteta, 1964. 234 p. Bibliography, p. 223-233.   23

About 310 references in alphabetical order, about 160 Soviet and 150 in foreign languages.

AGRANAT, Grigorii Abramovich. Zarubezhnyi Sever: opyt osvoeniia. (The North outside the USSR: lines of development). Moskva, Izdatel'stvo "Nauka," 1970. 414 p. Bibliography, p. 403-411. (Akademiia Nauk SSSR. Mezhduvedomstvennaia Komissiia po Problemam Severa Soveta po Izucheniiu Proizvoditel'nykh Sil pri Gosplane SSSR).   23

About 240 references.

AGRANAT, Grigorii Abramovich. Zarubezhnyi Sever: ocherki prirody, istorii, naseleniia i ekonomiki raionov (Foreign North: studies in its natural conditions, history, population, and economy of its regions). Moskva, Izdatel'stvo Akademii Nauk SSSR, 1957. 319 p. Bibliography, p. 302-310. (Akademiia Nauk SSSR. Nauchno-Populiarnaia Seriia).   23

About 230 references in alphabetical order, about 70 Soviet and about 160 in foreign languages.

LAPPO, Sergei Dmitrievich. Spravochnaia knizhka poliarnika: kratkie svedeniia ob okeanografii, klimate, zhivotnom mire i naselenii Arktiki. (Handbook for polar explorers: brief information on oceanography, climate, animal life, and population of the Arctic).   24

BIBLIOGRAFIIA (Bibliography), Problemy Arktiki (Glavsevmorput'. Vsesoiuznyi Arkticheskii Institut). Leningrad, 1938, no. 1, p. 181-191; no. 2, p. 231-239; no. 3, p. 147-155; no. 4, p. 163-167. N. F. Ustinova.   24

About 750 entries on the study of the Arctic organized by sections, with headings in both Russian and English: general part, history of exploration, and biography; aviation and air transport; astronomy, geodesy, and cartography; geology, paleontology, geomorphology, and glaciology; hydrology and hydrography; geophysics; biology; economics and the study of man; Soviet construction in the Far North; reindeer breeding; fiction.

___ . ____ . 1939, no. 1, p. 117-120; no. 2, p. 124-127; no. 3, p. 98-100;
.o. 4, p. 95-99; no. 5, p. 117-120; no. 6, p. 109-111; no. 7-8, p. 131-133;
.o. 9, p. 109-112; no. 10-11, p. 156-160; no. 12, p. 105-107.
707 entries.

___ . ____ . 1946, no. 1, p. 106-108.

69 entries for the period 1934-1944 on the Arctic, in Russian and other
.anguages, organized systematically: general problems, history of exploration,
.iography, hydrology, hydrography, geophysics, meteorology, geology, paleontology,
.eomorphology, biology, economics, Soviet construction in the Far North, liter-
.ture, and Antarctica.

The Antarctic

.AS Antarktiki, tom 2. Leningrad, Gidrometeorologicheskoe Izdatel'stvo, 1969.
.98 p. (Sovetskaia Antarkticheskaia Ekspeditsiia).   2402

Text volume with references at ends of sections, on geographical descrip-
.ion, p. 32 (48 entries); history of exploration, p. 114-117 (207 entries);
.eronomy and physics of the earth, p. 200-205 (275 entries); geology and relief,
.. 304-310 (274 entries); climate, p. 362-363 (67 entries); glaciation, p. 398-
.00 (90 entries); water and ice of the Southern Ocean, p. 471-473 (139 entries);
.iology, p. 536-541 (261 entries); and landscapes, p. 579-580 (101 entries): or
.,462 references in all. See annotation also in entry 228.

.ROVIN, Leonid Ivanovich, and PETROV, V. N. Nauchnye stantsii v Antarktike, 1882-
.963. (Scientific stations in Antarctica, 1882-1963). Leningrad, Gidrometeoizdat,
.967. 282 p. Bibliography, p. 272-282. (Arkticheskii i Antarkticheskii Nauchno-
.ssledovatel'skii Institut. Sovetskaia Antarkticheskaia Ekspeditsiia).   2403

391 entries, organized by country.

.VETSKAIA antarkticheskaia ekspeditsiia, 1955- . Trudy. Leningrad. 1959- .
.itle varies: "Materialy." (Arkticheskii i Antarkticheskii Nauchno-Issledovatel'-
.skii Institut).   2403a

.KOV, N. I., and TARASOVA, Zh. A. Desiat' let sovetskikh issledovanii v Antarktike.
.Bibliograficheskii ukazatel' otechestvennoi literatury za 1956-1965 gg. (Ten
.years of Soviet research in the Antarctica; bibliographical guide to the national
.literature for 1956-1965). Leningrad, Arkticheskii i Antarkticheskii Nauchno-
.Issledovatel'skii Institut, 1968. 167 p.   2404

2,204 entries of books, chapters in collected works, and periodical articles
.on Soviet research on Antarctica 1956-1965. Author index.

.BLIOGRAFICHESKII ukazatel' otechestvennoi literatury po Antarktike za 1966-1970 gg.
.(Bibliography of native literature on Antarctica in the years 1966-1970). Compiled
.by M. S. Dmitriev. Leningrad, 1973. 139 p. (Glavnoe Upravlenie Gidrometeoro-
.logicheskoi Sluzhby pri Sovete Ministrov. Arkticheskii i Antarkticheskii Nauchno-
.Issledovatel'skii Institut).   2405

PINA, I. Ia. O publikatsiiakh sovetskikh rabot po Antarktike, 1956-mai 1960 g.
(On the publications of Soviet work on the Antarctic 1956-May 1960). In the book:
Antarktika, vypusk 1. Moskva, Izdatel'stvo Akademii Nauk SSSR, 1961, p. 61-85
(Akademiia Nauk SSSR. Mezhduvedomstvennaia Komissiia po izucheniiu Antarktiki). 2406

About 470 entries arranged systematically: geophysics, glaciology, hydrology,
roughness, sea ice, geology, meteorology, geography, biology, medical and other
questions.

2402 - 2406

STAT'I, opublikovannye v no. 1-40 "Informatsionnogo biulletenia Sovetskoi
    antarkticheskoi ekspeditsii," (Articles published in nos. 1-40 of the Informa-
    tion bulletin of the Soviet Antarctic expedition), Informatsionnyi biulleten'
    Sovetskoi antarkticheskoi ekspeditsii, 1959-1963.   2
    .... v no. 1-10, 1959, no. 10, p. 43-48 (about 170 entries)
    .... v no. 11-20, 1960, no. 20, p. 56-59 (about 130 entries)
    .... v no. 21-30, 1961, no. 30, p. 40-44 (about 150 entries)
    .... v no. 31-40, 1963, no. 40, p. 57-61 (about 140 entries)

    Alphabetical lists of articles on the study of natural conditions in
    Antarctica.

UKAZATEL' statei, pomeshchënnykh v "Problemakh Arktiki" za 1937-1945 gg. (Guide
    to articles in Problemy Arktiki in 1937-1945), Problemy Arktiki, 1945, no. 5-6,
    p. 159-174.   2

    Alphabetical index with about 600 entries.

SPISOK izdanii Arkticheskogo nauchno-issledovatel'skogo instituta, 1920-1939,
    Problemy Arktiki, 1940, no. 3, p. 157-184.   2

    Covers articles and continuing publications of the Arctic Institute: Trudy
    Severnoi nauchno-promyslovoi ekspeditsii, from 1925; Trudy Instituta po izucheniiu
    Severa, 1920-1931; Trudy Arkticheskogo instituta, 1933-1939; Materialy po
    izucheniiu Arktiki, 1931-1935; Arctica, 1933-1936; Sovetskoe olenevodstvo, 1936-
    1937; Problemy Arktiki, 1937-1939; Biulleten' Arkticheskogo instituta, 1931-1936.
    Author index.

KOTLIAKOV, Vladimir Mikhailovich. Snezhnyi pokrov Antarktidy i ego rol' v sovremennom
    oledenenii materika (Snow cover in Antarctica and its role in contemporary
    glaciation of the continent). Moskva, Izdatel'stvo Akademii Nauk SSSR, 1961.
    246 p. Bibliography, p. 219-232 (Akademiia Nauk SSSR. Mezhduvedomstvennyi Geo-
    fizicheskii Komitet. Mezhdunarodnyi Geofizicheskii God 1957-1958. IX Razdel
    Programmy MGG. Gliatsiologiia, no. 7). Rezul'taty issledovanii po programme
    Mezhdunarodnogo Geofizicheskogo Goda. Available in English: The snow cover of
    the Antarctic and its role in the present-day glaciation of the continent.
    Jerusalem, Israel Program for Scientific Translation. 1966. 264 p. IPST 1474.   2

    413 references in alphabetical order, 210 Soviet and 203 in foreign languages.

AVER'IANOV, Viacheslav Grigor'evich. Tsentral'naia Antarktida: fiziko-geografiche-
    skaia kharakteristika (Central Antarctica: physical-geographic characteristics).
    Ia. Ia. Gakkel', ed. Leningrad, "Morskoi Transport," 1963. 147 p. Bibliography,
    p. 137-143. (Arkticheskii i Antarkticheskii Nauchno-Issledovatel'skii Institut.
    Mezhdunarodnyi Geofizicheskii God. Sovetskaia Antarkticheskaia Ekspeditsiia,
    Trudy. Tom 30).   2

    229 references in alphabetical order, 160 Soviet and 69 in foreign
    languages.

MARKOV, Konstantin Konstantinovich, BARDIN, Vladimir Igor'evich, and ORLOV, A. I.
    Fiziko-geograficheskaia kharakteristika beregovoi polosy Vostochnoi Antarktidy
    (Physical-geographic characteristics of the coastal zone of Eastern Antarctica).
    O. A. Borshchevskii, ed. Moskva, Izdatel'stvo Moskovskogo Universiteta, 1962.
    148 p. Bibliography, p. 135-138. (Mezhdunarodnyi Geofizicheskii God 1957-1959.
    Moskovskii Gosudarstvennyi Universitet. Sovetskaia Antarkticheskaia Ekspeditsiia).
    Summary in English. Available in English: The geography of Antarctica. Jerusalem,
    Israel Program for Scientific Translation, 1970. 376 p. IPST 5697.   2

    About 110 references in alphabetical order, about half Soviet and half in
    foreign languages.

OVALOV, Georgii Vasil'evich. Gliatsio-geomorfologicheskaia kharakteristika
apadnoi chasti Vostochnoi Antarktidy (Glacial-geomorphological characteristics
f the western part of Eastern Antarctica). E. S. Korotkevich, ed. Leningrad,
idrometeoizdat, 1971. 122 p. Bibliography, p. 112-121 (Glavnoe Upravlenie
idrometeorologicheskoi Sluzhby pri Sovete Ministrov SSSR. Arkticheskii i
ntarkticheskii Nauchno-Issledovatel'skii Institut).   2413

   242 references.

BER, Georgii Mikhailovich. Osnovnye cherty klimata i pogody [Antarktiki]
Basic characteristics of the climate and weather of Antarctica). L. F. Rudovits,
d. Leningrad, Gidrometeoizdat, 1956. 148 p. Bibliography, p. 143-147
Gosudarstvennyi Okeanograficheskii Institut. Antarktika. Chast' 1).   2414

   148 references in alphabetical order, 56 Soviet and 92 in foreign languages.

ONOV, I. M. Oazisy Vostochnoi Antarktidy (Oases of Eastern Antarctica).
u. A. Kruchinin, ed. Leningrad, Gidrometeoizdat, 1971. 176 p. Bibliography,
. 161-171. (Glavnoe Upravlenie Gidrometeorologicheskoi Sluzhby pri Sovete
inistrov SSSR. Arkticheskii i Antarkticheskii Nauchno-Issledovatel'skii Institut).
                                                                             2415

                              DESERTS

ROV, Mikhail Platonovich. Pustyni zemnogo shara (Deserts of the world).
eningrad, Izdatel'stvo "Nauka," Leningradskoe Otdelenie, 1973. 435 p. Bib-
iography, p. 400-418.   In Russian with supplementary table of contents in
nglish, p. 435. (Akademiia Nauk Turkmenskoi SSR. Institut Pustyn'. Lenin-
radskii Gosudarstvennyi Universitet. Geograficheskii Fakul'tet).

   About 700 references organized by the three major parts of the book:
1) physical features of the individual deserts of the world, (2) types of
nvironmental conditions in deserts and the adaptation of plants and animals
o these conditions, and (3) natural resources of deserts and prospects for
heir utilization and development. A fully international bibliography covering the relevant literature in Russian, English, French, German, and Spanish.
Indexes of geographical names and plant names.   2416

PART VI. BIBLIOGRAPHIES AND REFERENCE WORKS

ON THE SOVIET UNION IN WESTERN LANGUAGES

A. GENERAL BIBLIOGRAPHICAL AIDS

BIBLIOGRAPHIES OF BIBLIOGRAPHIES

Geographic

CHEL, Karol. Guide to Russian reference books. Vol. II. History, auxiliary
istorical sciences, ethnography, and geography, ed. by J. S. G. Simmons.
tanford, California, Stanford University, 1964. 297 p. (Hoover Institution.
ibliographical series 18).   2417

    Section F. Geography, p. 189-227, entries F1-F300, provides the most com-
rehensive English-language retrospective bibliography of bibliographies on the
eography of Russia and the Soviet Union (300 items). Extensive annotations
nclude English translation of titles and full information on the scope, contents,
nd value of each of the bibliographies. Includes general and selective bibliog-
aphies, dissertations, institutions, periodicals, abstracts and indexes, region-
l physical geographies, special subject bibliographies, cartography, maps and
lases, administrative regions, encyclopedias, dictionaries and gazetteers, biog-
aphy, and handbooks.

General

CHEL, Karol. Guide to Russian reference books. Stanford University, California,
he Hoover Institution on War, Revolution, and Peace, 1962- .

    v. 1. General bibliographies and reference books, ed. by J. S. G. Simmons.
962. 92 p. (Hoover Institution Bibliographical series, v. 10).   2418

    379 entries organized systematically: bibliographies of bibliographies;
ussian national bibliographies; national bibliographies of the Soviet republics;
ibliographies of Russian publications published outside the USSR; bibliographies
f non-Russian publications relating to Russia; library catalogs; selective bib-
iographies; bibliographies of dissertations; bibliographies of rare and illustra-
ed books; bibliographies of catalogs of manuscripts; publications of the Academy
f Sciences of the USSR and of the republican and branch academies; bibliographies
f translations; bibliographies of congresses, conferences, and meetings; bibliog-
aphies of periodicals and newspapers; indexes and abstracts; general encyclopedias;
iographical dictionaries; dictionaries of anonyma and pseudonyms; dictionaries of
bbreviations; language dictionaries; and handbooks of Russia and the USSR. Index.

    v. 2. History, auxiliary historical sciences, ethnography, and geography,
d. by J. S. G. Simmons. 1964. 297 p. (Hoover Institution Bibliographical
eries, v. 18).   2419

    Covers history of the USSR (633 entries, B1-B633, p. 33-122); world history
35 entries, C1-C135, p. 123-146); auxiliary historical sciences (174 entries,
-D174, p. 147-170); ethnography (124 entries, E1-E124, p. 171-188); and geog-
aphy and geology (370 entries, F1-F370, p. 189-236). Index of authors, compilers,
onsoring institutions, titles, and subjects. See separate listing for section
 geography.

v. 5. Science, technology, and medicine, with the assistance of B. J. Pooler and Rudolf Lednicky. 1967. 384 p. (Hoover Institution Bibliographical series, v. 32). 2

833 entries on science and technology and 233 on medicine, organized systematically. Science and technology include general science, general technology, agriculture, astronomy, biological sciences, chemistry, earth science (including physical geography, p. 135), engineering, food industry, forestry and forest products, mathematics, and physics. Index by author or compiler, sponsoring institution, title, and subject.

v. 3, social sciences, religion, and philosophy; v. 4, humanities; and v. 6, cumulative index and supplementary material, not yet published.

SIMMONS, J. S. G. Russian bibliography library and archives: a selective list of bibliographical references for students of Russian history, literature, political, social and philosophical thought, theology and linguistics. Twickenham, Middlesex, England, Anthony C. Hall, 1973. 76 p. 2

696 entries organized systematically: general, general bibliographies and reference works, historical, literary, political and social thought, philosophy, theology, and linguistic bibliography. Appendices of comparative table of systems of Russian transliteration, notes on the Granat encyclopedia, and table of reprints. Index in four parts: authors (Cyrillic), authors (non-Cyrillic), titles of title-entry works (Cyrillic), and (4) titles of title-entry works (non-Cyrillic). Limited to bibliographies proper but includes encyclopedias with extensive bibliographies. Notes library location of copies, primarily in Oxford or London. Informed and careful selection of the best bibliographies in the fields covered.

BESTERMAN, Theodore. Russia, in his A World bibliography of bibliographies. 4th ed., second printing, Totowa, New Jersey, Rowman and Littlefield, 1971, v. 4, cols. 5481-5508. 2

279 bibliographies on Russia and the Soviet Union by the following classification: bibliographies and periodicals (11); manuscripts (2); general (63); cartography (13); foreign relations (23); history (104); law and government (28); official publications (1); social life (7); topography (14); Northern Russia (2); Russia in Asia (2); South Russia (3); and miscellaneous (6).

See also: Ukraine, v. 4, cols. 6242-6244 (20 entries); White Russia, v. 4, cols. 6535-6536 (4 entries); Estonia, v. 2, cols. 2062-2063 (9 entries); Latvia, v. 3, cols. 3430-3431 (7 entries); Lithuania, v. 3, cols. 3567-3568 (10 entries); Moldavia, v. 3, col. 3993 (1 entry); Georgia, v. 2, col. 2507 (1 entry); Armenia, v. 1, cols. 513-514 (9 entries); Azerbaijan, v. 1, col. 642 (2 entries); Kazakhstan, v. 2, col. 3318 (4 entries); Khirgiztan, v. 2, col. 3334 (2 entries); Uzbekistan, v. 4, col. 6366 (1 entry); Tadzhikistan, v. 4, col. 5990 (1 entry); Siberia, v. 4, cols. 5743-5744 (12 entries); Far Eastern Province, v. 2, col. 2129 (1 entry); Central Asia, v. 1, col. 576 (3 entries); Ural region, v. 5, col. 6355 (1 entry); Moscow, v. 3, cols. 4026-4027 (9 entries); Leningrad, v. 3, cols. 3490-3491 (11 entries).

CURRENT BIBLIOGRAPHIES

GEORGE, Pierre, avec la collaboration de G. Krichevsky et Mlles. M.-M. Birot et Cl. Rondeau. U. R. S. S., Bibliographie géographique internationale. 1891- Annual. Paris, Centre National de la Recherche Scientifique. 2

The best annual bibliography in a Western language on the geography of the Soviet Union. Includes all major topical fields of physical and economic geography and regional studies. The 1970 volume contains 427 entries in section B. VI, p. 691-779, organized as follows: generalities (10 entries); general phy-

ical geography (1); geology, morphology, soils (31); climate, hydrology, bio-
eography (20); human geography: population, cities, toponymy (22); economy in
eneral and plans (23); agriculture, animal husbandry, fishing, and irrigation
38); raw materials, energy, and industry (17); commerce, circulation, ports,
nd tourism (8); regional studies: Baltic republics (19); Belorussia, Ukraine,
nd Moldavia (25); European part of the R.S.F.S.R. (68); Republics of the Caucasus
33); Kazakhstan and Soviet Middle Asia (41); Asiatic part of the R.S.F.S.R.,
Western Siberia, Central Siberia, Eastern Siberia, and the Soviet Far East (71
ntries). For retrospective volumes see entry 2422.

UMENTATIO geographica: geographische Zeitschriften- und Serien-Literatur.
- (1966-). Quarterly. Bonn-Bad Godesberg, Bundesforschungsanstalt für
andeskunde und Raumordnung.   2424

RENT geographical publications: additions to the research catalogue of the
merican Geographical Society of New York. 1- (1938-). Monthly except July
nd August. New York, American Geographical Society.   2425

    Provides a current supplement to the published volumes of the Research
atalogue of the American Geographical Society [2427].

                      RETROSPECTIVE BIBLIOGRAPHIES

                             GEOGRAPHIC

               Comprehensive Geographic Bibliographies

S. Library of Congress. Reference Department. Soviet geography: a bibliography,
d. by Nicholas R. Rodionoff. Washington, D.C., Library of Congress, 1951. 2 v.
68 p. Reprinted: New York, Greenwood Press, 1969.   2426

    Extensive bibliography of geography of Russia and the Soviet Union up to
950, covering geography as a science, general geography, exploration, historical
geography, physical geography, economic geography, political and military geog-
aphy, atlases and cartography, bibliography and biobibliography, and individual
egions of the Soviet Union in 4,421 entries. Indexes for authors and subjects.
ow somewhat dated, but still the most extensive bibliography in a Western lan-
uage of substantive works on Soviet geography.

EARCH catalogue of the American Geographical Society. Boston, Massachusetts,
. K. Hall and Co., 1962. First supplement [1962-1971]. Boston, Massachusetts,
. K. Hall and Co., 1972.   2427

    USSR as a whole, v. 9, p. 6607-6687, 1,673 entries, 1923-1961 on the geography
f the Soviet Union as a whole. 1st supplement, v. 2, p. 45-66, about 650 entries
or 1962-1971.

    European USSR, v. 9, p. 6687-6803, 2,429 entries, 1923-1961. 1st supplement,
. 2, p. 66-76, about 300 entries for 1962-1971.

    The Baltic States, v. 9, p. 6569-6572, 6583. 60 entries, 1923-1961. 1st
upplement, v. 2, p. 40-41, 14 entries for 1962-1971. Estonia, v. 9, p. 6572-
582, 229 entries; 1st supplement, v. 2, p. 41-43, 57 entries. Latvia, v. 9,
. 6583-6589, 137 entries; 1st supplement, v. 2, p. 43, 8 entries. Lithuania,
. 10, p. 6813-6819, 138 entries; 1st supplement, v. 2, p. 92-94, 37 entries.

    Armenia, v. 12, p. 8517-8522, 90 entries.

    Asiatic USSR, v. 12, p. 8693-8696. 64 entries; 1st supplement, v. 2, p. 76,
0 entries. Siberia and the Far East, v. 12, p. 8749-8820, 1,482 entries; 1st
upplement, v. 2, p. 83-91, about 250 entries. Sakhalin, v. 12, p. 8876-8877,
9 entries. Kuril Islands, v. 12, p. 8946-8947, 19 entries. Soviet Middle Asia,
. 12, p. 8697-8745, 1,004 entries; 1st supplement, v. 2, p. 76-83, about 220
ntries.

    About 8900 entries for the Soviet Union 1923-1971.

                                                        2424 - 2427

GEOGRAPHISCHES Jahrbuch, Gotha/Leipzig, VEB Hermann Haack Geographisch-Kartographis-
sche Anstalt. 1866-1967. The bibliographies on Russia and the Soviet Union are
as follows: 2|

European USSR

Schultz, Arved. "Europäisches Russland (1929-1936)," Geographisches Jahrbuch, v, 52
(1937), p. 75-248.   2|

   2,838 entries on geography of the European part of the Soviet Union, 1929-1936.
Organized both by systematic fields and by regions. Textual discussions of principal works for each field and region. Author index. A rich bibliography but
since Russian titles are translated into German, including titles of monographs
and even names of serials, identification and location of the references is sometimes difficult.

Friederichsen, Max. "Europäisches Russland (1918-1928) (ohne Kaukasus und Trans-
kaukasien)," Geographisches Jahrbuch, v. 43 (1928), p. 82-110.   2|

   Discussion of the literature on the geography of the European part of the
Soviet Union (excluding the Caucasus and the Transcaucasus) for the period 1918-
1928 with 388 bibliographical footnotes.

Friederichsen, Max. "Europäisches Russland (1912-1918)," Geographisches Jahrbuch,
v. 38 (1915-1918), p. 299-362; 1906-1911, v. 35 (1912), p. 455-475; 1894-1905,
v. 29 (1906), p. 148-208.   2|

   Discussion of the literature 1894-1918 on the geography of the European
Russia, with 1,437 bibliographical footnotes.

Anuchin, Dmitrii Nikolaevich. "Europäisches Russland," Geographisches Jahrbuch,
v. 17 (1894), 238-260.   2|

   Discussion of the literature on the geography of European Russia up to 1894,
with 75 bibliographical footnotes.

Giere, Werner. "Die Ostbaltischen Staaten, Litauen, Lettland, Estland (1928-1936),"
Geographisches Jahrbuch, v. 51 (1936), p. 358-418.   2|

   767 entries for 1928-1936 on the geography of the Baltic states, Lithuania,
Latvia, and Estonia, with 128 entries on the Baltic as a whole, 149 on Lithuania,
191 on Latvia, 286 on Estonia, and 13 on maps. Textual discussion on major works.
Author index.

Friederichsen, Max. "Die Randstaaten (Litauen, Lettland, Estland)," 1918-1927,
Geographisches Jahrbuch, v. 43 (1928), p. 66-81.   2|

   Discussion of the literature (1918-1927) on the geography of the Baltic States,
Lithuania, Latvia, and Estonia, with 164 bibliographical footnotes.

Asiatic USSR

Leimbach, Werner. "Nordasien, Westturkistan und Innerasien (1926-1937)," Geographisches Jahrbuch, v. 53, part 2 (1938), p. 437-565, v. 54, part 1 (1939),
p. 303-352, and v. 54, part 2 (1939), p. 555-596.   2|

   2,104 entries on the geography of the Asiatic part of the Soviet Union:
1,307 on Siberia and the Soviet Far East (in vol. 53) and 797 on Kakakhstan and
Soviet Middle Asia (in vol. 54). Within the two major regional divisions the
entries are arranged by topical categories. Author indexes. Titles given both in transliterated
Russian and in German translation. Pre-World War II library locations and call
numbers in Germany noted.

:ler, Paul. "Russisch-Asien and Zentralasien (1912-1925)," Geographisches
ahrbuch, v. 41 (1926), p. 309-360.   2436

Discussion of the literature 1912-1925 on the geography of Russian Asia and
entral Asia with 435 bibliographical footnotes.

:derichsen, Max. "Russisch-Asien 1905-1914," Geographisches Jahrbuch, v. 37
.914), p. 285-314; 1898-1904, v. 27 (1904), p. 376-425.   2437

Discussion of the literature 1898-1914 on the geography of Asiatic Russia
excluding the Trans-Caucasus), with 672 bibliographical footnotes.

:hin, Dmitrii Nikolaevich. "Geographische Erforschungen in Russisch-Asien 1896-
397," Geographisches Jahrbuch, v. 20 (1897), p. 409-424; 1893-1895, v. 18 (1895),
. 316-325; 1891-1892, v. 16 (1893), p. 398-421.   2438

Discussion of the literature on geographical expeditions to Asiatic Russia
391-1897, with 468 bibliographical footnotes.

:ies, Hans. "Geographische Erforschungen in Asien 1888-1890: Turan, Inner-Asien,
ord-Asien," Geographisches Jahrbuch, v. 14 (1890-1891), p. 325-340; 1885-1887,
. 12 (1888), p. 157-163, 176-177.   2439

Discussion of geographical exploration in Turan, the Pamirs, Inner Asia,
.beria, North Asia, and the Russian Far East 1885-1891, with bibliographical
ootnotes.

LIOGRAPHIE géographique internationale, Paris. 1891- . Annual. "Empire russe,"
3ussie," or "Russie d'Europe," v. 1-33 (1891-1923) and "Asia russe," v. 1-7
1891-1897), "Asie russe et Mantchourie," v. 8-13 (1898-1903), "Asie russe,"
. 14-29 (1904-1919), "Sibérie, Turkestan russe," v. 30-33 (1920-1923) and
Jaucasie, Géorgie," v. 30-33 (1920-1923); "U.R.S.S. (Europe)," v. 34-54 (1924-
944), and "U.R.S.S. (Sibérie. Turkestan russe)," v. 34-35 (1924-1925), and
J.R.S.S. (Caucasie. Géorgie)," v. 35 (1925), "U.R.S.S. (Sibérie, Pamir, Turke-
:an russe)," v. 36-54 (1926-1944), "U.R.S.S. (Caucasie, Arménie, Azerbeidjan,
3orgie)," v. 36-54 (1926-1944), "U.R.S.S.," as a whole, including both European
nd Asiatic USSR, v. 55- (1945-), and continuation.   2440

See also separate "Esthonie, Lettonie, Lituanie," v. 30-34 (1920-1924),
3stonie, Latvie, Lituanie," v. 35-54 (1925-1944).

Selective Guides to the Geographic Literature

?IS, Chauncy D. The land, chapter 2 in Basic Russian publications: an annotated
ibliography on Russia and the Soviet Union, ed. by Paul L. Horecky. Chicago and
ondon, University of Chicago Press, 1962, p. 25-48, entries 82-202.   2441

121 entries, highly selective and annotated, covering the key works in
ussian or other languages of the USSR on the geography of the Soviet Union
hrough 1961, organized systematically and regionally: bibliography, reference
ids, atlases, serials, general geography, physical geography, economic geog-
aphy, physical and economic regionalization, regions of the USSR, and history
f geography and geographical exploration. A somewhat expanded (193 entries)
ersion is the author's Geography, resources, and natural conditions in the Soviet
nion: an annotated bibliography of selected basic books in Russian. Chicago,
llinois, Department of Geography, University of Chicago, 1962. 45 p. mimeo-
raphed, o.p. with call numbers in the University of Chicago Library.

HARRIS, Chauncy D. The land, chapter 3 in Russia and the Soviet Union: a bibliographic guide to Western-language publications, ed. by Paul L. Horecky. Chicago
and London, University of Chicago Press, 1965, p. 46-54, entries 190-238.   2

 Annotated listing of 49 principal monographs or reference works on the Soviet
Union in Western languages, mainly English, French, or German, published up to
1964. Organized systematically: general reference aids, atlases, general geography, physical geography, economic geography, regions, exploration. Revisions
are prepared from time to time, the latest being the author's Geography of the
Soviet Union: an annotated bibliography of major books in Western languages.
Chicago, Department of Geography, University of Chicago, 1971. 15 p. mimeographed,
95 entries of works published up to 1970.

KUNSKÝ, Josef. Sovětský svaz (Soviet Union), in his book: Všeobecný zeměpis, I.
Úvod do studia: bibliografie (General geography, I. Introduction to the discipline:
bibliography). Praha, Nakladatelství Československé věd, 1960, p. 228-235.   2

 155 numbered entries (2429a-2584) on the geography of the Soviet Union,
mostly in Russian but including also major works in other languages as well.
Arranged by systematic fields for the country as a whole and by regions.

Bibliographies Contained in Geographies of the USSR as a Whole

COLE, John P., and GERMAN, F. C. A geography of the U.S.S.R.: the background to a
planned economy. 2nd ed. London, Butterworths, 1970. 324 p. Reference and
bibliographies at the ends of chapters, and a general bibliography, p. 316-317.
(1st ed., 1961. 290 p.).   2

 About 250 references in all.

GEORGE, Pierre. L'U.R.S.S. 2nd ed. Paris, Presses Universitaires de France, 1962.
497 p. Bibliographies, p. 5-6, 73-76, 128-132, 184-185, 203-209, 229-230, 265,
280-282, 438, 464, and 487.   2

 Several hundred references grouped by chapters or sections: general bibliographies, relief of plains and plateaus of Russia and Siberia, relief of mountains,
climatology and hydrology, seas, zones and types of vegetation, population, the
economy, cities, European regions, regions in Asia.

GREGORY, James S. Russian land Soviet people: a geographical approach to the
U.S.S.R. London, George G. Harrap, 1968. 947 p. Bibliography, p. 911-917.   2

 About 200 references arranged by major topics, separately for works in
English and in Russian.

HOOSON, David J. M. The Soviet Union [people and regions]. London, University of
London Press, Belmont, California, Wadsworth Publishing Co., 1966. 376 p. Bibliography, p. 356-363.   2

 111 references, arranged by chapters.

HOWE, G. Melvyn. The Soviet Union. London, MacDonald and Evans, 1968. 403 p.
Bibliography, p. 381-384.   2

 83 references in English.

LEIMBACH, Werner. Die Sowjetunion: Natur, Volk und Wirtschaft. Stuttgart,
Frankh'sche Verlagshandlung, 1950. 526 p. Bibliography, p. 481-506.   2

 200 references, arranged systematically.

LYDOLPH, Paul E. Geography of the USSR. 2nd ed. New York, London, John Wiley
and Sons, 1970. 683 p. (1st ed., 1964. 451 p.).   2

 604 references distributed in reading lists throughout the book: general
references, p. viii-xiii (101); historical and physical background, p. 26-27 (24);

egions, p. 31-32, 54, 65, 88, 122-123, 143-144, 161-162, 182-183, 214-215, 267-
8, 296, and 331 (215 references); population and economic geography, p. 394-
7, 458-459, 547-550, 589, and 649-651 (264 references).

OR, Roy E. H. Geography of the U.S.S.R. London, Macmillan, New York,
. Martin's Press, 1964. 403 p. Bibliography, p. 377-381.   2450

   141 references arranged by categories of material.

MEYER, Karl. Landeskunde der Sowjetunion. Frankfurt am Main, Bernard und
aefe Verlag für Wehrwesen, 1968. 218 p. Bibliography, p. 213-218.   2451

   166 references in alphabetical order.

GENERAL GUIDES

To Works Mainly in Russian

ECKY, Paul L., ed. Basic Russian publications: an annotated bibliography on
ussia and the Soviet Union. Chicago and London, University of Chicago Press,
962. 313 p.   2452

   1,396 entries on general reference aids and bibliographies, the land, people,
istory, state, economic and social structure, and intellectual life of the Soviet
nion, selected and annotated by 32 specialists on the country. An excellent over-
ll introduction to bibliographies, reference works, and substantive publications
n Russian and other languages of the USSR, valuable alike for its selectivity and
or the critical notations. Index of names of authors, compilers, editors, trans-
ators, and sponsoring organizations; titles of publications; and principal sub-
ect headings.

. Library of Congress. Monthly Index of Russian Accessions. v. 1-22, no. 3
pril 1948-March 1969). Monthly. Washington, D. C.: Government Printing Office.
.1-10, no. 9, 1948-1957 as Monthly List of Russian Accessions).   2453

   A union list of Russian-language books and serials received by the Library
 Congress and a group of cooperating libraries. Part A, monographs; part B,
riodicals with tables of contents; part C, subject index to articles in part B;
rt D, monographic works in Western languages on the Soviet Union or by Soviet
thors.

ARIDZÉ, David A. Russie-U.R.S.S. in L.-N. Malclès, Les sources du travail bib-
ographique, Genève, Libraire Droz, 1950-1958, réimpression 1965. v. 1, p. 302-
6, and Russie-R.S.F.S.R., v. 2, p. 684-749.   2454

   The first part is devoted mainly to general national bibliographies of the
SR, library catalogues, periodicals, encyclopedias, biographical directories,
d non-Russian republics in the USSR. The second part is devoted to language,
terature, and history of Russia proper.

To Works Mainly in Western Languages

ECKY, Paul L., ed. Russia and the Soviet Union: a bibliographic guide to
estern-language publications. Chicago and London, University of Chicago Press,
965. 473 p.   2455

   1,960 entries on the Soviet Union in Western languages, mainly English,
rench, and German, through 1964, selected and annotated by 31 specialists on
he country, organized systematically: general reference aids and bibliographies;
eneral and descriptive works; the land; the people (ethnic and demographic
eatures); the nations (civilization and politics); history; the state; the

economic and social structure; and intellectual and cultural life. Excellent inventory and evaluation of the major Western publications on the Soviet Union. Index of authors, compilers, editors, translators, and sponsoring organizations; titles of publications; and principal subject headings.

BYRNES, Robert F. and BACKOR, Joseph. Russia, in A select bibliography: Asia, Africa, Eastern Europe, Latin America, ed. by Phillips Talbot and Nelda S. Freeman. New York, American Universities Field Staff, Inc., 1960, p. 257-324.

    1,044 references (entries 4501-5544) on the Soviet Union, under the following principal headings: journals, land and peoples, general, history, Russia in world politics, government, law, economy, international communist movement, education and science, religion, philosophy, literature, and other arts. Includes the principal studies in English on the society and civilization of the Soviet Union, including the geography of the country.

BYRNES, Robert F. Russia in A select bibliography: Asia, Africa, Eastern Europe, Latin America. Cumulative supplement 1961-1971. New York, N. Y., and Hanover, New Hampshire, American Universities Field Staff, Inc., 1973, p. 146-184.

    633 references (entries 5545-6177) on the Soviet Union, 1961-1971, under the following headings: reference aids, general economics, education and science, history and description, land and peoples, literature, other arts, politics and government, philosophy and religion, Russia in world politics, military, and international communist movement. Valuable list of substantive works in English in the social sciences and humanities on the Soviet Union, including geographical works.

AMERICAN bibliography of Slavic and Eastern European studies for 1956. Edited by J. T. Shaw. Bloomington, Indiana, 1957. 89 p. (Indiana University publications. Slavic and East European series, v. 9). 807 entries. This volume, unlike later years, covers only language, literature, folklore and pedagogy.

____, 1957. 1958. 103 p. (____, v. 10). 1,363 entries.

____, 1958. 1959. 112 p. (____, v. 18). 1,380 entries.

____, 1959. Edited by J. T. Shaw and David Djaparidze. 1960. 134 p. (____, v. 21). 1,781 entries.

AMERICAN bibliography of Russian and East European studies for 1960. Edited by J. T. Shaw, Albert C. Todd, and Stephen Viederman. Bloomington, Indiana, 1962. 124 p. (Indiana University publications. Russian and East European series, v. 26). 1,615 entries.

____, 1961. Edited by Albert C. Todd and Stephen Viederman. 1963. 138 p. (____, v. 27). 1,863 entries.

____, 1962. Edited by Albert C. Todd and Stephen Viederman. Assisted by Donald Rowney. 1964. 173 p. (____, v. 29). 2,277 entries.

____, 1963. Edited by Fritz T. Epstein, Albert C. Todd, and Stephen Viederman. Assisted by Cynthia H. Whittaker. 1966. 116 p. (____, v. 32). 2,291 entries.

____, 1964. Fritz T. Epstein, editor. Cynthia H. Whittaker, associate editor. 1966. 119 p. (____, v. 34). 2,260 entries.

____, 1965. Edited by Fritz T. Epstein assisted by James Cairns Miller. 1968. 148 p. (____, v. 37), 2,369 entries.

____, 1966. Fritz T. Epstein, editor. Penelope H. Carson and Michael E. Shaw, associate editors. 1972. 148 p. (____, v. 40). 1,852 entries.

AMERICAN bibliography of Slavic and East European studies, 1967. Kenneth E. Naylor, editor. Craig N. Packard and Zorianna M. Paschyn, assistant editors. Assisted by Olga C. Shopay. 321 p. Columbus, Ohio, Ohio State University Press, 1972. 2,417 entries.

2456 - 2460

_. 1968 and 1969. Kenneth E. Naylor, editor. Timothy A. Duket and Timothy L.
nz, assistant editors. Assisted by Rosemary Stefanka. Columbus, Ohio, American
sociation for the Advancement of Slavic Studies, 1974. 173 p. 4,781 entries.

_. 1970-1972. James P. Scanlon, editor. Jeanne F. Eason, associate editor.
lumbus, Ohio, American Association for the Advancement of Slavic Studies. In
ess. About 8,300 entries. 3,000 book reviews.

   English-language publications regardless of country of origin, articles and
oks published in the United States or Canada, or by American abroad. Arranged
 subject fields. Through 1960 did not include British contributions.

LTHEISS, Thomas, ed. Russian studies 1941-1958. A cumulation of the annual
bliographies from the Russian Review. Ann Arbor, Michigan. The Pierian Press,
72. 395 p.   2461

   Arranged by years. Author and main entry index. Subject index with listings
r geography on p. 354-355.

ER, Harold H., ed. American research on Russia. Bloomington, Indiana, Indiana
iversity Press, 1959. 240 p. Index.   2462

   Bibliographic notes, p. 187-232. Geography, by W. A. Douglas Jackson, chapter
p. 113-122, 218-221.

EUR, Walter Z., and LABEDZ, Leopold, eds. The state of Soviet studies.
mbridge, Massachusetts, Massachusetts Institute of Technology Press, 1965.
7 p. Index.   2463

   Essays which originally appeared in Survey: a journal of Soviet and East
ropean studies, January and April 1965. Bibliographical footnotes.

RSON, Philip. Books on Soviet Russia, 1917-1942: a bibliography and guide to
ading. London, Methuen, 1943. 354 p.   2464

   About 2,500 books and pamphlets on the USSR published in Great Britain,
bruary 1917-June 1942. Soviet geography and exploration, p. 275-291,is organized
 topics: general works, European Russia, Asiatic Russia in general, Western
beria, the exploration and development of the Arctic regions, Eastern Siberia,
ssian Turkestan, Chinese Turkestan, and Outer Mongolia.

RSON, Philip. Books and pamphlets on Russia, 1942-45, Slavonic and East European
view, v. 24, no. 63 (January 1946), p. 133-147.   2465

_. _____ 1946-1947 [July 1945-December 1946], v. 25, no. 65 (April 1947), p.
8-517

_. _____ 1947, v. 26, no. 67 (April 1948), p. 512-518

_. _____ 1948, v. 27, no. 69 (May 1949), p. 556-562

_. _____ 1949, v. 28, no. 71 (April 1950), p. 486-492

_. _____ 1950, v. 29 (1950-1951), p. 550-557

   Books and pamphlets on the USSR published in Great Britain, from July 1942
rough 1950.

ET, East European, and Slavonic studies in Britain, 1971: a bibliography.
ublished by and distributed only to subscribers to:) ABSEES: Soviet and East
ropean abstract series, Glasgow, v. 3, no. 1 (35) (July 1972).   2466

_ 1972...v. 4, no. 2 (October 1973).

MARCOU, Lilly. L'Union soviétique. Paris, Armand Colin, 1971. 149 p. (Fondation Nationale des Sciences Politiques. Bibliographies françaises de sciences sociales Guides de recherches. 2).

    391 entries organized: 1. By sources: sources published in the Soviet Union; sources and collections of documents published in the West; memoirs and evidence; reference works; periodicals; specialized institutions. 2. State of research: historical studies; ideology and society; political institutions and life; the republics and the nationality question; international relations; cultural life; the economy. Author index. Index of periodicals cited.

L'U.R.S.S. dans les publications de la Documentation Française: Bibliographie 1945-1965. Paris, La Documentation Française, 1967. 172 p.

LHÉRITIER, Andrée. Bibliographie des travaux parus en France concernant la Russie et l'U.R.S.S. (Année 1962), Cahiers du monde russe et soviétique, v. 4, no. 1-2 (1963), p. 150-200.

    974 entries for publication in 1962, mainly periodical articles, arranged systematically.

_____. 1963, v. 5, no. 3 (1964), p. 329-405. 1,481 entries. Geology-geography, entries 90-113, p. 333-334.

_____. 1964, v. 6, no. 4 (1965), p. 586-657. 1,349 entries. 93-105, p. 591-592.

_____. 1965, v. 8, no. 1 (1967), p. 124-181, 1,112 entries.

_____. 1966, v. 9, no. 1 (1968), p. 70-137, 1,393 entries.

_____. 1967, v. 10, no. 2 (1969), p. 280-341, 1,174 entries.

_____. 1968, v. 11, no. 0 (1970), p. 000-000, 0,000 entries.

_____. 1969, v. 12, no. 3 (1971), p. 347-383, 656 entries.

AYMARD, Marguerite, and SEYDOUX, Marianne. Travaux et publications parus en français en 1970 sur la Russie et l'U.R.S.S. Domaine des sciences sociales, Cahiers du monde russe et soviétique, v. 13, no. 2 (1972), p. 286-315.

    Includes a list of accepted dissertations for 1970.

MEHNERT, Klaus. Die Sovet-Union 1917-1932. Systematische, mit Kommentaren versehen Bibliographie der 1917-1932 in deutscher Sprache ausserhalb der Sovet-Union veröffentlichten 1900 wichtigsten Bücher und Aufsätze über den Bolschewismus und die Sovet-Union. Königsberg, Berlin, Ost-Europa-Verlag, 1933. 186 p. (Deutsche Gesellschaft zum Studium Osteuropas).

    1900 entries organized in 10 major chapters, each subdivided into subject categories. Mainly German-language publications but one chapter devoted to English publications and another to French. Author index.

BEZER, Constance, A., ed. Russian and Soviet studies: a handbook. Columbus, Ohio, American Association for the Advancement of Slavic Studies, 1973. 219 p. (Columbia University. Russian Institute).

    Orientation by disciplines (geography by Robert A. Lewis, p. 38-42), librari and archives, research and study centers, services and sources, selected reference aids, and chronological list of Russian general bibliographies. Contains useful information but uneven in coverage and quality.

NERHOOD, Harry W., comp. To Russia and return: an annotated bibliography of travelers' English-language accounts of Russia from the ninth century to the present. Columbus, Ohio, Ohio State University Press, 1968. 367 p.

    1,473 accounts in English of travels to Muscovy, Tsarist Russia, and the Soviet Union.

UNG, Friedrich von. Kritisch-literärische Übersicht der Reisenden in Russland
s 1700, deren Berichte bekannt sind. St. Peterburg, Leipzig, 1846. 2 v.
Ɔ p., 430 p. Facsimile reprint, Amsterdam, 1960.   2474

266 references. Russian translation, Kritiko-literaturnoe obozrenie...
anslated by Aleksandr Klevanov. Moskva, 1864. Chast' 1. 264 p. 400 ref-
ences.

LIBRARY CATALOGUES

Specialized Slavic

, Library of Congress. Cyrillic union catalog. New York, Readex Microprint
rporation, 1963. 1,244 cards in 7 boxes.   2475

708,000 cards representing 178,226 titles for entries in Russian, Ukrainian,
Lorussian, Bulgarian, and Serbian reported to the Library of Congress by 185
jor research libraries of the United States and Canada up to March, 1956,
ranged in three parts: (1) author and added entry; (2) title with listing of
ɔrary locations, and (3) subject.

Box 1, author cards 1-199, A-Mel'gunov; box 2, author cards 200-399, Mel'gunov-
zykin; box 3, titles, cards 1-160, A. A. Aliab'ev-Opyt; box 4, titles, cards
L-321, Opyt (cont.)-Zyriansko; box 5, subjects, cards 1-175 Aaron-Historians,
ssia, box 6, subjects, cards 176-350, Historians, Russia (cont.)- Russia, his-
ry, Catherine II; box 7, subjects, cards 351-524, Russia, history, Catherine II
ɔnt.)-Zyrianov.

Box 5, cards 160-161 include geographers, geographical distribution, geo-
aphical societies, geography (bibliography, collected works, dictionaries,
story, periodicals, societies, study and teaching, textbooks), commercial geog-
ɔhy, economic geography, and historical geography.

YORK Public Library. Reference Department. Dictionary catalogue of the
avonic collection. Boston, Massachusetts, G. K. Hall, 1959. 26 v. 24,228 p.   2476

About 550,000 entries in Slavic or Baltic languages or on Slavic or Baltic
eas, representing 120,000 volumes. Photographic reproduction of card catalogue
tries by author and subject, in a single alphabetical order according to the
tin alphabet, in the transliteration scheme of the New York Public Library.
e proportion in various languages are Russian 65%, Polish 13%, Czech and Slovak
%, Serbo-Croation 4%, Ukrainian 3%, Latvian 3%, Bulgarian 2.5%, Lithuanian 1%,
ɔvenian 0.5%, Belorussian 0.5%, and others, including non-Slavic languages of
e USSR 1.5%.

Geography, v. 8, p. 7105-7133. Geography cross-references, v. 8, p. 7109.

A second edition in 44 volumes with 724,000 entries for 200,000 volumes is
press and has an expected publication date of 1974.

LOTHÈQUE Nationale. Catalogue général des livres imprimés: auteurs, collectiv-
és-auteurs, anonymes 1960-1969. Série 2. Caractères non latins. Tome 2.
ractères cyrilliques. Russe A-M. Paris, 1973. 1,037 p.   2477

Works in Russian and other languages written in Cyrillic received by the
ɔliothèque Nationale during the decade 1960-1969. With the publication of the
cond volume it will supersede Bibliothèque Nationale. Catalogue général des
vres imprimés: auteurs, collectivités-auteurs, anonymes 1960-1964. Tome 11.
vrages imprimés en caractères cyrilliques: russe, ukrainien, biélo russe,
Lgare. Paris, 1967. 940 p.

General

LIBRARY of Congress catalogs: National union catalog (Title 1956-1973: National
union catalog: a cumulative author list). 1956- . monthly, with quarterly,
annual, and quinquennial cumulations.

The 5-year cumulations are (a) National union catalog: a cumulative authors
list representing Library of Congress printed cards and titles reported by other
American libraries. Compiled by the Library of Congress with the cooperation of
the Committee on Resources of American Libraries of the American Library Association, 1958-1962. New York, Rowman and Littlefield, 1963. 54 v. This cumulation
includes titles from January 1, 1956. (b) National union catalog: a cumulative
authors list representing Library of Congress printed cards and titles reported
by other American libraries. Compiled, edited, and approved by the Library of
Congress with the cooperation of the Resources Committee of the Resources and
Technical Services Division, American Library Association, and published by J. W.
Edwards, Publisher, Inc., pursuant to an exclusive contract with the Library of
Congress, 1963-1967. Ann Arbor, Michigan, J. W. Edwards 1969. 72 v. (c) _____ .
_____ , 1968-1972. Ann Arbor, Michigan, J. W. Edwards, 1973-1974. 128 v.
(v. 1-104, Author list; 105-119, Register of additional locations; 5 v., Music
and phonorecords; 4 v. Motion pictures and filmstrips).

Preceded by U. S. Library of Congress. A catalog of books represented by
the Library of Congress printed cards issued to July 31, 1942. Ann Arbor, Michigan,
Edwards Brothers, 1942-1946. 167 v.; Supplement, cards issued August 1, 1942-
December 31, 1947. Ann Arbor, Michigan, J. W. Edwards, 1948. 42 v.; Library of
Congress author catalog, a cumulative list of works represented by the Library of
Congress cards, 1948-1952. Ann Arbor, Michigan, J. W. Edwards, 1953. 24 v.;
National union catalog: a cumulative author list representing Library of Congress
cards and titles reported by other American libraries, 1953-1957. Ann Arbor,
Michigan, J. W. Edwards, 1958. 28 v.

NATIONAL union catalog: pre-1956 imprints. A cumulative author list representing
Library of Congress printed cards and titles reported by other American libraries.
Compiled and edited with the cooperation of the Library of Congress and the
National Union Catalog Subcommittee of the Resources Committee of the Resources
and Technical Services Division, American Library Association. London, Mansell
Information/Publishing Ltd, v. 1- (1968-). In progress. To be about 610 volumes
(v. 1-344, 1968-1974, A-Ludwig).

BRITISH Museum. General catalogue of printed books, photolithographic edition to
1955. London, British Museum, 1959-1966. 263 v.

_____ . General catalogue of printed books. Ten-year supplement, 1956-1965. 1968.
50 v.

_____ . _____ . Five-year supplement, 1966-1970. 1971-1972. 26 v.

SERIALS

Inventories

HARRIS, Chauncy D., and FELLMANN, Jerome D. International list of geographical
serials, 2nd ed., Chicago, Illinois, University of Chicago, Department of
Geography, Research paper 138, 1971. 267 p. Union of Soviet Socialist Republics,
p. 149-190.

336 numbered entries (1684-2019) for geographical serials published in Russia
or the Soviet Union.

2477a - 2478

S, Rudolf, comp. <u>Half a century of Soviet serials, 1917-1968: a bibliography</u>
<u>d a union list of serials published in the USSR</u>. Washington, D. C., Government
inting Office, 1968. 2 v. 1,661 p. (Library of Congress. Reference Depart-
nt).   2479

    29,761 numbered serials from all fields, in alphabetical order, with listings
holdings of libraries in the United States and Canada. Invaluable for checking
titles and location of copies.

ORY, Winifred. <u>List of the serial publications of foreign governments 1815-1931</u>.
w York, H. W. Wilson Co., 1932. 720 p. (For the American Council of Learned
cieties, American Library Association, National Research Council. Committee:
T. Gerould, H. M. Lydenberg, H. H. B. Meyer).   2479a

    Russia, p. 577-716, includes about 3,500 titles. Compiled by Vladimir
ovski with the collaboration of George Novossiltzeff, of the Library of Congress.
jor subdivisions: (1) The Russian Empire, p. 580-630 (including Grand Duchy of
nland, p. 628-629, and the Kingdom of Poland, p. 629-630; (2) Russia under the
ovisional Government, February-October, 1917; (3) The Union of Soviet Socialist
publics (U.S.S.R.), p. 632-675 (including an appendix: Communist Party, p. 674-
5); (4) The Russian Socialist Federal Soviet Republic (R.S.F.S.R.), p. 675-693;
) Ukrainian Socialist Soviet Republic, p. 693-700; (6) White-Russian Socialist
viet Republic, p. 701-703; (7) The Transcaucasian Socialist Federal Republic;
e Socialist Republic of Armenia; the Socialist Republic of Georgia; and the
erbaijan Socialist Soviet Republic, p. 704-705; (8) Autonomous Turkestan Social-
t Soviet Republic, p. 705; Uzbek Socialist Soviet Republic, p. 706-707; Turkoman
cialist Soviet Republic, p. 707; The Tadjik Socialist Soviet Republic, p. 707.
cludes key to abbreviations, in Russian, p. 710-712 and in English, p. 713-716.

    Very helpful aid in tracing government publications of Imperial Russia 1815-
17, the provisional government, 1917, and the Soviet Union 1917-1931.

IOTHÈQUE Nationale. Département des Périodiques. <u>Catalogue collectif des</u>
<u>riodiques conservés dans les bibliothèques de Paris et dans les bibliothèques</u>
<u>iversitaires de France</u>. <u>Périodiques slaves en caractères cyrilliques</u>. <u>Etat</u>
<u>s collections en 1950</u>. Paris, Bibliothéque Nationale, 1956. 2 v. 873 p.   2480

_. _____. _____. _____. <u>Supplément 1951-1960</u>. 1963. 495 p.

_. _____. _____. _____. <u>Addenda et errata</u>. <u>Etat général des collections</u>
1960.

N, Peter. <u>Gesamtverzeichnis russischer und sowjetischer Periodika und</u>
<u>rienwerke in Bibliotheken der Bundesrepublik und West-Berlins</u>. Union list
Russian and Soviet periodicals and serial publications in libraries of the
deral Republic of Germany and West-Berlin. Herausgegeban von Werner Philipp.
erlin. Freie Universität. Osteuropa-Institut. <u>Bibliographische Mitteilungen</u>,
ft 3.) Wiesbaden, Otto Harrassowitz, 1960-1973. 3 v., 19 nos. 1843 p.   2481

    A union list of about 10,000 main entrie, 1702-1956, in alphabetical order,
th holdings of libraries. Vol. 3, no. 19 (1973) consists of addenda. Vol. 4,
dex, is in preparation.

5E, F. <u>Slawistik</u>. <u>Stand vom 1 Dezember 1967</u> (Deutsche Staatsbibliothek,
rlin. Zeitschriften-Bestandsverzeichnisse, 11). Berlin, 1968. 416 p.   2482

. Bureau of the Census. <u>Bibliography of social science periodicals and mono-</u>
aph series. U.S.S.R. 1950-1963. Washington, D. C., Government Printing Office,
55. 443 p. (Foreign social science bibliographies. Series P-92, no. 17).
                                                                                  2483
    2,313 periodical or monograph series arranged by fields. Human geography,
204-217. Indexes of subjects, titles, authors, and issuing agencies.

CKY, Paul L., and CARLTON, Robert C. <u>The USSR and Eastern Europe: periodicals</u>
<u>Western languages</u>. 3rd ed., Washington, D. C., Library of Congress, 1967.   2484

Geographical Serials

SOVIET geography: review and translation (American Geographical Society), New York
   v. 1- (1960-). Monthly, except July and August.

   English translations of selected significant articles from Soviet geographic
periodicals and serials. News notes. Section Current Literature lists tables of
contents of current issues of principal Soviet geographical periodicals, such as
Geograficheskoe Obshchestvo SSSR, Izvestiia; Akademiia Nauk SSSR, Izvestiia, seri
geograficheskaia; Moskva. Universitet. Vestnik, seriia 5. Geografiia; Leningra
Universitet. Vestnik. Seriia geologii i geografii; Akademiia Nauk SSSR, Sibirsko
Otdelenie. Institut Geografii Sibiri i Dal'nego Vostoka, Doklady; and Voprosy
geografii. Excerpts from translation journals. Excerpts from Current geographic
publications on the USSR.

Petermanns geographische Mitteilungen. Gotha. 1- (1855-). Quarterly.

   Since 1953 a special section or special attention has been devoted to Soviet
geography.

Other Serials of Geographic Interest

Abstract Journal

ABSEES: Soviet and East European abstract series (National Association for Soviet
   and East European Studies). Glasgow, Institute of Soviet and East European
   Studies, University of Glasgow. 1- (27) (July 1970-). quarterly.

   Divided into two parts USSR and East Europe by countries. The USSR section
consists of abstracts in English of Soviet materials arranged by topics such as:
agriculture; costing, prices, and planning; economic data; economic theory and
policy; history; industry; labour and wages; government; science; social data or
problems; standard of living; trade and services; and transportation and communic
tions. A separate section is devoted to abstracts of Soviet books in such fields
Occasional special bibliographies.

   Continues in parentheses numbering of Soviet studies: information supplement
nos. 1-26 (April 1964-April 1970), which contained similar abstracts.

   Includes an Information bulletin with news on research and theses on the USSI
and East Europe by British scholars and institutions, 1- (July 1966-), of which
nos. 1-10 (July 1966-January 1970) were in Soviet studies: information supplement
nos. 11-25 (July 1966-January 1970).

Translation Journal of General Interest

CURRENT digest of the Soviet press (American Association for the Advancement of
   Slavic Studies). Columbus, Ohio, v. 1- (1949-). Weekly.

   Translations of material of substantial political, cultural, or economic
interest from 60 major Soviet newspapers or periodicals arranged by subject matte
Includes weekly indexes to material in Pravda and Izvestiia.

Translation Serials in the Physical, Biological, and Earth Sciences

   The following translation journals are published by the American Geophysical
Union, 1707 L Street, N.W., Washington, D.C. 20036, U.S.A.

AKADEMIIA Nauk SSSR. Izvestiia. Atmospheric and oceanic physics. 1965- .
   Translation of Izvestiia. Fizika atmosfery i okeana.

ESY and aerophotography. 1962- . 6 nos. a year. Complete translation of
odeziia i aerofotos"ëmka.   2490

AGNETISM and aeronomy. 1961- . 6 nos. a year. Translations from Geomagnetizm
aeronomiia.   2491

ECTONICS. 1967- . 6 nos. a year. Translation of Geotektonika.   2492

NOLOGY. 1965- . 6 nos. a year. Complete translation of Okeanologiia. Super-
des Soviet oceanography, 1960-1964.   2493

ET hydrology: selected papers. 1962- . 6 nos. a year   2494

ET Antarctic expedition information bulletin. 1961- . Irregular.   2495

The American Geological Institute, 2201 M Street, N.W., Washington, D. C.
0037, publishes the following journals:

RNATIONAL geology review (American Geological Institute), Washington, D. C.
59- . monthly. Annual subject and author indexes. Articles from many Soviet
eriodicals.   2496

ADY: earth sciences sections. 1957- . Translation of the earth-science sec-
ons of Akademiia Nauk SSSR. Doklady.   2497

HEMISTRY international. 1964- . Cover-to-cover translation of Geokhimiia.   2498

The National Technical Information Service, U. S. Department of Commerce,
285 Port Royal Road, Springfield, Virginia 22151, U.S.A. issues:

OROLOGY and hydrology. 1957- . Irregular. Translation of Meteorologiia i
idrologiia.   2499

The American Institute of Biological Sciences and Scripta Technica, Inc.,
511 K Street, N. W., Washington, D. C. 20005, U.S.A. publish:

IET soil science. 1958- . 6 nos. a year. Translation of Pochvovedenie.   2500

The Consultants Bureau, Plenum Publishing Corporation, 227 West 17th Street,
ew York, N. Y. 10011, U.S.A. or Davis House, 8 Scrubs Lane, Harlesden, NW 10
SE, England, publishes:

SOVIET journal of ecology. 1- (1970-). 6 nos. a year. A cover-to-cover
ranslation of Ekologiia (Akademiia Nauk SSSR). Sverdlovsk. 1- (1970-). 6 nos.
year.   2501

The National Research Council of Canada, Ottawa, Canada, publishes:

BLEMS of the North (National Research Council of Canada). Ottawa. 1- (1958-).
rregular. Translation of Problemy severa.   2502

Translation Serials in the Social Sciences

The following translation journals are published by the International Arts
nd Sciences Press, Inc., 901 North Broadway, White Plains, New York 10603, U.S.A.:

BLEMS of economics: selected articles from Soviet economic journals in English
ranslation. 1958- . Monthly.   2503

IET and Eastern European foreign trade: a journal of translations (formerly
merican review of Soviet and Eastern European foreign trade). 1965- . Quarterly. 2504

IET anthropology and archeology: a journal of translations from Soviet sources.
962- . Quarterly.   2505

IET sociology: a journal of translations from scholarly Soviet sources. 1962- .
uarterly.   2506

2490 - 2506

Lists of Translations

BIBLIOGRAPHY--index to current US JPRS [United States Joint Publications Research
Service] translations: Soviet Union. Compiled and edited by Theodore E. Kyriak,
v. 1-v. 8, no. 11 (July 1962-June 1970). Published successively by Research
Microfilm, later Research and Microfilm Publications, Annapolis, Maryland, later
Washington, D. C., and then by CCM Information Corporation, a subsidiary of Crowell
Collier and Macmillan, Inc., New York, N. Y.

TRANSDEX: bibliography and index to United States Joint Publications Research Servic
(JPRS) translations, v. 9, no. 1- (June-December 1970-). New York, CCM Information Corporation, a subsidiary of Crowell-Collier and Macmillan, Inc. Preceded
by: Catalog cards in book form for United States Joint Publications Research Service, v. 1-v. 8 no. 2 (1957-1970) published by Research and Microfilm Publications
Annapolis, Maryland, later Washington, D. C.

GEOPUB, Tulatin, Oregon. Israel Program for Scientific Translations: geography and
the earth sciences. Tulatin, Oregon, Geopub Book Service, n.d. ca. 1973. 24 p.

162 Soviet scientific monographs, for which English translations have been
made by the Israel Program for Scientific Translations: 5 in cartography, 44 in
atmospheric sciences, 85 in geosciences, 25 in biosciences, and 3 miscellaneous.
Title in English, name of author, date of publication in Russian, date of translation, IPST number, number of pages, International Standard Book Number, and price
from Geopub Book Service. The contents of each volume are described in several
paragraphs.

General Scholarly Serials Originally in Western Languages

SLAVIC review: American quarterly of Soviet and East European studies (American
Association for the Advancement of Slavic Studies). Urbana, Illinois, University
of Illinois, 1940- . Quarterly.

SURVEY: a journal of East and West studies (International Association for Cultural
Freedom and Stanford University). London, Oxford University Press, 1956- .
Quarterly.

SOVIET studies: a quarterly journal on the USSR and Eastern Europe. Glasgow,
University of Glasgow Press, 1949- . Quarterly.

RUSSIAN review: an American journal devoted to Russia past and present. Stanford,
California, v. 1- (1941-). Quarterly.

Cumulative index to volumes 1-30 (1941-1971). Compiled by Virginia L. Close.
1972. 219 p. Supersedes earlier cumulative indexes, v. 1-10 (1941-1951) and v.
1-20 (1941-1961).

SLAVONIC and East European review. Cambridge University Press for the School of
Slavonic and East European Studies, University of London. 1922- . Semiannual.
Cumulative index v. 1-10 (1922-1932), in v. 11, no. 31 (July 1932), p. 223-
246.

CAHIERS du monde russe et soviétique (Ecole Pratique des Hautes Etudes. Sorbonne.
6ème Section: Sciences Economiques et Sociales. Centre d'Etudes sur l'U.R.S.S. et
les Pays Slaves). Paris and The Hague, Mouton, 1959- . Quarterly.

ÉCONOMIES et sociétés: cahiers de l'I.S.E.A. [Institut de Science économique
appliquée]. Série G. Economie planifiée. 1967- . Supersedes the institute's
Cahiers: Série G. Economie planifiée (1-7, 1960 as Série G. Economies des Démo-
craties populaires). Paris, Institut de Science économique appliquée; Genève,
Librairie Droz.

UROPA: Zeitschrift für Gegenwartsfragen des Ostens (Deutsche Gesellschaft für
teuropakunde). Stuttgart, Deutsche Verlags-Anstalt, 1951- . Monthly.   2517

UROPA: Wirtschaft (Deutsche Gesellschaft für Osteuropakunde). Stuttgart,
utsche Verlags-Anstalt, 1- (1956-). Quarterly. In German with English sum-
ries at end of each issue. Cumulative index, v. 1-6. (1956-1961).   2518

UROPA: Naturwissenschaft und Technik (Deutsche Gesellschaft für Osteuropakunde).
uttgart, Deutsche Verlags-Anstalt, 1- (1957-). Semiannual.   2519

   Includes a section, Osteuropäische Fachliteratur in westlichen Sprachen, in
ich geography is one division.

ITUTE for the Study of the USSR (Institut zur Erforschung der UdSSR). München.
lletin. v. 1-18 (1954-1971). Closed.   2520

   v. 18, no. 12 (December 1971), p. 48-59 contains a complete list of the pub-
cations of the Institute for the Study of the USSR, 1950-1971. In the same
sue, p. 5-12 there is a brief history of the Institute by Herman F. Achminow.

R record (Scott Polar Research Institute, Cambridge). Cambridge, Scott Polar
search Institute. 1931- . 3 nos. a year.   2521

   Recent polar literature, appearing separately with each issue, carries about
0 entries, or about 2700 entries a year, about a fourth of which concern the
viet Union. Comprehensive in scope covering all aspects of polar regions and
terature in all languages.

R-NORD: revue internationale d'études arctiques et nordiques; international
urnal of Arctic and Nordic studies. (Paris. École Pratique des Hautes Études.
e Section [Sorbonne]. Centre d'Études Arctiques). 1960- . Annual. (Title
ries). In French or English. Includes extensive material on Siberia and the
viet Arctic.   2522

RAL Asian review (Central Asian Research Centre). London. v. 1-16 no. 4
953-1968). Closed. Merged into Mizan (Central Asian Research Centre).   2523

DISSERTATIONS

SICK, Jesse J. Doctoral research on Russia and the Soviet Union. New York,
ew York University Press, 1960. 248 p.   2524

   Listing of 960 American, British, and Canadian doctoral dissertations on
ssia and the Soviet Union, arranged by fields for American and Canadian dis-
ertations: agriculture, archaeology and anthropology, art, architecture, com-
unications, drama and theater, music, ballet and dance, education, geography
d geology, language, literature, library studies, law, museology and numismatics,
ilosophy and religion, psychology, sociology, medicine and public health, home
onomics, mathematics and science, economics, history, political science and
ternational law and relations; and British dissertations. In addition to lists
 dissertation includes: Aids to further research in each of the fields, with
out 1,500 entries. Extensive background commentary. Geography, p. 64-69.

ICK, Jesse J. Doctoral dissertations on Russia and the Soviet Union accepted
 American, Canadian, and British universities 1960-1964, Slavic review, v. 23,
. 4 (December 1964), p. 797-812.   2525

_. Doctoral dissertations on Russia, the Soviet Union, and Eastern Europe
cepted by American, Canadian, and British universities 1964-1965, ____, v. 24,
. 4 (December 1965), p. 752-761.

_____· _____· 1965-1966, _____, v. 25, no. 4 (December 1966), p. 710-717.

_____· _____· 1966-1967, _____, v. 26, no. 4 (December 1967), p. 705-712.

_____· _____· 1967-1968, _____, v. 27, no. 4 (December 1968), p. 694-704.

_____· _____· 1968-1969, _____, v. 28, no. 4 (December 1969), p. 699-708.

_____· _____· 1969-1970, _____, v. 29, no. 4 (December 1970), p. 766-776.

_____· _____· 1970-1971, _____, v. 30, no. 4 (December 1971), p. 727-941.

_____· _____· 1971-1972, _____, v. 31, no. 4 (December 1972), p. 951-966.

_____· _____· 1972-1973, _____, v. 32, no. 4 (December 1973), p. 866-881.

DISSERTATIONS-in-progress in Slavic and East European studies. Washington, D. C., Slavic Bibliographic and Documentation Center, Association of Research Libraries, first list, 1971; second list, 1972, 60 p.

The 1972 list included 363 American and Canadian dissertations in progress from 58 universities and 414 German dissertations in history and Slavic literatures and languages.

SIMMONS, J. S. G. Theses in Slavonic studies approved for higher degrees by British universities, 1907-1966, Oxford Slavonic papers, v. 13 (1967), p. 133-160.

313 entries.

_____· _____ , 1967-1971, _____ , n. s. v. 6 (1973), p. 133-147.

161 entries.

SEYDOUX, Marianne, and BIESKIEKIERSKI, M. Répertoire des thèses concernant les études slaves, l'U.R.S.S. et les pays de l'Est européen et soutenues en France de 1824 à 1969 (Travaux publiés par l'Institut d'Etudes slaves, 30) Paris, Institut d'Études slaves, 1970. 157 p.

Continued for 1970 by Aymard and Seydoux, q. v. under General guides to works mainly in Western languages [2470].

B. REFERENCE WORKS AND ASSOCIATED BIBLIOGRAPHIES

MAPS AND ATLASES: BIBLIOGRAPHIES

XANDER, Gerard L. Guide to atlases: world, regional, national, thematic. An
nternational listing of atlases published since 1950. Metuchen, New Jersey, The
carecrow Press, 1971. 671 p.　　　　　　　　　　　　　　　　　　　　　　　2529

90 atlases of the USSR published in the years 1950-1970 arranged alphabetically
y publisher and for each publisher alphabetically by title, entries 3323-3412, p.
52-361.

EAR, Clara Egli. A list of geographical atlases in the Library of Congress,
ith bibliographical notes. v. 6. Titles 7624-10254. (A continuation of four
olumes by Philip Lee Phillips). Washington, D. C., Government Printing Office,
963. 681 p. U.S.S.R., p. 308-323, entries 9210-9301 (Library of Congress.
ap Division).　　　　　　　　　　　　　　　　　　　　　　　　　　　　　2530

92 atlases on the Soviet Union mostly published in the years 1920-1956 in
he collection of the Library of Congress in classed arrangement: special (agri-
ulture, climatology, economic geography, history, population, and railroads),
eneral, Kara Sea; Russia in Europe broken down into special (cities, economic
eography, ethnography, forests, geology, ice, linguistic geography, railroads,
nd religions), general, and provinces and regions (Caucasus, Crimea, Karelia,
eningrad Province, Moscow Province, Ukraine, and White Russia), cities (Lenin-
rad, Pushkin), and water bodies (Don River, Neva River, and White Sea); Russia
n Asia.

LLIPS, Philip Lee. A list of geographical atlases in the Library of Congress,
ith bibliographical notes. Washington, D. C. Government Printing Office, 1909-
920. 4 v. (Library of Congress. Division of Maps).　　　　　　　　　　2531

v. 1. 1909. Russia, p. 1149-1158, 1189, entries 3098-3126, 32227-32228;
. 3. 1914. Russia, p. 662-668, 685-686, entries 4055-4063 and 4080a; v. 4.
920. Russia, p. 487-490, 502, entries 5278-5283 and 5306.

48 atlases of Russia in the Library of Congress up to 1920.

3TERMAN, Theodore. Russia. Cartography, in his: A World bibliography of bibliographies. 4th ed., 2nd printing, Totowa, New Jersey, Rowman and Littlefield, 1971.
v. 4, p. 5488-5489.　　　　　　　　　　　　　　　　　　　　　　　　　　2532

13 bibliographies of maps and atlases of Russia and the Soviet Union, 1841-
1962.

3LIOGRAPHIE cartographique internationale (Centre National de la Recherche
5cientifique. Service de Documentation et de Cartographie Géographiques). Paris,
France, Librairie de la Faculté de Sciences, and Nendeln, Liechtenstein, Kraus-
Thomson Organization Ltd. 1946- . Annual. Name of issuing body varies.　　2533

In the 1971 annual volume, maps and atlases of the USSR, p. 279, entries
1050-1055, included only 5 maps and one atlas, all published outside the Soviet
Union. The 1970 volume, p. 285, includes 3 maps and one atlas. The 1969 volume,
p. 231 includes 1 map and 3 atlases. The 1968 volume, p. 235-236, includes 4
maps and 1 atlas.

'DENBERG, Leonid Arkad'evich. Russian maps and atlases as historical sources,
ranslated by James R. Gibson, Cartographica, monograph no. 3, Toronto, B. V.
utsell, 1971. 76 p. Bibliography, p. 48-76. Originally published as Russian
artographic materials of the 17th-18th centuries as an historical source and
heir classification, in Russian in Problemy istochnikovedeniia, no. 7 (1959),
. 296-347.　　　　　　　　　　　　　　　　　　　　　　　　　　　　　2534

193 references.

PREOBRAZHENSKII, Arkadii Ivanovich. Economic maps in pre-reform Russia, material
for a history of Russian economic cartography, translated by James R. Gibson,
Cartographica, monograph no. 7, Toronto, B. V. Gutsell, 1973. 46 p. Bibliography, p. 42-46. Originally published in Russian as Ekonomicheskie karty v
doreformennoi Rossii, Voprosy geografii, sbornik 17 (1950), p. 105-138, with
bibliographical footnotes.

    49 references.

ATLASES

THE TIMES atlas of the world, ed. by John Bartholomew. Mid-century ed. London,
The Times Publishing Co., 5 v. 1955-1959. 50 cm. v. II. South-West Asia and
Russia, 1959. Plates 38-47. Reissued in 1 v., 1967. 558 p. 46 cm. 2nd ed.,
1968. 558 p. 46 cm.

    Large-scale physical maps cover the Soviet Union as a whole at 1:15,000,000,
the Soviet Union by regions in seven plates at a scale of 1:5,000,000; and two
areas in greater detail at 1:2,500,000: the Moscow-Volga-Don-Dnepr region and the
Caucasus-Crimea.

THE ECONOMIST (London). The Oxford regional economic atlas: the U.S.S.R. and
Eastern Europe. Prepared by the Economist Intelligence Unit and the Cartographic Department of the Clarendon Press. London, Oxford University Press,
1956. 134 p. 27 cm. (Oxford Regional Economic Atlases). Reissued 1963. 1969.

    Primarily a topical atlas covering physical geography, agriculture, minerals,
manufacturing, and economic geography for the Soviet Union as a whole, with regional reference maps.

KISH, George, with the assistance of Ian M. Matley, Holly Fry, and Betty Bellaire.
Economic atlas of the Soviet Union, 2nd ed. Ann Arbor, Michigan, University of
Michigan Press, 1971. 90 p. 27 cm. Bibliography, p. 82-83. (1st ed., 1960. 96 p.)

    A regional atlas with four maps (agriculture and land use; mining and miner-
als; industry; transportation and cities) for each of 15 regions, and five general
maps for the country as a whole, with accompanying text. Index.

TAAFFE, Robert N. and KINGSBURY. Robert C. An atlas of Soviet affairs. New York,
Frederick A. Praeger, 1965. 143 p. 21 cm.

    65 black-and-white maps with accompanying text covering particularly the
economic geography of the Soviet Union but with maps also on distribution of
population and on political organization.

SOVIET Union in maps. Harold Fullard, ed. London, G. Philip, Chicago, Denoyer-
Geppert, 1961- . Revised frequently. 32 p. 25 cm. Editions 1942-1959, edited
by George Goodall. 1942-1943 editions entitled Soviet Russia in maps.

    Inexpensive paperbound atlas covering historical, physical, economic, and
regional geography.

PLUMMER, Thomas F., Jr., HANNE, William G., BRUNER, Edward F., and THUDIUM, Christian
C., Jr. Landscape atlas of the U.S.S.R. West Point, New York, Department of
Earth, Space and Graphic Sciences, United States Military Academy, 1971. 197 p. 45 cr

    Parts of 70 topographic maps on a scale of 1:250,000 selected to portray the
landscape contrasts in the Soviet Union, with textual explanation, diagrams, and
sketch maps showing regional location of topographic sheets selected.

KOVALEVSKY, Pierre. Atlas historique et culturel de la Russie et du monde slave.
Paris, Elsevier, 1961. 217 p. 35 cm. Bibliography. Index.

S of Siberia. Semyon U. Remezov. Facsimile edition, with an introduction by
o Bagrow. 's-Gravenhage, Mouton and Co. 1958. 17 p., 171 l. 30 cm. (Imago
ndi: a review of early cartography. Supplement 1).   2543
Facsimile reproduction of manuscript atlas in the Harvard University Library.

STATISTICS

IEWICZ, Ellen. Handbook of Soviet social science data. New York, Free Press,
ndon, Collier-Macmillan, 1973., 225 p. Lists of sources for tables.   2544
Statistical tables on the Soviet Union, often by union republics, for the
elds of demography, agriculture, production, health, housing, education, elite
ecruitment and mobilization, communications, and international interactions with
ntroductory text for each section by a specialist and with general textual fore-
ord and introduction.

IS, Chauncy D. Population of cities of the Soviet Union, 1897, 1926, 1939,
59, and 1967, tables, maps, and gazetteer, Soviet geography: review and transla-
on, v. 11, no. 5 (May, 1970), 138 p. (p. 307-444).   2545
Tabulation of population data for cities and towns of the Soviet Union from
e censuses of 1897, 1926, 1939, and 1959, and of estimates for 1967 for 1,247
viet cities and towns of more than 10,000 population in 1959 (15,000 in the
FSR); 104 other places of more than 15,000 population in 1967; 69 urban places
more than 10,000 in the 1897 or 1926 census but not in the 1959 census; 22
gional maps showing location and size of places listed in the tables; and index-
zetteer listing places in alphabetical order.

ENCYCLOPEDIAS

ENCYCLOPAEDIA Britannica. 15th ed. Chicago, London, Encyclopaedia Britannica,
974. 30 v.   2546
This new edition of the Encyclopaedia Britannica in its 19-volume Macropaedia
ortion includes many major signed articles on the Soviet Union, its union re-
ublics, cities, and physical features by Soviet geographers and other specialists:
he Soviet Union with sections on the land by V. M. Strygin, the people by V. V.
okshishevskii, and other aspects by I. V. Kozlov; Russian SFSR by S. A. Vodovozov,
krainian SSR by I. A. Erofeev, Estonian SSR by A. A. Keerna and V. J. Tarmisto,
atvian SSR by P. V. Gulian, Lithuanian SSR by K. A. Meskauskas, Belorussian SSR
y M. I. Rostovtsev, Moldavian SSR by F. N. Sukhopara, Georgian SSR by M. D.
zhibladze, Armenian SSR by A. A. Mints, Azerbaijan SSR by E. D. Silaev, Kazakh
SR by V. F. Kosov, Kirgiz SSR by S. N. Riazantsev, Uzbek SSR by V. Kopanev,
urkmen SSR by V. B. Zhmuida, and the Tadzhik SSR by A. I. Imshenetskii.

City articles include Moscow by I. I. Miachkin, Leningrad by E. M.
oroshinskaia, Kiev by L. A. Daen and P. I. Pozdniak, Novosibirsk by N. A. Meisak,
nd Vladivostok by F. V. Diakonov.

There are similar signed articles on mountain systems: the Urals by E. V.
astrebov, the Caucasus by N. A. Gvozdetskii and S. I. Bruk, Pamir mountain area
y T. K. Zakharova, Tien Shan (Tyan' Shan') by E. Ia. Rantsman and S. I. Bruk,
nd Altai by N. I. Mikhailov.

Major river articles are on the Dnepr, Dnestr, and Dvina (Western) by A. P.
omanitskii, Don and Dvina (Northern) by A. M. Gavrilov, Volga by P. S. Kuzin,
mu Darya by A. A. Kliukanova, Ob by L. K. Malik, Yenisey by C. G. Tikhotskii,
ena by I. V. Popov, and Amur by A. P. Muranov. There are articles on lakes
adoga and Onega by B. B. Bogoslovskii, Issyk-Kul by V. A. Blagoobrazov, and
aikal by G. G. Galazii.

The surrounding seas are well covered: the Black Sea by V. P. Goncharov and
. M. Fomin, Sea of Azov by A. M. Muromtsev, Caspian by A. N. Kosarev and O. K.

Leont'ev, Aral Sea by V. I. Lymarev, Barents Sea by M. M. Adrov, White Sea by
N. Ia. Arsen'eva, Kara Sea and Laptev Sea by E. G. Nikoforov and A. O. Shpaikher,
Bering Sea and Strait by A. P. Lisitsyn, and Sea of Okhotsk by T. I. Supranovich.

Other articles on physical features are on the Pripet Marches by A. M. Marinch
the Russian Steppe by F. N. Mil'kov, and the Kara-Kum (desert) by B. A. Fedorovich.

Two Soviet authors contribute to the 44-page monographic article on the
Continent of Asia: P. N. Kropotkin, the section on geological history, and Iu. K.
Efremov, the section on physical geography. Major geographical features of the
deep interior in Central Asia are treated in articles on the Karakorum Range by
G. D. Bessarabov, Kunlun Mountains by V. M. Sinitsyn, Koko Nor (lake) and Tarim Riv
by N. T. Kuznetsov, and Takla Makan Desert and Gobi (desert) by M. P. Petrov.

Soviet contributions in physical geography are reflected in articles on the
development of the continents by V. V. Beloussov, on physiographic effects of
tectonism by I. P. Gerasimov, on beaches and on gulfs and bays by V. P. Zenkovich,
and on water resources by G. P. Kalinin and V. D. Bykov.

Shorter unsigned articles on less-important physical features, administrative
units, regions, and cities in the Soviet Union are found in the 10-volume micro-
paedia portion of the encyclopedia.

GREAT Soviet encyclopedia. New York, Macmillan, Inc., London, Collier Macmillan
Publishers, 1973- . 30 v. In progress. [See entry 848].   2

A translation of Bol'shaia sovetskaia entsiklopediia, 3rd ed., Moskva, 1970- ,
in progress, "Although the encyclopedia is general in scope, it naturally concentrates on the Soviet Union." Omits certain articles that can be classified
simply as dictionary or gazetteer entries for places outside the USSR, but all
omissions are carefully noted. Some articles have been updated with the cooperation of the publishers of the Russian edition. All bibliographies that appear
in the Russian edition are fully listed in the English edition. Each article is
coded to show location in the Russian edition by volume, page, and column.

McGRAW-HILL encyclopedia of Russia and the Soviet Union. Editor: Michael T. Florin-
sky; consultants: Harry Schwartz, Theodore Shabad, and others. New York, London,
McGraw-Hill, 1961. 624 p. Included bibliographies.   2

3,500 entries in alphabetical order, by 91 contributors. Concentrates on the
Soviet period. Strong on economics.

UTECHIN, Sergej V. Everyman's concise encyclopaedia of Russia. London, J. M. Dent
and Sons, New York, E. P. Dutton, 1961. 623 p. (Everyman's Reference Library). 2

Handy, inexpensive, and informative small encyclopedia, covering the arts,
economy, education, ethnography, geography, history, government, literature,
military, politics, philosophy, religion, science and technology, and society of
Russia and the Soviet Union in alphabetical order, with list of articles by fields.
List of geography articles, p. xv-xix, includes about 650 entries for articles
on places in the USSR. 2,000 entries in all.

ENCYCLOPEDIA lituanica. Simas Sužredėlis and Vincas Rastenis, eds. Boston, Massachusetts, Juozas Kapočius, Encyclopedia Lituanica, 1970- . (v. 1-3, A-M, 1970-
1973-). In progress. To be 6 v. In English.   2

The article "Geographical research," v. 2, p. 302-305 treats the history of
geographical investigation in Lithuania. The article "Geography, physical, v. 2,
p. 305-311, by Antanas Bendorius, discusses the physical geography of Lithuania.
Other relevant articles on Lithuania are those on climate, fauna, flora, geology,
and lakes and rivers.

2547 - 2550

HANDBOOKS

WELL, Robert, ed. Information USSR; an authoritative encyclopaedia about the
nion of Soviet Socialist Republics. Oxford, New York, Pergamon Press, 1962.
32 p.   2551

    English translation of volume 50, USSR, of the Bol'shaia sovetskaia entsik-
opediia, 2nd ed., Moskva, 1957, 761 p., with some updating and appendices with
tatistical data, directory of institutions of higher learning, bibliography of
ecent books in English on the USSR, data on trade with the USSR, and resumé of
he 22nd Congress of the Communist Party of the USSR. Indexes of names and sub-
ects. [See entry 847].

A handbook for the Soviet Union. Co-authors: Eugene K. Keefe and 9 others.
ashington, D. C., Government Printing Office, 1971. 826 p. Bibliography,
. 763-805.   2552

    Four main sections: general survey, social, political, and economic. The
ibliography, with about 800 references, almost entirely in English, is organized
y these four sections, each further divided into recommended sources and other
ources used. The geographic content is modest.

ZSIMMONS, Thomas, MALOF, Peter, and FISKE, John C. USSR: its people, its society,
ts culture. New Haven, Connecticut, HRAF Press, 1960. 590 p. Bibliography,
. 521-548. (Survey of world culture).   2553

    About 500 references, nearly all in English, in alphabetical order.

ZSIMMONS, Thomas, ed. RSFSR, Russian Soviet Federated Socialist Republic.
lifford Barnett and others, eds. New Haven, Connecticut, Human Relations Area
iles, 1957. 2 v. 681 p. Bibliography, v. 2, p. 611-633. (Country survey
eries).   2554

    About 350 references, almost entirely in English on social and cultural,
olitical and economic aspects of the Russian part of the USSR.

TING, Kenneth R. The Soviet Union today. A concise handbook. rev. ed. New
ork, Frederick A. Praeger, 1966. 423 p. Bibliography, p. 395-412. Index.
1st ed., 1962. 405 p.).   2555

UAIRE de l'URSS: droit, économie, sociologie, politique, culture (Strasbourg.
niversité. Centre de Recherches sur l'URSS et les Pays de l'Est). Paris,
965- . annual. 1965-1969, Editions du Centre National de la Recherche Scientifi-
ue; 1970-1971, Istra. Index.   2556

: UdSSR: Enzyklopädie der Union der Sozialistischen Sowjetrepubliken.
erausgegeben von W. Fickenscher unter Mitwirkung von H. Becker, R. Rompe, W.
teinitz, u. a. Leipzig, Verlag Enzyklopädie, 1959. 1,104 p.   2557

    A German translation of volume 50, USSR, of the Bol'shaia sovetskaia ·
ntsiklopediia, 2nd ed., Moskva, 1957, 761 p., with some updating of statistics.
Jame and subject index. [See entry 847].

RKERT, Werner, ed. Sowjetunion. Köln, Böhlau, 1965. 587 p. (Osteuropa-
andbuch). Bibliography. Index.   2558

JETUNION (Koordinationsausschuss deutscher Osteuropa-Institute). München, Carl
anser Verlag, 1974. 356 p. Bibliography, p. 335-341. (Länderberichte Osteuropa
.).   2559

    155 references arranged by chapters: the area and its structure; population
nd social structure; political and judicial system; economic system; education,
cience, and culture. Seven folded colored maps.

UKRAINE: a concise encyclopaedia. Prepared by the Shevchenko Society. Edited by
Volodymyr Kubijovyč. Toronto, for the Ukrainian National Association by the
University of Toronto Press, 1963-1971. 2 v. v. 1, 1963. 1,185 p. v. 2, 1971.
1,394 p.   2

     v. 1 covers general information; physical geography and natural conditions
(history and present state of geographic and naturalistic studies of Ukraine;
geology; soils; relief and landforms; climate; the Black Sea and Sea of Azov;
inland waters; flora; fauna; Ukraine as a geographic entity and its sub-regions);
population, ethnography; language; history; culture; and literature.

     v. 2 covers law; the Ukrainian Church; scholarship; education and schools;
libraries, archives, museums; book publishing and the press; the arts; music and
choreography; theater and cinema; national economy; health and medical services
and physical culture; the armed forces; and Ukrainians abroad. Index.

     Based on Entsyklopediia ukraïnoznavstva (Ukrainian encyclopedia), published
in Ukrainian, München, 1949-1952, v. 1, parts 1-3 [entry 860].

                                  GAZETTEERS

COLUMBIA Lippincott gazetteer of the world. Edited by Leon E. Seltzer with the geographical research staff of Columbia University Press and with the cooperation of
the American Geographical Society. New York, Columbia University Press by arrange-
ments with J. B. Lippincott Co., 1952. 2,148 p.   2

     The best English-language gazetteer for information on regions, places, and
features of the Soviet Union. The entries for the Soviet Union were prepared by
Theodore Shabad, who was assistant editor of the gazetteer.

WESTERMANNS Lexikon der Geographie, ed. by Wolf Tietze. Braunschweig, Georg
Westermann Verlag, 1968-1972. 5 v.   2

     About 1500 entries on physical features, major regions, and cities in the
Soviet Union, many with bibliographies. The entries on the European Soviet Union
by Adolf Karger, 206 cols., and on the Asiatic part of the Soviet Union by Erich
Thiel and Otto Plaschka, 121 cols.,were also separately issued as reprints in 1967.

MEYERS Kontinente und Meere in 8 Bänden: Sowjetunion. Daten. Bilder, Karten.
Herausgegeben vom Geographisch-Kartographischen Institut Meyer. Bearbeitet von
Dr. Werner Jopp. Mannheim, Wien, Zürich, Bibliographisches Institut, Geographisch-
Kartographisches Institut Meyer, 1969. 340 p.   2

     Gazetteer, p. 83-338, has about 4,000 entries. Handbook section, p. 9-82
contains articles on geology, geomorphology, climate, vegetation, soil, peoples,
economy, foreign trade, government, history, and natural zones. The bibliography,
p. 339-340 includes about 200 references. Colored maps and colored and black-and-
white photographs.

U. S. Board on Geographic Names. Gazetteer no. 42. U.S.S.R. 2nd ed., Washington,
D. C., Government Printing Office, 1970. 7 v. (Preliminary ed., 1951; 1st ed.,
1959, 7 v. 362,000 names).   2

     About 400,000 names of places and features in the Soviet Union, with latitude
and longitude, identification, and map references. The most comprehensive list of
places in the USSR.

TELBERG, Ina. Russian-English geographical encyclopedia. New York, Telberg Book Co.,
1960. 142 p. mimeographed.   2

     Gazetteer based on English translation of information in Bodnarskii, M.S.,
Slovar' geograficheskikh nazvanii. Moskva, 2nd ed., 1958.

RUGINE, T., translator. Dictionary of Russian geographical names. New York,
Telberg Book Co., 1958. 82 p.   2565

English translation of Volostnova, M. B. Slovar' russkoi transkriptsii
geograficheskikh nazvanii. Chast' I. Geograficheskie nazvaniia na territorii
SSR (Dictionary of Russian transcription of geographical names. Part I.
Geographical names on the territory of the USSR), 1955.

Includes place names of the USSR, administrative divisions, and natural
features, such as mountains, lakes, and peninsulas. In alphabetical order of
the Cyrillic alphabet as in the original [entry 897].

KELEIN, Wolfgang. Ortsumbenennungen und -neugründungen im europäischen Teil
der Sowjetunion. Nach dem Stand der Jahre 1910/1938/1951 mit einem Nachtrag für
Ostpreussen 1953. Berlin, Duncker and Humbolt, 1955. 135 p. (Berlin. Freie
Universität. Osteuropa-Institut. Wirtschaftswissenschaftliche Veröffentlichungen,
Band 2). Bibliography.   2566

Listing with analysis of 1,363 name changes and newly founded places in the
European part of the Soviet Union between 1910 and 1951, with listing of names
as of 1910, 1938, and 1951.

MES index-gazetteer. London, Times Publishing Co., 1965. 959 p.   2567

As the most detailed one-volume gazetteer of the world, it contains many
thousand entries for the Soviet Union.

DICTIONARIES

Bibliographies

YNEN, Emil. Bibliography of mono- and multilingual dictionaries and glossaries
of technical terms used in geography as well as in related natural and social
sciences. Wiesbaden, Franz Steiner Verlag, 1974. 246 p. (International Geographical Union. Commission on International Geographical Terminology).   2568

3,211 dictionaries in geography and related fields. In section II, Multilingual glossaries, dictionaries and lexicons of technical terms in geography,
p. 4-7, multilingual geographical dictionaries for languages of the Soviet Union
are listed as follows, mostly from Russian into the stated language (with entry
number in parentheses): Ukrainian (67, 83), Belorussian (69), Azerbaijani (75),
Kazakh (52), Kirgiz (73, 80), Uzbek (54, 59), Turkmen (72), Tajik (82), Yiddish
(59, 67, 83), Tatar (60), Bashkir (57), Osset (63-64), Tuvinian (78), Buryat
(65), and Digor and Iron (70). Russian with English or French (62, 84, 86).
Part I, Monolingual explanatory dictionaries of geographical terms lists four in
Russian (37-40). In addition there are many dictionaries in languages of the
Soviet Union, especially Russian, listed in Part III, special fields of geog-
raphy and related branches of science.

WANSKI, Richard C. A bibliography of Slavic dictionaries. 2nd ed. Bologna,
Editrice Compositori, in press, 4 v. (Johns Hopkins University. Bologna Center
Library). (1st ed., New York, New York Public Library, 1959-1963. 3 v.).   2568a

Lists 11,200 dictionaries, including about 5,320 Russian, 680 Ukrainian,
100 Belorussian, and 350 Bulgarian, 1,350 Czech, 115 Lusatian, 45 Macedonian,
100 Old Church Slavic, 1,760 Polish, 660 Serbocroatian, 440 Slovak, and 230
Slovenian. Includes alphabets of national minority groups of the Soviet Union.

BLIOGRAPHIE der Wörterbücher erschienen in der Deutschen Demokratischen Republik,
Rumänischen Volksrepublik, Tschechoslowakischen Sozialistischen Republik, Ungari-
schen Volksrepublik, Union der Sozialistischen Sowjetrepubliken, Volksrepublik
Bulgarien, Volksrepublik China, Volksrepublik Polen, 1945-1961. Bibliography of
dictionaries published in Bulgarian People's Republic, Chinese People's Republic,
Czechoslovak Socialist Republic, German Democratic Republic, Hungarian's People's
Republic, Polish People's Republic, Rumanian People's Republic, Union of Soviet

Socialist Republics, 1945-1961. Title also in Polish and Russian: Bibliografia
słowników...Bibkiografiia slovarei... D. Rymsza-Zalewska, I. Siedlecka, eds.
Warszawa,Wydawnictwa Naukowo-Techniczne, 1965. 249 p. Entries 1-2249 with 2
additions.

2251 dictionaries divided into two groups, monolingual and bi- or polylingual and by universal decimal classification within each group. Geographical
dictionaries, monolingual, p. 53-55, entries 554-569, and bi- or polylingual,
p. 202-205, entries 2228-2248. Indexes of languages, of names and titles, and of
subjects.

_____. 1962-1964. 1968. 167 p. Entries 2250-3333. Geography, entries 2528-
2532, and 3327-3333.

_____. 1965-1966. 1969. 111 p. Entries 3334-3980. Geography, entries 3537-
3540 and 3973-3976.

_____. 1967-1968. 1970. Entries 3981-4581.

_____. 1969-1970. 1972. 133 p. Entries 4582-5338. Geography, entries 4842-4857
and 5332-5337.

ZAUMÜLLER, Wolfram. Bibliographisches Handbuch der Sprachwörterbücher. Stuggart,
Anton Hiersemann, 1958. 496 cols.

An annotated international bibliography of 5600 dictionaries, 1460-1958 in
more than 500 languages and dialects: Russian, p. 323-333; Ukrainian, p. 393-395;
Belorussian, p. 402-403; Estonian, p. 123-124; Latvian, p. 256-258; Lithuanian,
p. 258-260; Moldavian, p. 272; Georgian, p. 171-172; Armenian, p. 19-21; Azerbaijani, p. 22; Kazakh and Kirgiz, p. 225-226; Uzbek, p. 399, Turkmen, p. 390; and
Tajik, p. 373.

Geographical

SARNA, Andrei. Russian-English dictionary of geographical terms. V. G. Telberg,
ed. New York, Telberg, 1962-1965. 2 v. 73, 87 p. Mimeographed.

Russian geographical terms with English equivalents, based on the stock of
terms and definitions in Kratkaia geograficheskaia entsiklopediia, v. 1-4, 1960-
1964. About 3,000 terms. [See entries 845 and 916].

BURGUNKER, Mark E. Russian-English dictionary of earth sciences. New York.
Telberg, 1961. 94 p. Mimeographed.

Russian terms in physical geography with English equivalents. Based on stock
of terms in A. S. Barkov, Slovar'-spravochnik po fizicheskoi geografii, 3rd ed.
Moskva, 1954. 1,900 terms. [See entry 915].

U. S. Army Map Service. Russian glossary. 2nd ed. Washington, D. C. 1951.
881 p. (AMS technical manual, no. 12) (1st ed., 1946 with title: Russian map
terms).

English equivalents of nearly 30,000 terms in Russian found on maps, including abbreviations.

Related Fields

CALLAHAM, Ludmilla Ignatiev. Russian-English chemical and polytechnical dictionary.
2nd ed. New York and London, John Wiley and Sons, 1962. 892 p. (1st ed., 1947,
794 p., as Russian-English technical and chemical dictionary).

Particularly useful in physical geography and related fields in the physical
and biological sciences.

FORD, M. H. T., and ALFORD, V. L. Russian-English scientific and technical
dictionary. Oxford, Pergamon Press, 1970. 2 v. 1423 p.   2575

  More than 100,000 entries, particularly from the physical and biological
sciences and engineering.

ITH, R. E. F. A Russian-English dictionary of social science terms. London,
Butterworths, 1962. 493 p.   2576

  Particularly useful in economic geography and in explanation of technical
terms used in planning, in government agencies, and in programs.

General

FORD Russian-English dictionary, by Marcus Wheeler. B. O. Unbegaun, general
editor, with the assistance of D. P. Costello and W. F. Ryan. Oxford, Clarendon
Press, 1972. 918 p.   2577

  About 70,000 entries.

PLACE-NAME DICTIONARIES

CHHARDT, Rosemarie (with contributions from M. Vasmer and B. O. Unbegaun). Bibliographie zur russischen Namenforschung (Bibliography on Russian name research),
Onoma (Louvain), v. 5 (1954), p. 1-77.   2578

  926 books and articles, of which more than 750 are concerned with place
names, published in Russian, other languages of the USSR, or in other languages,
systematically organized, from the 19th century to 1953. For publications from
_954 see annual bibliographies in Onoma.

ASMER, Max, ed. Wörterbuch der russischen Gewässernamen. Bearbeitet von A. Kerndl',
R. Richhardt, and W. Eisold. Wiesbaden, Otto Harrassowitz, v. 1-5, no. 1-14, 1961-
1969. no. 15, 1973. Supplement. (Freie Universität, Berlin. Osteuropa-Institut.
Abteilung für Slavische Sprachen und Literaturer. Veröffentlichungen, Band 22). 2579

  Exhaustive multi-volume inventory of hydronyms of Slavic origin in the
European part of the Soviet Union, with description of each river, location, and
source of information. Names in Cyrillic form.

USSISCHES geographisches Namenbuch. Begrundet von Max Vasmer. Herausgegeben von
Herbert Bräuer. Bearbeitet von Ingrid Coper, Ingeborg Doerfer, Marit Podeschwik
Jürgen Prinz, Georg Viktor Schulz, and Rita Siegmann. Wiesbaden, Otto Harrasso-
witz, 1962- . In progress. (Akademie der Wissenschaften und der Literatur,
Mainz).   2580

  A detailed inventory of about 600,000 place names in the European part of
the Soviet Union. Indicates sources from which information is taken. An important contribution to place-name research. v. 1-7 no. 1. A. Poliule, 1962-1974
are published.

BIOGRAPHICAL DIRECTORY

RECTORY of Soviet geographers, comp. by Theodore Shabad, Soviet geography: review
and translation, v. 8, no. 7 (September, 1967), p. 508-610.   2581

  About 2,000 names of Soviet geographers active in publication in geographical
periodicals and monographs during the period 1955-1967. The most complete biographical directory of contemporary Soviet geographers. About one-fourth of the
names are also listed in the biographical section of the Kratkaia geograficheskaia
entsiklopediia (Short geographical encyclopedia) v. 5, p. 410-544, but the other
three-fourths are based on an examination of the Soviet geographical literature
for the stated period.

  Information provided includes full name, with given name and patronymic;
academic degrees; fields of specialization and publication; institutional affilia-
tions; and, sometimes, year of birth.

C. SYSTEMATIC FIELDS OF GEOGRAPHY

GENERAL

)VIET geography: accomplishments and tasks. A symposium of 50 chapters, contributed by 56 leading Soviet geographers and edited by a committee of-the Geographic Society of the USSR, Academy of Sciences of the USSR. I. P. Gerasimov,
chairman, G. M. Ignat'yev, secretary, S. V. Kalesnik, O. A. Konstantinov, E. M.
Murzayev, K. A. Salishchev. Translated from the Russian by Lawrence Ecker.
English edition edited by Chauncy D. Harris. New York, American Geographical
Society, 1962. 409 p. (Occasional publication no. 1).   2582

English translation of Akademiia Nauk SSSR. Geograficheskoe Obshchestvo
Soiuza SSR. Sovetskaia geografiia: itogi i zadachi. Moskva, Geografgiz, 1960.
635 p. Extensive bibliographies at ends of chapters on all principal fields of
geography as developed in the Soviet Union. For details see entry under Russian
title [23 and 24].

PHYSICAL GEOGRAPHY

Comprehensive Physical Geographic Regions

ERG, Lev Semënovich. Die geographischen Zonen der Sowjetunion. Leipzig, B. G.
Teubner, 2 v. 1958-1959. 437 p. 604 p. See annotation under Russian Title [976].
                                                                               2583

ERG, Semënovich. Natural regions of the USSR. Translated from the Russian by
Olga Adler Titelbaum. Edited by John A. Morrison and C. C. Nikiforoff. New York,
Macmillan, 1950. 436 p. See annotation under Russian title [977].   2584

USLOV, Sergei Petrovich. Physical geography of Asiatic Russia. Translated from
Russian by Noah D. Gershevsky. Edited by Joseph E. Williams. San Francisco,
W. H. Freeman, 1961. 594 p. See annotation under Russian title [987].   2585

Minerals and Energy
(See also entry 2598)

HIMKIN, Demitri B. Minerals: a key to Soviet power. Cambridge, Massachusetts,
Harvard University Press, 1953. 452 p. Bibliography, p. 393-418.   2586

About 550 references, arranged alphabetically.

ODGKINS, Jordan A. Soviet power: energy resources, production and potentials.
Englewood Cliffs, New Jersey, Prentice-Hall, 1961. 190 p. Bibliographical notes
at end of chapters.   2587

Geomorphology

'OSTER, Harold C. The changing focus of geomorphology, Soviet geography: review
and translation, v. 13, no. 6 (June, 1972), p. 337-343. Bibliography, p. 341-343.
                                                                               2588

41 bibliographical citations arranged in order of discussion of comparative
work in the Soviet Union and in Western literature.

Climatology and Meteorology

ORISOV, Anatolii Aleksandrovich. Climates of the USSR. Edited by Cyril A. Halstead.
Translated by R. A. Ledward. Edinburgh, London, Oliver and Boyd; Chicago, Illinois,
Aldine, 1965. 255 p. Bibliography, p. 240-242. Translation of: Klimaty SSSR.
Moskva, Uchpedgiz, 1959. (1st ed. 1948).   2589

70 references, all in Russian, given in English translations.

LYDOLPH, Paul E. Soviet work and writing in climatology, Soviet geography: review and translation, v. 12, no. 10 (December, 1971), p. 637-666. Bibliography, p. 645-660.

    207 references to selected Soviet publications in climatology, with notations of which are also available in English. Arranged by major fields of climatology. List of 50 papers on climatology, translated into English in Soviet geography: review and translation, arranged systematically.

ZIKEEV, Nikolai Tikhonovich and DOUMANI, George A. Weather modification in the Soviet Union, 1946-1966; a selected annotated bibliography. Washington, D. C., Library of Congress. Reference Department, Science and Technology Division, 1967. 78 p.

    503 references.

Hydrology

MICKLIN, Philip P. Dimensions of the Caspian Sea Problem, Soviet geography: review and translation, v. 13, no. 9 (November, 1972), p. 589-603. Bibliography, p. 598-603.

    68 references on the Caspian sea problem and related alleviatory schemes, of which 14 are available also in English.

Biogeography

KÜCHLER, August Wilhelm, comp., Vegetation maps of the Union of Soviet Socialist Republics, Asia, and Australia. Lawrence, Kansas, University of Kansas Libraries, 1968. 389 p. International bibliography of vegetation maps, v. 3 (University of Kansas publications. Library series, no. 29).

    Union of Soviet Socialist Republics. p. 1-93, by D. V. Lebedev. Arranged by large regions: Soviet Union, European part of the Soviet Union, Ukraine, Caucasus, Regions of the Soviet Union east of the Urals, and within each region by date. Provides full legends of each map .

Conservation and Resources

PRYDE, Philip R. Conservation in the Soviet Union. Cambridge, University Press, 1972. 301 p. Bibliography, p. 264-294.

    About 500 references, organized by the 9 chapters of the book.

GERASIMOV, I. P., ARMAND, D. L., and YEFRON, K. M., eds. Natural resources of the Soviet Union: their use and renewal. Translated from the Russian by Jacek I. Romanowski. English edition edited by W. A. Douglas Jackson. San Francisco, W. H. Freeman Co., 1971. 349 p. Bibliographies at ends of articles. [For Russian edition see entry 1648].

JENSEN, Robert G. Land evaluation and regional pricing in the Soviet Union, Soviet geography: review and translation, v. 9, no. 3 (March, 1968), p. 145-153. Bibliography, p. 148-153.

    45 references in order of discussion in text.

ECONIMIC, POPULATION, AND URBAN GEOGRAPHY

Economic Geography: Agriculture

)NS, Leslie. Russian agriculture: a geographic survey. London, G. Bell and
)ns, 1972. 348 p. Bibliography, p. 319-329.   2597

    About 200 references, arranged alphabetically.

Economic Geography: Manufacturing and Industrial Resources

3AD, Theodore. Basic industrial resources of the U.S.S.R. New York and London,
)lumbia University Press, 1969. 393 p. Bibliography, p. 347-358.   2598

    About 175 references arranged systematically: general references, statistical
.blications, atlases, economic geography of the USSR, Soviet periodicals, minerals
;eneral), fuels (general), coal, petroleum and natural gas, peat, electric power,
ron and steel, other metals and nonmetallics, chemical industries, regional pub-
.cations.

\ES, Leslie. Locational factors and locational developments in the Soviet chem-
:al industry. Chicago, Illinois, University of Chicago, Department of Geography,
≥search paper no. 119. 1969. 262 p. Bibliography, p. 243-262.   2599

    309 references arranged by categories: public documents; books, monographs,
ıd pamphlets; articles, periodicals, and essays; abstracts; atlases; and other
ources. Within each category entries are listed alphabetically.

4, Alfred. Industriegeographie der Sowjetunion. Berlin, VEB Deutscher Verlag
≥r Wissenschaften, 1963. 226 p. Bibliography, p. 224-226.   2600

    74 references.

₹, Brenton M. The Soviet wood-processing industry: a linear programming analysis
f the role of transportation costs in location and flow patterns. Toronto,
.iversity of Toronto Press, 1970. 135 p. Bibliography, p. 129-135 (University
f Toronto Department of Geography. Research publication no. 5).   2601

    125 references.

Population Geography

3HISHEVSKII, Vadim Viacheslavovich, ed. Evaluation of the Soviet population
≥nsus 1970, Geoforum: journal of physical, human and regional geosciences, no. 9
1972), p. 3-60. Bibliographical footnotes.   2602

    Includes: Population of the USSR--changes in its demographic, social, and
thnic structure by S. I. Bruk; Urbanization in the USSR by V. V. Pokshishevskii;
ransformation of rural settlements in the Soviet Union by S. A. Kovalev, Ethnic
rocesses in the USSR by V. I. Kozlov; and Development of the peoples of the north
.ring the Soviet epoch by A. V. Smoliak.

IS, Robert A. The postwar study of internal migration, Soviet geography: review
ıd translation, v. 10, no. 4 (April, 1969), p. 157-166. Bibliography, p. 161-
56.   2603

    81 references arranged in order of discussion.

IMER, Frank. The population of the Soviet Union: history and prospects. Pre-
ared by the Office of Population Research, Princeton University for the League
f Nations. Geneva, League of Nations, 1946. 289 p. (League of Nations. Series
f publications. II. Economic and financial. 1946. II.A.3). Includes 22 plates
f colored maps. 28 x 44 cm. Bibliography, p. 259-284.   2604

    512 references arranged by type of material.

Urban Geography

HARRIS, Chauncy D. Cities of the Soviet Union: studies in their functions, size,
  density, and growth. Washington, D. C., Association of American Geographers,
  1970. 484 p. Bibliography, p. 412-468. (Monograph series, no. 5).

  More than 700 references, mostly in Russian. Index.

Settlement Geography

FISCHER, Dora. Siedlungsgeographie in der Sowjetunion, Erdkunde, v. 20, no. 3
  (1966), p. 211-227.

  50 references plus bibliographical footnotes. List of Soviet institutions
  active in geography of settlement, population, and planning.

HAHN, Roland. Jüngere Veränderungen der ländlichen Siedlungen im europäischen Teil
  der Sowjetunion, Stuttgart, Geographisches Institut der Universität, 1970. 146 p.
  Bibliography, p. 113-146 (Stuttgarter geographische Studien, Band 79).

  About 500 references.

Cultural Geography and Ethnography

JAKOBSON, Roman, HÜTTL-WORTH, Gerta, and BEEBE, John Fred. Paleosiberian peoples
  and languages: a bibliographical guide. New Haven, Connecticut, Human Relations
  Area Files Press, 1957. 224 p.

  1,904 references, arranged by ethnic groups.

HALPERN, Joel, ed Bibliography of anthropological and sociological publications
  on Eastern Europe and the USSR (English language sources). Los Angeles, California,
  University of California Russian and East European Studies Center, 1961. 142 p.
  (Russian and East European Studies Center series, v. 1, no. 2).

City and Regional Planning

PARKINS, Maurice F. City planning in Soviet Russia: with an interpretative bibliography. Chicago, Illinois, University of Chicago Press, 1953. 257 p. Bibliography, p. 127-240.

  More than 800 entries, arranged by subjects.

CHAMBRE, Henri. L'Aménagement du territoire en U.R.S.S. Introduction à l'étude
  des régions économiques soviétiques. Paris, Mouton, 1959. 250 p. Bibliographical
  footnotes. (École Pratique des Hautes Études - Sorbonne. VIe Section: Sciences
  Économiques et Sociales. Études sur l'Economie et la Sociologie des Pays Slaves,
  IV).

2605 - 2610

HISTORICAL GEOGRAPHY AND RELATED FIELDS

Historical Geography

NCH, Richard Anthony. Historical geography in the USSR, in <u>Progress in historical geography</u>, ed. by Alan R. H. Baker. Newton Abbot, England, David and Charles, 972, p. 111-128. Bibliography, p. 237-242. 2611

95 references in order of discussion in the text.

NCH, Richard Anthony. Historical geography in the USSR, <u>Soviet geography: review nd translation</u>, v. 9, no. 7 (September, 1968), p. 551-561. Bibliography, p. 558-61. 2612

50 bibliographical footnotes arranged in order of discussion.

KER, W. H. <u>An historical geography of Russia</u>. London, University of London ress, Ltd., 1968. 416 p. Bibliography, p. 381-391. 2613

About 275 references, arranged alphabetically.

KO, George J. <u>The Russian colonization of Kazakhstan: 1896-1916</u>. Bloomington, ndiana, <u>Indiana University Publications, Uralic and Altaic series</u>, v. 99, 1969. 71 p. Bibliography, p. 259-271. 2614

122 references arranged by type of material: articles in periodicals, books nd monographs, government publications, reports, and unpublished material, and isted alphabetically by author within each type. Also, 301 bibliographical notes, . 226-251.

OR, Thomas S. <u>Patterns of urban growth in the Russian Empire during the nineteenth entury</u>. Chicago, Illinois, University of Chicago, Department of Geography, Re-earch paper no. 163. In press. 2615

177 references arranged by: basic statistical sources (publications of the sarist government; other statistical sources) and general sources. Within each ategories entries are listed in alphabetical order.

SON, James R. <u>Feeding the Russian fur trade: provisionment of the Okhotsk sea-oard and the Kamchatka peninsula 1639-1856</u>. Madison, Wisconsin, University of isconsin Press, 1969. 337 p. Bibliography, p. 285-310. 2616

About 400 references. Also bibliographical notes, p. 229-284.

History of Exploration

G, Lev Semenovich. <u>Geschichte der russischen geographischen Entdeckungen</u>. esammelte Aufsätze. Leipzig, VEB Bibliographisches Institut, 1954. 283 p. ee annotation under Russian title [1846]. 2617

History and Historical Sources

MSTED, Patricia Kennedy. <u>Archives and manuscript repositories in the USSR:</u> oscow and Leningrad. Princeton, New Jersey, Princeton University Press, 1972. 36 p. (Columbia University. Studies of the Russian Institute). Author-title nd subject indexes. 2618

PIRO, David. <u>A select bibliography of works in English on Russian history 1801-917</u>. Oxford, Basil Blackwell, 1962. 106 p. 2619

1070 entries arranged by topics. Author index.

MEYER, Klaus, with the collaboration of John.L. H. Keep, Klaus Manfrass, and Arthur Peetre. Bibliographie zur osteuropäischen Geschichte: Verzeichnis der zwischen 1939 und 1964 veröffentlichten Literatur in westeuropäischen Sprachen zur osteuropäischen Geschichte bis 1945. Herausgegeben von Werner Philipp. Wiesbaden, Otto Harrassowitz, 1972. 649 p. (Berlin. Freie Universität. Osteuropa-Institut. Bibliographische Mitteilungen, Het 10).

  2

12,152 unannotated entries on publications in Western European languages 1939-1964 on the history of Russia, the Soviet Union, and Poland up to 1945, arranged in six main chapters: Eastern Europe as a whole; Russia and the Soviet Union; Finland and the Baltic countries; Ukraine and Belorussia; Asiatic Russia; and Poland. Each chapter is divided by disciplinary, topical, and chronological categories. Alphabetical index of authors and editors.

MEYER, Klaus. Bibliographie der Arbeiten zur osteuropäischen Geschichte aus den deutschsprachigen Fachzeitschriften 1858-1964. Herausgegeben von Werner Philipp. Wiesbaden, Otto Harrassowitz, 1966. 314 p. (Berlin. Freie Universität. Osteuropa-Institut. Bibliographische Mitteilungen, Heft 9).

  2

2597 entries of articles on the history of Eastern Europe in German technical periodicals 1858-1964, arranged in seven main chapters: Eastern Europe in general; Russia and the Soviet Union; Baltic countries; Ukraine and Belorussia; Caucasus and Russian Asia; Poland; and Southeast Europe. The chapter on the Soviet Union divided into general and related disciplines, bibliographies, and historical periods. Author index. List of periodicals.

LADA-MOCARSKI, Valerian. Bibliography of books on Alaska published before 1868. New Haven and London, Yale University Press, 1969. 567 p.

  2

Includes facsimile title page and analysis of contents of more than 160 books with English translation of Russian titles.

METHODS IN GEOGRAPHY

BRUNER, Edward F. A selected bibliography of mathematical articles in Soviet economic geography (1960-1968), Soviet geography: review and translation, v. 10, no. 9 (November, 1969), p. 559-562.

  2

43 selected articles, listed by years, with notes on the mathematical techniques employed, the source of data if used, and citations of American or British authors. English translations, if available, are also noted.

VINOGRADOV, Boris Veniaminovich. Main trends in the application of airphoto methods to geographical research in the U.S.S.R.: a review of publications, 1962-1964. Photogrammetria, v. 23, no. 3 (May, 1968), p. 77-94. Bibliography, p. 88-94.

  2

123 references in alphabetical order.

2620 - 2624

D. REGIONAL GEOGRAPHY

BIBLIOGRAPHY OF BIBLIOGRAPHIES

CHEL, Karol. Guide to Russian reference books, v. 2. History, historical
ciences, ethnography, geography. Stanford, California, The Hoover Institution
n War, Revolution, and Peace, 1964. 297 p.   2625

"Regional physical geography bibliographies," p. 197-202, entries F52-F93,
"Regional handbooks," p. 222-227, entries F256-F300, and "Regional history and
opography," p. 78-111, entries B313-B550, include some 325 regional and local
ibliographies, many of geographic value, extending from the earliest periods
p to 1962.

UKRAINE AND BELORUSSIA

L, Jiří. Geografický bibliografie Podkarpatské Rusi (Geographical bibliography
f Subcarpathian Ruthenia). Praha, Geograficky ústav Karlovy Universita, 1923-
928. 2 v. (Travaux Géographiques Tchèques, no. 11,13). (French title: Bib-
iographie géographique de la Russie Subcarpathique). In Czech.   2626

2,156 entries on the Transcarpathian Ukraine from the second half of the
9th century to 1926. Each volume is divided into four sections: bibliographies;
hysical geography and related fields; human geography and related fields; and
aps and atlases. Author index for each volume.

AR, Nicholas P. A bibliographical guide to Belorussia. Cambridge, Massachusetts.
arvard University Press, 1956. 63 p. Multigraphed. (Russian Research Center
tudies, 22). Index.   2627

SOVIET MIDDLE ASIA

RCE, Richard A. Soviet Central Asia: a bibliography. Berkeley, California,
niversity of California, Center for Slavic and East European Studies, 1966.
parts. Mimeographed.   2628

part 1. 1558-1866. 52 p. 497 entries.

part 2. 1867-1917. 66 p. 715 entries.

part 3. 1917-1966. 71 p. 797 entries.

Each organized systematically and each with a separate author index.

TRAL Asian Research Centre, London. Bibliography of recent Soviet source material
n Soviet Central Asia and its borderlands. 1957-1962. 2 nos. a year, except
957. Closed. Title varies slightly.   2629

Earlier bibliographies included in the Central Asian review, up to v. 5, no.
(1957, no. 1). Supplement to the Central Asian review.

Entries briefly annotated. Organized systematically for Soviet Middle Asia
nd by country for bordering areas.

WORTH, Edward, ed. Soviet Asia bibliographies: the Iranian, Mongolian, and Turkic
ationalities. New York, Praeger, 1974.   2630

About 4,000 entries grouped alphabetically and by discipline in 35 nationality
nd subregional categories, cross-indexed. Social science and humanities fields
ith coverage of Tsarist, Soviet, and Western sources, 1850-1968.

TAAFE, Robert N. Rail transportation and the economic development of Soviet
Central Asia. Chicago, Illinois, University of Chicago, Department of Geography,
Research paper no. 64, 1960. 186 p. Bibliography, p. 181-186.

    91 references.

SIBERIA, SOVIET FAR EAST, AND THE NORTH

STEWART, John M. Siberia and the Soviet Far East (a short bibliography of Soviet
publications, 1959-1969), ABSEES: Soviet and East European abstracts series, no.
4 (30), April, 1971, Special section, p. i-xxiii.

KERNER, Robert J. Northeastern Asia: a selected bibliography. Contributions to
the bibliography of the relations of China, Russia, and Japan, with special reference to Korea, Manchuria, Mongolia, and eastern Siberia, in Oriental and
European languages. Berkeley, California, University of California Press, 1939.
2 v. v. 1. 675 p. v. 2. 621 p. Photoprinted. (Publications of the Northeastern Asia Seminar of the University of California).

    13,884 entries. Asia...: geography and cartography, v. 1, p. 14-17, entries
148-181; China: geography and cartography, p. 118-127, entries 1235-1345; The
Japanese Empire: geology, geography, and natural resources, v. 2, p. 14-25,
entries 7384-7463; The Russian Empire and the Soviet Union in Asia and on the
Pacific: geography and cartography, v. 2, p. 292-300, entries 10,521-10,583.
Sections also on the Soviet Far East, v. 2, p. 482-515; Buryat-Mongolism ASSR,
v. 2, p. 515-524; Yakut ASSR, v. 2, p. 525-534; Siberia, v. 2, p. 534-547;
Western Siberia, v. 2). p. 547-560, and Soviet Middle Asia, v. 2, p. 560-607.

GRIER, Mary C., comp. Oceanography of the North Pacific Ocean, Bering Sea and
Bering Strait: a contribution toward a bibliography. Seattle, Washington,
University of Washington, 1941. 290 p. (University of Washington Publications,
Library series, v. 2. Reprinted: New York, Greenwood Press, 1969. 290 p.

    2,929 entries. Index of authors and personal names.

THIEL, Erich. The Soviet Far East: a survey of its physical and economic geography.
Translated by Annelie and Ralph M. Rookwood. New York, Frederick A. Praeger, 1957.
388 p. Bibliography, p. 363-377. In German: Sowjet-Fernost: eine landes- und
wirtschaftskundliche Übersicht. München, 1953. 329 p. Bibliography, p. 309-319
(München. Osteuropa Institut. Veröffentlichung, Band 1).

    240 references.

ARMSTRONG, Terence. Russian settlement in the north. Cambridge, University Press,
1965. 224 p. Bibliography, p. 198-217.

    About 275 references arranged alphabetically.

.T VII.   A NOTE ON THE SOURCES UTILIZED IN THE COMPILATION OF THIS BIBLIOGRAPHY

Although in aggregate more than a hundred different sources were consulted
the compilation of this bibliography, a substantial fraction of the titles and
the bibliographical information were derived from a few key references as fol-
rs:

(1) Bibliografiia sovetskoi bibliografii 1939, 1946-1970. (Vsesoiuznaia Knizh-
a Palata). Moskva, Izdatel'stvo "Kniga," 1941, 1948-1972. Annual volumes.
s superb systematically organized inventory of Soviet bibliographies, both
se published separately and those included incidentally in books and articles,
by far the best source by virtue of its comprehensiveness. It listed, however,
e bibliographies than the present work, since it covers items with more than
references, whereas the present work generally is limited to titles with at
st 100-200 references. It listed 1,894 bibliographies in the general section
geography and thousands more in closely related fields [entry 10, p. 3].

(2) Referativnyi zhurnal: geografiia. Svodnyi tom (Vsesoiuznyi Institut
.chnoi i Tekhnicheskoi Informatsii). Moskva, 1954-1972. Monthly. The annual
.ject index lists for most of the fields covered the main bibliographical items,
ugh under diverse headings. This source is world-wide in coverage; only those
ms in Russian or on the Soviet Union are included here [entry 16, p. 7-9].

(3) Fizicheskaia geografiia: annotirovannyi perechen' otechestvennykh biblio-
fii, izdannykh v 1810-1966 gg. M. N. Morozova and E. A. Stepanova, comps.
V. Klevenskaia, ed. L. G. Kamanin, scientific consultant. Moskva, Izdatel'-
o "Kniga," 1968. 309 p. (Gosudarstvennaia Biblioteka SSSR imeni Lenina.
titut Geografii Akademii Nauk SSSR. Sektor Seti Spetsial'nykh Bibliotek
demii Nauk SSSR). 1,336 entries. Excellent comprehensive source with exact
liographical information and notes for the fields and dates covered [entry 1,
1].

(4) Bibliografii po geografii, vyshedshie v SSSR (1946-1961 gg.). S. A.
rilova, I. I. Parkhomenko, and N. K. Kuznetsova, comps. In Geograficheskii
rnik [v. 1]. Moskva, Proizvodstvenno-Izdatel'skii Kombinat VINITI, 1963,
32-45. (Institut Nauchnoi Informatsii Akademii Nauk SSSR. Raboty po
reticheskim i Spetsial'nym Voprosam Nauchno-Tekhnicheskoi Informatsii). 266
ries. It includes only separate works that are primarily bibliographies and
s excludes incidental bibliographies in books or articles. It lacks annota-
ns [entry 2, p. 1].

(5) Literatura po geografii i smezhnym naukam. Ezhemesiachnyi informatsionnyi
lleten' (izdaetsia s 1969 goda). (Akademiia Nauk SSSR. Tsentral'naia Biblioteka
Estestvennym Naukam. Biblioteka Instituta Geografii). Moskva. 1969- .
thly. This was the principal source for recent works, since it has only a
rt time lag before publication, the material is already selected for signifi-
ce in geographical research, the bibliographical information is full, and the

presence and extent of bibliographies is noted. By scanning the bottom of the card for each entry, I could ascertain which works contained extensive bibliographies. Only a small fraction of the works listed, however, contained bibliographies of a scope to be included in the present volume [entry 17, p. 10].

Since information on titles from any any single source was likely to incomplete, most entries had to be checked against three different reference volumes. Regardless of ultimate source, most titles were checked against Bibliografiia sovetskoi bibliografii [10] for information on size and organization of bibliographies and appropriate form, such as whether a work should be listed under the names of the compilers or the institution which sponsored the publication. For completeness I have tried to include both since a work might be listed under onl one or the other in any given library. Unfortunately Bibliografiia sovetskoi bibliografii, invaluable in its information on number of entries and arrangement of each bibliography, generally does not include the name of the publisher or number of pages. To complete the information in an entry one must turn to the standard inventories of Soviet book publications:

(6) Ezhegodnik knigi SSSR: sistematicheskii ukazatel' (Vsesoiuznaia Knizhnaia Palata). Moskva, Izdatel'stvo "Kniga," 1941- . Annual. Typically information is sought through the name index, which includes authors, compilers, and editors, or through the separate index of titles, which lists alphabeticall by title works published primarily under a title or ones for which authors, compilers, or editors are not shown on the title page, as is typical for many major works sponsored by scientific institutions [entry 22, p. 12].

(7) Knizhnaia letopis': organ gosudarstvennoi bibliografii SSSR (Vsesoiuzna Knizhnaia Palata). Moskva, Izdatel'stvo "Kniga," 1907- . Weekly, with quarter: author indexes. This is more comprehensive than Ezhegodnik knigi SSSR and is published with a much shorter time lag, but is less easy to use [entry 20, p. 1:

Because many of the titles here listed are to be found only in very large research and reference libraries and the card catalogues of such libraries are enormous and are likely to have many authors with the same last name and even with the same initials for given name and patronymic, I have tried to include i full both the given name and patronymic. These are not given in most Soviet bibliographic sources, though they are included in the cards prepared by the Vsesoiuznaia Knizhnaia Palata or in the card catalogue of the Gosudarstvennaia Biblioteka SSSR imeni Lenina. Compiling this list in Chicago, I did not have direct access to either of them. My two principal sources for such names were:

(8) Directory of Soviet Geographers, prepared by Theodore Shabad, Soviet geography: review and translation, v. 8, no. 7 (September, 1967), p. 508-610 [entry 2581, p. 379].

(9) National union catalog: a cumulative author list representing Library of Congress printed cards and titles reported by other American libraries, 1958 1962, 1963-1967, 1968-1972 [entry 2477a, p. 364].

Recent volumes in the National union catalog, based on cards prepared by the
oiuznaia Knizhnaia Palata as well as on inspection of information contained
ndividual publications, include much fuller information on given names and
onymics, than do earlier volumes.

National union catalog. Pre-1956 imprints [entry 2477b], and the British
um, General catalogue of printed books to 1955, with supplements, 1956-1965
1966-1970 [entry 2477c], were also useful in checking.

(10) Kartograficheskaia letopis'; organ gosudarstvennoi bibliografii SSSR
soiuznaia Knizhnaia Palata). Moskva, "Kniga, 1954-1972. Annual. The sec-
"Nauchno-spravochnye atlasy" in each annual volume provided the best biblio-
hical control on atlases considered for inclusion in this Guide [entry 112,
3].

In summary, identification of works with substantial bibliographies and
ription of the characteristics of these bibliographies are based primarily
ources 1-5 above; confirmation of data on publication and publisher, on
ces 6 and 7 above; and information on the full name of the author, on sources
d 9 above; and atlases on source 10, above.

Number of entries: 2660

In the final editing two numbers were omitted, 413 and 1736, and 26 were
ed: 11a, 926a,1075a, 1275a, 1334a, 1461a, 1503a, 1524a, 1642a, 1723a, 1756a,
la, 1918a, 1934a, 1936a, 2196a, 2289a, 2403a, 2442a, 2477a, 2477b, 2477c,
9a, 2562a, 2568a, 2608a. To the last numbered entry, 2636, must therefore be
ed 24 to give the total number of entries, 2660.

I N D E X

Listed here in a single alphabetical order are names of all authors, editors,
and compilers; institutional sponsors; and titles of books and relevant articles,
sections or bibliographies in collected works or serials. In addition, entries are
provided in English for general subject categories and major regions.

Numbers refer to entries, not to pages. Titles, many of which have been con-
densed, appear in italics.

A

THE UNIVERSITY OF CHICAGO
DEPARTMENT OF GEOGRAPHY
RESEARCH PAPERS (Lithographed, 6×9 Inches)

(Available from Department of Geography, The University of Chicago, 5828 S. University Ave. Chicago, Illinois 60637. Price: $5.00 each; by series subscription, $4.00 each.)

62. GINSBURG, NORTON, editor. *Essays on Geography and Economic Development* 1960. 196 pp.
71. GILBERT, E. W. *The University Town in England and West Germany*
    1961. 79 pp. 4 plates. 30 maps and diagrams. (Free to new subscribers)
72. BOXER, BARUCH. *Ocean Shipping in the Evolution of Hong Kong* 1961. 108 pp.
74. TROTTER, JOHN E. *State Park System in Illinois* 1966. 152 pp.
84. KANSKY, K. J. *Structure of Transportation Networks: Relationships between Network Geometr
    and Regional Characteristics* 1963. 155 pp.
91. HILL, A. DAVID. *The Changing Landscape of a Mexican Municipio, Villa Las Rosas, Chiapas*
    NAS-NRC Foreign Field Research Program Report No. 26. 1964. 121 pp.
94. MC MANIS, DOUGLAS R. *The Initial Evaluation and Utilization of the Illinois Prairies, 1815–184C
    1964. 109 pp.
97. BOWDEN, LEONARD W. *Diffusion of the Decision To Irrigate: Simulation of the Spread of a Nev
    Resource Management Practice in the Colorado Northern High Plains* 1965. 146 pp.
98. KATES, ROBERT W. *Industrial Flood Losses: Damage Estimation in the Lehigh Valley*
    1965. 76 pp.
102. AHMAD, QAZI. *Indian Cities: Characteristics and Correlates* 1965. 184 pp.
103. BARNUM, H. GARDINER. *Market Centers and Hinterlands in Baden-Württemberg* 1966. 172 pp.
105. SEWELL, W. R. DERRICK, et al. *Human Dimensions of Weather Modification* 1966. 423 pp.
106. SAARINEN, THOMAS F. *Perception of the Drought Hazard on the Great Plains* 1966. 183 pp.
107. SOLZMAN, DAVID M. *Waterway Industrial Sites: A Chicago Case Study* 1967. 138 pp.
108. KASPERSON, ROGER E. *The Dodecanese: Diversity and Unity in Island Politics* 1967. 184 pp.
109. LOWENTHAL, DAVID, et al. *Environmental Perception and Behavior.* 1967. 88 pp.
110. REED, WALLACE E. *Areal Interaction in India: Commodity Flows of the Bengal-Bihar Industria
     Area* 1967. 210 pp.
112. BOURNE, LARRY S. *Private Redevelopment of the Central City: Spatial Processes of Structura
     Change in the City of Toronto* 1967. 199 pp.
113. BRUSH, JOHN E., and GAUTHIER, HOWARD L., JR. *Service Centers and Consumer Trips: Studie
     on the Philadelphia Metropolitan Fringe* 1968. 182 pp.
114. CLARKSON, JAMES D. *The Cultural Ecology of a Chinese Village: Cameron Highlands, Malaysi
     1968. 174 pp.
115. BURTON, IAN; KATES, ROBERT W.; and SNEAD, RODMAN E. *The Human Ecology of Coastal Floo
     Hazard in Megalopolis* 1968. 196 pp.
117. WONG, SHUE TUCK. *Perception of Choice and Factors Affecting Industrial Water Supply Deci
     sions in Northeastern Illinois* 1968. 96 pp.
118. JOHNSON, DOUGLAS L. *The Nature of Nomadism* 1969. 200 pp.
119. DIENES, LESLIE. *Locational Factors and Locational Developments in the Soviet Chemical Industr
     1969. 285 pp.
120. MIHELIC, DUSAN. *The Political Element in the Port Geography of Trieste* 1969. 104 pp.
121. BAUMANN, DUANE. *The Recreational Use of Domestic Water Supply Reservoirs: Perception an
     Choice* 1969. 125 pp.
122. LIND, AULIS O. *Coastal Landforms of Cat Island, Bahamas: A Study of Holocene Accretionar
     Topography and Sea-Level Change* 1969. 156 pp.
123. WHITNEY, JOSEPH. *China: Area, Administration and Nation Building* 1970. 198 pp.
124. EARICKSON, ROBERT. *The Spatial Behavior of Hospital Patients: A Behavioral Approach to Spatia
     Interaction in Metropolitan Chicago* 1970. 198 pp.
125. DAY, JOHN C. *Managing the Lower Rio Grande: An Experience in International River Develop
     ment* 1970. 277 pp.
126. MAC IVER, IAN. *Urban Water Supply Alternatives: Perception and Choice in the Grand Basin
     Ontario* 1970. 178 pp.
127. GOHEEN, PETER G. *Victorian Toronto, 1850 to 1900: Pattern and Process of Growth* 1970. 278 pp

8. GOOD, CHARLES M. *Rural Markets and Trade in East Africa* 1970. 252 pp.

9. MEYER, DAVID R. *Spatial Variation of Black Urban Households* 1970. 127 pp.

0. GLADFELTER, BRUCE. *Meseta and Campiña Landforms in Central Spain: A Geomorphology of the Alto Henares Basin* 1971. 204 pp.

1. NEILS, ELAINE M. *Reservation to City: Indian Urbanization and Federal Relocation* 1971. 200 pp.

2. MOLINE, NORMAN T. *Mobility and the Small Town, 1900–1930* 1971. 169 pp.

3. SCHWIND, PAUL J. *Migration and Regional Development in the United States, 1950–1960* 1971. 170 pp.

4. PYLE, GERALD F. *Heart Disease, Cancer and Stroke in Chicago: A Geographical Analysis with Facilities Plans for 1980* 1971. 292 pp.

5. JOHNSON, JAMES F. *Renovated Waste Water: An Alternative Source of Municipal Water Supply in the U.S.* 1971. 155 pp.

6. BUTZER, KARL W. *Recent History of an Ethiopian Delta: The Omo River and the Level of Lake Rudolf* 1971. 184 pp.

7. HARRIS, CHAUNCY D. *Annotated World List of Selected Current Geographical Serials in English, French, and German* 3rd edition 1971. 77 pp.

8. HARRIS, CHAUNCY D., and FELLMANN, JEROME D. *International List of Geographical Serials* 2nd edition 1971. 267 pp.

9. MC MANIS, DOUGLAS R. *European Impressions of the New England Coast, 1497–1620* 1972. 147 pp.

0. COHEN, YEHOSHUA S. *Diffusion of an Innovation in an Urban System: The Spread of Planned Regional Shopping Centers in the United States, 1949–1968* 1972. 136 pp.

1. MITCHELL, NORA. *The Indian Hill-Station: Kodaikanal* 1972. 199 pp.

2. PLATT, RUTHERFORD H. *The Open Space Decision Process: Spatial Allocation of Costs and Benefits* 1972. 189 pp.

3. GOLANT, STEPHEN M. *The Residential Location and Spatial Behavior of the Elderly: A Canadian Example* 1972. 226 pp.

4. PANNELL, CLIFTON W. *T'ai-chung, T'ai-wan: Structure and Function* 1973. 200 pp.

5. LANKFORD, PHILIP M. *Regional Incomes in the United States, 1929–1967: Level, Distribution, Stability, and Growth* 1972. 137 pp.

6. FREEMAN, DONALD B. *International Trade, Migration, and Capital Flows: A Quantitative Analysis of Spatial Economic Interaction* 1973. 202 pp.

7. MYERS, SARAH K. *Language Shift Among Migrants to Lima, Peru* 1973. 204 pp.

8. JOHNSON, DOUGLAS L. *Jabal al-Akhdar, Cyrenaica: An Historical Geography of Settlement and Livelihood* 1973. 240 pp.

9. YEUNG, YUE-MAN. *National Development Policy and Urban Transformation in Singapore: A Study of Public Housing and the Marketing System* 1973. 204 pp.

0. HALL, FRED L. *Location Criteria for High Schools: Student Transportation and Racial Integration* 1973. 156 pp.

1. ROSENBERG, TERRY J. *Residence, Employment, and Mobility of Puerto Ricans in New York City* 1974. 230 pp.

2. MIKESELL, MARVIN W., editor. *Geographers Abroad: Essays on the Problems and Prospects of Research in Foreign Areas* 1973. 296 pp.

3. OSBORN, JAMES. *Area, Development Policy, and the Middle City in Malaysia* 1974. 273 pp.

4. WACHT, WALTER F. *The Domestic Air Transportation Network of the United States* 1974. 98 pp.

5. BERRY, BRIAN, J. L., et al. *Land Use, Urban Form and Environmental Quality* 1974. 464 pp.

6. MITCHELL, JAMES K. *Community Response to Coastal Erosion: Individual and Collective Adjustments to Hazard on the Atlantic Shore* 1974. 209 pp.

7. COOK, GILLIAN P. *Spatial Dynamics of Business Growth in the Witwatersrand* 1975. 143 pp.

8. STARR, JOHN T., JR. *The Evolution of Unit Train Operations in the United States: 1960–1969—A Decade of Experience* 1975.

9. PYLE, GERALD F. *The Spatial Dynamics of Crime* 1974. 220 pp.

0. MEYER, JUDITH W. *Diffusion of an American Montessori Education* 1975.

1. SCHMID, JAMES A. *Urban Vegetation: A Review and Chicago Case Study*

2. LAMB, RICHARD. *Metropolitan Impacts on Rural America*

3. FEDOR, THOMAS. *Patterns of Urban Growth in the Russian Empire during the Nineteenth Century*

4. HARRIS, CHAUNCY D. *Guide to Geographical Bibliographies and Reference Works in Russian or on the Soviet Union* 1975. 496 pp.

ADMINISTRATIVE AND STATISTICAL AREAS OF THE SOVIET UNION ARRANGED ALPHABETICALLY

Key to numbers and letters on the maps

Abkhaz ASSR. Trans-Caucasus, Georgian SSR, 1
Adygey AO. RSFSR, North Caucasus region, 42
Adzhar ASSR. Trans-Caucasus, Georgian SSR, 2
Aktyubinsk oblast. Kazakh SSR, 3
Alma-Ata city. Kazakh SSR, AA
Alma-Ata oblast. Kazakh SSR, 16
Altay kray. RSFSR, West Siberian region, 60
Amur oblast. RSFSR, Far Eastern region, 68
Andizhan oblast. Middle Asian region, Uzbek SSR, 8
Archangel oblast. RSFSR, Northwest region, 7
Arkhangel'sk oblast. RSFSR, Northwest region, 7
Armenian SSR. Trans-Caucasus region, Ar
Ashkhabad city. Middle Asian region, Turkmen SSR, Ash.
Astrakhan' oblast. RSFSR, Volga region, 38
Azerbaijan. See Azerbaydzhan
Azerbaydzhan SSR. Trans-Caucasus, Az. Consists of areas directly under administration of the republic plus two special areas, nos. 1-2 and Baku city, B.

Baku city. Trans-Caucasus region. Azerbaydzhan SSR, B
Baltic region, includes three union republics, Estonian SSR (Es), Latvian SSR (La), Lithuanian SSR (Li), and Kaliningrad oblast of the RSFSR (76)
Bashkir ASSR. RSFSR, Volga region, 32
Belgorod oblast. RSFSR, Central Black-Earth region, 27
Belorussian region, same as Belorussian SSR, Be.
Belorussian SSR, Be, consists of 6 oblasts, nos. 1-6, and one city oblast, Minsk, M.
Black-Earth Center. See Central Black-Earth region
Brest oblast. Belorussian SSR, 5
Bryansk oblast. RSFSR, Central region, 12
Bukhara oblast. Middle Asian region, Uzbek SSR, 3
Buryat ASSR. RSFSR, East Siberian region, 66

Center. See Central region
Central Black-Earth region. RSFSR, IV, nos. 26-30
Central region. RSFSR, II, nos. 9-20
Chardzhou oblast. Middle Asian region, Turkmen SSR, 3
Chechen-Ingush ASSR. RSFSR, North Caucasus region, 47
Chelyabinsk oblast. RSFSR, Urals region, 52
Cherkassy oblast. Ukrainian SSR, Southwestern region, 18
Chernigov oblast. Ukrainian SSR, Southwestern region, 13

Chernovtsy oblast. Ukrainian SSR. Southwestern region, 21
Chimkent oblast. Kazakh SSR, 14
Chita oblast. RSFSR, East Siberian region, 67
Chuvash ASSR. RSFSR, Volga-Vyatka region, 24
Crimean oblast. Ukrainian SSR, Southern region, 25

Dagestan ASSR. RSFSR, North Caucasus region, 48
Dnepropetrovsk oblast. Ukrainian SSR, Donets-Dnepr region, 4
Donets-Dnepr region. Ukrainian SSR, Uk-1, nos. 1-8
Donetsk oblast. Ukrainian SSR. Donets Dnepr region, 6
Dushanbe city. Middle Asian region, Tadzhik SSR, D.
Dzhambul oblast. Kazakh SSR, 15

East Kazakhstan oblast. Kazakh SSR, 1
East-Siberian region. RSFSR, IX, nos. 62-67
Estonian SSR. Baltic region, Es.

Far-Eastern region. RSFSR, X, nos. 68-75
Fergana oblast. Middle Asian region. Uzbek SSR, 9
Frunze city. Middle Asian region, Kirgiz SSR, F.

Georgian SSR. Trans-Caucasus region, Ge. Consists of area directly under administration of the republic, three special areas, nos. 1-3, and the city of Tbilisi, Tb.
Gomel' oblast. Belorussian SSR, 4
Gor'kiy oblast. RSFSR, Volga-Vyatka region, 21
Gorno-Altay AO. RSFSR, West Siberian region, 61
Gorno-Badakhshan AO. Middle Asian region, Tadzhik SSR, 2
Grodno oblast. Belorussian SSR, 6
Gruzinskaya SSR. See Georgian SSR.
Gur'yev oblast, Kazakh SSR, 2

Irkutsk oblast. RSFSR, East Siberian region, 65
Issyk-Kul' oblast. Middle Asian region, Kirgiz SSR, 1
Ivano-Frankovsk oblast. Ukrainian SSR, Southwestern region, 20
Ivanovo oblast. RSFSR, Central region, 18

ADMINISTRATIVE AND STATISTICAL AREAS OF THE SOVIET UNION ARRANGED GEOGRAPHICALLY

Key to numbers and letters on the maps

R.S.F.S.R.: RUSSIAN SOVIET FEDERATED SOCIALIST REPUBLIC, the largest of the 15 union republics, is divided into 10 large regions and 78 smaller units.

I. Northwest region
1 Murmansk oblast
2 Karelian ASSR
3 Leningrad oblast
L Leningrad city
4 Pskov oblast
5 Novgorod oblast
6 Vologda oblast
7 Archangel oblast
8 Komi ASSR

II. Central region
9 Kalinin oblast
10 Moscow oblast
M Moscow city
11 Smolensk oblast
12 Bryansk oblast
13 Kaluga oblast
14 Orël oblast
15 Tula oblast
16 Ryazan' oblast
17 Vladimir oblast
18 Ivanovo oblast
19 Yaroslavl' oblast
20 Kostroma oblast

III. Volga-Vyatka region
21 Gor'kiy oblast
22 Kirov oblast
23 Mordva ASSR
24 Chuvash ASSR
25 Mari ASSR

IV. Central Black-Earth region
26 Kursk oblast
27 Belgorod oblast
28 Voronezh oblast
29 Lipetsk oblast
30 Tambov oblast

V. Volga region
31 Tatar ASSR
32 Bashkir ASSR
33 Ul'yanovsk oblast
34 Kuybyshev oblast
35 Penza oblast
36 Saratov oblast
37 Volgograd oblast
38 Astrakhan' oblast
39 Kalmyk ASSR

VI. North Caucasus region
40 Rostov oblast
41 Krasnodar kray
42 Adygey AO
43 Stavropol' kray
44 Karachayevo-Cherkess AO
45 Kabardino-Balkar ASSR
46 North Osset ASSR
47 Chechen-Ingush ASSR
48 Dagestan ASSR

VII. Urals region
49 Udmurt ASSR
50 Perm oblast
51 Sverdlovsk oblast
52 Chelyabinsk oblast
53 Kurgan oblast
54 Orenburg oblast

VIII. West-Siberian region
55 Tyumen' oblast
56 Omsk oblast
57 Novosibirsk oblast
58 Tomsk oblast
59 Kemerovo oblast
60 Altay kray
61 Gorno-Altay AO

IX. East-Siberian region
62 Krasnoyarsk kray
63 Khakass AO
64 Tuva ASSR
65 Irkutsk oblast
66 Buryat ASSR
67 Chita oblast

X. Far-Eastern region
68 Amur oblast
69 Khabarovsk kray
70 Jewish AO
71 Maritime (Primorskiy) kray
72 Sakhalin oblast
73 Yakut ASSR
74 Magadan oblast
75 Kamchatka oblast

(Baltic region)
76 Kaliningrad oblast

UKRAINIAN SSR

Uk-1 Donets-Dnepr region
    1 Sumy oblast
    2 Poltava oblast
    3 Kirovograd oblast
    4 Dnepropetrovsk oblast
    5 Khar'kov oblast
    6 Donetsk oblast
    7 Voroshilovgrad oblast
    8 Zaporozh'ye oblast

UK-2 Southwestern region
    9 Volyn' (Volhynian) oblast
  10 Rovno oblast
  11 Zhitomir oblast
  12 Kiev oblast
   K Kiev city
  13 Chernigov oblast
  14 L'vov oblast
  15 Ternopol' oblast
  16 Khmel'nitskiy oblast
  17 Vinnitsa oblast
  18 Cherkassy oblast
  19 Trans-Carpathian oblast
  20 Ivano-Frankovsk oblast
  21 Chernovtsy oblast

Uk-3 Southern region
  22 Odessa oblast
  23 Nikolayev oblast
  24 Kherson oblast
  25 Crimean oblast

BALTIC REGION
  Es Estonian SSR
  La Latvian SSR
  Li Lithuanian SSR
  76 Kaliningrad oblast (RSFSR)

Be BELORUSSIAN REGION. Belorussian SSR
    1 Minsk oblast
  Mi Minsk city
    2 Vitebsk oblast
    3 Mogilëv oblast
    4 Gomel' oblast
    5 Brest oblast
    6 Grodno oblast

Mo MOLDAVIAN SSR

TRANS-CAUCASUS REGION

Ge Georgian SSR
  Tb Tbilisi city
    1 Abkhaz ASSR
    2 Adzhar ASSR
    3 South Osset AO

Ar Armenian SSR

Az Azerbaydzhan SSR
   B Baku city
   1 Nakhichevan' ASSR
   2 Nagorno-Karabakh AO

Ka KAZAKHSTAN REGION. Kazakh SSR
    1 Ural'sk oblast
    2 Gur'yev oblast
    3 Aktyubinsk oblast
    4 Kustanay oblast
    5 North Kazakhstan oblast
    6 Kokchetav oblast
    7 Turgay oblast
    8 Tselinograd oblast
    9 Pavlodar oblast
  10 Karaganda oblast
  11 Semipalatinsk oblast
  12 East Kazakhstan oblast
  13 Kzyl-Orda oblast
  14 Chimkent oblast
  15 Dzhambul oblast
  16 Alma-Ata oblast
  AA Alma-Ata city
  17 Taldy-Kurgan oblast

MIDDLE ASIAN REGION

Ki Kirgiz SSR
   F Frunze city
   1 Issyk-Kul' oblast
   2 Naryn oblast
   3 Osh oblast

Uz Uzbek SSR
    1 Karakalpak ASSR
    2 Khorezm oblast
    3 Bukhara oblast
    4 Samarkand oblast
    5 Syrdar'ya oblast
    6 Tashkent oblast
   T Tashkent city
    7 Namangan oblast
    8 Andizhan oblast
    9 Fergana oblast
  10 Kashkadar'ya oblast
  11 Surkhandar'ya oblast

Tu Turkmen SSR
  Ash Ashkhabad city
    1 Tashauz oblast
    2 Mary oblast
    3 Chardzou oblast

Ta Tadzhik SSR
   D Dushanbe city
   1 Leninabad oblast
   2 Gorno-Badakhshan AO

Based on: USSR. Tsentral'noe Statisticheskoe Upravlenie. Itogi vsesoiuznoi perepisi naseleniia 1970 goda. Tom 1. Moskva, "Statistika," 1972. p. 10-21.

473

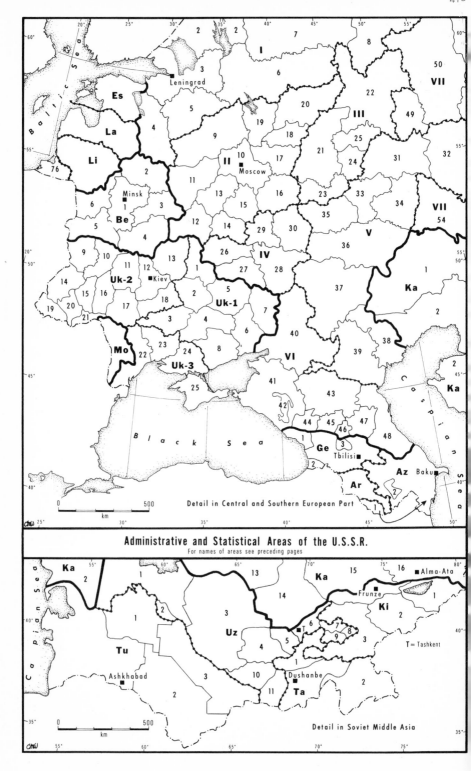

Administrative and Statistical Areas of the U.S.S.R.

For names of areas see preceding pages

Detail in Central and Southern European Part

Detail in Soviet Middle Asia

T = Tashkent